Robotics Science

System Development Foundation Benchmark Series

Robotics Science, 1989
Michael Brady, editor

Current Directions in Computer Music Research, 1989
Max. V. Mathews and John R. Pierce, editors

Robotics Science

edited by
Michael Brady

System Development Foundation Benchmark Series

The MIT Press
Cambridge, Massachusetts
London, England

This book was set in Times Roman by Asco Trade Typesetting Ltd., Hong Kong, and printed and bound by Halliday Lithograph in the United States of America.

Library of Congress Cataloging-in-Publication Data

Robotics science/edited by Michael Brady.
p. cm.—(System Development Foundation benchmark series; 1)
Includes bibliographies and index.
ISBN 0-262-02284-2
1. Robotics. I. Brady, Michael, 1945–. II. Series.
TJ211.R5683 1989
629.8′92—dc19 88-33682
CIP

Contents

Preface

This book consists of sixteen invited original chapters by prominent robotics researchers. It is intended as an assessment by the authors of the current state of the art and way forward in robotics. The introduction, by the editor, is a personal list of thirty problems, problem areas, and issues that seem to lie on the critical path to the development of a *Science of Robotics*.

The book marks the termination of the Systems Development Foundation, whose grants contributed significantly to the development of robotics in the United States during the 1980s. Of the sixteen chapters in this book, the work reported in ten of them was generously supported by the System Development Foundation. In addition, the SDF contributed to the establishment of *The International Journal of Robotics Research* and provided substantial funding for the International Symposia of Robotics Research, the proceedings of which are also in the MIT Press Artificial Intelligence Series.

The contributing authors and chapters were selected by a steering committee, consisting of the editor, Haruhiko Asada, Gerd Hirzinger, Steven Jacobsen, Tomás Lozano-Pérez, Marc Raibert, and Russell H. Taylor. Contributing authors were asked to prepare chapters according to the following brief:

> We invite you to contribute an article that describes your work. It is *not* intended that your article should be a survey. Rather, the article should take stock of the current state of your work and the problems it addresses, and it should point the way forward. We ask that you describe the context of your work rather more fully than is normal in a journal article. What problems are you trying to solve? What was known about those problems when you embarked upon your research? Similarly, there should be an extended conclusion that details the outstanding problems in your field, and sketches possible approaches to solving those problems. We hope that your article will excite interest in your field and encourage young scientists and engineers to attack the outstanding problems that you describe.

The draft chapters were read by two members of the steering committee, who then discussed their comments with the authors. The chapters were also discussed at a meeting that immediately followed the Fourth International Symposium on Robotics Research, held at the University of California at Santa Cruz in August 1987. Every author responded well to suggestions, comments, and criticisms.

Projects such as this depend upon the collective efforts of many people. The list of acknowledgments is consequently rather long. First, and foremost, thanks are due to the authors for entering the spirit of the venture by producing chapters that are special and different, and keeping (mostly) to time, after the initial shock of being asked for yet another version. Special thanks are also extended to the steering Committee, for its work in selection, careful reading of drafts, and discussions with the authors.

The book would not have been written without the considerable administrative and financial support of the System Development Foundation, particularly the members of the board, Carl York and Roberta Ishihara, whose quiet efficiency and understanding smoothed the entire process. Charles S. Smith, onetime director of programs for the System Development Foundation, championed the support of robotics as a science and originally suggested that it ought to be possible to write a book along these lines. Of course, he would tear the current text apart and start over to meet his exacting standards. Nevertheless, I hope this goes at least some way towards redressing the debt that many of the authors, myself in particular, owe Charles.

I thank Sarah Harrington, the administrator of Oxford University's Robotics Research Group, for her continuing heroic support in keeping me (roughly) up to date despite the tangled web of committees in which Oxford threatens to snare me, by defending my phone like Horatio on the bridge, and by ensuring that I am in approximately the right place at something like the right time.

Finally, Terry Ehling of MIT Press showed considerable patience when term intervened halfway through the writing of my chapter. Bob Bolles and Bernie Roth set up the facilities, interpreted broadly, for the Fourth ISRR at which the draft chapters were discussed and informally presented.

Michael Brady
Oxford
October 1988

Robotics Science

1 Introduction: The Problems of Robotics

Michael Brady

1.1 Introduction

This chapter is meant to serve a dual role. First, it introduces the other chapters in the book. Second, it cuts across them to offer a personal list of 30 problems that need to be solved if we are to develop a *science of robotics*.

Robotics is a broad, interdisciplinary subject that involves various facets of sensing and manipulation, as well as thinking that integrates sensing with action. "Thinking" takes place at many different levels. At its lowest levels, thinking is largely reflexive. Controlled responses to sensed changes in the environment are necessary, for example, for protective autonomy of a robot [Steer 1988], and they typically need to be fast and unconscious. Naturally, reflexive thinking involves only rudimentary descriptions of the environment. At higher levels, intelligent thought enables the robot to cope with a complex, uncertain, changing world. To be intelligent, robotics must rely upon artificial intelligence. An earlier essay [Brady 1985] explored this relationship. Here our emphasis is less on the state of the art in robotics and more on outstanding problems, problem areas, and issues that seem critical in the development of the subject. Where appropriate, we refer to the chapter(s) in the volume in which the problems are discussed more thoroughly or from a different perspective.

The discussion has been grouped (in no particular order) under the following headings:

- sensing, other than vision,
- vision,
- mobility,
- design,
- control,
- manipulation operations,
- reasoning,
- geometric reasoning, and
- systems integration.

We conclude the paper with a list of *nonproblems*, in the hope that by doing so we might encourage researchers starting out in robotics to work on the multitude of difficult open problems. Unfortunately, the nonproblems do appear to be quite tempting to researchers.

1.2 Sensing

Over the past 25 years of robotics research, many different types of sensors have been experimented with, to varying degrees of success. As we noted in Brady [1985], robot perception is only a special case of computer perception insofar as there are occasional opportunities for engineering solutions to what are, in their general form, difficult problems. As robots move outdoors, and away from the "blocks world" of many manufacturing environments, the opportunities for shortcuts become fewer.

Conventionally, sensors are classified into a number of different categories:

• *Internal state sensors:* including potentiometers and other position encoders, tachometers and Doppler shift sensors, accelerometers, (gyro)compasses, and odometers. Position and velocity sensors are widely available, cheap, and reliable, though 16-bit position encoders have only recently emerged from military applications. Remarkably, the use of compasses and odometers is still rare on wheeled and legged robots despite the recent effort on those topics.

• *Contact sensors:* including contact switches, touch, force, proximity, and slip. Touch sensors and touch sensing are reviewed in the papers by Dario (this volume), Hollerbach [1987], and in Brady [1985, section 6.2]. While there has been substantial progress in the past 4 years, it is still unfortunately the case that commercially available touch sensors have low spatial resolution, limited dynamic range, are prone to wear and tear, and have poor hysteresis. Tactile sensors are hampering the development of tactile interpretation algorithms.

• *Noncontact sensors:* including ultrasound, active infrared rangers, structured illumination, radar, and vision. As Horn [1986] has noted, vision is our most powerful sense; but it is also our most complex. Vision has attracted more research than all the other sensors combined, so it is discussed more fully in the next section. Many of the experimental structured illumination sensors developed for robotics applications are closely analogous to visual processes. For example, light striping is based on the same geometry as binocular stereo, while Moiré fringe sensors are analogous to trinocular stereo (see Faugeras's chapter in this volume for a discussion of vision and trinocular stereo). Radar sensors have been applied in Europe to autonomous vehicle navigation.

Ultrasound sensors have been popular for many years, not least because they are cheap, rugged, and provide a low-bandwidth channel for directly sensing range. The papers by Brown [1986], as well as by Steer, by Flynn and by Richardson and Marsh in a special issue of the *International Journal of Robotics Research* (Vol. 7, No. 6, December 1988) are representative of current robot projects using ultrasound. Problems associated with sonar sensors include false depth readings caused by specularities arising from the long wavelength of sonar; excessive rounding of features such as corners because of the beam spread needed to get sufficient signal-to-noise; and the attenuation of sonar in air.

• *Application-dependent sensors:* including smoke alarms, temperature sensors, smell sensors, and microphones/sound and speech sensors. Dario's [1986] book is a representative selection of sensors for Robotics.

Each of the sensors enumerated above is appropriate for a certain range of tasks; which sensor is appropriate for which task is currently decided heuristically by experimenters on the grounds of cost, speed, and operating conditions. Different sensors provide different kinds of information, and they all have characteristic failure modes such as those elaborated above for sonar.

Since no sensor/sensor interpretation algorithm works perfectly in all circumstances, the idea has developed of having multiple sensors, perhaps of different types, and pooling the results for more reliable, more accurate sensor readings. The problem is complicated since different sensors yield different kinds of information. It is clear that simple averaging of sensor measurements is ineffective and that a more robust statistical combination of measurements is called for. Durrant-Whyte [1986] is a good introduction to the issues of multisensor data fusion, including the problem of sensors that are distributed in space. Several projects have experimented with temporal filters such as Extended Kalman Filters to integrate uncertain measurements over time. Faugeras's chapter in this volume is a good introduction to the application of the technique to vision. The books by Gelb [1974] and by Bar-Shalom and Fortmann [1988] cover the ground more thoroughly for simpler sensors.

1.2.1 Problems in Sensing

Problem 1: The continuing need for better sensors

Tactile sensors that overcome the problems enumerated above, and that compute shear, stress, and strain, have been slow to emerge from the research laboratory. Until better sensors are available, theories for interpreting touch data will continue to languish.

As robots begin to move around factories [Brady et. al. 1988] or outdoors, they confront an awkward hiatus in direct range sensing. Almost all structured light range sensors have been developed for the range 0–2 m. Because the magnitude of the speed of light implies high (gigahertz) rates for time-of-flight sampling, robust, active electromagnetic range sensors are currently only available for ranges in excess of 50–100 m. One partial exception is the ERIM sensor used in the CMU Navlab project; but it has several other problems (see Hebert). Factory mobile robots need range sensors that operate in the range 2–10 m, and they are currently not commercially available. Clearly, there is a limit to the amount of laser power than can be beamed around a primarily metallic environment that contains people. Bell Laboratories have experimental prototypes of an infrared sensor that measures range up to 15 m.

Continued development of all kinds of sensors, including TV cameras, is a problem for robotics.

Problem 2: Better use of sensor data

Currently, the choice of which sensor(s) to use in an application is based on experience or nontechnical considerations such as availability or cost. More rational choice demands sensors models that use a better description of a sensor's abilities and limitations. Similarly, statistically combining uncertain sensor measurements requires sensor models to be developed that make explicit the type of information provided by the sensor, the error distribution in that measurement, and the sensor's characteristic failure modes. Sensor models may be quantitative, such as the corrupted Gaussian probabilistic models championed by Durrant-Whyte [1986], or they may be qualitative, such as the logical sensor concept pioneered by Henderson and Shilcrat [1984]. Models of both sorts are required, together with functions that relate them, but how this should best be done remains an open problem.

Problem 3: Sensor control

Uncertainty is only one of many problems associated with sensors. An equally important problem stems from the fact that many sensors (including force, touch, and even certain visual processes) only deliver information that is local relative to the overall size of the object being sensed. This raises the unsolved sensor control problem: how to compute the most useful place to look next given what one has sensed previously and given prior expectations in the form of models.

Problem 4: Architectures for real-time sensory processing
Sensors are generally distributed in space, and they yield information at different rates, of different sorts, that needs to be interpreted in the light of past sensing and expectations. For all but the simplest sensors and tasks, serial hardware is too slow to support real-time sensor data interpretation and control. Parallel architectures are needed; but few compelling suggestions have been proposed. For example, early blackboard architectures depended critically on the time to process information being long compared to the time to write or access that information from a shared block of memory. This is unlikely to be true for a sensory processes distributed over a local area network. It has recently been suggested that MIMD architectures based on processors such as transputers are appropriate for implementing schemes for sensor integration; but the suggestion has yet to be realised in practice.

1.3 Vision

Over the past 20 years there has probably been more research on computer vision than on all the sensors discussed in the previous section combined. In part, this reflects the importance of vision as man's most powerful sense. It is also the case that vision sensors have, since the early 1970s, been relatively cheap and of high and improving standard. Also, unlike the immediacy of visual perception, humans have difficulty interpreting force, radar, ultrasound, and even touch data. Of course, while the results of visual processing are relatively easy to assess qualitatively, it is difficult to quantify them.

Faugeras's chapter in this volume is a fine overview of some of the best current work in computer vision, and he provides his own list of major open problems. Horn's [1986] book *Robot Vision* provides a good introduction to the subject, and a previous essay by the author [Brady 1986] overviews intelligent vision systems. There has been considerable progress in computer vision over the past 20 years, yet there is plainly even more to do.

Among the key advances have been

• *Intensity change detection:* Studies of step changes in intensity have led to filters such as the nondirectional difference-of-Gaussians (as well as variants such as the Laplacian of a Gaussian, or the difference of offset

Gaussians), and directional first and second derivatives championed by Canny and Haralick.

Despite the intensive research effort applied to intensity change detection, existing filters are only effective for low curvature (i.e., straight), isolated (no texture), static, monochrome, step changes. Recent work has aimed to overcome these limitations (Fleck [1988], Noble [1988], Ponce and Brady [1986], Buxton and Buxton [1983]); but there are many shortcomings and open problems.

• *Early motion:* The apparent motion caused by changing brightness patterns has been much studied. There have been dozens of proposed algorithms for computing this optic flow. Most of them are about equally effective in densely textured regions of an image, and they are all about equally ineffective at motion boundaries. Hildreth [1984] suggested that motion should be computed only at boundaries, an idea investigated further by Buxton and Buxton [1983]. Nagel has noted that the motion constraint equation can be solved at what he calls image corners. Gong [1988] has recently developed this idea further. Scott [1987] has developed an algorithm that adapts to the local image structure, resembling optic flow algorithms in densely textured regions and Hildreth's proposal at motion boundaries.

Algorithms for computing 3-dimensional structure from motion have been developed and investigated theoretically. The work is less advanced than the study of optic flow, with Charnley and Blisset [1988] being representative of the state of the art.

• *Stereo:* Several stereo algorithms, embodying many different constraints, have been implemented and carefully assessed. Most algorithms consist of constrained matching of intensity changes. Fewer algorithms attempt to match textured regions of images, while fewer still match both regions and edges. Recently there has been a spate of multiimage stereo matching algorithms, mostly using three cameras, in an attempt to restrict the importance of the correspondence problem of stereo matching. Faugeras's chapter in this volume gives details. In particular, he describes work aimed at temporal integration of partial or noisy stereo matches using an Extended Kalman Filter (EKF). Following Faugeras and his group, the computer vision community has discovered EKFs.

• *Shape from shading:* Image formation is, to a good approximation [Horn 1986], governed by a first-order partial differential equation called the image irradiance equation. A number of practical techniques, including photometric stereo, have been developed to compute visible surface geometry from one or more brightness measurements. Since image measurements are noisy, Horn and Ikeuchi pioneered the addition of smoothness terms in a least squares computation. Such terms have since been dubbed regularizers.

• *Model-based vision:* Instances of known objects have been located in brightness or range images by constrained search techniques. Approaches differ mainly on the representations of image structure used and in the technique(s) adopted for search.

Since no single visual process is yet nearly completely understood, improvements to all visual processes are open problems. The next subsection highlights just a few of them.

1.3.1 Problems in Vision

Problem 5: Applied optics
There is a considerable body of heuristic knowledge about what might be called applied optics. This includes the placement of light sources and cameras, and the kinds of effects that are clearly visible (even to a computer) in images of various wavelengths and states of polarisation. An application that seems hard to make completely reliable or efficient may often be solved easily if polarized or ultraviolet light is used, or two images are superimposed to overcome shadowing effects.

A corpus of knowledge does not constitute a theory, and the largely ad hoc nature of such knowledge means that it is rarely written down in books. All too often the tricks have to learned afresh by engineers. Vision would be more widely adopted in robotics if processes worked reliably all of the time. This is the single most important reason why edge detection, for example, continues to be avoided in industrial vision. Speed is of lesser importance; it can always be increased if it is considered worthwhile and to do so is financially feasible.

Problem 6: Seeds of perception
Not all visual information is created equal [Brady 1987]. Intensity changes provide more information than points in the middle of regions of smoothly changing or constant brightness. Since they provide more information, they offer tighter constraint. Corners offer even more information than edges, and so they in turn offer tighter constraint. Of course, it is harder to determine reliably image locations where the intensity changes significantly

than locations where it barely changes at all. With care, however, it can usually be done. In turn, it is harder to determine image corners than edges; but again it can mostly be done. As Nagel, Gong, and Harris [1988] have shown, it is worth the effort to do so. We refer to corners, and similar locations of tight constraint as *seeds of perception*, and they can be found in processes as diverse as stereo, motion, shape from contour, and lightness.

Seeds of perception suggest a process in which interpretations are grown from locations of tightest constraint to those of decreasing constraint. Such a regime is implicit in a number of algorithms; but it currently falls some way short of a theory. The problem is to develop seeds of perception so that it involves (i) a precise model of the information or constraint associated with a feature, (ii) robust techniques for computing seeds, and (iii) a mechanism for combining the constraints associated with spatially distributed seeds.

Problem 7: Model-based vision

Current model-based vision systems are narrow minded. Some argue that since objects may be overlapped, the visibility of local features cannot be guaranteed. This leads to the pessimistic stance that assumes that those features will never be visible, and programs based on such a view fail to take advantage of distinct local features if they are visible. In consequence of their pessimism, such programs are ineffective when there is too much extraneous clutter.

Others argue for image structures such as parallelism, junctions, or salient features, and their programs collapse unceremoniously if those features are not visible. Graceful degradation in performance on the one hand, coupled with opportunism and the ability to recognise model instances in a sea of clutter on the other, may be possible if there is a hierarchy of representations of image structure, at one level making parallelism, symmetry, and the like explicit, at a lower level simply consisting of a piecewise linear approximation to image edges. It is an open problem whether opportunistic programs that degrade gracefully can be built, and whether or not such a hierarchy of image structure representations holds a part of the key.

Problem 8: Dynamic vision

Few computer vision programs have a continuing perceptual existence. Instead, they gaze at one, or at most a small number, of images until their processing is done. This often takes time, though the author often consoles us that the processing could, in principle, be done in "real time." Even research in visual motion has worked on short image sequences, often consisting of as few as two images. Recently, several programs that aim at a continuing existencc have been constructed, including [Khatib, 1986; Andersson 1988; Dickmanns 1988; Ayache and Faugeras 1988; Charnley and Blissett 1988]. Andersson, for example, argues persuasively that building a vision program that had to work continuously and in real time qualitatively changed the problems that he confronted.

Many of the systems described in the papers cited are based on ideas from control theory, particularly the extended Kalman filter. More generally, the relationship between control theory and vision needs to be explored more thoroughly, including (high-bandwidth) hardware architectures for achieving vision-based control. The term "dynamic vision" has recently been coined for this overlap between control based on vision and continuing vision based on ideas from control theory. Anticipating the discussion of control (section 1.6), there is no easy off-the-shelf solution to dynamic vision: ideas about hierarchical control are rudimentary; and since the matrices involved in EKFs are not banded, it is difficult to implement EKF schemes in real time. In any case, the optimality results of the Kalman filter do not apply to the EKF. The development of dynamic vision systems, together with appropriate architectures is timely in view of the increasing availability of parallel hardware.

Problem 9: Shape

As Winston [1985] has noted, the primary role of a representation is to make information explicit and/or to expose the constraints that are exploited in recognition and reasoning. A system capable of performing a variety of tasks needs to make a variety of sorts of information explicit, hence needs to be equipped with a variety of representations. Nearly all vision programs have been constructed for a single goal, and they have used/championed a single representation that was chosen to be more or less well suited to the information required to achieve that goal.

One well-studied goal has been to determine the position and orientation of an object. The position and orientation of a rigid object can be specified by giving the position of one point—for example, the center of mass—and a set of three characteristic directions—for example, the principal axes of the object. Since the center of mass and principal axes are defined by integrals over the object, representations that accumulate information can be used to achieve the goal. Examples include the Hough transform and the Extended Gaussian Image.

More detailed representations are required if the goal is to recognize, inspect, or grasp an object, or to navigate among stationary objects. Yet more is required if the object is flexible or articulated and has movable parts—for example, a robot arm. Polyhedral approximations have been used for recognising objects, even when they are overlapped, and for some automatic assembly operations. The later section on reasoning explains why polyhedral representations are of limited use in automatic planning of assemblies that involve parts such as shafts, gears, keyways, and cams.

Other proposed representations try to make explicit commonly occurring geometric regularities, such as translational, reflectional, or rotational symmetries; surface discontinuities; and local geometric or topological surface features. Yet others attempt to represent the hierarchical decomposition of a shape into parts, potentially enabling them to relate function to form. A useful set of criteria by which to judge shape representations has been identified by Marr [1982], and augmented by other authors.

Still, there is no elixir for shape, no single representation that is useful and effective in all situations. It is probably illusory to chase such an all-embracing representation given the diversity of tasks that might confront an intelligent robot. However, no vision-guided system can mobilize a variety of representations, intelligently choosing between them to achieve different goals. And though we have progressed far in vision, no system can compute rich and useful descriptions of the complex, curved, textured, multicoloured objects that populate our everyday world.

1.4 Mobility

This section is concerned with robots that are capable of moving around their environment. We restrict attention to land vehicles, yet we omit discussion of tele-operation [Hollier 1986], and of the ambitious European PROMETHEUS project, which aims to integrate advanced information technology into automobiles to aid the driver.

Wheeled vehicles demand a continuous path with limited grades that is not excessively rough relative to the smoothing filter provided by the suspension, tyres, and wheel radius. Wheeled vehicles deal poorly (or not at all) with obstacles even as small as steps. However, particularly on specially prepared terrain, such as roads and in factories, wheeled vehicles substantially outperform tracked and legged vehicles.

Most automated guided vehicles in industrial use today exploit additional modifications to the environment in the form of guide wires or stripes that simplify steering control [Hollier 1986]. Simplified control is purchased at the cost of flexibility and excessive use of space; this is analogous to the difference between road and rail transportation. Nevertheless, automated guided vehicles suffice for many applications, particularly those in which payloads need to be shuttled around between permanently fixed locations. Automated airport shuttles (e.g., those operating at Dallas-Fort Worth and at London Gatwick) illustrate this.

Work at Oxford on mobile robots is based on two vehicles, one of which is a research version (the Turtle) of a commercial free-ranging autonomous guided vehicle (see [Brady et al. 1987; Brady et al. 1988] for details) originally constructed by GEC-Caterpillar. The Turtle has a scanning infrared laser that reads bar-coded targets attached to the wall at known locations. Triangulation from a number of visible targets yields an estimate of the Turtle's position. The industrial GEM-80 controller follows a user-supplied trajectory by applying a Kalman filter to the laser data and to odometry. The Turtle's estimate of its *own* position is very accurate. However, it cannot sense the positions of movable objects in its environment, and this provides the context for the Oxford work [Brady et al. 1988], which is aimed at acquiring pallets from a large unstructured area and at real-time obstacle avoidance. The Oxford work emphasizes sensor integration (the Turtle is being equipped with an array of 16 sonar sensors, a steerable digital sonar sensor, a time-of-flight infrared ranging sensor, a structured light laser sensor that is effective in the range 2–5 m, a trinocular stereo head, and model-based vision), path planning, and systems integration.

Over the past 5 years there has been a spate of wheeled vehicle projects whose shared goal appears to be to follow roads. The chapters by Herman and Kanade and by Bolles and Pentland in this volume describe work on two of the major US projects. Dickmann [1988] is representative of European work on image-based control of steering. The emphasis of most recent projects is on 3-dimensional vision, though both chapters in this volume describe the ERIM range sensor. In other work, Charnley and Blissett [1988] have demonstrated a 3-dimensional vision system capable of computing "drivable" regions in image sequences of curved and rough roads.

The limitations of wheels are partly overcome by tracked vehicles. Military tracked vehicles can overcome remarkably rough terrain and surprisingly high obstacles. The second mobile robot at Oxford is a tracked vehicle, originally designed for teleoperated bomb disposal. Current work aims to equip it with a controller based on a hardware/software architecture for real-time sensor fusion.

Raibert's chapter surveys legged robots. He points out that legs are active suspension units unrivaled by those available on current automobiles. The practical implication is that the payload can move along a smooth path despite rough terrain. He also notes that legged vehicles do not require a continuous path to move along: feet can be placed at discrete locations to provide traction and support. Some preliminary work has investigated techniques for computing foot placements after sensing the terrain. For example, the OSU Hexapod [McGhee, Ozguner, and Tsai 1984] has a simplified stereo laser system and, based on the information it provides, has crawled over a pile of railway ties. Raibert draws attention to a number of lessons learned in research on legged locomotion:

- Mechanical designs can finesse control, as exemplified by Hirose's pantograph mechanism, which is directly controlled in Cartesian coordinates.
- The "virtual leg" concept supports elegant hierarchical control systems, building from a one-legged system to one with two legs, and then to four legs.
- Symmetry is a feature of legged locomotion.

1.4.1 Problems in Mobility

Problem 10: Systems architecture for mobility
Mobile robots focus attention on many of the problems discussed elsewhere in this essay: sensor integration, dynamic sensing and vision, sensor control, and navigation. These components aside, mobile robots also force attention on systems integration (see section 1.10). The appropriate hardware and software architecture for a multisensor, purposive, mobile vehicle remains an open problem, though there has been some preliminary work [Brooks (this volume); Giralt et. al. 1985; Harmon 1987].

Problem 11: High-bandwidth control
Advanced autonomous guided vehicles, such as that developed and marketed by GEC-Caterpillar, are based on low-bandwidth industrial controllers. As noted earlier, the GEM-80 controller applies a Kalman filter to a sequence of trajectory set points using sensed data from the vehicle's odometers and laser ranging system.

As AGVs move into more complex, less structured environments, typified by outdoors applications such as stockyarding, the sensors will, of necessity, need to provide richer information about the environment. Vision and range sensors produce information at much faster rates than sonar and infrared sensors used in current AGVs. In consequence, high-bandwidth controllers will need to be developed, based on more sophisticated nonlinear filters than the EKF. The design of such filters and controllers will be a major challenge for advanced robotics, and for computer vision.

Problem 12: Sensor-based steering and foot placement

Legged robots barely know where to place their feet. Notwithstanding the extraordinary achievements by Raibert's group in making a biped robot perform gymnastics and a quadruped robot trot, no legged robot could reliably go for a jog through a park. Wheeled robots do not fare much better. They may be able to (slowly) circumnavigate obstacles; but they cannot judge the quality of the terrain and so choose which way to steer locally.

There are several dimensions to local steering control: first, the sensing problems of locating small rocks and holes and estimating the mechanical surface properties of the ground; second, given a perceptual representation of the terrain immediately in front of the vehicle, deciding in real time which way to steer or where to place the feet to provide stability, traction, and support.

Problem 13: Energy and legged locomotion

Why does a horse change from a trot to a canter, from a canter to a gallop? Part of the answer seems to be to minimize the energy expended for a given speed, terrain, and payload. Such energy considerations have yet to be studied in research on legged robots.

1.5 Design

Several of the chapters in this volume are concerned with facets of robot design:

• Whitney presents a historical account of techniques for manipulation and assembly. The difficulties posed by stiff, overdesigned robot arms and the need for fast coordinate conversion (resolved motion rate control) are discussed. His equation (9) gives an expression for the ratio between the torque required of a hand to that required of an arm, and he notes that this is achieved in human manipulation by a small dextrous hand. The remote-centre compliance device for compliant assembly is described.

• Hollerbach reviews progress in modeling and computing the kinematics and dynamics of robot arms, and explores the implications for controlling those arms. The solvability of spherical wrists, singularities of 6-degrees-of-freedom arms, and redundant manipulators are discussed. The problems caused by gears are recalled and recent experimentation with a direct-drive robot arm reviewed.

• Salisbury summarizes the considerations that informed the design of a three-fingered hand. The kinematics was chosen because it was minimally capable of imparting arbitrary small forces in arbitrary directions, while allowing interfinger forces necessary to grasp objects. The degree of manipulability of an object at a point in the workspace of the hand was estimated using the condition number of the Jacobian of the coordinate transformation into hand coordinates. This idea is similar to the *measure of manipulability* developed by Yoshikawa.

• Asada sketches an approach to arm design in which the mechanical construction is devised in tandem with the controller for the arm. He suggests that a geometric quantity, defined throughout the workspace, called the *generalized inertia ellipsoid*, can help arm design. He illustrates the use of the generalized inertia ellipsoid in the design of both a serial and a parallel arm, showing how decoupled dynamics can be accomplished (his theorems 1 and 2).

• Roth's chapter sketches the use of constraint equations to design open-loop systems that achieve specified output forces, velocities, or first-order path derivatives. The equations relate those entities that define the desired capabilities of the system and a set of unknowns that correspond to structural dimensions. The technique can be applied, for example, to determine appropriate dimensions for robot arms, dextrous hands, and walking machines. He illustrates the approach for the space of designs of robot arms consisting of two revolute joints, and suggests how it might be extended to three and more degrees of freedom.

• Jacobsen and his colleagues propose an approach to the design of robot effectors that starts from a definition of system behaviors and ends with a set of parametric constraint equations. En route a set of eleven "quantitative performance criteria" are defined for a linear effector model, including "structural smoothness," "structure/actuator stability assurance," "quickness," "strength," and "positional accuracy."

• Raibert's chapter uncovers a set of design criteria for legged robots (see above).

1.5.1 Problems in Design

Problem 14: Designing a head, knuckle, or knee

Roth's constraint equations, Jacobsen's acuator model and parametric design, Asada's generalized inertia ellipsoid, and Salisbury's grasp stability all point the way toward a theory of design for robot arms, end effectors, and legs. Necessarily the theories are, to varying extents, abstract. The power of Roth's approach, for example, precisely derives from postulating idealizations such as instantaneous quantities at a point.

Though each of the authors referred to in this section has designed and built innovative robot structures, applying the currently fragmentary theory of design to produce real devices poses a considerable challenge. The title of the problem enumerates three specific devices to which the theory of design might be applied:

- The control of attention is a problem for visually guided control of robots as there is too much to look at in an image. Directing attention involves controlling head and eye movements. A number of "head" mechanisms have been built for this purpose, but they are clumsy when compared with the beautiful structures implemented by the authors referred to above. A study of head designs for two (and three) camera vision systems would be of immediate interest to vision researchers. Coincidentally, the physiology and control of the human eye-head system has been closely studied, though this should not necessarily constrain the design.
- The fingers of a human hand approximate RRRR kinematic chains. This is not quite exactly the case, as the distal knuckle joint cannot move independently of the proximal joint. Wood, for example, has investigated the intricate tendon structure of human fingers. Knuckles are largely driven remotely by an extraordinary tangle of tendons, though finger muscles to provide additional actuation. While Jacobsen, Wood, and their colleagues have done great work on robot hand design, the knuckle remains a paradigm of the challenge of building

joints and (see below) actuators that are light, small, accurate, and fast.

• The knee is one of the most complex joints in the human body. It achieves remarkable performance in active suspension during running, jumping, and carrying loads. Prosthetic knees have been developed with some success, though their designs are but pale imitations of real, human knees. The mechanical design of a human knee equips it for the dynamic challenges that it must achieve, yet even the geometry of how it does this is scantily understood. The knee joint is a paradigm of relating mechanical design to complex control and dynamics for active suspension.

Problem 15: Actuators and mechatronics

Actuation is all too often the Achilles heel of robotics. Available actuators are either too big, too difficult to control (most hydraulics), too slow, not powerful enough, or have to overcome enormous friction from backlash, cogging, and other non-linear effects of gearing (see Hollerbach's chapter). Choosing an actuator tends to be driven heuristically by a single facet of the design; then the further consequences of that choice have somehow to be overcome. For example, if strength or power-to-weight ratio is a key design requirement, hydraulic or pneumatic actuation is often chosen; but then the limited (relative to electric motors) possibilities for fine control have to be faced.

The deficiencies of current actuators have retarded the development of high-performance robot arms. Hollerbach notes that "the advent of direct drive technology has been a big boost to experimental robotics, since meaningful control studies incorporating dynamic models are being carried out." But direct drive motors are also plagued with problems—for example, the significant levels of torque ripple (despite manufacturers' published claims to the contrary). Torque ripple severely degrades the signal-to-noise ratio in the signal emanating from the motor; hence it limits control. Other actuator technologies have been explored, including arrays of tiny electrostatic motors implemented on silicon chips and shape memory alloys.

A recurrent theme in the chapters referred to in this section is the need to design a mechanical structure in tandem with the control. This is in essence one of the key ideas of *mechatronics*. Conventionally, a motor is connected to a computer that is at some distance removed from the motor. The electrical characteristics of the motor are fixed; it is the computer's thankless task to try to make the motor appear other than it really is, in effect trying to make a silk purse out of a sow's ear. A fundamental idea of mechatronics is to put the computer *inside* the motor, and to make the electronics of the motor define its performance characteristics. The idea is that simply by changing EPROMs, and the software they contain, one might change the velocity or torque profiles of the motor. In view of critical role played by actuators, realizing this dream is a key problem for robotics.

Problem 16: Parallelism, flexibility, and Robotworld

The potentially enormous space of designs for robotics has barely been charted. The vast majority of arms are serial kinematic chains. Key contributions will be made by studying relatively uncharted areas of the design space:

- Asada's chapter discusses some of the advantages of parallel structures such as pantograph mechanisms, as does Raibert's in reviewing Hirose's quadruped. Asada's parallel direct drive arm should encourage further efforts.
- Most arms are supposed to be rigid, though all have some (unmodeled, uncontrolled) flexibility in both joints and links. Flexible arms can be lighter and faster than rigid arms, and they can support much greater payload-to-weight ratios. The downside is that flexible arms are considerably more difficult to control. Daniel [1987, 1988] has designed flexible 1-, 2-, and 3-degrees-of-freedom arms. Torque ripple of the direct drive motors powering the 3-degrees-of-freedom Rotabot currently poses a problem for the effective control of the device. The design and control of flexible arms remains an open problem.
- Scheimann has proposed and implemented a totally integrated system of sensors, actuators, and fixtures called Robotworld. As the name implies, the entire design was oriented toward the needs of the robot and away from the needs or biases of the human designer putting it together. In essence, it works out the Gedanken exercise: "If I were a robot, what would I need to know, to what accuracy?"

1.6 Control

Robotics is a tough nut for control theory and implementation. Among the more difficult challenges are

• *Multiple-degrees-of-freedom devices:* Most industrial robot arms have between 4 and 6 degrees of freedom. Often, multiple arms are required to cooperate—for example, in painting the inside of a vehicle. Existing robot hands have as many as 32 degrees of freedom.
• *Complex rapidly changing dynamics:* The equations of motion of a robot arm are usually modeled by coupled second-order differential equations (see Hollerbach's chapter). High link speeds mean that velocity dependent torques and inter-joint reaction torques become significant.
• *Nonlinearities:* There are many sources of nonlinearity, including cogging, backlash, saturation, digital approximations to analog quantities, and thresholds used to remove sensor noise. Some sources of nonlinearity can be overcome using actuators (like direct drive motors) without gears; but nonlinearity is inevitable and significant for high-performance robots.
• *Many sources of error to be controlled:* Whitney's figure 8.9 (this volume) illustrates the diversity of error sources (he enumerates seven distinct types) that inevitably arise for generic applications such as peg-in-hole insertion.
• *Discontinuous changes:* for example, in inertial loading when objects are picked up or let go, or the formation of closed kinematic chains when the robot contacts the environment.
• *Plant models that are a complex function of state:* such as the Jacobian transformation used in resolved motion rate control, and Asada's generalized inertia ellipsoid measure that varies throughout the workspace.
• *Large-amplitude, high-frequency changes:* generated by hard-on-hard contacts that are characteristic of assembly or surface processing.
• *No colocated sensors for force control:* Few robot arms can supply reasonably accurate force measurements at their joints, so force control strategies quickly become unstable.

Arimoto's chapter notes that in current practice, control design for robot arms ignores most (typically, all) of the above considerations. He suggests that despite the complications of nonlinearities and coupling torques, classical proportional-derivative techniques can work well in practice, at least for the low-performance arms currently available commercially. Essentially, this is done by increasing the loop gains sufficiently. He further suggests a design in which this classical servo is augmented by modern control techniques to compensate for nonlinearities and coupling torques.

Many advanced control techniques have been proposed, occasionally simulated, and sometimes implemented in practice, to counter the problems enumerated above. Among these are (see Arimoto's chapter or Craig [1987] for references) resolved motion rate control, inverse kinematic control, computed torque control, hybrid force-position control, generalized stiffness and impedance control, nonlinear decoupling, model-reference control, stochastic control, and self-tuning adaptive control. Stability results are currently only available for PID control. Robust control and multivariable control that tolerates uncertainty are all promising, but experimental evaluation of all these advanced techniques lags behind theoretical developments by a large margin.

1.6.1 Problems in Control

Problem 17: Experimental evaluation of modern control techniques
The actual problems posed by advanced robotics for the design of control systems were listed above. In a sense, they are already being worked on. However, too many papers that are backed up only by software simulation continue to be written on the application of modern control techniques to robotics. To be sure, simulation has an important place in the development of control algorithms—for example, in analyses of stability and in checking one's proper use of numerical methods. However, the real world is hard to simulate. For example, Gaussian noise models are only an approximation (occasionally poor) to reality.

In view of the problems for control design enumerated at the beginning of this section, it is important that extensive experimental evaluation be made of the relevance and applicability to robotics of the techniques of modern control theory. Hollerbach (this volume) notes that this is not at all easy; he expresses the opinion that direct drive motors have only recently made feasible experimentation with control based on dynamic models. Nevertheless, experimental evaluation is imperative if the claims for modern control techniques are to be taken seriously and become routine in robotics practice. Here are two specific examples:

- *On-line parameter estimation:* High-performance arms with discontinuous changes in inertial loading implies accurate, on-line parameter estimation. Arimoto emphasises the importance of "learning" for accurate trajectory following. Hollerbach reports some encouraging initial results for the MIT direct-drive arm. On-line parameter estimation is essential to self-tuning adaptive control. Daniel [1987] reports some initial findings with the GPC algorithm. Clearly, considerably more work needs to be done.

- *Computed-torque control:* This is a promising technique for achieving force control of a single joint with a colocated force sensor. It has yet to be convincingly demonstrated in practice, however.
- *Multiple-input/multiple-output control:* Joints 2 and 3 of the Unimation PUMA robot, and joints 2 and 3 of a SCARA robot have parallel axes. Interaction torques experience at joint 2 (for example), caused by the motion of joint 3, can be the dominant considerations in its motion. The independent joint control techniques that predominate in robotics simply reject the interaction torque as a disturbance. An alternative approach is to view joints 2 and 3 as a two-input/two-output system, in which the interaction torques are (approximately) linear. Multivariable control techniques have, however, rarely been applied to robots.

Problem 18: Multiprocessor architecture for controlling a mobile vehicle or a robot hand

In a later section on systems integration, we note that the software architecture of robot systems inevitably amounts to a real-time operating system. A number of multiprocessor architectures for control have been designed and implemented—for example, based on recursive formulations of the dynamics of an open kinematic chain, a custom built bus, or on transputer arrays. Brooks's chapter argues for a different approach to systems architecture for robot control that he calls a *subsumption architecture.* He presents some initial results with an implementation of the approach.

It is too early to know which of these multiprocessor architectures will be most appropriate for which aspects of robot control. Multiprocessor control of a sensor-guided vehicle, or of a multifingered hand, would provide good experimental data for judging the merits of different architectures.

1.7 Manipulation Operations

In his historical survey of manipulation and assembly, Whitney (this volume) concludes that "consideration of what robots *are* has evolved over time into consideration of what robots *do* ... this fact forces us to think of robot and task together." Similarly, Mason (this volume) focuses on the mechanics of primitive manipulation operations, to address issues such as "What are the fundamental mechanical processes of manipulation?" and "What is the correspondence between manipulation task domains, manipulator operations, and automatic planning procedures?" Both are led to investigate operations that might be considered *generic* in robot manipulation: peg-in-hole insertion, planar pushing, squeezing, parts orienting, and grinding.

Whitney begins by discussing compliant assembly, and shows that insertion is an important, typical example. He contrasts *active* compliance, which is, perhaps inevitably, computationally and sensing intensive, with *passive* compliance, in which the mechanical structure significantly reduces, and occasionally eliminates, the need for computing and sensing. He describes the evolution of remote-center compliance (RCC) for peg-in-hole insertion and generalizes it into a mathematical theory of parts mating that shows close agreement with observations. The mathematical theory underpins the design of parts mating devices—for example, if a particular dimensionless measure of the insertion depth exceeds the coefficient of friction, then elastic deformation of the parts leading to "wedging" cannot occur.

Mason also emphasizes sensorless assembly, and illustrates his approach with discussions of planar pushing, squeezing, and orienting objects using a tilt tray. The operation of pushing—for example, with a finger—has two variables that must be determined by a planning procedure: the finger orientation relative to the object and the direction of finger motion. Choosing values for these variables is simplified by a representation called the *pushing space diagram*; to plan a pushing motion, any point can be chosen within a region of the pushing space diagram and the corresponding pushing motion executed, leading to predictable object orientation. Squeeze-grasping begins with no contact, then pushing with one finger, finally pushing with two fingers. The squeeze-grasping diagram can be constructed automatically from a pair of pushing space diagrams.

The analyses of the manipulation operations described by Mason and Whitney support important practical applications: fast parts insertion by RCC devices and parts positioning and orientation on conveyor belts using guide rails whose placement can be derived from Mason's pushing space diagrams.

Despite the enormous potential of the work of Mason, Whitney, and others, caution is in order. Mason notes that a "troublesome aspect of [his] approach is that the task domains are terribly narrow. Mechanical analysis of a task requires a list of assumptions that severely limit the applicability of the results." Similarly, Whitney introduces his work on robot grinding with the caution that "we do not understand the process very well and we cannot predict perfectly what will happen, even if we could control the robot and the grinder perfectly." His approach is to rely upon stochastic dynamic programming to develop a plan that minimizes the time taken to do the job, avoids overgrinding, and does not overheat the work or leave a rough finish.

1.7.1 Problems in Manipulation Operations

Problem 19: Increasing the repertoire
Operations with a reasonable claim to be generic that have barely been studied include (i) *impacts*, such as tapping, striking, dropping, and choppings, and (ii) *twisting*, such as screwing, spinning, and drilling. No doubt others can be identified. More complex geometries such as L-shaped parts and T-sections pose additional problems.

The RCC is designed to accommodate two different events that may occur during chamfer crossing: a sideways translation and pure rotation. That is, the RCC blends the *two* manipulation operations of pushing and twisting, in the context of chamfered insertion. The axis of the twist is orthogonal to the direction of push. Bayonet fitting fixtures (such as some lightbulbs) also involve the simultaneous application of a push and a twist; but here the axis of the twist is parallel to the direction of push. In short, there is a hierarchy of operations in manipulation and assembly. However, it is unclear how different representations of the sort proposed by Mason can be combined (subject to temporal constraints) to build representations automatically for these "higher-order" processes, in the way that pushing leads to squeeze-grasping.

Problem 20: Adding sensing
Passive compliance has the advantage of speed, as computation and sensing are significantly reduced and occasionally eliminated. The price paid is adaptability—software control algorithms are easier to change, perhaps on-line, than are mechanical structures. A descendant of the RCC was instrumented for force control to make the device applicable to a wider range of tasks. Recently, Mason and his colleagues have begun to investigate the trade-offs between (i) a sequence of sensorless operations, such as allowing a part to collide with the sides of a tray, and (ii) using a sensor to constrain the configuration of the part. Such analyses have great potential throughout robotics. The heart of the problem is to relate the constraint on the state of an object as a result of a sensing step with the result of a sensor-less operation. This is relatively straightforward in a 2-dimensional,

rectangular environment. More generally it involves several other problems discussed in this chapter—for example sensor models, logical sensors, and shape representations.

Problem 21: Qualitative process models

The applied mechanics of most practical instances of any manipulation operation are likely to be too complex to carry out analyses of the sort described by Mason and Whitney. Significantly, Mason's work emphasizes representations that form the basis for automatic planning of pushing, squeezing, and orienting parts. This leads to the problem of developing qualitative accounts for manipulation operations, perhaps based on the idea of qualitative reasoning (naive physics). The next section explains this.

1.8 Reasoning

The relationship between artificial intelligence (AI) and robotics is the subject of an earlier essay [Brady 1985]. The relationship is necessarily two-way:

• AI provides techniques important to intelligent robotics. Examples abound: *search* processes are central to Brooks's [1982] work on symbolic reasoning with uncertainty, to most robot path planners, and to Lozano-Pérez, Mason, and Taylor's [1985] work on automating compliant assembly; *constraint propagation* is the basis of Popplestone's [1986] work on design-to-product; *expert systems* have been developed for the design of control systems and for workcell layout; and *neural networks* have been used to learn robot trajectories [Pabon and Gossard 1988].
• Conversely, robotics continues to be a fruitful if demanding domain in which to develop AI. AI issues that are ubiquitous in robotics include time-critical planning and dealing with uncertainty. Many robotics applications are characterized by *competence intensity*—humans know a vast amount about even the tiniest aspect of a craft skill [Brady et al. 1984]. This can be contrasted with the shallow knowledge of most rule-based expert systems.

Robotics and artificial intelligence have developed symbiotically. Naturally, the central problems in robotics reasoning are central problems in AI.

1.8.1 Problems in Robotic Reasoning

Problem 22: Real-world planning

Early work on AI planning concentrated on off-line planning of simple actions in an idealized, static, toy world. Although this represents a considerable abstraction of planning, improvising, and reasoning about the real world, a number of important issues were uncovered and studied, including the frame problem, the need for hierarchical planning, and the need to represent dependences and constraints. After something of a hiatus in planning research there is a resurgence of interest, this time on issues that are clearly important in any practical application:

- *Temporal planning* of time-critical events (in particular, exploring the relationship between planning and classical control). Temporal planners must satisfice, in the sense of being able to offer their best suggestions for a plan when a resource such as time is exhausted.
- *Planning in the face of uncertainty* including sensor-based planning.
- *Reactive planning and situated action*, in which a planner responds quickly to a changing world in which changes invalidate or render irrelevant a plan conceived in a previous world state. Agre and Chapman [1987] describe a program that plays a video game in which the hostile environment is continually changing. They suggest an indexical-functional representation for variables as opposed to that used in conventional robotics programs.

Despite this resurgence of effort, each of these topics requires considerable research.

Problem 23: Qualitative process theory and naive physics

Hopcroft and Krafft [1984] point out that conventional engineering is founded upon a variety of representations, each one of which supports some operations but makes others infeasible. Differential equations and finite element approximations have been used with great success to model heat and stress distributions in solids and in processes such as injection moulding. Such models are inappropriate for macroscopic processes such as assembly. Classical mechanics, though nondeterministic because of friction, is the basis for most current work in assembly and robot dynamics and control. Geometric modeling systems support computer-aided design and the simulation of the kinematics of robot motion. But none of these representations is appropriate for reasoning about whether or not a rusty screwdriver with a slightly chipped blade might be adequate for removing a partly worn screw.

The lesson is that modeling the physics of a situation is not necessarily the way to solve a problem of robotics. On occasion, perhaps more often than we might initially suppose, a *qualitative*, symbolic, "naive" representation might be more appropriate. The past few years have seen a rapid growth in work on qualitative reasoning, together with its attendant technologies of truth-maintenance systems and sign algebras. [Bobrow 1985] is a good introduction.

Qualitative reasoning occasionally needs to be refined to be quantitative, but the existing qualitative reasoning techniques support this poorly or not at all ([Williams 1988] addresses this issue in his work on sign algebras). A second qualitative reasoning issue in need of research is covered in the next topic: shape and dynamics: existing qualitative reasoning systems use only the simplest representations of objects, and so their application to practical problems such as arise in, for example, production engineering is extremely limited.

Problem 24: Shape (and dynamics)

Some time ago we introduced a project called the *Mechanic's Mate* [Brady et al. 1984] that aimed to reason about simple problems of assembly and disassembly—for example, those that arise in woodwork. It turns out that there are many different ways to extract a nail or screw; which method is used depends upon the fine details of the shapes of the objects involved. We proposed to apply the smoothed local symmetry representation of shape since, based on it, we had implemented a system that generated and learned semantic network representations of shape starting from images. The project has been in abeyance since the author moved to Oxford; but the issues that it raised have not diminished in importance or timeliness. In related work, Winston et al. [1985] explored the relationship between function and form in robotics reasoning; their work was based on the generalized cylinder representation of shape.

By and large, however, robotics reasoning systems ignore shape; at most they rely on polyhedral approximations. Yet shape and symbolic representation of motion and dynamics are central to our ability to reason about the world. "If you put that too near the edge of the bench, it will topple over." "If you put that on top of the structure it will likely collapse, unless you first shore it up with a scaffold" [Fahlman 1973], and "If you put that milk on top of the eggs they will break because they are too delicate."

Related problems arise in building (perhaps automatically) representations of large-scale space for mobile robots. As we shall see in the next section, representations of shape (for example, the symmetric axis transform, the Delaunay triangulation, and generalized cylinders) originally developed in computer vision have been used in geometric reasoning and path planning.

Work on shape representations for practical reasoning and for robot perception continues. Recently, for example, Fleck [1988] has noted a number of problems with representations of shape and time that correspond to subsets of $\mathfrak{R}^n$. She proposes instead a generalization of regular cell complexes to include boundaries. Considerably more work on the development of shape representations for practical reasoning is required.

Devising representations of shape that suffice for reasoning about how to assemble or troubleshoot one of a family of gear boxes would raise most of the important issues.

Problem 25: Design-to-product

In the vast majority of manufacturing applications, design takes little or no account of the processes of production. Production engineers complain about the often unnecessary challenges that result and, more often than not, change the specified design *force majeure* to make their own job simpler. Recent AI projects have aimed to close the loop between design and production.

One such project was conceived by Popplestone, and involved modeling shafts, transmissions, and motors parametrically, and then propagating constraints between these components to derive a design. The project subsequently became a demonstrator project for the UK Alvey program [Fehrenbach and Smithers 1987]. Other work [Fox and Kempf 1985] has addressed important issues such as tolerancing.

Although such pioneering projects point the way toward closing the loop between design and production, they barely scratch the surface of what is required in order to transform one of the blackest arts of manufacturing into a science.

1.9 Geometric Reasoning

Reasoning about problems involving geometry has been a continuing application of, and challenge to, AI reasoning technology. The two most studied problems areas have been (i) the derivation of sequences of steps to achieve assemblies and (ii) the determination of sequences of moves to navigate a robot around a multiroomed environment. Early work concentrated almost entirely on models of reasoning such as resolution theorem proving; the trivial representations of geometry simply modeled objects as logical variables, and the layouts of rooms or blocks by simple "nonlogical" axioms. The goal-achieving sequence of paths planned for the Shakey mobile robot were derived using the STRIPS theorem provng system.

Following Shakey/STRIPS, each successive increase in the power of reasoning systems has been applied to assembly and navigation, with increasingly complex representations of geometry. Popplestone and Ambler's RAPT system, for example, emphasizes the representation and specification of geometric goals and relational descriptions of objects. Descriptions are compositions of (homogeneous) matrix transformations that place a shaft in a hole, or constrain a block to lie somewhere on the surface of another. An early version of RAPT was based on the technology of equation solving and constraint propagation.

Recently, there has been a further advance in techniques for geometric reasoning, including Wu's method, Gröbner bases, and residual techniques. A special issue of the *Artificial Intelligence Journal* (Vol. 37, Nos. 1–3, December 1988), edited by Kapur and Mundy, summarizes the techniques and progress. Mundy has applied Wu's method to the geometry of perspective projection and has automatically derived constraints published in a research article only 4 years earlier. Residual techniques have been applied to path planning (q.v.) by Canny in his ACM award winning thesis.

In our earlier essay [Brady 1985, section 5.2], we recalled Taylor's and Brooks's work on dealing with uncertainty. The errors in joint angles, and in the positions and orientations of fixtures and of tools relative to the hand, combine according to complex trigonometric expressions to yield uncertainties that threaten the success of an assembly plan. Taylor's approach was to assume numerical bounds for the errors and to apply linear programming techniques to bound errors in parameters crucial to achieving the goal. Brooks's approach was to deduce error bounds for critical parameters using an expression bounding program applied to the symbolic trigonometric expressions defining those parameters. Brooks demonstrated that the bounds that he achieved were comparable to those obtained by Taylor; his approach had the advantage that the bounds could be used in several different ways—for example, to work backwards from a goal tolerance to bounds on the tool size guaranteed to meet the specification. The advantage of explicit structural, symbolic representation of constraints over unstructured sets of equations was also demonstrated by Fahlman

[1973]. He showed that a symbolic representation of support and stability in a pile of blocks allowed new forces and torques to be propagated quickly as blocks were added or removed. In contrast, a linear programming technique had to be run afresh every time a block is added or removed.

Two particularly fundamental geometric reasoning problems have attracted wide interest: *findpath* and *findspace*. Lozano-Pérez has noted the equivalence between these problems, and papers tend to concentrate on findpath. Lozano-Pérez and Taylor (this volume) discuss geometrical aspects of robot task planning, with particular emphasis on findpath. They distinguish (i) planning with complete prior knowledge, (ii) incomplete prior knowledge but complete execution knowledge, and (iii) fine motion planning. The chapter elegantly sets out the current state of the art and has an extensive set of references to the literature.

1.9.1 Problems in Geometric Reasoning

Problem 26: Knowledge of assembly

Despite the attention that has been paid to systems that reason about assembly, we still do not know much about how to plan automatically real assembly operations such as the construction of electric motors, gearboxes, televisions, and washing machines. Viewed as abstract geometric problems that involve such convoluted shapes as gears and cams, they are probably infeasible so far as general solutions to findpath are concerned.

Fortunately, we know a considerable amount about real assemblies—electronics boards have connecting sides that slot into backplanes, gears mesh with other gears and do so easily if they are spun together, and motors have housings with designated places that enable them to be fit to brackets. Some of our knowledge of assembly involves the shapes of objects and the constraints that shape representations suggest in practice for the motions of objects. Rotationally symmetric objects tend to be moved along or twisted about their axes; we know in practice that combining these motions is an effective technique for inserting shafts and adding sleeves. A keyway on a shaft must be lined up with the keyway on the hole. It is surely unnecessary to *derive* these motions *formally* from abstract geometric representations. In fact, the numerical complexity and inaccuracies of boundary representations of the surfaces of shafts, sleeves, and keyways may well preclude such motions from being derived at all.

It seems necessary to mobilize both (i) the general knowledge of geometry and dynamics that are the topics of the chapters by Mason and by Lozano-

Pérez and Taylor and (ii) commonsense knowledge about real assemblies like that enumerated in the previous paragraph. The development of techniques that can effectively mobilize both sorts of knowledge poses a well-known challenge to AI; but such techniques are crucial if we are to develop useful assembly systems.

Problem 27: Fine motion planning

The work reported in section 6.6 of Lozano-Pérez and Taylor's chapter and in Mason's chapter makes a good start at automating and reasoning about the crucial problems involved in fine motions. To an increasing extent, robotics is concerned with controlled collisions, and this poses as much of a challenge to reasoning as it does to design and control. As we noted in our discussion of manipulation problems, there is much to do. Mason draws attention to the severe assumptions that need to be made for applied mechanics analysis to be feasible. Lozano-Pérez and Taylor draw attention to the need to add more realistic dynamics to the operations they discuss.

Problem 28: Clash detection and its progeny

Geometric problems other than findpath have been studied. For example, Cameron [1988] has proposed a fast, complete algorithm for *clash detection*, which asks if two possibly complex shapes overlap. Such a technique may be useful in path and assembly planning if crucial points in a rough, putative assembly plan can quickly be identified and checked for collisions. Demonstrating the feasibility of such an approach remains an open problem. More generally, a list of practically important, generic, geometric reasoning tasks would be extremely valuable.

1.10 Systems Integration

Building a robot system to perform a challenging task is difficult in part because of the many separate facets of the project that need attention [Harmon 1987]. Consider, by way of illustration, the western Australian sheep shearing robot [Trevelyan 1985, 1986, 1987, 1988]. A novel wrist mechanism had to be invented, as did proximity capacitance sensors for the cutters. Trajectories were planned using a "software sheep" model consisting of a cubic spline parametric mesh defining curves on the model sheep body. The software sheep had to be transformed to fit it to the actual sheep and the trajectories defined by the mesh curves modified in real time as the sheep breathed and moved. A clamp had to be invented to hold the sheep, as did a hydraulic arm to transport the wrist and cutters. But developing all of these individual facets was simply the first stage in realizing the sheep shearing robot; at least as much work has gone into putting all these facets together to make a reliable working whole. This poses tough systems integration challenges, and it is with such issues that we are concerned in this section.

Trevelyan's experience is by no means unique. Other inspiring demonstrations as diverse as Andersson's [1987] ping pong playing robot, Miura's *Collie* system, and Horn and Winston's *Copydemo* all focused attention on systems integration. Of course, robot systems almost invariably involve large amounts of software and hardware, and building any large piece of software poses systems engineering challenges. However, robot systems are subject to additional demands:

- they need to work in (near) real time,
- they should gracefully degrade or upgrade as sensors or actuators are added, removed, or changed, and
- they are purposive.

In short, robot systems tend to be (more or less) intelligent, real-time operating systems. Systems integration software naturally assumes the form of communicating sequential processes. Every major robotics project has discovered this (for example, Miura [1985], Trevelyan [1988], Khatib [1987], and Andersson [1987]).

Worse, some robot systems consist of many processors that are distributed over a local area network. The MAP protocol layer has gained a large measure of acceptance, though there are many dark corners and parts that are not as yet specified.

1.10.1 Problems in Systems Integration

Problem 29: Communicating processes

From a software engineering perspective, there is an open problem concerning the type of asynchronous software architecture that is most appropriate to a given project. Some projects have explored object-oriented programming [Henderson and Shilcrat 1984]; others have provided semaphore guards to a shared data structure [Thorpe et al. 1988]; while others have proposed using communicating sets of processes either based on the CSP-Occam-transputer model [Dickenson 1988], on the Ada model [Volz et al. 1986], or on a parallel version of C [Cox and Gehani 1989]. At the moment, it is not known how to gauge the appropriateness of a software architecture for a robot system, or what the design principles should be. A problem consequent upon the development of such architectures is the need to develop debugging tools to speed their development.

Problem 30: Distributed hierarchical organization

Albus [1987] has begun to investigate the distribution of control over large systems. Some processors, such as those cooperating at a workcell, need to communicate quickly, while others may communicate mostly at low bit rates, apart from alarms and exceptions. Mapping asynchronous software architectures onto networked hardware architectures is a difficult problem that faces new challenges in robotics.

1.11 Concluding Remarks

Plainly there is much work to do if we are to realize a *Science of Robotics.* This chapter has enumerated a number of what we consider to be important problems that need to be solved, or at least explored. To conclude in a lighthearted vein, we now enumerate a number of "nonproblems," some (but by no means all) of them spoofs, that we hope to deflect people from working on. Any subject has its "pure scholarship" syndrome; problems that once were important or timely take on a life of their own and keep a band of dedicated scholars busy long after they become less important, even irrelevant. There are still far too few researchers working in robotics for the effort to be diluted by nonproblems. The following is a tentative list of nonproblems:

• *Yet another formulation of rigid body dynamics* applied, for its own sake, to an open kinematic chain such as a robot arm waving in the breeze. Of course, a judicial choice of representation can critically determine the effectiveness of some process such as computing the dynamics of closed kinematic chains (e.g., a hand grasping an object, or the legs of a robot contacting the ground), or computing dynamics in parallel; but it is the use of the representation that is of primary interest. Hollerbach (this volume) discusses this nonproblem in more detail.

• *Kinematic trajectory planning* is of limited interest. The papers in [Brady et al. 1983] pose a number of problems, particularly concerning knot placement in spline trajectories; but the entire approach has been overtaken by the invention of trajectory planning algorithms that incorporate the robot dynamics (see, for example, [Tan and Potts 1988]).

• *Gratuitous complexity results.* Complexity results about spatial reasoning problems—for example, *the piano mover's problem*—have provided deep insight into the intrinsic difficulty of many of the problems that are confronted regularly in robotics. However, some problems have assumed a life of their own with no obvious application to robotics. We caricature this with a spoof paper title: "The Tethered Piano Tuner's Problem I: A Single Tuning Fork."
• *Resurrecting well-known limited techniques.* The needs of a particular application may dictate the use of a well-known technique whose (typically severe) limitations are also well known. Reasons for applying such techniques include speed, cost, processing power, and memory limitations. Applying such techniques may be expedient; but the fact that the application is in robotics does not make it a contribution to robotics science. Examples include (i) using the Sobel operator to find edges quickly or applying the Hough transform to inspect widgets and (ii) using (chronological) backtracking for robotic reasoning or planning.
• *Applying a conventional SIMD architecture to a simple task in vision and/or control.* This is another instantiation of the previous nonproblem.

Acknowledgments

Thanks to Chris Brown, Ron Daniel, Hugh Durrant-Whyte, John Leonard, and Barry Steer for reading a draft of this chapter. Thanks especially to my daughter Sharon, first for getting sick, so that I could spend time at home nursing her and writing this chapter, and then for getting well again. Thanks to Terry Ehling for patience when I delivered the chapter two months late. Thanks to the SDF for support, and for the administrative assistance of Carl York and Roberta Ishihara. Thanks finally to Charles S. Smith, who provoked me 5 years ago to write this chapter, though I doubt he would be satisfied with it!

References

Agre, P., and Chapman, David. 1987. "Pengi: an implementation of a theory of activity," Proc. AAAI, Seattle.

Albus, James S. 1988. "The central nervous system as a low and high level control system," in Dario ed. 1988. *Sensors and Sensory Systems for Advanced Robots*, NATO ASI series, Springer-Verlag.

Andersson, Russell L. 1988. *A Robot Ping-Pong Player*, MIT Press.

Ayache, Nicholas, and Faugeras, Olivier D. 1988. "Maintaining representations of the environment of a mobile robot," in Bolles and Roth (eds.), *Fourth Int. Symp. Rob. Res.*, MIT Press, 337–350.

Bar-Shalom, Yaakov, and Fortmann, Thomas E. 1988. *Tracking and Data Association.*

Bobrow, Daniel G. 1985. *Qualitative Reasoning about Physical Systems*, MIT Press.

Brady, Michael. 1985. "Artificial Intelligence and Robotics," *Artificial Intelligence*, 26, 79–121.

Brady, Michael. 1986. "Intelligent vision," in Grimson and Patil (eds.), *AI in the 1980s and Beyond*, MIT Press, 201–242.

Brady, Michael. 1987. "Seeds of perception," Proc. 3rd Alvey Vis. Conf. Cambridge, UK.

Brady et al. 1984. "The mechanic's mate," O'Shea (ed.) *ECAI84: Advances in Artifical Intelligence*, North-Holland, 681–696.

Brady et al. 1987. "Progress towards a system that can acquire pallets and clean warehouses," in Bolles and Roth (eds.), *Fourth Int. Symp. Rob. Res.*, MIT Press, 359–374.

Brady et al. 1988. "Vision and the Oxford AGV," Proc. Image Processing '88 Conf., London.

Brooks, R. A. 1985. "Symbolic error analysis and robot planning," *Int. J. Rob. Res.*, 1(4), 29–68.

Brown, M. K. 1986. "The extraction of curved surface features with generic range sensors," *Int. J. Rob. Res.* 5(1), 3–18.

Buxton, B. F., and Buxton, H. 1983, "Monocular depth perception from optical flow using space time signal processing," Proc. R. Soc. Lon. B, Vol. 218, pp 27–47.

Cameron, S. A. 1988. "Efficient intersection tests for objects defined constructively," *Int. J. Rob. Res.*, 8.

Charnley, D., and Blissett, R. J. 1988, "Surface reconstruction from outdoor image sequences," Proc. Alvey Vision Conf., 153–158.

Cox, I. J., and Gehani, N. J. 1989. "Concurrent programming and robotics," *Int. J. Rob. Res.* (to appear)

Craig, John J. 1987. *Adaptive Control of Mechanical Manipulators*, Addison-Wesley.

Daniel, R. W. 1988 "A compliant direct drive industrial robot arm," Proc. RoManSy, Sept, Udine, Italy.

Daniel, R. W., Irving, M. A., Fraser, A. R., and Lambert, M. 1987. "The control of compliant manipulator arms," in Bolles and Roth (eds.), *Fourth Int. Symp. Rob. Res.*, MIT Press, 119–125.

Dario, P. 1988. *Sensors and Sensory Systems for Advanced Robots*, NATO ASI series, Springer-Verlag.

Dickenson, Mark. 1988. "A parallel-processing controller architecture for force control of an industrial robot," Proc. Second Workshop on Parallel Processing and Control, Bangor, UK.

Dickmanns, E. D. 1988. "4D-dynamic scene analysis with integral spatio-temporal models," in Bolles and Roth (eds.), *Fourth Int. Symp. Rob. Res.*, MIT Press, 311–318.

Durrant-Whyte, H. F. 1988. *Integration, Coordination, and Control of Multisensor Robot Systems*, Kluwer Academic Publishers.

Fahlman, Scott. 1974. "A planning system for robot construction tasks," *Artif. Intell.* 5(1), 1–49.

Fehrenbach, Paul, and Smithers, Tim. 1987. "Design and sensor-based robotic assembly in *design to product* project," in Bolles and Roth (eds.), *Fourth Int. Symp. Rob. Res.*, MIT Press, 391–399.

Fleck, Margaret M. 1988 *Boundaries and Topological Algorithms*, PhD thesis, MIT Artificial Intelligence Laboratory.

Fox, B. R., and Kempf, K. G. 1985. "A representation for opportunistic scheduling," in Faugeras and Giralt (eds.), *Third Int. Symp. Rob. Res.*, MIT Press, 109–116.

Gelb, Arthur. 1974. *Applied Optimal Estimation*, MIT Press.

Giralt et al. 1984. "An integrated navigation and motion control system for autonomous multisensory mobile robots," in Brady and Paul (eds.), *First Int. Symp. of Robotics Research*, MIT Press, 191–214.

Gong, S. 1988. "Improved local flow," Proc. 4th Alvey Vis. Conf., Manchester, UK, 129–134.

Harmon, S. Y. 1987. "The ground surveillance robot (GSR): an autonomous vehicle designed to transit unknown terrain," *IEEE J. Rob. and Aut.* RA-3 (3), 266–279.

Harris, C., and Stephens, M. 1988. "A combined corner and edge detector," Proc. 4th Alvey Vis. Conf., Manchester, UK, 147–152.

Henderson, T. C., and Shilcrat, E. 1984. "Logical sensor systems," *J. Robotics Systems*, 1(2), 169–193.

Henderson, T., Mitiche, A., Hansen, C., and Weitz, E. 1988. "Multi-sensor knowledge systems: interpreting three-dimensional structure," *Int. J. Rob. Res.*, 7(6).

Hildreth, E. C. 1984. *The Measurement of Visual Motion*, MIT Press.

Hollerbach, John M. 1987. "Robot hands and tactile sensing," in Grimson and Patil (eds.), *AI in the 1980s and Beyond*, MIT Press, 317–342.

Hollier, B. 1986. *Automated Guided Vehicles*, IFS Press, UK.

Hopcroft, John E., and Krafft, Dean B. 1984. "The challenge of Robotics for Computer Science," Cornell University Internal Memo.

Horn, B. K. P. 1986. *Robot Vision*, MIT Press.

Khatib, O. 1986. "Real time obstacle avoidance for manipulators and mobile robots," *Int. J. Rob. Res.*, 5(1), 90–98.

Lozano-Pérez, T, Mason, M. T., and Taylor, R. H. 1984. "Automatic synthesis of fine motion strategies for robots," in Brady and Paul (eds.), *First Int. Symp. of Robotics Research*, MIT Press, 65–96.

Marr, D. 1982. *Vision*, Freeman.

McGhee, R. B., Ozguner, F., and Tsai, S. J. 1984. "Rough terrain locomotion by a hexapod using a binocular ranging system," in *First Int. Symp. of Robotics Research*, MIT Press, 227–252.

Miura et al. 1985. "Dynamical walk of quadruped robot (COLLIE-1)," in Hanafusa and Inoue (eds.), *Second Int. Symp. Rob. Res.*, MIT Press, 317–324.

Noble, J. A. 1988. "Morphological feature detection," Proc. 4th Alvey Vis. Conf., Manchester, UK, 203–210.

Pabon, Jahir, and Gossard, David. 1988. "Connectionist networks for learning coordinated motion in autonomous systems," Proc. Seventh AAAI Conf., St. Louis, 791–795.

Ponce, Jean, and Brady, Michael. 1986. "Towards a surface primal sketch," Proc IEEE Conf. Robotics and Automation, St. Louis, 420–425.

Popplestone, Robin J., 1986. "An integrated design system for engineering," in Faugeras and Giralt (eds.), *Third Int. Symp. Rob. Res.*, MIT Press, 397–404.

Scott, G. L. 1987. "'Four line' method of locally estimating optic flow," *Im. and Vis. Comp.*, 5(2), 67–72.

Steer, B. 1988. "Protective autonomy," Second Int. Workshop on Manipulators, Sensors, and Steps toward Mobility, Univ. Salford, UK, Oct 24–26.

Tan, H. H., and Potts, R. B. 1988. "Minimum time trajectory planner for the discrete dynamic robot model with dynamic constraints," *IEEE J. Rob. and Aut.* RA-4(2), 174–185.

Thorpe, C., Hebert, M. H., Kanade, T., and Shafer, S. A. 1988. "Vision and navigation for the Carnegie-Mellon Navlab," *IEEE Trans. PAMI*, 10(3), 361–372.

Trevelyan, J. 1985. "Skills for a shearing robot: dexterity and sensing," in Hanafusa and Inoue (eds.), *Second Int. Symp. Rob. Res.*, MIT Press, 273–280.

Trevelyan, J., Kovesi, P., and Ong, M. 1984. "Motion control for a sheep shearing robot," in Brady and Paul (eds.), *First Int. Symp. of Robotics Research*, MIT Press, 175.

Trevelyan, J., Nelson, Mark, and Kovesi, Peter. 1988. "Adaptive motion sequencing for process robots," in Bolles and Roth (eds.), *Fourth Int. Symp. Rob. Res.*, MIT Press, 445–453.

Volz, R. A., Mudge, T. N., Naylor, A. W., and Brosgol, B. 1986. "Ada in a manufacturing environment," Proc. Control Eng. Conf. and Expo., Rosemont Ill, 433–440.

Williams, Brian C. 1988. "MINIMA: a symbolic approach to qualitative algebraic reasoning," Proc. Seventh AAAI Conf., St. Louis, 264–269.

Winston, Patrick Henry. 1985. *Artificial Intelligence*, 2nd ed., Addison-Wesley.

Winston et al. 1984. "Learning physical descriptions from functional definitions, examples, and precedents," in Brady and Paul (eds.), *First Int. Symp. of Robotics Research*, MIT Press, 117–135.

I PERCEPTION

2 A Few Steps toward Artificial 3-D Vision

Olivier D. Faugeras

2.1 Introduction

This chapter examines a number of issues that I think should be worked on in order for 3-D vision to become available on a large scale for practical applications. For this to become true, basic scientific questions have to be answered. We need to understand what has to be computed from images [MAR82] in order to make it possible for robots to adapt to an unknown or imperfectly known environment; also, we need to derive robust realizations of these computations and to implement them in such a way that real time constraints are met at reasonable costs.

I believe that this ambitious program can be achieved in our lifetimes; the mathematical tools are here for us to use, computer vision is coming of age, and the hardware to make it work is also becoming increasingly available. This is not to say that all the problems have been solved, but that they have been fairly well identified and that their solution is mostly a question of attacking them with the right tools and the right people. This is also not saying that they are easy problems, but I do not believe that cracking them implies major theoretical breakthroughs. Solving the vision problem is more to me like putting a man on the moon (an extremely difficult technological problem) than proving Fermat's theorem (an extremely difficult mathematical problem). This caveat is to keep all of us vision scientists honest.

In this chapter, I discuss three main topics: first, how we obtain basic 3-D data from sensors; second, how we organize these data and what we compute with; and third, how we use them to recognize and locate objects or places. For each of these three topics, I have tried to organize the discussion along three paths:

- what needs to be computed;
- how can it be computed; and
- what we think are the most challenging problems to be worked on.

2.2 Obtaining 3-D Data

There are many ways of acquiring 3-D data from the environment, depending on how far we are from the objects of interest and on how accurate

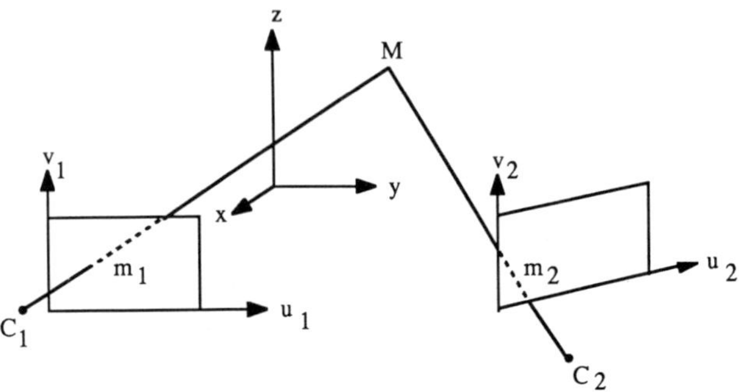

Figure 2.1
The problem of 3-D vision.

we want our measurements to be. I do not discuss here techniques such as contact sensing provided, for example, by tactile sensors and concentrate on noncontact sensing.

2.2.1 Calibration

From here on, we can describe the first problem to be solved in 3-D vision as in figure 2.1. Two optical systems, which we leave unspecified at the moment, yield two images $I_1(u_1, v_1)$ and $I_2(u_2, v_2)$ of the scene that is referenced to a coordinate system $(Oxyz)$. Two questions need to be answered:

- First, given a point m_1 of coordinates (u_1, v_1) in image 1, to which point m_2 in image 2 does it correspond, in the sense that they are the images of the same physical point M?
- Second, given m_1 and m_2, how do we compute the coordinates of M in the coordinate system $(Oxyz)$?

2.2.1.1 Geometry In order to answer these questions, we have to define models of the optical systems. The model that is universally used in the computer vision community is the pinhole model shown in figure 2.2. The pinhole model in its simplest form assumes that the system is defined by its optical axis perpendicular to the image plane and that the image m in that plane of a point M is formed by perspective projection with respect to the optical center C located on the optical axis at a distance f (the focal length) of the image plane. The image plane is referred to a system of coordinates (ouv), whereas space is referred to the coordinate system $(Oxyz)$. It is also convenient to consider the coordinate system $(CXYZ)$, where CZ is parallel

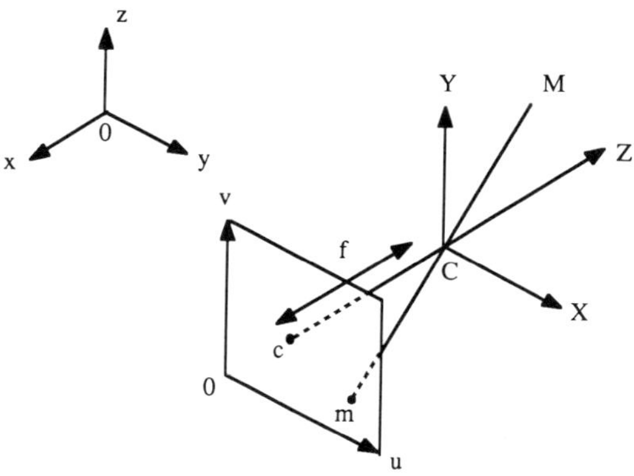

Figure 2.2
The pinhole camera model.

to the optical axis and CX and CY are parallel to ou and ov, respectively. The pinhole model is then defined by its extrinsic parameters [position and orientation of $(CXYZ)$ with respect to $(Oxyz)$] and its intrinsic parameters: the coordinates in (ouv) of the point c where the optical axis pierces the image plane and the focal length f. The problem of estimating these parameters by processing images is called the optical calibration problem and has been studied fairly widely in at least two communities:

• photogrammetry ([AK71, AK74, BRO66, BRO71, FAI75], [KAR79, KOL74, LIN72], and [MAL71 OKA81, OKA84, WON75]) and
• computer vision ([FT86, GEN79, HTMS82, IPS85, MBK81, STR84, SUT74] and [TSA85a, TSA85b, TSA86, YC78]).

One basic idea is to use a number of reference points M_i of known coordinates in $(Oxyz)$, read their image coordinates, and use the simple analytical relation (1) between space and image projective coordinates to estimate matrix M by least-squares techniques:

$$\begin{bmatrix} U \\ V \\ T \end{bmatrix} = \mathbf{M} \begin{bmatrix} x \\ y \\ z \\ 1 \end{bmatrix}. \tag{1}$$

When T is different from 0, the actual image coordinates are given by $u = U/T$ and $v = V/T$. From the 3×4 matrix $\mathbf{M}$, the unknown extrinsic and intrinsic parameters can be easily computed [GAN84, FT86, TSA86].

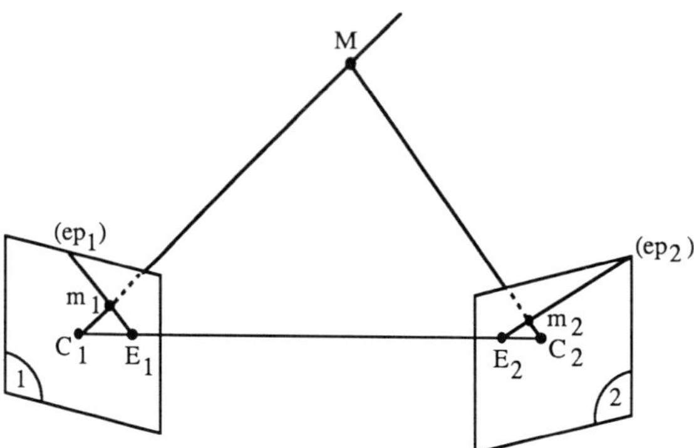

Figure 2.3
Epipolar geometry.

Before we explore any further the calibration problem, let us go back for a moment to our original questions. Considering now as in figure 2.3 a system of two pinhole cameras, the first question was, given a point m_1, image of an unknown 3-D point M, how do we decide which point m_2 in image 2 it corresponds to? At first glance, m_2 can be anywhere in image 2, but if we look a little closer at the problem, with our simple pinhole camera model we can discover some geometric properties that will help us limit our search. Indeed, the line $C_1 C_2$ intersects the image planes 1 and 2 at two points E_1 and E_2 (which are either at finite or infinite distances) called the epipoles. Notice that E_1 (respectively, E_2) can be considered as the "image" of C_2 (respectively, C_1) by camera 1. Point M is located on the half-line $C_1 m_1$; therefore point m_2 is located on the image by camera 2 of this half-line. This image is a half-line starting at E_2. These purely geometrical considerations allow us to reduce considerably the search space from two to one dimension.

The second question was, given two corresponding points m_1 and m_2, how do we compute the 3-D coordinates of the point M that they are the images of? At first glance, the answer seems to be the intersection of the two half-lines $C_1 m_1$ and $C_2 m_2$. In fact, because of the uncertainties and noise, they may very well not intersect. What are then the coordinates of M? A possible answer is to take M as the midpoint of the shortest segment between the half-lines $C_1 m_1$ and $C_2 m_2$ (see figure 2.4). We return to this problem in section 2.2.7 on motion.

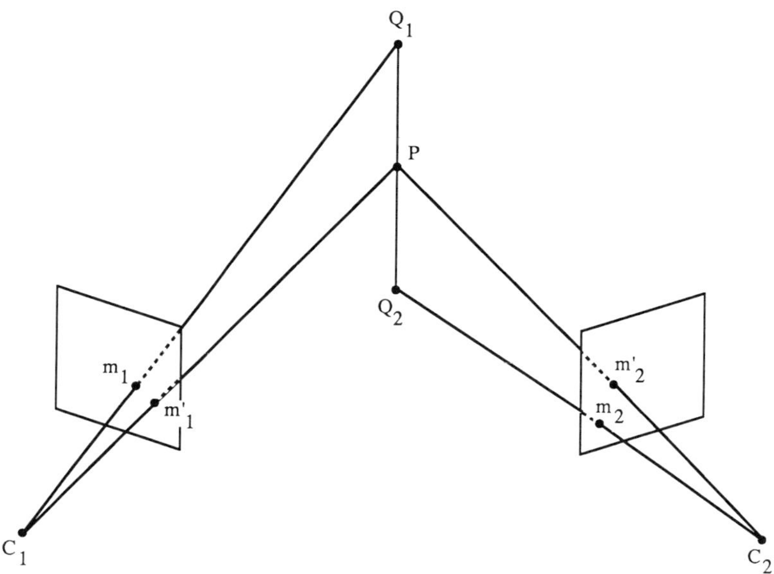

Figure 2.4
Reconstructing a 3-D point from its two images.

At this point, from a simple pinhole camera model, we have inferred a powerful geometric constraint, the epipolar constraint, which allows us to reduce the search for corresponding points. Several new questions must then be asked:

1. What is the validity of the pinhole camera model?
2. How accurately can it be recovered by optical calibration?
3. How can the geometric primitives that are necessary to enforce the constraints be computed from the model [i.e., for example, how do we compute the epipolar line of a point m_1 of coordinates (u_1, v_1)?]?

The answer to the first question is that for most of the commercially available Vidicon, *CID*, or *CCD* cameras and lenses, the model is inadequate and needs to be refined in order to include effects such as lens and off-center distortions. This implies that equation (1) does not hold anymore and nonlinearities have to be introduced. A related question is the stability in time of the model parameters. Very little is known on this subject at present. The usual solution to this nonlinearity problem is to model the real camera as the concatenation of a pinhole system followed by a non-

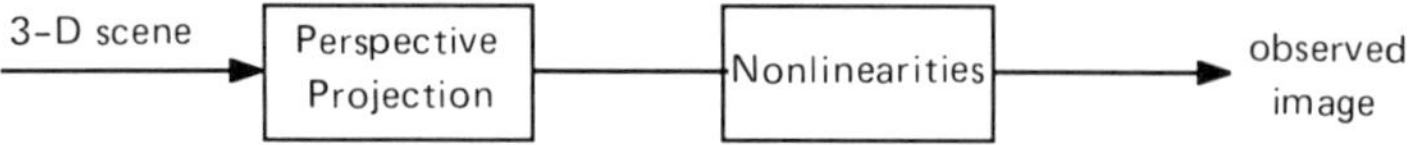

Figure 2.5
Realistic camera model.

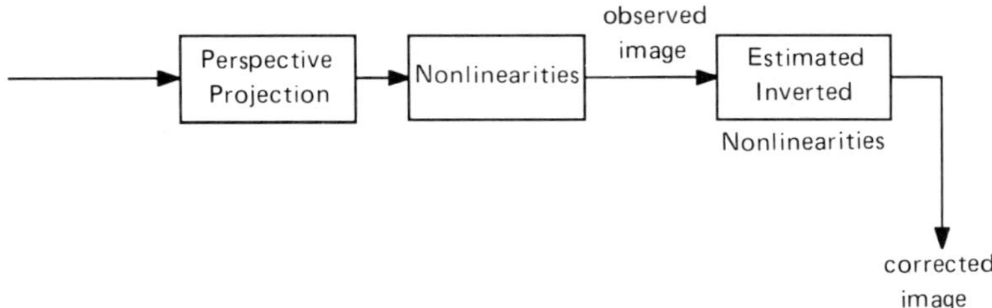

Figure 2.6
Pinhole camera model after eliminating the nonlinearity.

linear system (figure 2.5). The nonlinearity is estimated and inverted and the resulting system is shown in figure 2.6.

The second question is equivalent to asking how accurately do we have to measure the 3-D coordinates of our reference points and their image coordinates in order to obtain a reasonable estimation of the pinhole camera model parameters. Again, very few things are known on this subject at present (but see [FT86, TSA86]).

2.2.1.2 Computing the Epipolar Geometry Now, coming to the third question, we see that in order to answer it we need to find the coordinates of the optical centers C_1 and C_2, given the perspective matrices $\mathbf{M}_1$ and $\mathbf{M}_2$. These coordinates are obtained by solving the two systems of three linear equations in three unknowns:

$$\mathbf{M}_i\mathbf{C}_i = \mathbf{0}, \qquad i = 1, 2,$$

where $\mathbf{C}_i = [x_i, y_i, z_i, 1]^T$. From there, the image coordinates of the epipoles E_1 and E_2 are computed as $\mathbf{M}_1\mathbf{C}_2$ and $\mathbf{M}_2\mathbf{C}_1$, respectively.

Let us get now to the question of computing the epipolar lines. The camera model that is usually assumed in the computer vision community is that of figure 2.7, where the line C_1C_2 is parallel to the image rows. Both epipoles are at infinity and the epipolar lines are parallel to the scanlines,

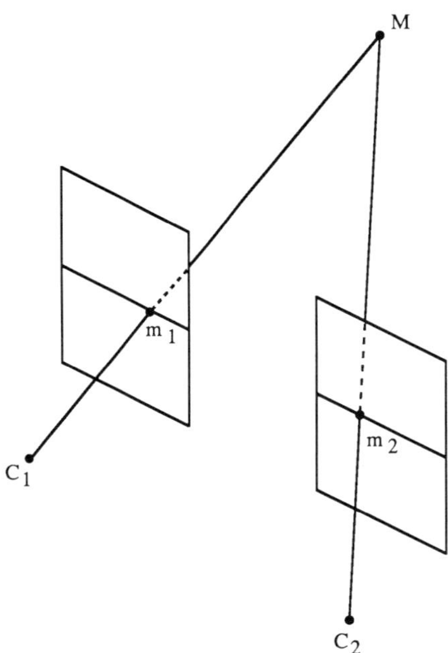

Figure 2.7
Simplified camera geometry: epipolar lines are image rows.

thus making the search process even simpler. In practice, it is very difficult to guarantee mechanically that this condition is satisfied; therefore the assumption is not reasonable. We must assume a general position for the cameras. In that situation, one popular approach is to rectify the images to make them appear the way they would if the camera geometry was that of figure 2.7. This computationally expensive procedure, called the epipolar transformation, involves interpolation and accuracy is likely to be lost.

Let us study what is meant by the epipolar transformation. We have seen in the previous analysis of the epipolar geometry that in order for the epipoles to be both at infinity, the retina planes had to be parallel to the line joining the camera optical centers $C_1 C_2$. This ensures that the epipolar lines are parallel. Our further condition is that they are parallel to the original scanlines. Therefore, one way to define an epipolar transformation is to project the two images onto a plane such as shown in figure 2.8. Since there are many such planes, we need a criterion to choose one. Indeed, the only condition is that the plane is parallel to $C_1 C_2$. One such possible

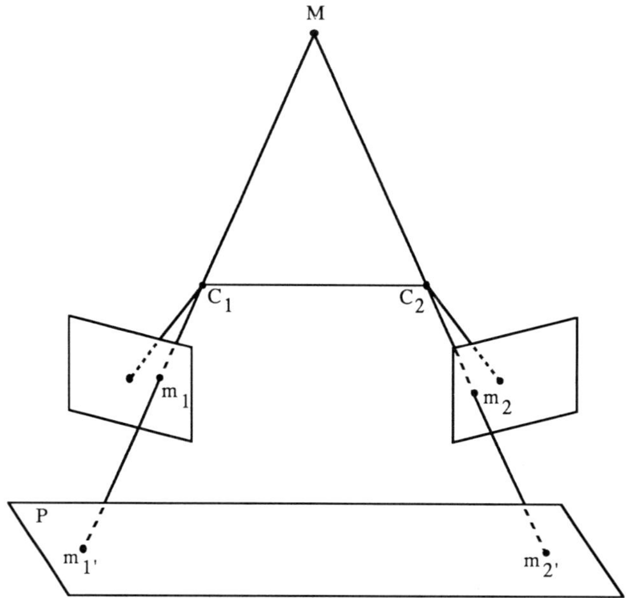

Figure 2.8
Epipolar transformation: plane P is parallel to $C_1 C_2$ and therefore the epipoles are at infinity.

criterion is to minimize the distortion between the actual images and the projected ones. In the case where the two retina planes are not parallel, this probably leads to a plane parallel to $C_1 C_2$ and to their intersection. Some work is required to understand more fully the epipolar transformation. At the moment, we believe that it is simpler to compute the equation of the epipolar line each time it is needed, especially for approaches to stereo based on symbolic matching rather than on numerical correlation (see sections 2.2.4–2.2.6). We show in appendix A how this can be efficiently done.

2.2.2 Tokens and Features

Having discussed the problems of camera modeling, we go back to our original subject of getting 3-D data. There are two main techniques for obtaining such data: stereo and structure from motion. Stereo involves a number of cameras with known positions with respect to each other; structure from motion, in the simplest case, involves one camera moving in an otherwise static environment with some unknown or imperfectly known motion. In both cases, the main question to be answered is the following:

Given a token in image 1, what is the corresponding token in image 2?

This matching problem has been shown to be ambiguous, and the previous question raises a number of other questions, such as which tokens, features, and constraints to use to reduce this ambiguity. The discussion for tokens is common to stereo and motion. Constraints are in general different.

Tokens must be such that they can be reliably extracted from images. There are a number of possible candidates.

• The simplest one is the pixel used in the so-called intensity-based correlation techniques [KMM77, GEN80, MOR80]. In order to identify pixels we have to attach features to them. What features should we compute in both images to help the correspondence problem? one condition is certainly that if pixels m and n match, then their features should be the same or approximately the same. Since these features are computed from the measured light intensities, we have to analyze their invariance with respect to the perspective transformation.

The light reflected by the surface of an object is a function of the positions of the sources, the orientation of the surface, and the viewing direction. This behavior can be summarized in the reflectance function [HOR74, HOR75, HOR77]. For a Lambertian surface (a completely matte surface), the reflected light is the same in all directions; therefore the intensities at two corresponding points are the same. Very few surfaces are Lambertian, though, the extreme case being a purely glossy one acting like a mirror for which the intensities at two corresponding points are most likely to be different. In between these two extremes there is a whole class of surfaces for which the reflectance varies slowly with the direction of viewing and therefore for which the intensities at corresponding points are fairly close if the base line is small with respect to the viewing distance. A lot of the original techniques for computing stereo matches are based on the idea of correlating the left and right intensity images. This works well when the reflectance functions of objects satisfy the previous condition and when the disparity range is not too large to prevent large correlation distances.

• The next simplest token is the edge pixel. We can also think of grouping edge pixels together to form curves or portions of curves. The simplest curve invariant by perspective projection is the straight line. We show in section 2.2.4 how to use line segments as tokens in the stereo matching process. Segments can have both geometric and intensity based features attached to them such as length, orientation, and average constrast across them. Care should be taken when using these features to guide the matching since they may not be invariant by perspective projection [BIN84, NP84].

An important related point is the classification of edges. Edges in images can be classified along several dimensions. The first dimension is the shape of the intensity function at the edge; step edges are by far the most commonly used either explicitly or implicitly, but there are other types of edges such as roof, shoulder, etc. (see section 2.3.2 and [PB85] for an application of this idea to depth images). A second dimension for classifying edges is their physical origin, i.e, reflectance edges, shadow edges, depth edges, texture edges, etc. If we could label every edge in the image by the class to which it belongs, then the ambiguity in matching could be considerably reduced. Indeed, edges in images are produced by physical events that fall broadly into two classes: first, markings on the surface of objects, such as textures, changes in reflectance, and shadows, produce edges that are invariant with respect to perspective and can therefore be reliably used for stereo matching; second, depth discontinuities may also produce edges that are invariant if the surface does not recede smoothly. If it does, the edge in the right image will be displaced by an amount that is a function of the curvature of the surface and the viewing distance. Except for this case, edges seem to be a reliable source of information to guide the stereo matching process. Edge-pixels-based stereo algorithms are by far the most common [BB81, NIS84, OK85].

• Last, we can use image regions as tokens to be matched in the stereo process. Depending on how these regions are extracted, their shapes and the intensity-based features attached to them may or may not be invariant to perspective transformation [CG87].

2.2.3 Constraints

We have already discussed the epipolar constraint. There are others that can also be put to use, which we now examine.

2.2.3.1 Uniqueness If we limit ourselves to opaque objects, one point in the first image should have at most one match in the second image. This is not true for transparent objects.

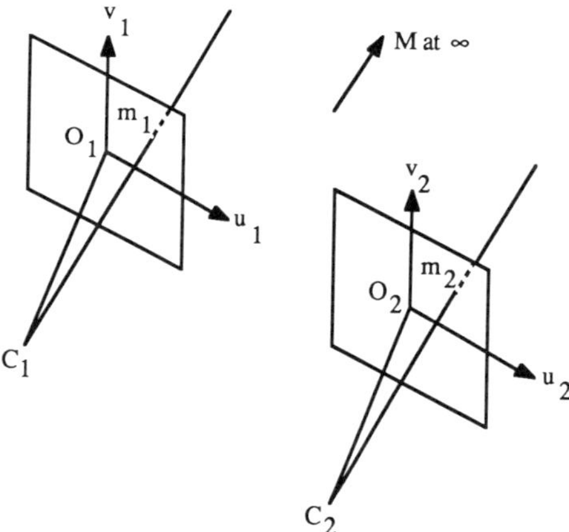

Figure 2.9
Simple disparity definition.

2.2.3.2 Continuity The basic idea of this constraint is that the world is mostly made of objects with smooth surfaces. This means that the reconstruction function that assigns to a pair of matched points a 3-D point M is smooth almost everywhere. This reconstruction function is, for historical reasons, usually summarized as a function $z = f(d)$, where z is the distance of M to the cameras and d is the so-called disparity. In order to define precisely the continuity constraint, we have to define disparity.

Given a point m_1 of coordinates (u_1, v_1) in image 1, and its corresponding point m_2 of coordinates (u_2, v_2) in image 2, disparity is defined as $u_1 - u_2$. This definition implicitly assumes the camera geometry of figure 2.9, where the two retina planes are the same. A disparity of 0 implies that the 3-D point M is at infinity. If we bring point M to m_1 along the half-infinite line $C_1 m_1$, the disparity will decrease from 0 to $-d_{12}$, where d_{12} is the distance between C_1 and C_2. There is also a simple relationship between disparity and the distance of the 3-D point M, distance measured from the two camera planes (figure 2.10):

$$d = u_1 - u_2 = d_{12} f/z.$$

Frontoparallel planes are locus of points with constant disparity.

When we assume the general camera position of figure 2.11, there is no longer a simple relationship between the coordinates (u_1, v_1) and (u_2, u_2) of the two corresponding points when M is at infinity. To find out what

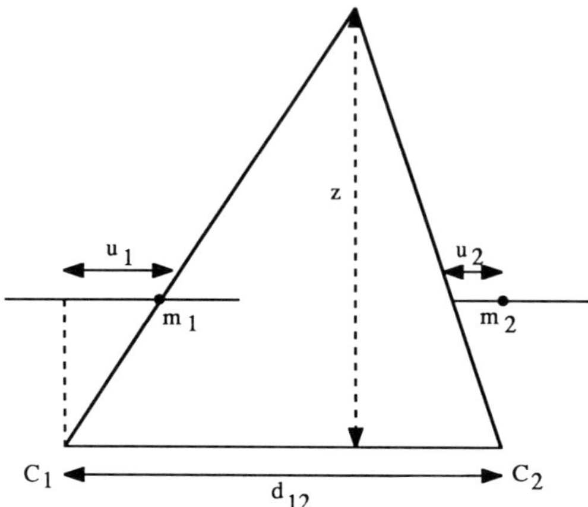

Figure 2.10
Relation between depth and disparity.

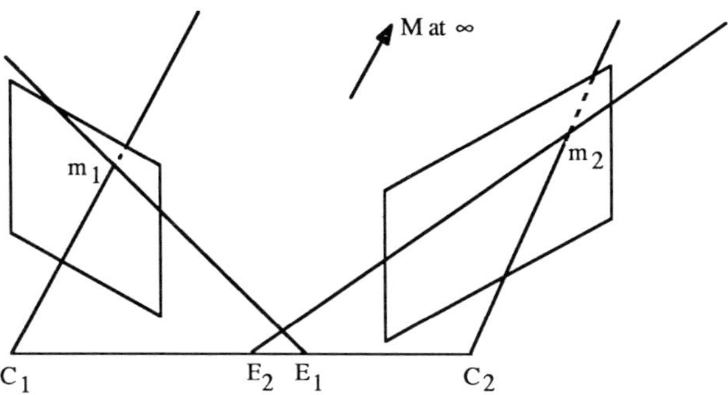

Figure 2.11
General camera position.

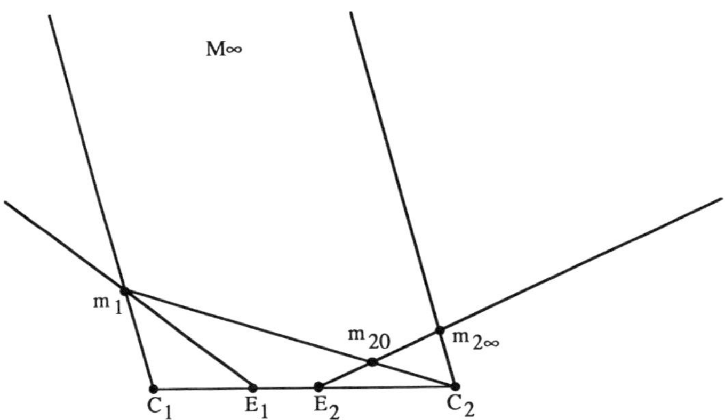

Figure 2.12
Epipolar plane configuration.

happens, let us draw the figure in the epipolar plane $C_1C_2m_1m_2$ (figure 2.12). When M travels from infinity to m_1 along the line C_1m_1, m_2 varies from $m_{2\infty}$ to m_{20} (m_{20} is the point of intersection of line C_2m_1 and the epipolar line $E_2m_{2\infty}$). What is important for the matching process is that the point corresponding to m_1 is to be found on the epipolar segment $m_{20}m_{2\infty}$.

Nonetheless, it is possible to define a notion of disparity for a point M with the help of figure 2.13. Indeed, there is a known relationship between the abscissas u_1 and u_2 of m_1 and m_2 along the epipolar lines and the coordinates (x, z) of M, where z is the distance to the axis C_1C_2 [this relation is the 2-D analog of equation (1)]:

$$u_1 = (a_1x + b_1z + c_1)/(d_1x + e_1z + f_1).$$

Similarly,

$$u_2 = (a_2x + b_2z + c_2)/(d_2x + e_2z + f_2),$$

where the coefficients $a_i, b_i, c_i, d_i, e_i, f_i, i = 1, 2$, depend only on the geometry of the two epipolar lines and the cameras centers. Eliminating x between these two equations yields

$$z = (c_2 - u_2f_2)(u_1d_1 - a_1) - (c_1 - u_1f_1)(u_2d_2 - a_2)$$
$$/(b_1 - u_1e_1)(u_2d_2 - a_2) - (b_2 - u_2e_2)(u_1d_1 - a_1).$$

Disparity can then be defined as $d = 1/z$. The origins on the u_1 and u_2 axes

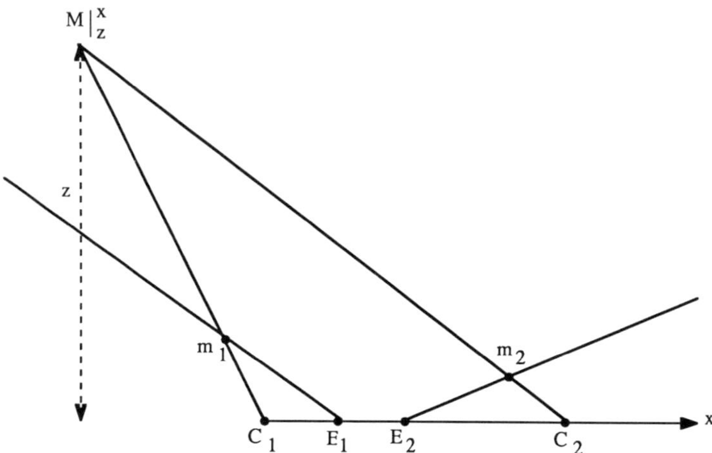

Figure 2.13
General definition of disparity.

can be chosen at the epipoles E_1 and E_2, respectively. Constant disparity now means constant z. Points of equal disparities are on circular cylinders of axis $C_1 C_2$.

If we assume that the objects that we are looking at are smooth almost everywhere, then the disparity constraint can be introduced in the following manner. Let M be a point in 3-D space with projections m_1 and m_2 on retinas 1 and 2 with a disparity d as previously defined. Then a neighbor n_1 of m_1 in retina 1 should find a match n_2 in retina 2 with a disparity close to d.

2.2.3.3 Ordering Let us consider a 3-D point M_1 and its projections m_1 and n_1 in retinas 1 and 2, respectively. Figure 2.14 shows the corresponding epipolar plane configuration. Now let us choose a point M_2 in the cone defined by M_1, C_1, C_2 containing the base line $C_1 C_2$ (lined in figure 2.14). M_2 has images m_2 and n_2 in retinas 1 and 2, respectively. Let us now look at the order of the images along the epipolar lines: we have (E_1, m_2, m_1) for retina 1 and (E_2, n_2, n_1) for retina 2. In other words, the images of M_1 and M_2 appear in the same order along the epipolar lines when M_2 is in the previous cone. It is easy to see that the converse is true when M_2 varies in the other cone defined by M_1, C_1, and C_2 (the one that is not lined in figure 2.14).

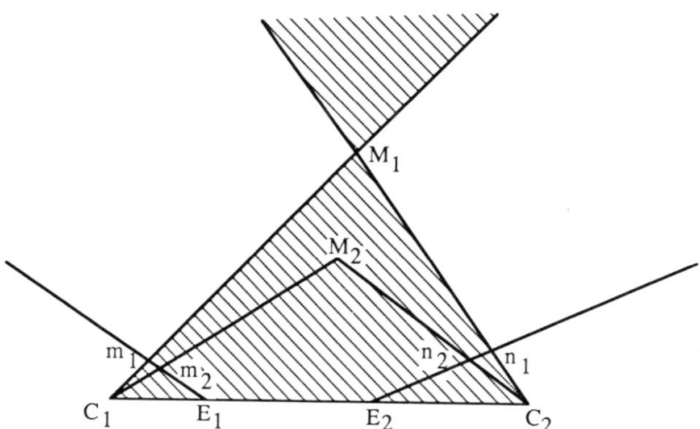

Figure 2.14
Forbidden zone attached to M.

There are several arguments for calling the lined zone in figure 2.14 the forbidden zone attached to M_1:

1. If the distance of M_1 to the base line C_1C_2 is large with respect to the length of C_1C_2, then the angle $C_1M_1C_2$ is small and the probability for a point M_2 to fall into it is quite small but nonzero.
2. If we assume that M_1 and M_2 are situated on an opaque object of nonzero thickness as shown in figure 2.15, then M_1 and M_2 cannot possibly be seen simultaneously by retinas 1 and 2.

Since the fact that M_2 is in the forbidden zone defined by M_1 can be easily checked by checking the ordering of their images along the epipolar lines, this constraint can be used to eliminate matches for m_2, given the match (m_1, n_1).

In practice, it is difficult to eliminate the whole cone of figure 2.15, because of configurations such as the one depicted in figure 2.16. In this figure, the points M_1 and M_2 are seen to belong to two different objects for which the ordering constraint does not apply. It seems therefore more reasonable to force only the neighbors of M_1 within a small neighborhood (for example, a disc of radius ε) to belong to the nonforbidden zone (figure 2.17).

2.2.3.4 Geometric Constraints In some applications, we may know a priori that the surfaces of objects have locally some known analytic form. The simplest form of this assumption is to assume that objects are locally/

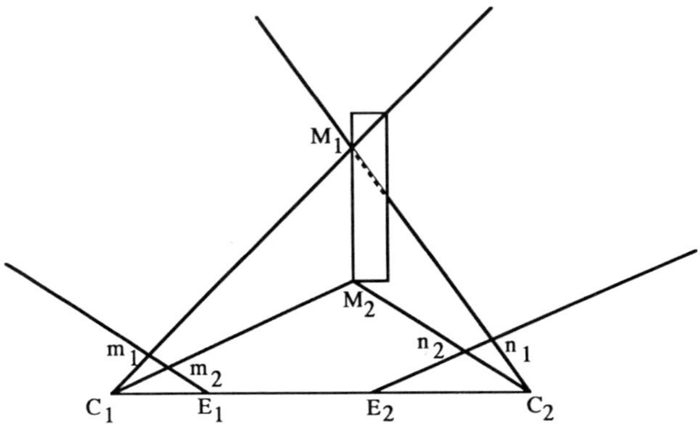

Figure 2.15
M_1 and M_2 are not both visible from retina 2.

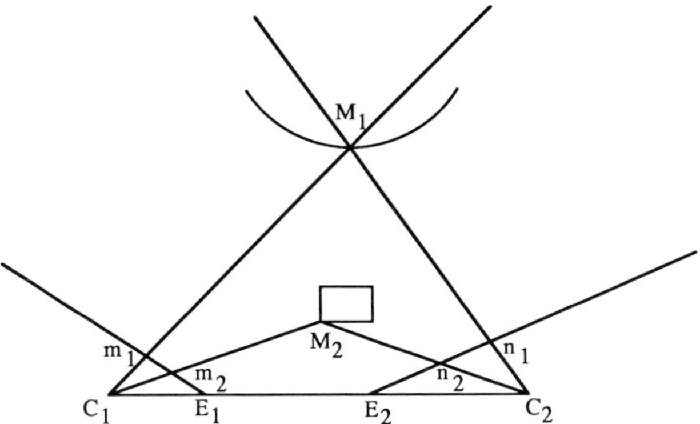

Figure 2.16
Breaking the ordering constraint.

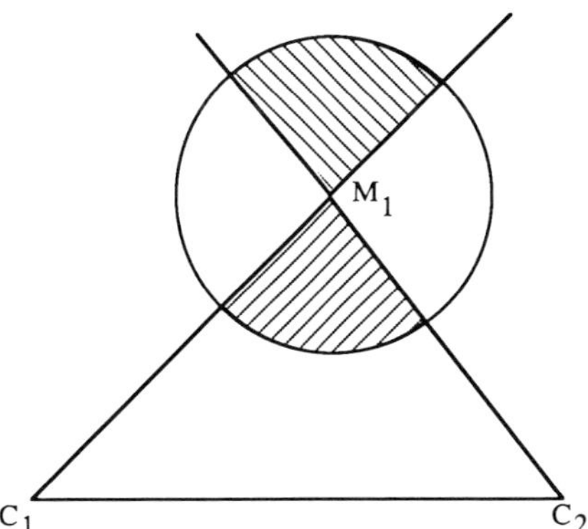

Figure 2.17
Actual forbidden zone.

piecewise planar, i.e, are well approximated by their tangent planes almost everywhere—that is, except at discontinuities. We show in section 2.2.5 how to use this extra constraint.

2.2.4 Finding the Stereo Matches

There have been a large number of algorithms published to solve the stereo matching problem, and we do not intend to review all of them. The interested reader is referred to the following references: [MP76, MP79, BB81, GRI81, OK85]. We want to discuss here an algorithm developed at INRIA by Ayache and Faverjon [AF85] and Ayache and Lustman [AL87] because it is a good example of the application of a powerful technique, hypothesis prediction and verification, to the matching of symbolic primitives, a technique that we have used successfully in a number of other applications (see section 2.4).

The symbolic primitives are line segments described by a number of features. The algorithm uses four constraints to reduce the size of the search space:

- the epipolar constraint, suitably modified for the case of line segments;
- the geometric similarity constraint to forbid matches between dissimilar segments;
- the continuity constraint implemented as a disparity gradient constraint; and
- the uniqueness constraint.

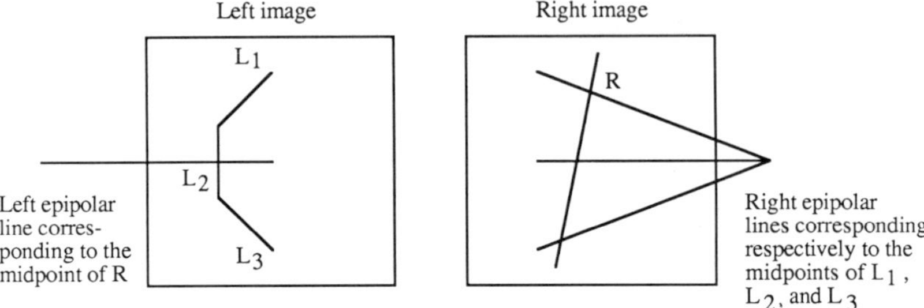

Figure 2.18
Multiple segment matches.

2.2.4.1 Building the Symbolic Descriptions Both images are represented as sets of line segments. These segments are extracted by first finding edge points, and then approximating those points by straight line segments. Each segment is represented by a number of features, both geometric (coordinates of midpoint, length, orientation) and intensity based (contrast across the segment, average intensity of image gradient along the segment, etc.). A neighborhood structure is also introduced on the two sets of segments by using buckets. Each image is divided into a number of non-overlapping square windows. To each window is attached the list of segments intersecting it, and to each segment is attached the list of windows it intersects. These buckets give fast access to the segments that are close to a given segment and are used to speed up the processes of generating and propagating the hypotheses.

2.2.4.2 Defining Matches A match is a pair (L, R) or (R, L) of left and right segments satisfying the epipolar constraint and the feature similarity constraint. The epipolar constraint for line segments implies that homologous segments have at least one analogous point. To make things simpler, this point is chosen to be the midpoint of the segment; i.e, we impose that the epipolar line associated with the midpoint of segment L intersects segment R for the match (L, R) to be possibly acceptable. Notice that this definition is not symmetric with respect to L and R, which has the advantage of potentially permitting to globally match contours that have been approximated by a different number of segments (see figure 2.18).

The disparity of a match (L, R) or (R, L) is defined in a symmetric way (see figure 2.19). We consider the end points A and B of L [C and D of R in the case of an (R, L) match], their corresponding epipolar lines intersect

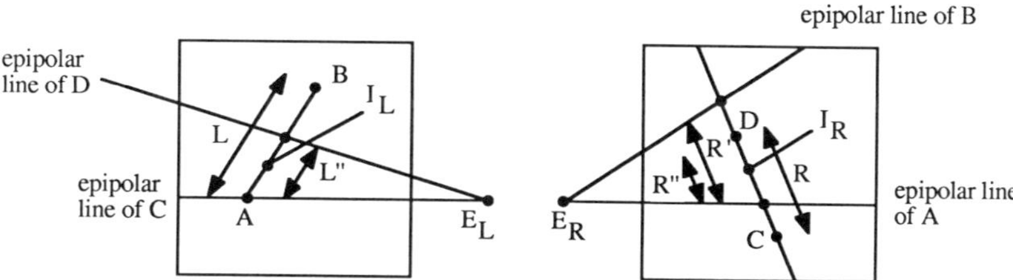

Figure 2.19
Disparity of two segments.

the line supporting R (L) along a segment R' (L'). The intersection R'' of R' and R (L'' of L' and L) is not empty, by definition of an acceptable match; using a similar procedure, the end points of R'' yield L'' included in L (R'' included in R). The segments L'' and R'' have all of their points in correspondence. The disparity of the two matches (L, R) and (R, L) is then defined as the average disparity along L''.

2.2.4.3 Making Hypotheses A number of hypothetical matches are made uniformly over the whole image by randomly selecting segments in the left image. In order to reduce the number of false matches, only segments that are neither too small (poor estimation of the orientation) nor too long (likely to be broken in the right image) and with an orientation not too close to that of the epipolar lines are selected. The thresholds for comparing segments features are set quite low at this stage, to eliminate as many false matches as possible.

2.2.4.4 Verifying Hypotheses Once hypotheses have been made, the algorithm attempts to propagate them by traversing a graph called the disparity graph, whose use is to enforce the disparity gradient constraint (to be defined below). Vertexes of this graph are matches (L, R). Two vertexes are connected by an arc if and only if the corresponding disparity gradient (the difference between the disparities of the two matches) is less than a threshold. An important point is that this threshold is variable and depends on the position in 3-D space of the line segments reconstructed from the matches. This threshold is set to tolerate a constant variation in depth; therefore it can be considered as a constraint on the roughness of the reconstructed objects. It is computed as follows. Given a match (L, R) and the corresponding disparity $DISP$, let M be the reconstructed point from I_L and I_R (see figure 2.20). A maximum variation of depth of ε around M cor-

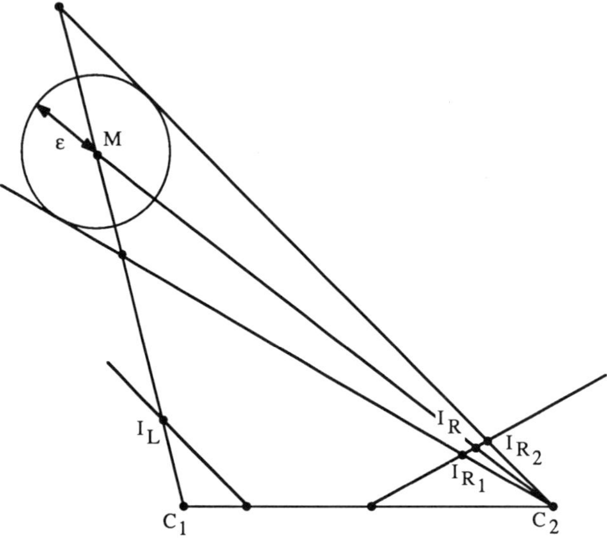

Figure 2.20
Disparity gradient.

responds to points I_{R1} and I_{R2} on the epipolar line of I_L, with disparities $DISP1$ and $DISP2$. A match (L', R') is connected to (L, R) if and only if its disparity $DISP'$ is such that $DISP' - DISP$ lies in the interval $[\mathrm{Min}(DISP1 - DISP, DISP2 - DISP), \mathrm{Max}(DISP1 - DISP, DISP2 - DISP)]$. The propagation of hypotheses can be considered as the building of connected components of the disparity graph. This is implemented by the following two procedures:

```
procedure propagate_left (left_segment, predicted_disparity)
begin
  if not already_visited (left_segment) then
    begin
      (right_segment, actual_disparity):= match_left (left_segment,
      predicted_disparity)
      if not (right_segment = NIL) then
        for each neighbor of right_segment
          propagate_right (right_segment, actual_disparity)
    end
end
```

```
procedure propagate_right (right_segment, predicted_disparity)
begin
  if not already_visited (right_segment) then
    begin
      (left_segment, actual_disparity):= match_right (right_segment,
      predicted_disparity)
      if not (left_segment = NIL) then
        for each neighbor of left_segment
          propagate_left (left_segment, actual_disparity)
    end
end
```

To propagate an hypothesis (L, R) with disparity $DISP$, the procedure **propagate_right** is activated with the parameters R and $DISP$. It first checks whether R has already been visited; if not, it calls the procedure **match_right**, which selects among the left segments those whose disparity with R is in the range $DISP1$, $DISP2$, and whose features are sufficently close to those of R. If several candidates segments remain, the one with disparity closest to $DISP$ is chosen.

If no match is found, the procedure **match_right** returns NIL and then **propagate_right** stops; otherwise, when a match L' with disparity $DISP'$ has been found, **propagate_right** calls the symmetric procedure **propagate_left** for each neighbor of L' with the disparity $DISP'$, etc.

2.2.4.5 Handling Conflicts A conflict is defined as a pair of matches (L, R) and (L, R') or (L, R) and (L', R). Conflicts occurring in the propagation phase are handled immediately by choosing the best match in terms of the features. Conflicts occurring between two distinct components of the disparity graph are solved by comparing the sizes of the two components and erasing from the smallest one the conflicting matches. To avoid a dependence on the order in which the components are compared, the size comparison is made on the initial size of the components.

2.2.5 Adding the Planarity Constraint

In many human made environments, we are bound to encounter many flat surfaces such as walls, floors, ceilings, doors. Those surfaces are planar. This in turn implies a very strong constraint on the kind of correspondence which occurs between the two images of a stereo pair. In fact it turns out that if the visual features which are matched are emanating from a plane, there is no ambiguity in the matching since there exists an analytic transformation from the left image coordinates to the right image coordinates. This analytic transformation is a function of the rotation and translation from camera 1 camera 2 and of the plane equation.

2.2.5.1 Analytic Transformation between Left and Right Images In order to derive this transformation, we use the notations of section 2.2 and figure 2.21. Let $\mathbf{M}_1$ and $\mathbf{M}_2$ be the perspective transformations for the two cameras. $\mathbf{R}$ the rotation matrix from the left camera to the right, and $\mathbf{t}$ the translation vector. It can be easily shown, that the relationship between $\mathbf{M}_1$ and $\mathbf{M}_2$ is

$$\mathbf{M}_2 = \mathbf{M}_1 \begin{bmatrix} \mathbf{R} & \mathbf{t} \\ \mathbf{0} & 1 \end{bmatrix}.$$

Let us suppose now that these two cameras are looking at a plane given by its normal $\mathbf{n} = [a, b, c]^T$ and its distance to the origin d (we assume that the normal is of length 1), and let m_1 of coordinates (u_1, v_1) be a point in the camera plane of camera 1. The line $C_1 m_1$ is described by the points of coordinates $r\mathbf{N}_1$ for $-\infty < r < +\infty$, with $\mathbf{N}_1$ the vector $[u_1, v_1, f]^T$, where f is the focal length.

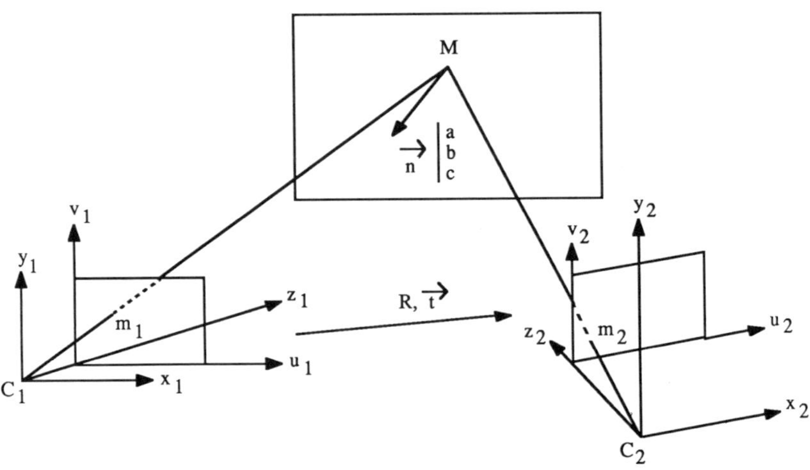

Figure 2.21
Looking at a plane. An arrow over a symbol indicates that the symbol is a vector.

The value of r corresponding to the intersection of line $C_1 m_1$ with the plane is obtained by writing

$$\mathbf{n}.r\mathbf{N}_1 = d.$$

This yields the object point M. The coordinates of the corresponding point m_2 in the second camera plane are given by

$$\begin{bmatrix} U_2 \\ V_2 \\ T_2 \end{bmatrix} = \mathbf{M}_2 \begin{bmatrix} (d/\mathbf{n}.\mathbf{N}_1)\mathbf{N}_1 \\ 1 \end{bmatrix}.$$

This is the same as

$$\begin{bmatrix} U_2 \\ V_2 \\ T_2 \end{bmatrix} = \mathbf{M}_2 \begin{bmatrix} d\mathbf{N}_1 \\ \mathbf{n}.\mathbf{N}_1 \end{bmatrix}.$$

Since $\mathbf{n}.\mathbf{N}_1$ is equal to $au_1 + bv_1 + cf$, this can be rewritten as

$$\begin{bmatrix} U_2 \\ V_2 \\ T_2 \end{bmatrix} = \mathbf{M}_2 \begin{bmatrix} d\mathbf{I} & & \\ a & b & c \end{bmatrix} \begin{bmatrix} U_1 \\ V_1 \\ T_1 \end{bmatrix}, \tag{2}$$

where $\mathbf{I}$ is the 3×3 identity matrix. Therefore, the relationship between the point m_1 and the point m_2 is linear in projective space when the object point M moves in a plane. Let us rewrite it more compactly as

$$\mathbf{m}_2 = \mathbf{A}\mathbf{m}_1. \tag{3}$$

Equations (2) and (3) describe completely the relationship between features in the left and right images.

Two questions can be asked at this point. First, once we know matrix $\mathbf{A}$, can we recover the plane equation? Second, how can we use this information to help solve the stereo problem?

2.2.5.2 Recovering the Plane Equation Let us answer the first question first and denote by β_i, $i = 1, \ldots, 4$, the 4-column vectors of size 3 of matrix $\mathbf{M}_2$. We have to solve the following equation:

$$\mathbf{M}_2 \begin{bmatrix} d\mathbf{I} & & \\ a & b & c \end{bmatrix} = \mathbf{A},$$

where the unknowns are the plane parameters a, b, c, d. This is a linear system in those unknowns that can be solved by classical mean-square techniques. Here we find, denoting by $\mathbf{p}$ the vector $[a, b, c, d]^T$ and by $\mathbf{a}_i$, $i = 1, \ldots, 3$, the three column vectors of size 3 of matrix $\mathbf{A}$;

$$\mathbf{Mp} = \mathbf{a},$$

and therefore

$$\mathbf{p} = (\mathbf{M}^T\mathbf{M})^{-1}\mathbf{M}^T\mathbf{a}$$

where $\mathbf{M}$ is the 9×4 matrix

$$\begin{bmatrix} \beta_4 & \mathbf{0} & \mathbf{0} & \beta_1 \\ \mathbf{0} & \beta_4 & \mathbf{0} & \beta_2 \\ \mathbf{0} & \mathbf{0} & \beta_4 & \beta_3 \end{bmatrix}$$

and $\mathbf{a}$ is the 9-dimensional vector $[\mathbf{a}_1^T, \mathbf{a}_2^T, \mathbf{a}_3^T]^T$. We do not need to take into account the constraint that the normal to the plane is of unit norm since matrices $\mathbf{M}$ and $\mathbf{A}$ are defined up to a scale factor.

2.2.5.3 Estimating Matrix A Another important related question is the estimation of matrix $\mathbf{A} = [a_{ij}]$, $i, j = 1, 2, 3$, from matches between the left and right images. So far we have assumed that we were only using points, but as we show next we can also use line segments or rather the infinite lines supporting those line segments. Let us consider the two cases separately.

Matching points: From equation (3) we get

$$T_2 = T_1(a_{31}u_1 + a_{32}v_1 + a_{33}) \tag{4}$$

and therefore

$$a_{11}u_1 + a_{12}v_1 + a_{13} = u_2(a_{31}u_1 + a_{32}v_1 + a_{33}), \tag{5}$$

$$a_{21}u_1 + a_{22}v_1 + a_{23} = v_2(a_{31}u_1 + a_{32}v_1 + a_{33}). \tag{6}$$

Since $\mathbf{A}$ is defined up to a scale factor, we can impose a condition on the coefficients a_{ij}. A possible and often used condition is to set $a_{33} = 1$. It must be kept in mind that of course this prevents a_{33} from being equal to zero. This has a corresponding geometric interpretation: if $a_{33} = 0$, the condition for the point m_2 to be at infinity can be written, according to equation (4), $a_{31}u_1 + a_{32}v_1 = 0$. This is the equation of a line in the first camera plane that goes through the origin. This line is the image by the first camera of the line intersection of the plane being looked at and the focal plane of the second camera. Therefore, if we set $a_{33} = 1$, this prevents the previous situation from happening; i.e., the image by the first camera of this line cannot go through the origin. In practice, this is not a problem. Let us now define

$$\mathbf{b} = [u_1, v_1, 1, 0, 0, 0, -u_2 u_1, -u_2 v_1]^T,$$

$$\mathbf{c} = [0, 0, 0, u_1, v_1, 1, -v_2 u_1, -v_2 v_1]^T,$$

$$\mathbf{a} = [a_{11}, a_{12}, a_{13}, a_{21}, a_{22}, a_{23}, a_{31}, a_{32}]^T.$$

Equations (5) and (6) can then be rewritten as

$$\mathbf{b}^T \mathbf{a} = u_2,$$

$$\mathbf{c}^T \mathbf{a} = v_2.$$

This is for one pair of points; for n pairs the corresponding mean-square problem is to minimize

$$C = \sum_n (\mathbf{b}_n^T \mathbf{a} - u_{2n})^2 + (\mathbf{c}_n^T \mathbf{a} - v_{2n})^2,$$

or

$$C = \mathbf{a}^T \mathbf{H} \mathbf{a} - 2\mathbf{h}^T \mathbf{a} + r,$$

where

$$\mathbf{H} = \sum (\mathbf{b}_n \mathbf{b}_n^T + \mathbf{c}_n \mathbf{c}_n^T),$$

$$\mathbf{h} = \sum (u_{2n} \mathbf{b}_n + v_{2n} \mathbf{c}_n),$$

$$r = \sum (u_{2n}^2 + v_{2n}^2).$$

The solution $\mathbf{a} = 2\mathbf{H}^{-1}\mathbf{h}$ can be computed iteratively using standard recursive least squares techniques.

Matching lines: It turns out that this is very similar to the matching of points, something to be expected since in planar projective geometry points and lines play the same role. Indeed, let L_1 be a straight line in the first camera plane defined by its equation

$$\mathbf{n}_1^T \mathbf{m}_1 = 0$$

and let L_2 be the corresponding line in the second camera plane defined by its equation,

$$\mathbf{n}_2^T \mathbf{m}_2 = 0.$$

Using equation (3) we get

$$\mathbf{n}_1 = \mathbf{A}^T \mathbf{n}_2. \tag{7}$$

Therefore the previous procedure can also be applied to lines modulo the difference between equations (3) and (7).

How many matches are necessary in order to estimate matrix **A** since it depends only on eight coefficients, four points in general position, i.e, such that not three of them are aligned are necessary, or two segments? Two segments can be seen to yield four points in general position by choosing two on each line supporting the segments, and using the epipolar geometry to compute their correspondents in the second image.

It is now possible to modify the algorithm of section 2.2.4 in such a way that it takes into account the planarity constraint. The result is an algorithm that solves the stereo matching problem and at the same time finds planes.

A hypothesis is defined to be two pairs (L_1, R_1) and (L_2, R_2) of acceptable matches in the sense of the previous section that are also such that the corresponding 3-D lines are coplanar. If this hypothesis is correct, then it can be propagated by first estimating the corresponding matrix **A**, and second applying it to the unmatched neighbors of L_1 and L_2. Let L be such a neighbor; then $\mathbf{A}L$ is a segment in the right image that should find a matching segment with similar features in its immediate neighborhood if the hypothesis is right and none if it is not.

When such a match has been found, estimation of matrix **A** can be updated and then applied to more segments until no more matches can be found. Once a plane has been found by propagating a hypothesis, others can be searched for by considering different hypotheses involving yet unmatched segments. Results are shown in figures 2.22 and 2.23.

2.2.6 Using Three Cameras

The idea of using three cameras ([GPSL86, GDS86, II86a, II86b, OWI86, PH86, YAC86] and [YKK86, AL87]) is that of making fuller use of the epipolar constraint. Indeed, let us consider the three cameras system of

A

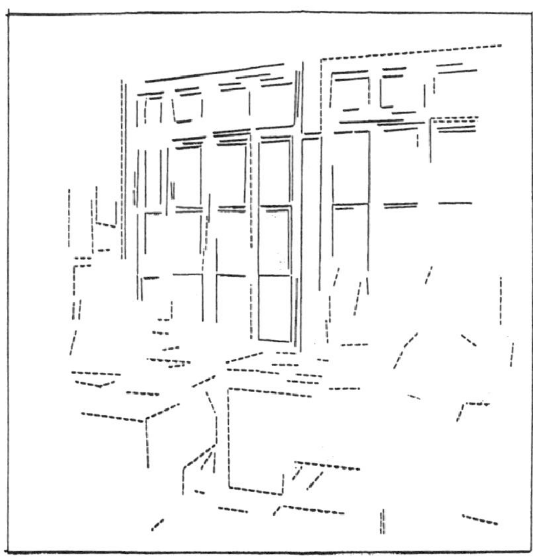

B

Figure 2.22
(A) Stereo pair of an office scene. (B) Segments found to be in the window plane.

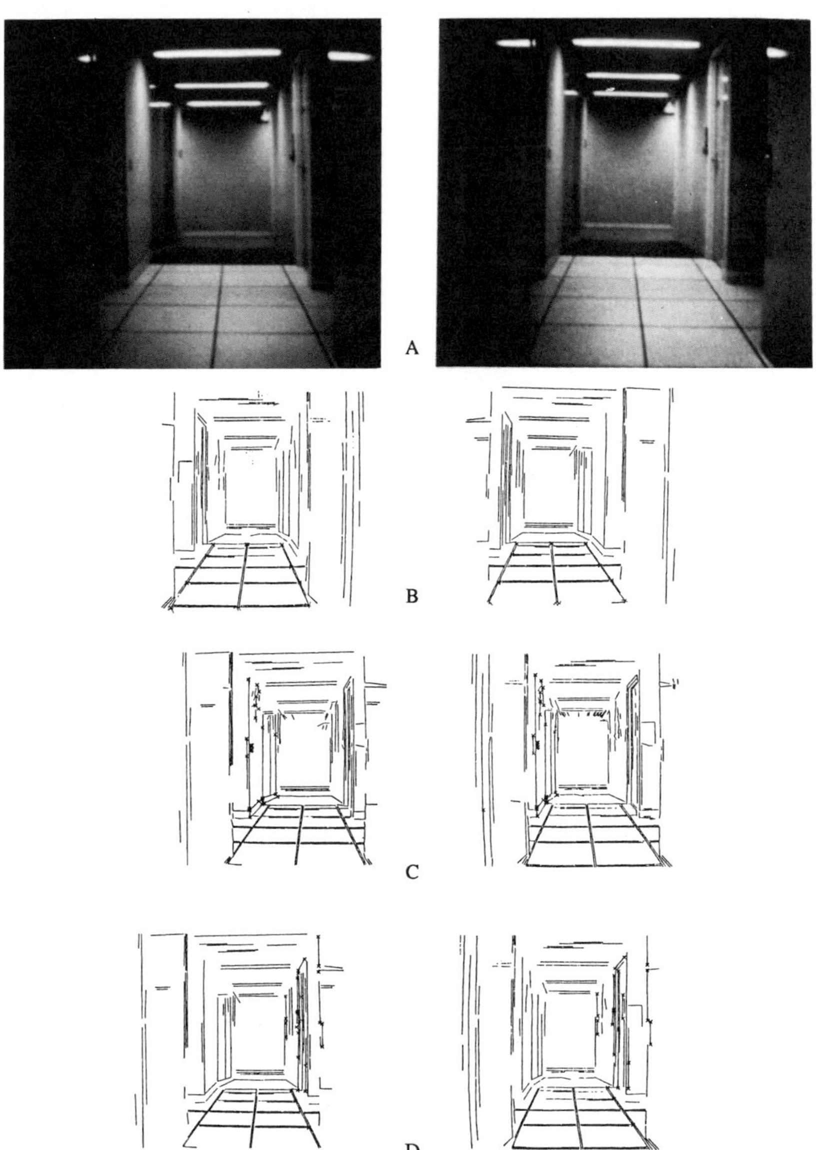

Figure 2.23
(A) Stereo pair of a hallway. (B) Points found to be on the floor plane. (C) Points found to be on the left wall plane. (D) Points found to be on the right wall plane.

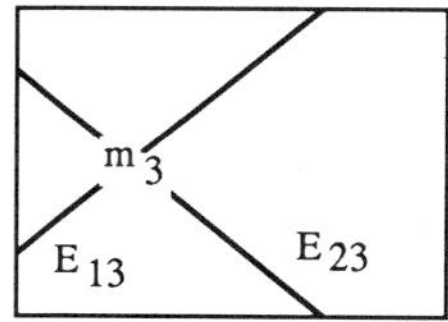

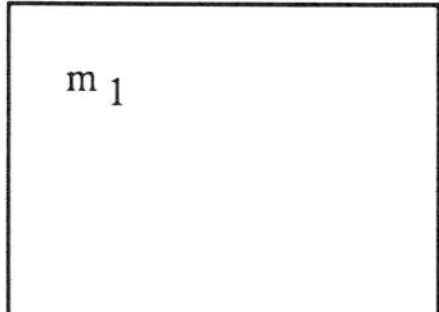

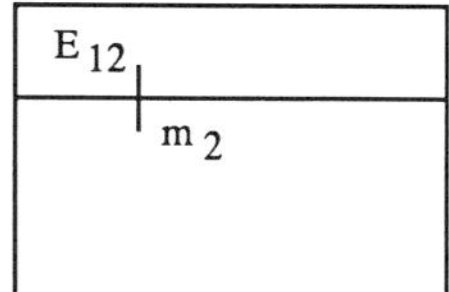

Figure 2.24
Three-camera stereo.

figure 2.24. We give the idea for a point token, understanding that it can be extended to other tokens as well.

Let $\mathbf{m}_1$ be a point in camera 1. Its potential matches in cameras 2 and 3 are located on the epipolar lines E_{12} and E_{13} of $\mathbf{m}_1$. Let $\mathbf{m}_2$ be a candidate match on E_{12}; $\mathbf{m}_2$ has an epipolar line E_{23} in camera 3 that intersects E_{13} at a point $\mathbf{m}_3$. Therefore, if the match $(\mathbf{m}_1, \mathbf{m}_2)$ is correct, there should be an image feature at $\mathbf{m}_3$ that is similar to that at $\mathbf{m}_1$ and $\mathbf{m}_2$. This test is quite simple when the epipolar geometry is known and drastically reduces the complexity of the search for correspondences.

The algorithm is then as follows (for line segments):

```
For each segment S1 in image 1
  Find S2 in image 2
        Intersecting E12
        Similar to S1
    For each such S2 in image 2
      Find S3 in image 3
        Intersecting the intersection of E13 and E23
          Similar to S1 and S2
```

The verification phase of section 2.2.4.4 has now disappeared and is replaced by a simple elimination of potential false triplets (S_1, S_2, S_3) by applying to the reconstructed 3-D segments a procedure enforcing local

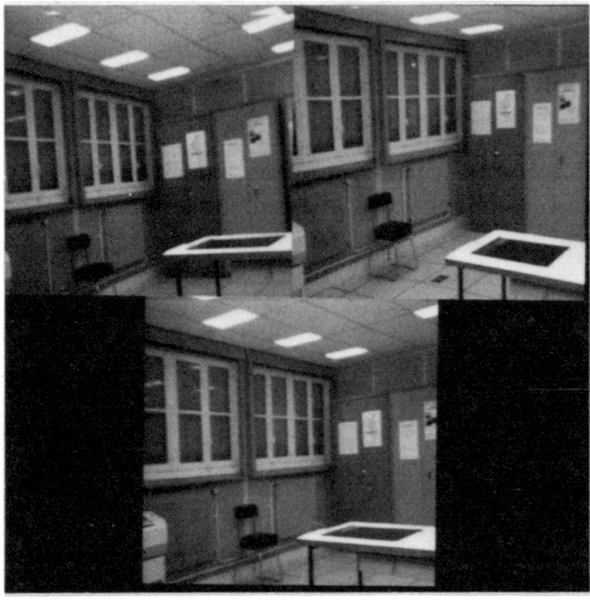

Figure 2.25
Stereo triplet of images.

compatibility [AL87]. The result is a more accurate, more reliable, and faster algorithm. Figure 2.25, 2.26, and 2.27 show some results.

2.2.7 Motion and Structure from Motion

Motion analysis is also a way of recovering 3-D data that we believe is important and may not have been exploited enough in the past. The motion work we present here is a simplification of the general problem in that we assume that only the camera is moving and that the environment is otherwise static. Under this simplifying assumption, it is well-known that the motion of the camera and the 3-D structure of the environment can be recovered up to a scale factor. This scale factor can then be recovered if the absolute size of some object in the scene is known.

Traditionally there have been two main approaches to the analysis of motion. The first approach involves computing the optical flow or image displacement field, assuming that it is the projection of the actual 3-D motion field, and recovering the 3-D field from it. The second approach involves matching tokens between frames and recovering 3-D motion and

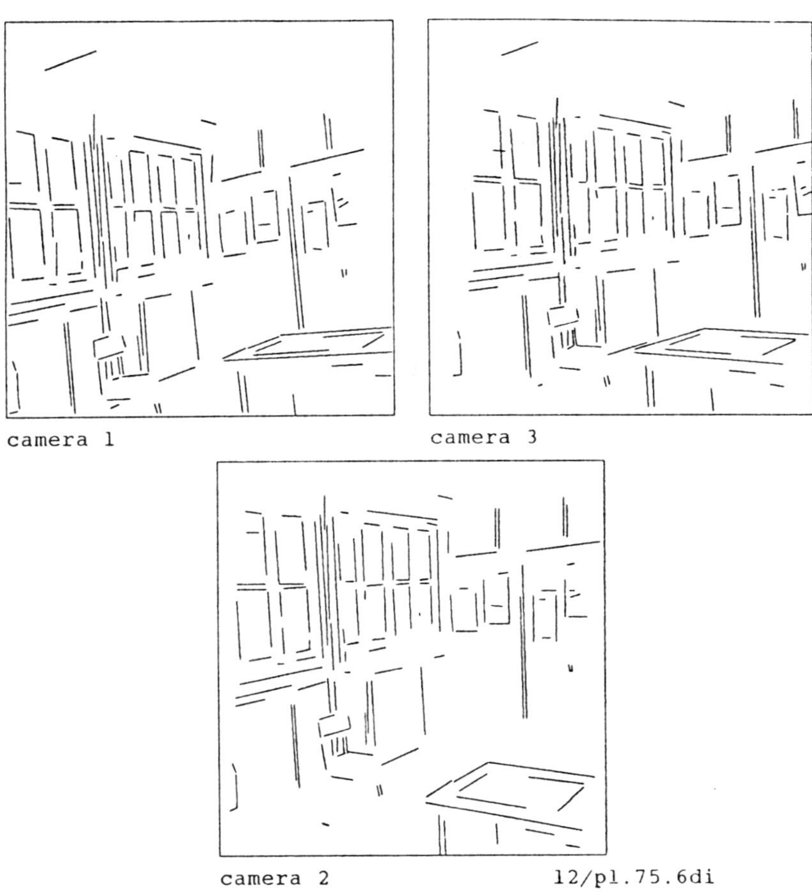

Figure 2.26
Matched segments in the three images.

structure from those matches. For an excellent recent review, see Nagel [NAG86].

Problems with the first approach are

1. the flow cannot be recovered entirely at every pixel (window effect, points with no gradient) [HIL84] and
2. the optical flow is usually not the projection of the 3-D field [VP87].

The main problem with the second approach is that finding the matches may not be simple. A general problem with both approaches is that the results are believed to be quite sensitive to noise and errors. Some recent work conducted at INRIA [FLT87] seems to indicate that 3-D motion and structure can be robustly estimated from token matches. This work points to the fact that the sensitivity to noise is not as a big a problem as we used

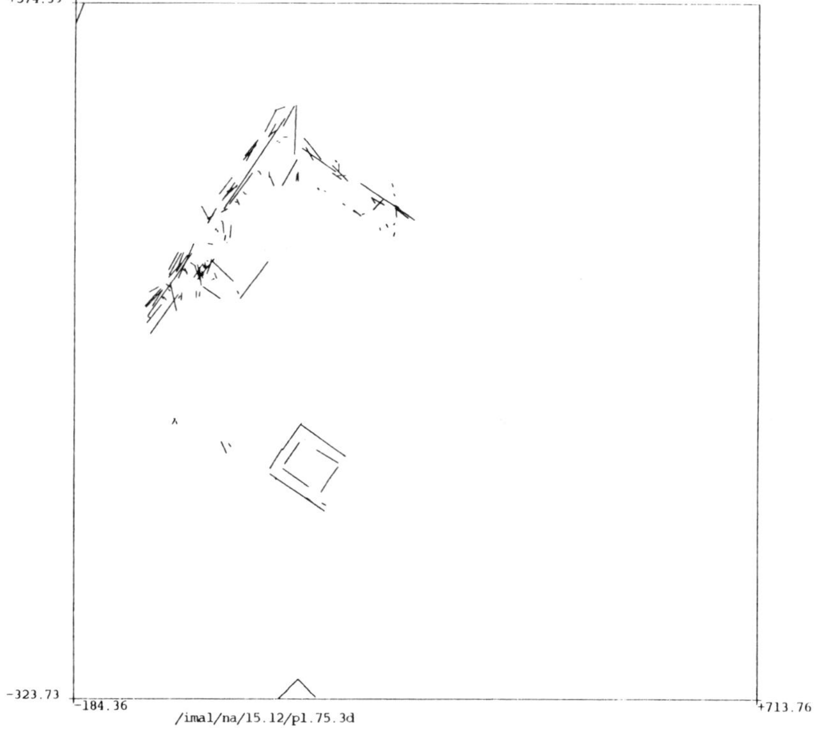

Figure 2.27
Orthographic projection in a horizontal plane of the reconstructed 3-D segments.

to think it was. We also think that in most robotics applications, given fast enough processing, the matching problem becomes essentially trivial because objects have not moved much from one frame to the next if the sampling frequency along the time axis is high enough.

We have investigated two kinds of tokens: points and lines. Points are corners, i.e., intersections of lines. Matching two points yields the whole 2-D displacement (optical flow), whereas matching two lines does not.

2.2.7.1 Point Matches The problem of recovering motion and structure from point matches has been studied by Longuet-Higgins [LON84, LON81], who proposed an algorithm based on 8 point matches. This algorithm fails when the points are not in general configuration—and in particular when they are coplanar. It is also quite sensitive to noise. In the planar case, a method developed by Tsai and Huang [TH82] requires 4 point matches. It has also been reported to be quite sensitive to noise.

In the two cases, the use of mean-square techniques (linear and nonlinear) has allowed us to become quite insensitive to noise. We discuss here the case of nonplanar points. The interested reader is referred to [FL87, FLT87] for the case of planar points.

The problem with points in general positions has been solved by Longuet-Higgins [LON84] in the exact case, assuming no noise. When noise is present, the method is quite sensitive [HUA86]. We propose here a method that can cope more robustly with the noise problem. It is based on the exploitation of the planarity constraint used by Longuet-Higgins.

Let us look at figure 2.28, where m_1 and m_2 are the images of the physical

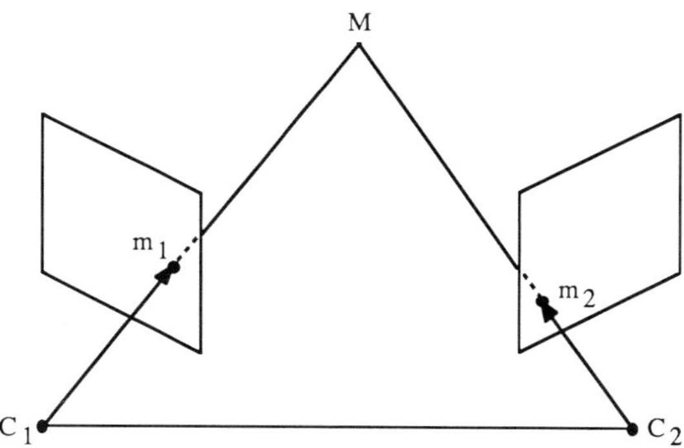

Figure 2.28
Planarity constraint in motion estimation.

point M in the two cameras. The three vectors $\mathbf{C}_1\mathbf{m}_1$, $\mathbf{t}$, and $\mathbf{C}_2\mathbf{m}_2$ are coplanar and therefore their determinant is zero. What we can measure are the coordinates of $\mathbf{C}_1\mathbf{m}_1$ and $\mathbf{C}_2\mathbf{m}_2$ in the coordinate systems $C_1x_1y_1z_1$ and $C_2x_2y_2z_2$ of figure 2.21. $\mathbf{C}_2\mathbf{m}_2$ expressed in the coordinate system 1 is equal to $\mathbf{RC}_2\mathbf{m}_2$. Therefore, we have, in the coordinate system 1,

$$\det(\mathbf{C}_1\mathbf{m}_1, \mathbf{t}, \mathbf{RC}_2\mathbf{m}_2) = 0,$$

or, more geometrically,

$$\mathbf{C}_1\mathbf{m}_1.(\mathbf{t} \wedge \mathbf{RC}_2\mathbf{m}_2) = 0. \tag{8}$$

Introducing the antisymmetric matrix $\mathbf{T}$,

$$\mathbf{T} = \begin{bmatrix} 0 & -t_z & t_y \\ t_z & 0 & -t_x \\ -t_y & t_x & 0 \end{bmatrix}.$$

Equation (8) can be rewritten as

$$\mathbf{C}_1\mathbf{m}_1^T\mathbf{EC}_2\mathbf{m}_2 = 0, \tag{9}$$

with

$$\mathbf{E} = \mathbf{TR}.$$

Since only the direction of $\mathbf{t}$ can be recovered, $\mathbf{E}$ is defined up to a scale factor. We normalize things by assuming that $\|\mathbf{t}\| = 1$. This implies that $\|\mathbf{E}\|^2 = 2$.

Notice that equation (9) is a linear equation in the coefficients of matrix $\mathbf{E}$. Denoting its row vectors by $\mathbf{X}_1^T$, $\mathbf{X}_2^T$, $\mathbf{X}_3^T$, and by $\mathbf{X}$ the 9×1 vector $[\mathbf{X}_1^T, \mathbf{X}_2^T, \mathbf{X}_3^T]^T$, we can write equation (9) as

$$\mathbf{a}^T\mathbf{X} = 0,$$

where

$$\mathbf{a}^\mathbf{T} = [x_1x_2, x_1y_2, x_1, y_1x_2, y_1y_2, y_1, x_2, y_2, 1]^T$$

and the x_i's and y_i's are the coordinates of $\mathbf{C}_1\mathbf{m}_1$ and $\mathbf{C}_2\mathbf{m}_2$ in coordinate systems 1 and 2 (we assume a focal length of 1).

If we have n matches, we have n such equations, and a linear system

$$\mathbf{A}_n\mathbf{X} = 0, \tag{10}$$

where $\mathbf{A}_n$ is an $n \times 9$ matrix. Longuet-Higgins solves this system for $n = 8$ when the rank of $\mathbf{A}_8$ is 8. The null space is then of dimension 1 and $\mathbf{X}$ is the vector of norm $\sqrt{2}$ in the null space. Degenerate cases, which occur when rank($\mathbf{A}_8$) < 8, are analyzed in Longuet-Higgins [LON81] and Zhuang and Haralick [ZH85]. We discuss in [FLT87] the case where the 3-D points are coplanar, which also yields a degenerate matrix $\mathbf{A}_8$.

When $rank(\mathbf{A}_n) = 8$, our technique works in three steps:

1. *Compute* **X**: Solve equation 10 by mean square. Noticing that $\|\mathbf{X}\|^2 = 2\|\mathbf{t}\|^2 = 2$, this is equivalent to solving

$$\min_{\mathbf{X}} \|\mathbf{A}_n\mathbf{X}\|^2 \qquad \text{subject to} \quad \|\mathbf{X}\|^2 = 2,$$

which is a standard problem.
2. *Compute* **t**: We then notice [NAG86] that $\mathbf{t}^T\mathbf{E} = \mathbf{t}^T\mathbf{TR} = \mathbf{0}$. Therefore we obtain **t** as the solution of

$$\min_{\mathbf{t}} \|\mathbf{E}^T\mathbf{t}\|^2 \qquad \text{subject to} \quad \|\mathbf{t}\|^2 = 1,$$

another standard problem
3. *Compute* **R**: **R** is then computed as the solution of

$$\min_{\mathbf{R}} \|\mathbf{E} - \mathbf{TR}\|^2 \qquad \text{subject to} \quad \mathbf{R}^T\mathbf{R} = \mathbf{I}.$$

This is easily solved using, for example, the quaternion representation of rotation matrixes (see section 2.4, appendix 2B, and [FH86, FT86]).

We have tested this algorithm on synthetic and real data. The image of the 3-D test scene we have used is shown in figure 2.29. It is made of 10 points situated at a distance of 4 m from our camera. The size of the object is also about 4 m. The perspective transformation matrix we use to project the 3-D points has been obtained by calibration of a real camera. We have simulated a number of camera displacements and noise in pixel measurements. The resolution is 512×512 pixels, and noise varies from 1 to 5 pixels. Once we have estimated the rotation **R** and the direction of the translation **t**, we can reconstruct the 3-D points by mean square and project them back in the two retinas before and after motion. We then compute the average distance between the actual and the reconstructed pixels. The sum of these two measurements constitute our measure of quality.

Table 2.1 shows some typical results for two techniques (the original Longuet-Higgins technique, improved to take into account the noise when

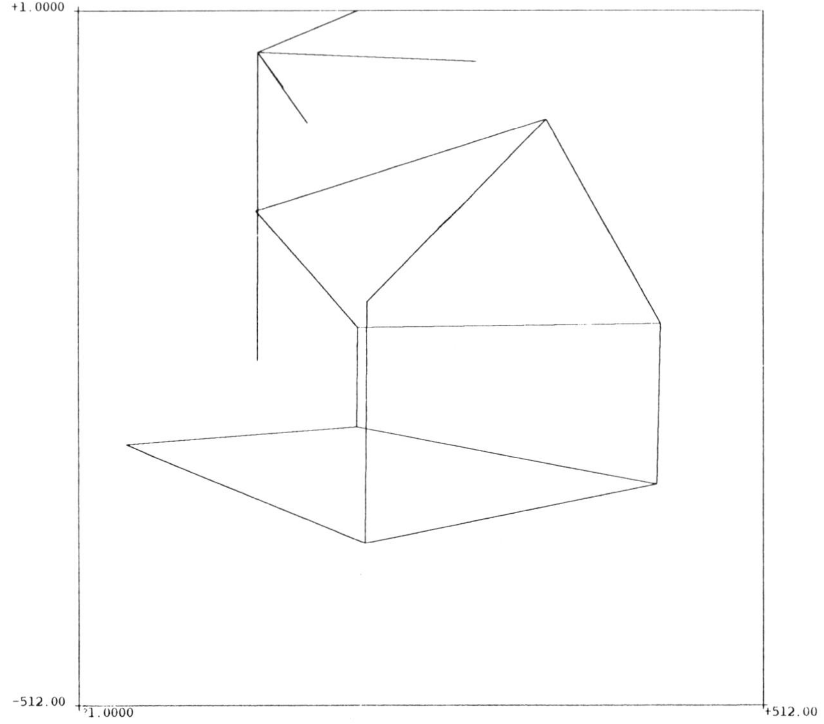

Figure 2.29
3-D test scene.

estimating **E**, and ours). Two main remarks can be made at this point:

1. Both techniques appear to be quite robust to quite large noise distortions, contrary to what has been reported in the literature so far.
2. Our method performs at least as well as the Longuet-Higgins one, and significantly better in many cases.

Figure 2.30 shows the projected reconstructed points in the two retinas for our method in the case of line 2 of table 2.1.

We have also experimented with real data that have been used in camera calibration experiments. We have taken four stereo views of a special pattern displayed in figure 2.31 made of lines painted black on a sheet of white paper at distances of 3, 3.5, 4, and 4.5 m. We have extracted automatically those lines from the four images and computed with great accuracy the

Table 2.1
Synthetic data

θ	t	∂p	av	sd	θ'	$\theta\%$	α	β	
					5.89	17.8	24.59	26.24	(1)
5	10	1	57.8	7.2	5.36	7.2	11.48	27.58	(2)
					5.21	4.2	2.60	4.12	(3)
					15.18	203.6	42.21	4.53	(1)
5	50	1	91.8	14.3	5.72	14.5	6.78	1.78	(2)
					5.25	5.0	3.31	0.78	(3)
					11.03	120.7	37.31	3.09	(1)
5	100	1	139.6	31.1	5.74	14.8	6.30	0.82	(2)
					5.25	5.1	3.38	0.38	(3)
					6.06	21.2	47.34	72.43	(1)
5	10	5	57.7	8.1	5.04	0.8	16.94	72.61	(2)
					5.14	2.8	8.61	71.91	(3)
					20.01	300.2	115.43	24.59	(1)
5	50	5	91.5	14.7	8.37	67.3	37.77	11.48	(2)
					5.99	19.8	10.12	4.31	(3)
					55.90	1018	121	23.2	(1)
5	100	5	139.4	31.1	9.09	82	22.3	5.06	(2)
					6.17	23.4	11.81	2.11	(3)
					34.6	130.7	35.6	6.68	(1)
15	50	5	189.1	20.4	19.5	30.3	8.5	7.84	(2)
					15.05	0.14	2.78	5.05	(3)
					25.15	67.6	24.98	4.19	(1)
15	100	5	232.3	30.3	19.73	31.6	5.37	1.93	(2)
					15.24	1.6	2.63	2.64	(3)

Axis of rotation: (1, 1, 1)

θ = angle of rotation (degrees)

t = magnitude of exact translation (cm)

∂p = measurement noise (pixels)

av = average of image displacement (pixels)

sd = standard deviation of image displacement (pixels)

θ' = angle of computed rotation

Direction of translation: (1, 0, 0)

$\theta\%$ = rotation angle error (in %)

α = angle between exact and computed axes of rotation

β = angle between exact and computed direction of translation

(1) = improved Longuet-Higgins

(2) = new linear technique

(3) = reconstruction and reprojection

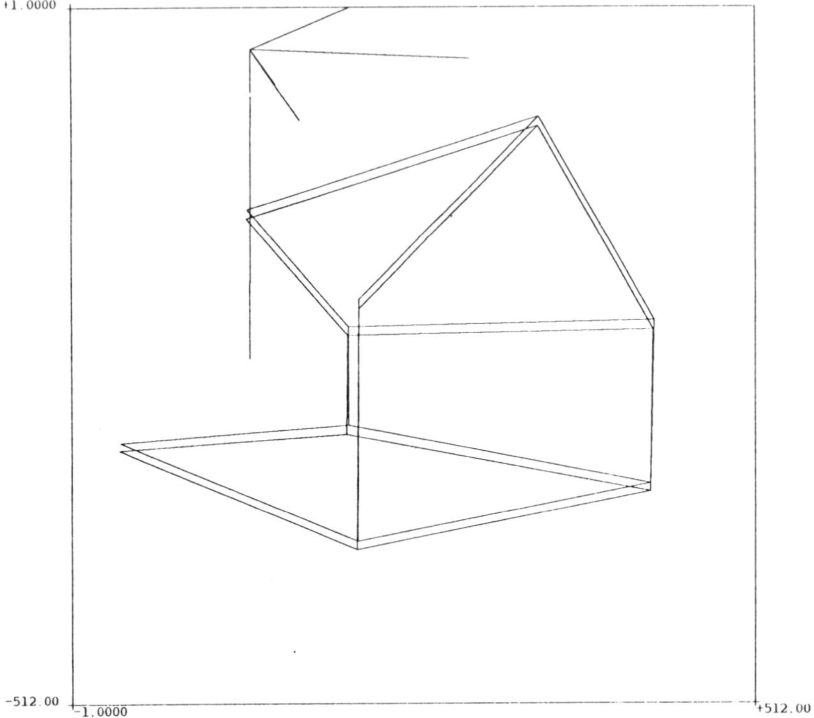

Figure 2.30
Projected reconstructed points.

pixel coordinates of their intersections, thus yielding 113 points in each image. Our method of stereo calibration described in [FT86] gives us the displacement (rotation **R** and translation **t**) from camera 1 to camera 2 by using at least two planes of data. We have then used our motion estimation technique on the points in the other two planes and estimated **R** and the direction of translation **t**. Results are shown in table 2.2 for two types of camera. They show that this technique for determining camera motion yields excellent results.

Remarks The coplanarity constraint of equation (8) is only one way of approaching the problem. There are several other possibilities that yield more complicated equations but may in practice yield better and more robust results. We discuss now two such possibilities. The first is to consider the shortest distance between the lines C_1m_1 and C_2m_2 as a function of **R**

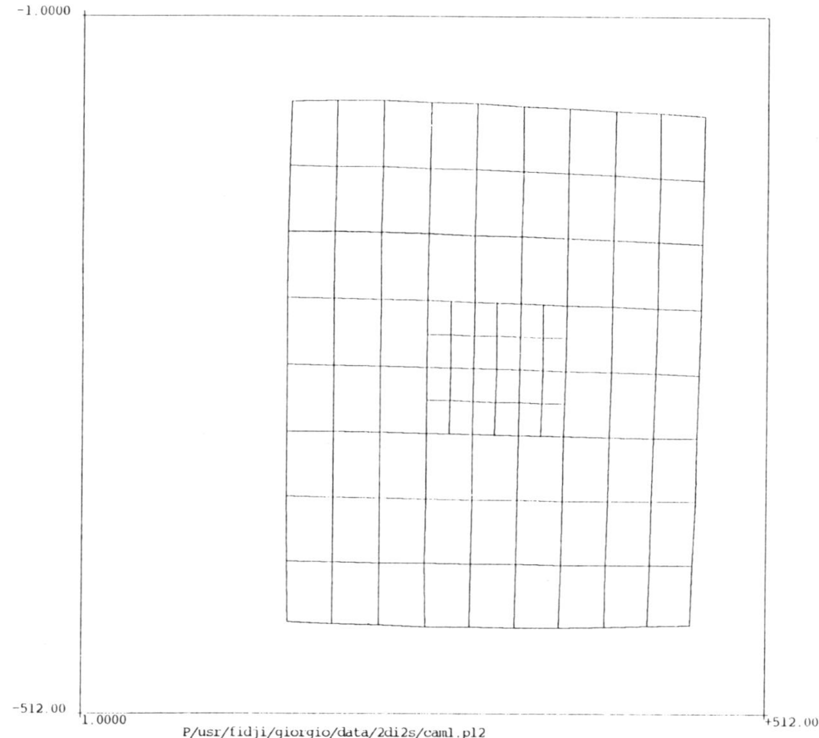

Figure 2.31
Special grid pattern.

and **t**. It can be easily proved that

$$d = \left| \frac{\mathbf{C}_1\mathbf{m}_1.(\mathbf{t} \wedge \mathbf{R}\mathbf{C}_2\mathbf{m}_2)}{\|\mathbf{C}_1\mathbf{m}_1 \wedge \mathbf{R}\mathbf{C}_2\mathbf{m}_2\|} \right|.$$

Therefore minimizing the sum over the pairs of matched points of the d^2 is the same as minimizing $\|\mathbf{A}_n\mathbf{X}\|^2$ but with the normalization by $\|\mathbf{C}_1\mathbf{m}_1 \wedge \mathbf{R}\mathbf{C}_2\mathbf{m}_2\|$, which is the sine squared of the angle between the lines C_1m_1 and C_2m_2.

The second is to minimize the image reconstruction error as follows. Let us denote by Q_1 and Q_2 the end points on C_1m_1 and C_2m_2 of their common perpendicular (see figure 2.4), and suppose we choose their midpoint P as the reconstructed 3-D point from m_1 and m_2. An alternative candidate is the point obtained by solving for the intersection of C_1m_1 and C_2m_2 by

Table 2.2
Real data

	θ	t	av	sd	θ'	$\theta\%$	α	β	
					3.36	3.3	0.96	0.93	(1)
I2S	3.48	8	13.4	6.2	3.37	3.1	0.75	0.91	(2)
					3.47	0.2	0.81	0.70	(3)
					14.64	175	58.1	7.25	(1)
I2S	5.32	44	20.6	5.2	5.13	3.6	0.41	0.76	(2)
					5.14	3.5	0.66	0.31	(3)
					4.60	2.0	4.16	2.18	(1)
I2S	4.69	37	24.9	7.2	4.58	2.4	2.40	2.17	(2)
					4.81	2.5	0.53	0.92	(3)
					9.55	1.5	3.10	0.52	(1)
Pul	9.71	53	26.2	16.4	9.68	0.2	0.32	0.02	(2)
					9.69	0.2	0.80	0.008	(3)
					9.43	3.0	1.36	0.22	(1)
Pul	9.15	49	20.7	12.2	9.06	1.0	0.61	0.22	(2)
					9.03	1.3	0.22	0.23	(3)
					8.13	1.6	2.74	0.31	(1)
Pul	8.27	45	18.8	12.4	8.14	1.5	0.65	0.01	(2)
					8.29	0.25	0.02	0.02	(3)

θ = angle of reference rotation
t = magnitude of reference translation (cm)
av = average of image displacement (pixels)
sd = standard derivation of image displacement (pixels)
θ' = angle of computed rotation
$\theta\%$ = rotation angle error (in %)
α = angle between reference and computed axes of rotation
β = angle between reference and computed direction of translation
(1) = improved Longuet-Higgins
(2) = new linear technique
(3) = reconstruction and reprojection
I2S = I2S CCD camera
Pul = Pulnix CCD camera

mean square. We can then project P as m_1' and m_2' on the two retinas and compute the distances $d(m_1, m_1')$ and $d(m_2, m_2')$ as functions of $\mathbf{R}$ and $\mathbf{t}$. Summing those squared distances over all pairs of matched points yields a criterion that can be minimized. This criterion weighs differently points that are far and points that are close since it accepts a larger 3-D reconstruction error for far points than for close points. We have experimented with this idea and the preliminary results are more accurate and robust to noise than the ones obtained by the previous technique [TF87]. Results also appear in tables 2.1 and 2.2. A similar idea has recently been used by Harris [Har87].

2.2.7.2 Line Matches Lines are tokens that can be extracted very reliably from many kinds of images and are therefore good candidates as a basis

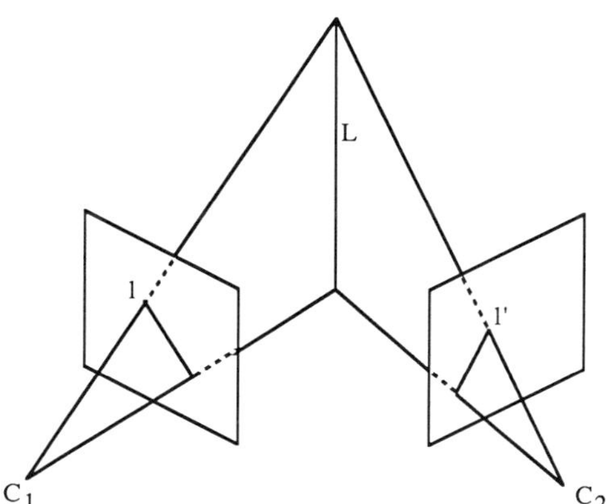

Figure 2.32
Matching two lines in two frames does not put any constraint on the motion.

for matches. Some work in this area has been reported by Mitiche et al. and Liu and Huang [MSA86a, MSA86b, LH86b, LH86a]. The difference between the two approaches consists in the order in which the problems are solved. Mitiche et al. first solve for structure and then foı motion, Liu and Huang first solve for motion and recover structure next. We have extended the work of Liu and Huang and shown that fairly robust estimates of motion and structure can be obtained from line matches [FLT87].

The characteristic of line matches when compared to point matches is that two views are not sufficient, but that three are needed to recover both motion and structure. Indeed, as shown in figure 2.32, matching l and l' simply defines a 3-D line L but puts no constraint on C_1 and C_2 (compare with figure 2.28). If we now consider figure 2.33, the fact that l, l', and l'' are the images at three different times of the same 3-D line L imposes constraints on the motion of the camera between positions 1, 2, and 3.

These constraints have their roots in the fact that the three planes defined by l, l', and l'' and that the camera optical centers meet along L: they belong to the same pencil of planes. Given a plane of equation $ax + by + cz + d = 0$, we associate a 4-dimensional vector $\mathbf{P} = [a, b, c, d]^T$. Expressing the fact that the three planes belong to he same pencil is equivalent to saying that the 4×3 matrix formed by their associated vectors $\mathbf{P}_1$, $\mathbf{P}_2$, and $\mathbf{P}_3$ is of rank less than or equal to two. It can be shown that this provides two constraints on the two motions. Since the two motions depend on 11 parameters (6 for the two rotations and 5 for the two translations), a simple

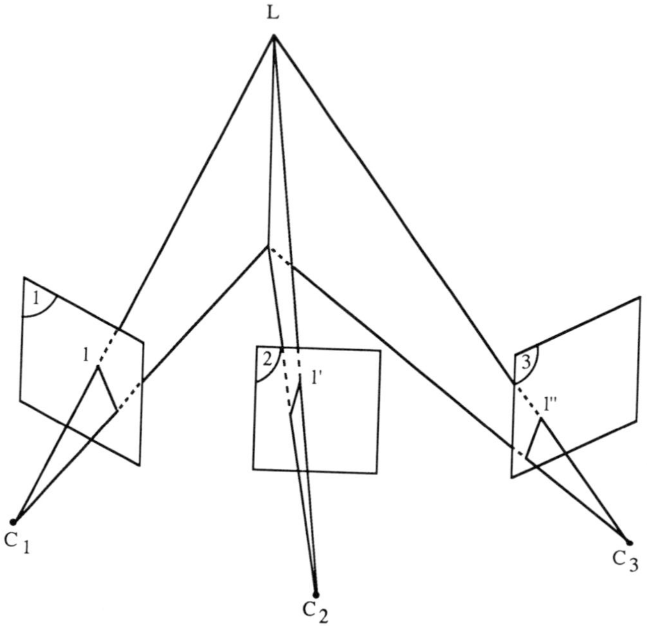

Figure 2.33
Three lines in three frames.

counting argument shows that in order to hope to solve for the unknowns, at least 6 line matches are necessary. Because the constraints on the rotations are nonlinear, there is no guarantee that the solution is unique or even that there is a finite number of solutions. We have implemented the motion computation using the extended Kalman filtering approach, which we briefly mention in section 2.4.3 and is developed in [AF87].

We have also implemented this technique on both synthetic and real data. Our synthetic data are the same as those used in the previous experiment, except that instead of using the points, we have used some of the lines defined by them. Again, we have simulated a number of camera motions and measurement noise. The noise is added on the pixels defining the end points of the projected segments and varies between 0.1 and 1 pixel. Results are presented in table 2.3.

With a 3-camera system commandable in rotation we took 7 stereo triplets of the same scene (one such triplet is shown in figure 2.25), each one differing from the previous by a rotation of 5° around a vertical axis. The

Table 2.3[a]

1	2	3	4	5	6	7	8	9	10
5	5	0.1	0.61	0.081	0.47	1.5	0.064	0.29	0.44
5	5	0.5	4.4	1.1	3.2	10	0.71	1.8	3.4
5	5	1	12	9.4	7.8	24	2.4	4.9	10
30	5	0.1	0.029	0.16	0.027	0.58	0.059	0.24	0.24
30	5	0.5	0.32	1.2	0.26	5.2	0.33	1.8	2.9
30	5	1	1.1	5.4	0.79	15	1.9	6	10
30	30	0.5	0.39	0.51	0.29	1.1	0.1	1.4	2.5
30	30	1	1.1	1.2	0.82	2.7	1.1	4	7.5
30	30	3	3.8	5.8	2.8	8.9	5.4	23	41
10	20	0.5	2.4	0.61	1.7	2.7	0.74	1.7	3.3
10	20	1	6.1	2	4.2	6.4	2.4	4.6	9.1

a. translation 1, 50 0 25 (cms); translation 2, 0 100 0 (cms); axis of rotation 1, 1 1 1; axis of rotation 2, 0 1 1; starting point, 0 for all the parameters. 1 (respectively, 2), angle rotation 1 (respectively, 2) in degrees; 3, pixel error eps (uniform noise in an interval of length 2 eps); 4 (respectively, 5), relative error on angle 1 (respectively, 2) in %; 6 (respectively, 7), absolute error on angle of axis 1 (respectively, 2) in degrees; 8 (respectively, 9), absolute error on direction of translation 1 (respectively, 2) in degrees; 10, relative error on ratio of norms of translations 1 and 2.

matches were obtained by using a 3-D matching procedure between two views following the trinocular stereo algorithm described in [AL87]. The results appear in table 2.4. The deviation of the angle from its predicted value of 5° and of the axis from a perfect verticality being quite stable, we believe that our computations are more accurate than the command of the system. Furthermore, it indicates that our method can cope with noisy data.

They somewhat contradict the results reported by Liu and Huang [LH86b, LH86a], who reported a large sensitivity of their method to noise and convergence toward false minima. The Extended Kalman Filtering approach seems to be much more immune to these problems since in all cases our initial guess was the identity matrix for the rotation. Of course, this needs to be tested further. Another big advantage of this method is that it allow us to easily inject any a priori knowledge about the rotation as constraints on the initial covariance matrix S_0.

Remark A remark similar to the one made in the case of points applies here. Indeed, we could compute the reconstructed 3-D line L as a function of the motion D_1 from camera 1 to camera 2, and the motion D_2 from camera 2 to camera 3 (see [FLT87] for a way of doing this), project it back on the three retinas as l_1, l'_1, and l''_1, and consider the distances $d(l, l_1)$,

Table 2.4[a]

1	2	3			4	5
0-1-2	206	0.02168	0.996340	−0.85454	5.2	4.9
		0.03186	0.996462	−0.83988	5.4	4.8
1-2-3	158	0.001352	0.997401	−0.072036	5.3	4.1
		0.000696	0.993526	−0.113600	5.2	6.5
2-3-4	108	0.002404	0.995785	−0.091683	5.3	5.3
		0.002502	0.997182	−0.074976	5.3	4.3
3-4-5	63	−0.039143	0.997057	−0.065924	5.3	4.4
		0.002230	0.995984	−0.089499	5.3	5.1
4-5-6	10	0.001226	0.996201	−0.087078	5.2	5
		0.002527	0.997028	−0.077000	5.2	4.4
5-6-7	13	0.014293	0.998154	−0.059027	5.3	3.5
		0.003504	0.995266	−0.097128	5.2	5.6

a. 1, triplet of images used; 2, number of matched segments; 3, axis of rotation (first line from image 1 to image 2, second line from image 2 to image 3); 4, angle of rotation; 5, angle of axis relative to the vertical axis of the absolute coordinate system.

$d(l', l'_1)$, and $d(l'', l''_1)$. The sum over all matches of these distances is a function of D_1 and D_2, which could be minimized. Again, this has the advantage of weighting less the remote lines than the close ones.

2.2.7.3 Finding the Matches The main difference in the case of stereo is that since the motion of the camera is a priori unknown, it is not possible to use the epipolar constraint to reduce the size of the search space for matches. Nor can other constraints, such as continuity (based on the computation of disparity) and ordering (based on epipolar geometry), be used.

It looks as though the matching problem may be very serious. It may not be true for a number of reasons. One such reason is that if the system that extracts tokens and features works fast enough as compared to the image velocity of the tokens, the matching from frame to frame becomes simple. Motion estimation from these matches may be quite inaccurate, as pointed out by Nagel [NAG86]. Nonetheless, if we track the motion over several frames and wait until the equivalent baseline is long enough, then we may be able to obtain much more robust estimates. A second reason is that in some cases (such as with robots), we may have some good a priori knowledge of the camera motion. This in turns implies that some of the epipolar geometry is known, thus simplifying the matching problem. Nonetheless, we think that this is an area where a lot of useful research could be performed.

2.2.8 Active 3-D Vision

We have seen that the problem of matching tokens may be difficult in stereo and motion (ambiguity for large disparities or fast motions). Moreover, if no contrast is present, then few tokens will be extracted and the 3-D maps obtained may be too sparse for the application considered. For these two reasons, systems have been designed that use extra lighting to reduce the matching ambiguity and increase the density of the resulting depth maps.

Examples are the stereo matcher developed by K. Nishihara at MIT [NIS84], where a texture is projected onto the scene to increase the density of features and therefore that of the resulting depth map, or the laser range finders developed at INRIA [FGK*83] and MIT [BRO84], and now by a number of companies. Another possibility is the ERIM laser range finder, which measures the time of flight of a laser beam to compute distances.

Laser-based range-finding techniques have the obvious advantages of making the matching in general simpler and increasing the density of data. They have the disadvantage that they do not work well at depth edges, for surfaces that have a large glossy reflectance component, and that they require an extra source of illumination, implying extra power and additional mechanical constraints. Another disadvantage is that lasers emit monochromatic lights, so that the richness of the whole visible spectrum is lost.

Despite these disadvantages, we believe that in a number of constrained industrial environments they are quite enticing and can in general provide another source of 3-D information for systems that use passive vision. We think nonetheless that because of their flexibility and simplicity, passive vision systems will be the main sources of 3-D data for the robotics systems of the future.

2.2.9 What Needs to Be Worked On

Here are a number of areas in which we believe considerable progress must and can be achieved:

1. **Classify edges for stereo and motion.** A potential source of errors in stereo and motion matching in the current algorithms is the small number of stable features attached to the tokens that they use. As we discussed in section 2.2.2, it may be possible to reliably extract features that would allow us to classify edges from the standpoint of either their physical origin or the shape of the intensity function in the vicinity of those edges.

2. **Understand matching errors in stereo and motion algorithms.** Indeed, these errors are the source of countless problems when the results are used to build higher-level representations than depth maps. On the other hand, no errors seem to be made by the human visual system. The main difference between computer vision stereo algorithms and human perception seems to us to come from the fact that human perception is essentially dynamic: our eyeballs are continuously in motion and so is our head. We believe that this is one key to eliminating errors in stereo matching by comparing views taken from slightly different viewpoints for inconsistencies.
3. **Understand fully the number of solutions for motion.** In sections 2.2.7.1 and 2.2.7.2 we discussed a number of algorithms for solving for the motion parameters from point and line matches in the case where noise is present. The computation of the number of solutions of the corresponding equations and the analysis of the degenerate cases have only been partially done, and a lot remains to be done. We believe this is important in order to understand the next problem.
4. **Understand the errors in motion estimation as functions of the geometry.** Indeed, when computing motion parameters using, for example, the algorithms discussed in sections 2.2.7.1 and 2.2.7.2 on synthetic data where we control both the 3-D position of the tokens that we project in the image plane and the amount of noise that we add to these projections, we realize that the error on the motion parameters is a complicated function of the geometry of the problem. Understanding this relation is the key to more robust algorithms.
5. **Find the matches in motion estimation.** There is still a large potential for good work on the token tracking problem.
6. **Combine stereo and motion.** This is a good example of a class of sensory problems where several sensors or sources of information are available and there is need to combine them. The whole theory of sensory integration remains mostly to be done even though some work has recently been devoted to this problem [CRO86, FAF86, AF87, DUR86, SC86]. On the specific problem of integrating stereo and motion (for example, to correct errors), very little has been done.

2.3 Shape Representation

Now that we have dealt with the problem of obtaining rough 3-D data, we would like to use them for a number of application tasks. One such application is the navigation of a mobile system in an unknown environment; another one is the recognition and localization of objects.

Even if these tasks do not have exactly the same requirements in terms of the kinds of representations they use, we believe that they should use a common representation with the following properties:

1. **Object centered so that the effect of rigid motions can be readily assessed**: Indeed, since our representations are intended to support tasks such as recognition from various viewpoints—localization, navigation, manipulation—they should behave simply with respect to the group of 3-D rigid motions.
2. **Volume and surface oriented**: In robotics applications, we are both interested in a volume representation of free space and in a surface representation of the objects and obstacles present.
3. **Work for both sparse and dense data**: We would like our representations to have good convergence properties when the density of measurements increases.
4. **Work for both dense and sparse data**: If many measurements are available, and if the geometry of the environment is simple, we would like to be have the possibility of simplifying the representations.
5. **Easy update when new data become available**: We would like the representations to be incremental in the sense that adding new measurements does not imply recomputing everything from scratch.
6. **Efficient computation**: This is connected to requirement (5) but not equivalent to it.

A representation satisfying these requirements having been built, it can then be used to construct other representations more tailored to specific applications. We concentrate first on this representation and propose a fairly general scheme for building it.

2.3.1 Interpolation through the Delaunay Triangulation

In a number of publications, Boissonnat [BOI84, BOI85, BOI86] has proposed roughly to use the Delaunay triangulation to interpolate 3-D data obtained, for example, from a laser range finder. Due to lack of space, we cannot describe here the Delaunay triangulation and refer the interested reader to [BOW81, PS85, WAT81].

We describe here the application of this technique to the interpolation of stereo data—more precisely to the construction, from the output of the stereo matchers discussed in sections 2.2.4–2.2.6 that produce line segments

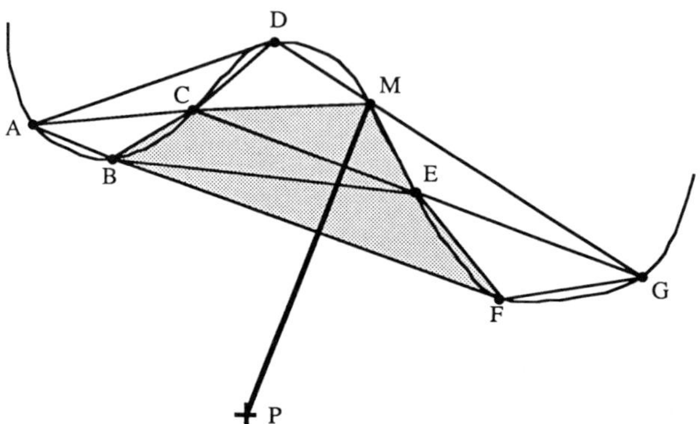

Figure 2.34
Basic idea in 2-D for volume and surface representation.

in 3-D space of a polyhedral approximation of the objects and obstacles present in the scene and of a volume representation of free space.

Figure 2.34 shows a two-dimensional example of the basic principle of the method. Point P represents one of the cameras optical centers; points A, B, C, D, M, E, F, and G are points in a set S, measured by the stereo process. We have computed the Delaunay triangulation of the points in S and we are considering point M. This point is by definition visible from P, and therefore all triangles in the Delaunay triangulation intersected by the segment PM should be marked as empty space. These are the shadowed triangles in figure 2.34. By considering more points and the second camera optical center, more triangles can be marked as empty space.

In three dimensions, things are very similar, except for the fact that we are now dealing with line segments in 3-D rather than points. We assume that the Delaunay triangulation is built for the set S of the segments end points.

As shown in figure 2.35, for a given stereo segment AB, the visibility property means that all tetrahedra intersected by triangle PAB can be marked as empty space. For example, the triangle PAB intersects tetrahedron $QRST$, the intersection being triangle HIJ. Therefore, tetrahedron $QRST$ can be marked as empty space.

The next question then is, given the Delaunay triangulation of the end points of the stereo segments, how can we efficiently detect those tetrahedra that are intersected by at least one triangle PAB, where P is one of the cameras optical center and AB one of the stereo segments? The obvious

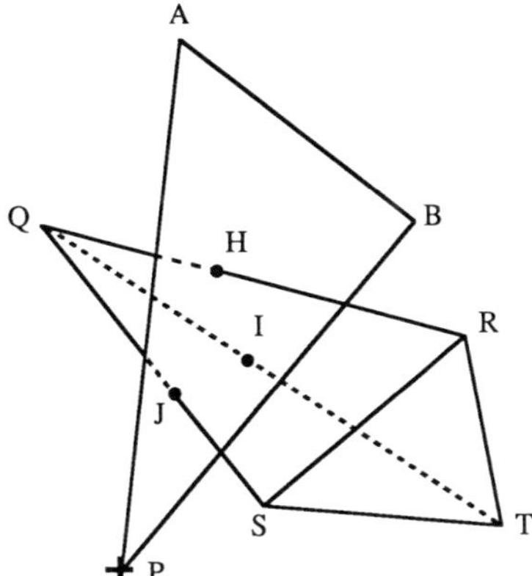

Figure 2.35
Same idea in 3-D.

way of going is by performing the following algorithm:

For all stereo segments AB
  For all unmarked tetrahedra T
    Determine if the intersection of triangle PAB with tetrahedron T is not empty
      If it is not empty then mark tetrahedron T

Clearly, the computation time of this algorithm depends heavily on the number of intersections that are going to be computed. In order to minimize this number, there are several ways of proceeding, all of which imply some preprocessing of the data. The obvious idea is to use bucketing techniques, not unlike what was done in section 2.2.4.1. For details, the interested reader is referred to [BFLB87].

Assuming that empty tetrahedra have been marked, we can identify two difficulties with this approach. The first one is that since the output of the stereo algorithm is a set of line segments in 3-D space, there is no guarantee that, in the Delaunay triangulation of their end points, the segments will be Delaunay edges. This may be a problem if we wish the reconstructed surface to be a good approximation of the real one. We show next that this can be guaranteed if we allow ourselves to add more points on some of the segments.

2.3.1.1 Constrained Delaunay Triangulation Indeed, one problem with the Delaunay triangulation is that it is not immediately obvious how to modify it so that is possible to include line segments, triangles, or even tetrahedra while making sure that these primitives will be part of the final triangulation. This is a problem known in computation geometry as the constrained triangulation problem. A paper by Lee and Lin [LL86] defines a constrained Delaunay triangulation for any planar straight line graph that can be computed in $O(n^2 \log n)$ time. We present here an algorithm that runs in $O(n^2)$ in the 2-dimensional worst case, can be generalized to any dimension, and is very efficient in practice.

We modify the input data by adding more points, in such a way that the resulting Delaunay triangulation is guaranteed to contain, say, a set of initial edges. Indeed, from the definition of the Delaunay triangulation, a given set of line segments will be Delaunay edges of the Delaunay triangulation of their end points if the disks having those segments as diameters contain in their interiors no other points than those of the segment.

Let S be such a disk attached to a segment s, and $s_1, s_2, \ldots, s_n$ be the segments intersecting that disk. It is possible to partition s in a finite number of subsegments such that none of the disks attached to those segments intersect any of the s_i's. Indeed, let d be the smallest distance of the s_i's to segment s. The way to compute the distance of a segment to another segment is shown in figure 2.36.

Assuming that segments do not cross, d is strictly positive. If D is the length of s, we can divide s into $\lceil D/2d \rceil$ intervals of length less or equal $2d$. This corresponds to adding $\lceil D/2d \rceil - 1$ points, the end points of these segments. Therefore, as shown in figure 2.37, the disk having those segments as diameters do not contain any points on other segments. In this figure, points A_1 and A_2 have to be added to the original set of points.

In the case where AB is connected to one segment BC, we consider the plane perpendicular to AB in B. If BC and AB are not on the same side of this plane, then we can ignore BC (in fact, we can ignore all the line seg-

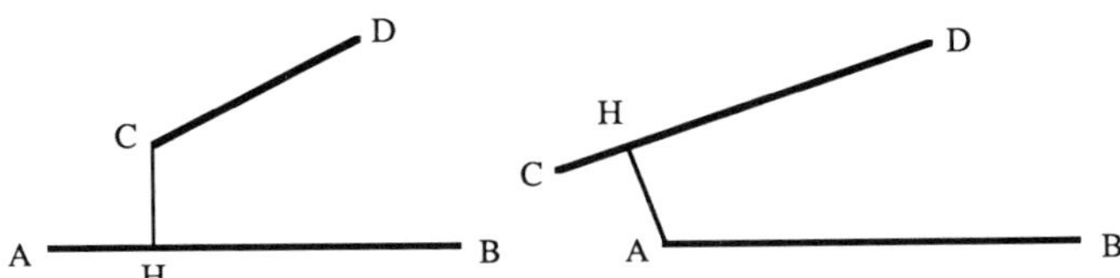

Figure 2.36
Computing distances between segments.

ments on the same side of the plane) and the previous technique for splitting AB can be used (see figure 2.38).

When they are on the same side, there exists a sphere passing through point B, a point A_1 of AB, and a point C_1 of BC, and not intersecting any other segment. This is shown in figure 2.39. Points A_1 and C_1 have to be added to the original set, and the previous technique can then be applied to segments AA_1 and CC_1, if necessary.

After this preprocessing, the resulting augmented set of line segments satisfies the condition that they are Delaunay edges of the Delaunay triangulation of their end points. This preprocessing has been described in the 2-dimensional case; it can be extended to any dimension in a straightforward manner.

An interesting question at this point is, how many points do we add in this process? the answer is that the worst case is $O(n)$. Indeed, it can be seen

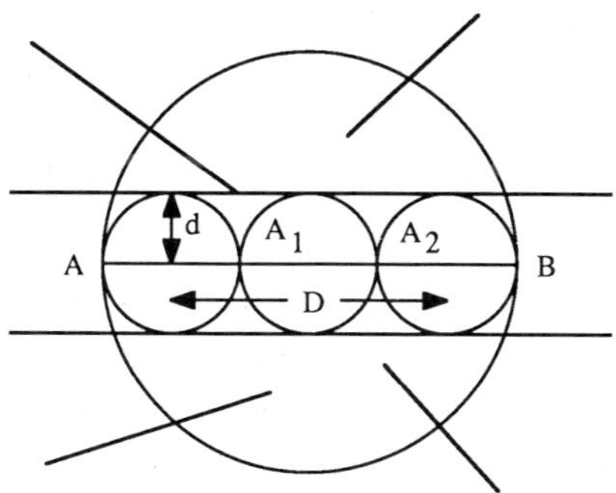

Figure 2.37
Adding new points on the segments to make them Delaunay edges.

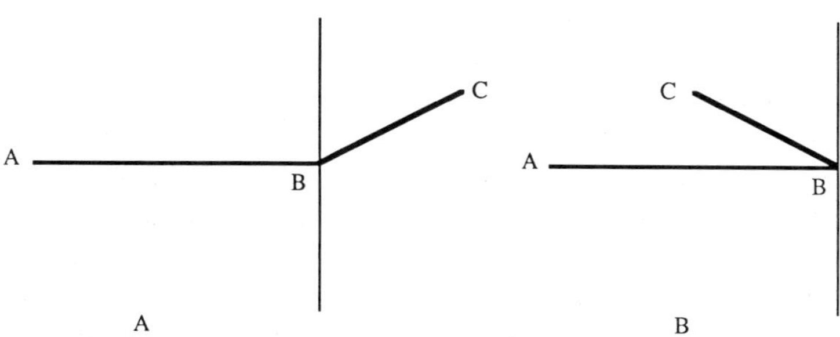

Figure 2.38
Segments connected to AB: (A) do not cause problems; (B) require special processing.

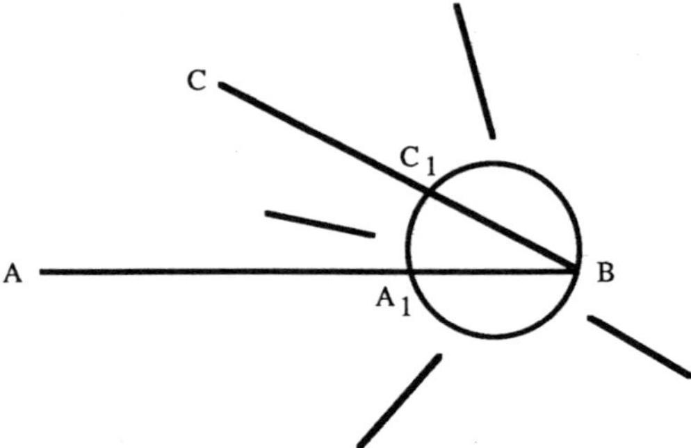

Figure 2.39
Special processing for connected segments.

from figure 2.37 that for a given segment AB of length D, there exists a number $d > 0$ such that in the band parallel to AB of width $2d$ there are no other points belonging to the other segments. We then only need to add $\lceil D/2d \rceil - 1$ points to segment AB so that the corresponding segments thus created satisfy the constraint.

We present now the algorithm, based on the above ideas, that computes a Delaunay triangulation constrained to contain a set of edges E:

Compute the Delaunay triangulation of the end points of the edges in E
For any edge, say AB, in E that is not an edge of the triangulation do
  Find $i \in [1, n]$ such that the distance d_i between s_i and AB is minimal.
  Add the corresponding new points (as described above) in a list of new points.
Update the triangulation by introducing the new points.

The computation time of this algorithm is $O(n^2)$ in the planar case and $O(n^3)$ in the 3-D case, if we use the sequential algorithm of the Delaunay triangulation. Let us do the analysis for the 3-D case. The complexity comes from the Delaunay triangulation, which is $O(n^3)$ worst-case, and from step (a). Since we may have $O(n^2)$ tetrahedra in the worst case in the triangulation, we may have to compare AB with $O(n^2)$ edges, thus adding another $O(n^3)$ term to the complexity. In practice, computation of the Delaunay triangulation by the iterative technique is linear. In order to reduce the second source of complexity, we use *bucketing techniques*. Indeed, we do not need to compare, for each edge AB, the distance between AB and every

Figure 2.40
3-D office scene.

other segment. The segments that must be considered are those intersecting the sphere whose diameter is the segment AB.

The bucketing technique consists here in dividing the space into buckets that are cubes or parallelepipeds and computing for the sphere the buckets that cover it and for each segment the buckets intersected.

We then compute the distances between the segment AB and each segment that intersects the sphere, i.e., the segments that have at least a bucket in common with the sphere. In practice, this makes the complexity of step (a) linear at the cost of preprocessing the data to compute the buckets.

Figure 2.40 shows an example of a 3-D office scene, Figure 2.41 shows a cross section by a horizontal plane of the original constrained triangulation before marking tetrahedra, and figure 2.42 shows the same cross section of the remaining tetrahedra after exploiting the visibility property.

2.3.2 Finding Important Curves on a Surface

The methods we describe next apply more to the case where the previous triangulation has been built from a dense range map such as the one provided by a laser range finder. We draw heavily on the work of Brady and his colleagues [BPYA85, PB85].

Curves on a surface are a useful source of constraints on the geometry of the surface and can be used as a basis for its description. There are three

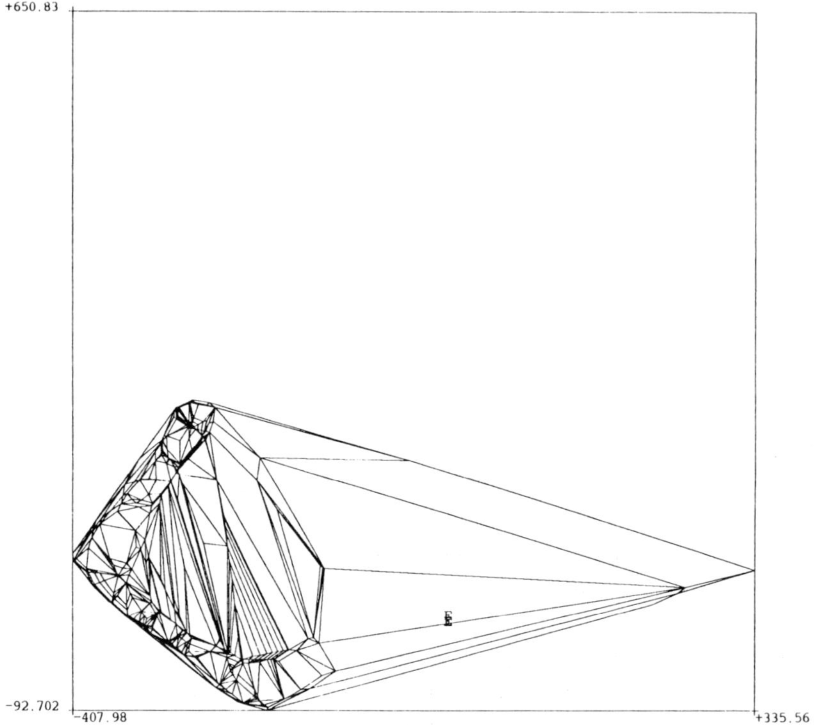

Figure 2.41
Cross section by a horizontal plane of the original constrained Delaunay triangulation.

types of curves we are interested in detecting and describing:

1. bounding contours,
2. lines of curvature, geodesics, and parabolic lines, and
3. curves where significant surface changes occur.

Bounding contours are either extremal, i.e., places where the surface normal turns smoothly away from the viewer, or mark discontinuities of some order of the surface.

2.3.2.1 Bounding Contours For extremal boundary contours, there exists an important relationship between the curvature of the projection on the retina of the boundary curve and the Gaussian curvature of the surface at points along the boundary curve. This relationship is the following [KOE84, BPYA85]:

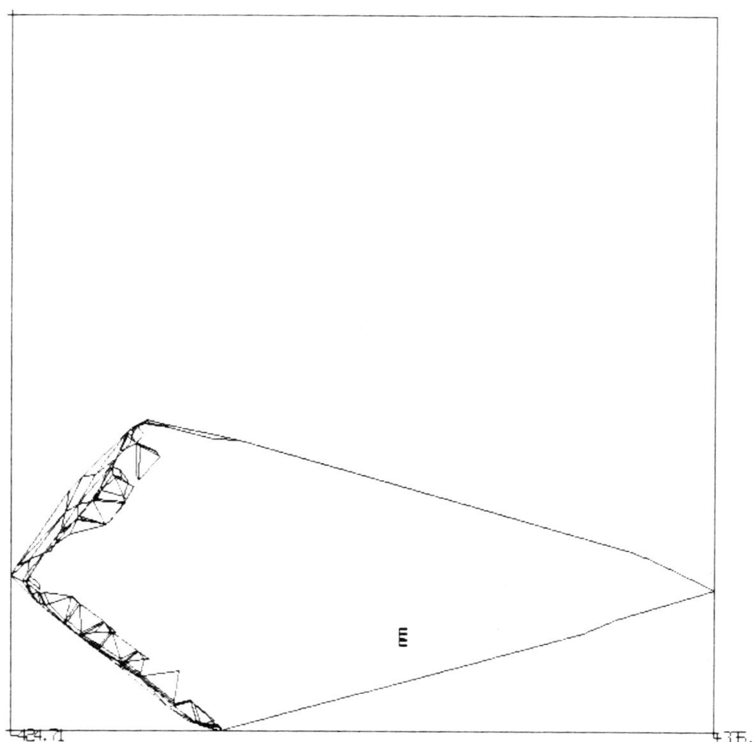

Figure 2.42
Cross section by a horizontal plane of the remaining tetrahedra after using the visibility constraint.

The sign of the Gaussian curvature of the surface at points along the boundary curve is the same as the sign of the curvature of the projection of the boundary curve.

The proof of this result is a little bit involved and we refer the interested reader to the previous references.

2.3.2.2 Lines of Curvature, Geodesics Lines of curvature give the locally flattest coordinate system intrinsic to the surface. They are local statements about the surface. In practice, we aim at describing surfaces at a larger scale. This is done by linking the local directions to form smooth curves. As pointed out in [BPYA85], there are generally infinitely many such curves and we want to make explicit in our representation only a small finite subset of those.

Figure 2.43
Lines of curvature for an ellipsoid.

Figure 2.44
Results of linking the principal directions on the surface of an oil bottle and a coffee mug [PB85].

One way of obtaining such a subset is by adding constraints to the curves. Possible constraints are

1. the curve is planar, and
2. the descriptors along the curve are constant.

Indeed, as it is shown in figure 2.43, there are only three planar lines of curvature for an ellipsoid, and they are the intersections of the surface with its planes of symmetry. The surface is effectively described by these curves. Brady [BPYA85] has developed a program that, given a dense depth map of an object, smoothes the depth map by Gaussian filtering, and computes at every point the principal directions and the surface curvatures. Lines of curvature can be found by linking the principal directions at the individual surface points. Results are shown in figure 2.44 for an oil bottle and a coffee mug.

Another set of potential good candidates is the set of geodesics. However, potential disadvantages of using the geodesics as descriptors are that there are too many of them (one through every point on a surface in every direction) and that if we add the constraint that the geodesics be planar, then they are lines of curvature.

2.3.2.3 Finding Surface Discontinuities Another potentially very attractive way of describing a surface is to detect the places where it changes abruptly. This is very similar to the edge detection problem with one very important difference. The difference is that edge detectors usually operate on intensity images where the measured intensity is a mixture of object surface geometry, object surface reflectance, and illumination. The goal of image analysis is to extract from such an image a set of intrinsic features that characterize the geometry of the objects in the scene. Examples of such features include surface normals, lines of curvature, and discontinuities in depth, in the normals, etc.

We are facing here a very different situation, and apparently a simpler one, in the sense that, unlike the case of an intensity image, range is available to us. Therefore edge models are really surface models and we do not have such things as reflectance, texture, shadow edges.

In order to detect the surface changes, we need to characterize the type of changes we are interested in; that is, we must define quantitative models for them. We must also propose algorithms that reliably detect those models in range data, taking into account the inevitable presence of noise. Ponce and Brady [PB85] consider three types of models:

- steps where the surface height function is discontinuous,
- roofs where the surface is continuous but its normal is discontinuous, and
- shoulders, which consist of two roofs and correspond to a step viewed obliquely.

The three models are shown in figures 2.45 and 2.46. They are one-dimensional models and therefore the surface is modeled locally as cylindrical.

The basic idea is to smooth the surface data, using Gaussian smoothing, at different scales, extract significant features and use them to **identify** the presence of the models—track these features at different scales to **localize** the identified model on the surface. It turns out that roofs consist of extrema of the dominant curvature; on the other hand, steps and shoulders consist of parabolic points, that is, zero crossings of the Gaussian curvature, and can be distinguished by their behavior at different scales.

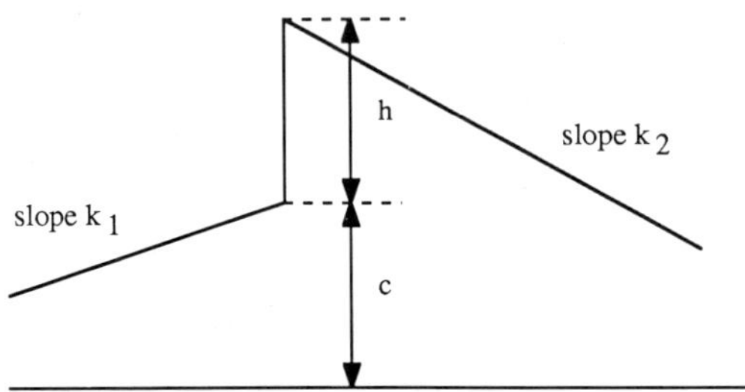

Figure 2.45
Step model.

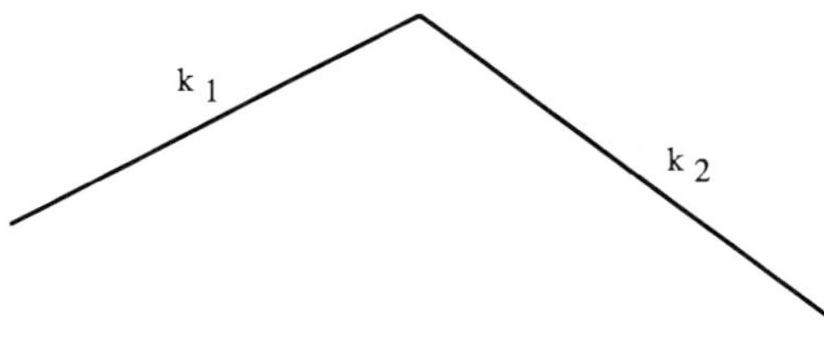

Figure 2.46
Roof model.

The key to using the combination of cylindrical models and detection of lines of curvature is in the following theorem [BPYA85]:

The convolution of a cylindrical surface with a Gaussian distribution is cylindrical.

Since we also know that the lines of curvature of a cylinder are in the plane of the curve and parallel to the generating line, then we know that this property is conserved after Gaussian filtering, thus allowing us to detect the models by computing the derivatives of the principal curvatures in the direction of corresponding line of curvature.

For a planar curve given by its equation $y = f(x)$, the curvature is given by the formula

$$\kappa(x) = \frac{f''}{(1 + f'^2)^{3/2}}. \quad (11)$$

From formula (11) and the equations describing the step model, it is easy to derive the fact that for a step smoothed by a Gaussian of variance σ, the

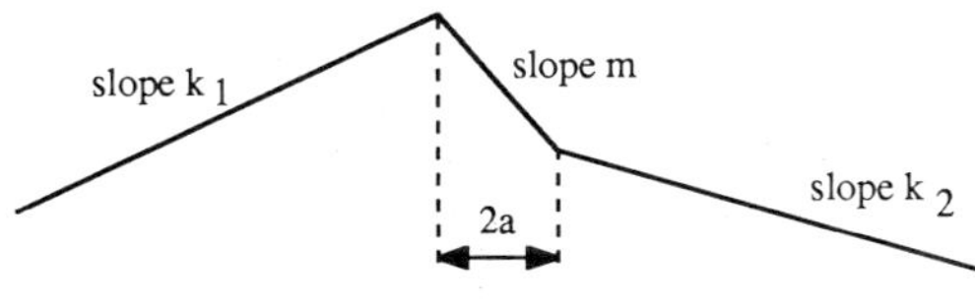

Figure 2.47
Shoulder model.

curvature κ has a zero crossing at the point $x_\sigma = \sigma^2 \delta / h$ (h is defined in figure 2.45 and $\delta = k_2 - k_1$; see figure 2.47). Also, using the fact that the second derivative of f_σ (f smoothed by a Gaussian of variance σ^2) is 0 at x_σ, it is also easy to show that

$$\frac{\kappa''(x_\sigma)}{\kappa'(x_\sigma)} = -2\delta/h,$$

which says that the ratio of the second and first derivatives of the curvature at the zero crossing is constant over the scales.

For the roof model, we can specialize the previous analysis to the case where $h = 0$ and show that the curvature satisfies the following relation:

$$\kappa(x, \mu\sigma) = \kappa(x/\mu, \sigma).$$

From this, it follows that the extremum value of κ is proportional to $1/\sigma$, and that its distance from the origin is proportional to σ.

Finally, for the shoulder model, it can be shown that there exists a zero crossing of the curvature at the point $x_\sigma = (\sigma^2/2a)\log(\delta_1/\delta_2)$. It is then straightforward to show that

$$\frac{\kappa''(x_\sigma)}{\kappa'(x_\sigma)} = -\frac{1}{a}\log\left(\frac{\delta_1}{\delta_2}\right),$$

so the ratio of the first and second derivatives at the zero crossing is constant over scale.

These three models can now be used to detect and localize surface intersections by tracking extrema of curvatures (roofs) and parabolic points (steps and shoulders). The technique goes as follows.

- First, smooth the original depth map with a Gaussian distribution at a variety of scales σ.

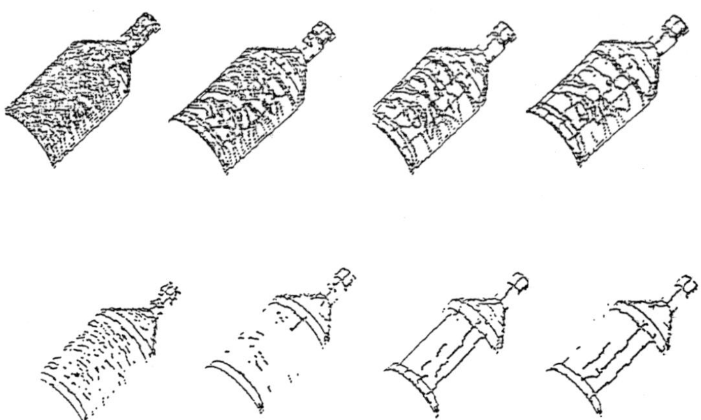

Figure 2.48
Detection of discontinuities on the surface of an oil bottle [PB85].

• Second, compute at each scale the principal curvatures and their directions, as well as the first and second derivatives of the principal curvatures in their associated directions.
• Third, mark parabolic points (zero crossings of the Gaussian curvature) and (directional) maxima (minima) of the maximum (minimum) curvature.
• Fourth, track the extrema and zero crossings across scales, from coarse to fine. The result is a forest of points, a "fingerprint" [YP83a, YP83b]. A path in the forest corresponds to a feature point.
• Fifth, eliminate points at the finest scale that have no ancestor at the coarsest scale on their path (detection). The remaining points are located from the finest scale.
• Sixth, label each remaining path as roof, step, or shoulder, according to the behavior of its curvature and its first and second derivatives across scales, using the previous characteristic relations. Paths that do not follow the properties of any of those models are eliminated.

Examples and results are shown in figure 2.48 taken from Ponce and Brady [PB85].

2.3.3 Finding Surface Patches

So far, we have seen how to represent a set of measured points by finding a simplicial polyhedron that interpolates these points. When the density of the points is high, this polyhedron is a very fine approximation of the set of points that may not be suitable for many applications. We have seen in section 2.3.2 how to use this structure to make explicit important curves, mostly from the computation of lines of curvature.

In this section, we want to develop a complementary approach to that of finding curves, namely, that of finding surface patches. The kind of surface primitives that can be used is highly constrained by their accessibility from the raw data. More precisely, we want them to be

1. robust to measurement noise and
2. viewer independent.

Condition (1) implies that we cannot use high-degree polynomials since they will tend to "follow" the noise. Condition (2) implies that if we approximate the points in two different coordinate systems, then the approximations should relate to each other by the same transformation as that which relates the two coordinate systems. These considerations have prompted us to use first- and second-degree polynomials (planes and quadrics) to approximate surface patches.

2.3.3.1 Representing Planes and Quadrics A plane is represented by its unit normal vector **n** and its distance to the origin d (see section 2.2.5.2). The plane equation is given by

$$\mathbf{n}^T\mathbf{x} = d, \qquad \text{where} \quad \mathbf{x} = [x, y, z]^T.$$

This implies that a plane has two equivalent representations $(\mathbf{n}, d)$ and $(-\mathbf{n}, -d)$.

A standard representation for a quadric is a symmetric 3×3 matrix **A**, a 3×1 vector **v**, and a number d; the equation of the quadric is then given by

$$\mathbf{x}^T\mathbf{A}\mathbf{x} + \mathbf{x}^T\mathbf{v} + d = 0.$$

Notice that the representation $(\mathbf{A}, \mathbf{v}, d)$ is defined up to a scale factor; we show in the next section how to impose a constraint on matrix **A** to reduce the ambiguity of the representation to two.

2.3.3.2 Surface Fitting When constructing the regions of the object that are best approximated by planes or quadrics, we have to solve repeatedly the following problem. Given a set of points forming a region, find the plane (quadric) that minizes the error measure,

$$E = \sum d^2(M_i, S), \tag{12}$$

where the M_i's are the points in the region and d is the distance of M_i to the surface S. In the case of a plane $(\mathbf{n}, d)$, it is well-known that $d(M_i, P) = \mathbf{n}^T\mathbf{x}_i - d$. In the case of a quadric Q, there is no such simple formula relating the representation of Q and the distance d. In fact, it can be shown, that for a given point M, its distance to a quadric is found by solving an equation of degree 3. For the sake of simplicity, and by analogy to the planar case, we define as the point error measure the value

$$d^2(M_i, Q) = \mathbf{x}_i^T\mathbf{A}\mathbf{x}_i + \mathbf{x}_i^T\mathbf{v} + d.$$

In the case of a quadric with a center, this is equal to the square of the euclidian distance between M_i and m_i, the intersection of the line CM_i (C the center of the quadric) with the quadric.

The problem is now to minimize criterion E in the case of a plane and a quadric. The case of a plane is classic [DH73]: the normal to the best plane is the eigenvector of length one, $\mathbf{n}_{\min}$, of the covariance matrix $\mathbf{C}$ of the points in the region, associated with the smallest eigenvalue λ that is also the minimum error $E_{\min}$. The covariance matrix is given by

$$\mathbf{C} = \sum \mathbf{A}_i\mathbf{A}_i^T, \qquad \mathbf{A}_i = \mathbf{x}_i - \mathbf{x}_g, \qquad \text{and} \qquad \mathbf{x}_g = 1/N \sum \mathbf{x}_i.$$

The distance to the best approximating plane is given by

$$d_{\min} = -1/N \sum \mathbf{n}_{\min}^T \mathbf{x}_i.$$

In the case of a quadric, we represent it, for the purpose of approximation, by ten numbers $a_{1 \cdots 10}$ related to the previous description by the following relationships:

$$\mathbf{A} = \begin{bmatrix} a_1 & a_4/\sqrt{2} & a_5/\sqrt{2} \\ a_4/\sqrt{2} & a_2 & a_6/\sqrt{2} \\ a_5/\sqrt{2} & a_6/\sqrt{2} & a_3 \end{bmatrix},$$

$$\mathbf{v} = [a_7, a_8, a_9]^T,$$

$$d = a_{10}.$$

As we noted earlier, the representation for a quadric is not scale invariant, and therefore so is criterion E. This implies that if we do not add a constraint on the a_i's, the minimum of E is 0, attained for $a_i = 0$ for all i's. Since there does not exist a "natural" constraint like $\|\mathbf{n}\| = 1$ in the case of a quadric, several constraints can be considered. Among those proposed in the literature,

1. $\sum a_i^2 = 1$,
2. $a_{10} = 1$, and
3. $\mathrm{Tr}(\mathbf{A}\mathbf{A}^T) = \sum a_i^2 = 1$,

we select the third one since it is the only one invariant with respect to rotations and translations. This implies that if we use (1) or (2), the solution of the minimization of E in two different coordinate systems will not be the same quadric!

Let us define the vector $\mathbf{p} = [a_1, \ldots, a_{10}]^T$. Criterion E can be rewritten as

$$E = \sum \mathbf{p}^T \mathbf{M}_i^T \mathbf{M}_i \mathbf{p} = \mathbf{p}^T \mathbf{M} \mathbf{p},$$

where $\mathbf{M}$ is a 10×10 symmetric matrix and $\mathbf{M}_i = [x_i^2, y_i^2, z_i^2, x_i y_i, x_i z_i, y_i z_i, x_i, y_i, z_i, 1]^T$. The constraint (3) can be written as

$$\|\mathbf{B}\mathbf{p}\|^2 = 1,$$

where $\mathbf{B}$ is a 6×10 matrix that "selects" the first six elements of vector $\mathbf{p}$. The solution to the optimization problem can be found in [FH86].

2.3.3.3 Region Growing In the computer vision literature there are traditionnally two ways of segmenting a picture, a curve, an object [PAV78]. The first method, called the split technique, starts from the entire object—let us say checks if it is homogeneous according to some criterion, and a number of subobjects if it is not. The process is then applied recursively until the objects become either too small or homogeneous. The major question where to split, has not received any satisfying answer in more than one dimension. So far, the only splitting schemes have been defined with respect to a fixed coordinate system, usually the discrete grid on which the object is defined, as in quadtrees or octtrees. This is quite disrupting since we insist on having representations intrinsic to the object rather than on its frame of reference.

In this part, we focus on a region-growing paradigm that starts from the simplicial polyhedron obtained, for example, by the techniques of section 2.3.1, and starts merging the initial triangular regions into larger ones best approximated either by planes or by quadrics. The process stops when either only one region is left or when the total approximation error, as measured by criterion E above, reaches some tolerance threshold. The scheme can easily incorporate the information provided by the results of section 2.3.2 about curves on the surface where significant changes occur. This is simply achieved by forbidding two regions to merge if their union contains in its interior one of the previous curves.

The merge procedure works then as follows. At every iteration, the pair of regions R_i and R_j that produce the smallest error $E\ (R_i \cup R_j)$ and such that $R_i \cup R_j$ does not contain any of the lines computed in section 2.3.2 are merged. This guarantees that the total error E grows as slowly as possible and can be easily implemented by maintaining a heap of all pairs of adjacent regions [AHU74]; the best pair is then always at the top of the heap. The merge consists in updating the region adjacency graph (*RAG*) and the heap. The algorithm can be described as follows:

INPUT: a simplicial polyhedron interpolating the data point a tolerance T on the maximum approximation error

OUTPUT: a segmentation of the points into regions best approximated by a plane or a quadric

Initialization: Build the region adjacency graph of the set of initial triangles.

Initialize the approximation error for each region to 0.

For each pair of adjacent regions, compute the error for the best approximating plane (quadric).

Use these values to build a heap such that the top of the heap is the pair of adjacent regions R_i, R_j with the smallest error E_{ij}.

Iteration: Update the total error: $E^n = E^{n-1} + E_{ij} - E_i - E_j$

If E^n is less than the tolerance T

then

remove the top of the heap and update it (this involves recomputing the errors E_{ik} and E_{jl} for all regions R_k and R_l adjacent to R_i or R_j and reinserting the corresponding pairs in the heap)

Update the *RAG*

else *STOP*

A few points need to be clarified in this algorithm. When we compute the best approximating plane (quadric) for the union of two adjacent regions R_i and R_j, we consider a plane or a quadric only if both R_i and R_j are currently approximated by a plane and only a quadric in the other cases. In the first case, the type of surface that yields the smallest error is selected. Also, when using the curves of discontinuity obtained in section 2.3.2, we

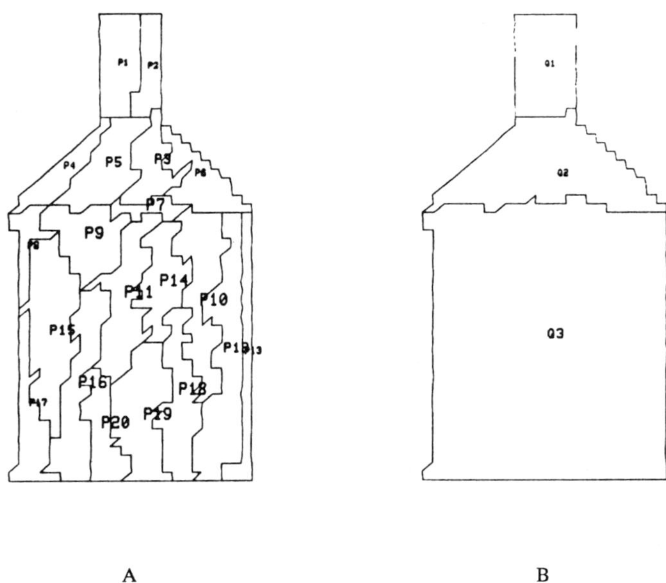

Figure 2.49
Segmentation of the surface of the oil bottle: (A) using only planar patches; (B) using quadric patches.

do not insert in the heap pairs of regions such that their union contains part of one of these curves. Results of this segmentation program are shown on the oil bottle in figure 2.49.

2.3.4 What Needs to Be Worked On

The question of shape representation cannot be separated from the applications. We focus in this chapter (see the next section) on applications such as recognition and localization of rigid objects. The kinds of representations that are needed for these applications may be different from those needed for manipulating objects, or building them, or designing them for specific tasks. And this is only for manufactured objects; but if we extend our realm to the class of natural objects, such as rocks, rivers, and flames, or even living beings, such as plants, trees, animals, and humans, then it is likely that the kind of models needed will be very different. In particular, they should incorporate the notion of growth, of dynamic evolution [THO61, THO72, ARN86]. This delimits a very large research territory, which is mostly unexplored and where we believe that some of the future efforts should be invested in the long term. Closer to our everyday preoccupations, we think that the following areas should be considered:

1. The question of symmetry is important in general for many objects and needs more attention, even though some preliminary attempts have been made at understanding the problem [BA84, TWK87].
2. The relationships between computational geometry [PS85] and shape representation should also be investigated further, in connection with other issues discussed in section 2.3.4.
3. The relationship between shape and function is extremely important if we want to be able to design flexible vision systems. Very little is known on the subject, and a lot more needs to be done [WBKL84, CB87].

2.4 Shape Recognition/Localization

2.4.1 Position of the Problem

We have seen in the previous sections a number of ways for constructing 3-D representations of objects in terms of geometric primitives. We have studied ways of reliably extracting these primitives from a number of sensors outputs, and we would like now to address the question of how to use these representations for recognizing and localizing objects or for the navigation of a robot. In this section we concentrate on the first of these two applications, referring the reader interested in the second to [AF87].

There have been many solutions proposed in the last twenty years or so for solving this problem in a various number of different contexts, and the difficulty of the problem is a function of many variables. We present here a number of these variables:

- First, objects can be one-, two-, or three-dimensional. Examples of one-dimensional objects are waveforms of various kinds; examples of two-dimensional objects are flat parts, such as keys, coins, etc., for which the third dimension is negligible with respect to the other two. Examples of three-dimensional objects can be found by looking around.
- Second, objects can be completely visible (observed) or partially hidden. It is almost always the case in robotics applications that the objects to be dealt with are partially hidden. The fact that parts of the object may be missing seriously adds to the difficulty of the recognition problem.

- Third, objects can be rigid or not. Nonrigid objects range from articulated bodies, such as a robot manipulator, to continuously deformable objects, such as clouds.
- Fourth, and related to this distinction between rigid and nonrigid objects, is the distinction between recognizing one instance of a specific shape, such as a given chair of known shape and dimensions, and recognizing classes of objects, such as the class of chairs, for example. An articulated object is somewhat in between in the sense that there exists an analytical parametrization of all its possible configurations, whereas very likely such a parametrization does not exist for the class of chairs.
- Fifth is the question of noise. Typically, measurements made at the scene (and sometimes also when building the model) are noisy. The amount of noise makes the problem more or less difficult.

In this section we concentrate exclusively on the problem of recognizing and localizing specific instances of partially overlapping rigid 3-D objects. We assume therefore that every object is defined by an accurate model $\mathcal{M} = (M_1, \ldots, M_n)$ composed of number of geometric primitives M_i of the types we have described in section 2.3 and that a number of visual sensors provide us with a similar description $\mathcal{S} = (S_1, \ldots, S_p)$ for the observed scene.

The recognition problem is then to produce a list of pairs of scene/model primitives $\mathcal{R}_n = ((M_1, S_{i1}), \ldots, (M_n, S_{in}))$, where the S_{ij}'s are either scene primitives or the special symbol NIL, indicating that the model primitive M_j is not present in the scene.

The localization problem is to find the best rigid transformation **T** that, when applied to the identified model primitives, brings them as close as possible to their corresponding scene primitives. A rigid transformation is the product of a translation and a rotation. We shall make heavy use of the rigidity constraint to find quickly the best matches between model and scene primitives.

2.4.2 Search Approach to the Problem

The complexity of the recognition problem depends directly on the number of scene (p) and model (n) primitives since the number of sequences $\mathcal{R}$ is n^p. This becomes rapidly extremely large when n and p increase, and it is therefore impossible to explore all possibilities. Nonetheless, the brute force approach consisting in exploring the interpretation tree of figure 2.50 yields

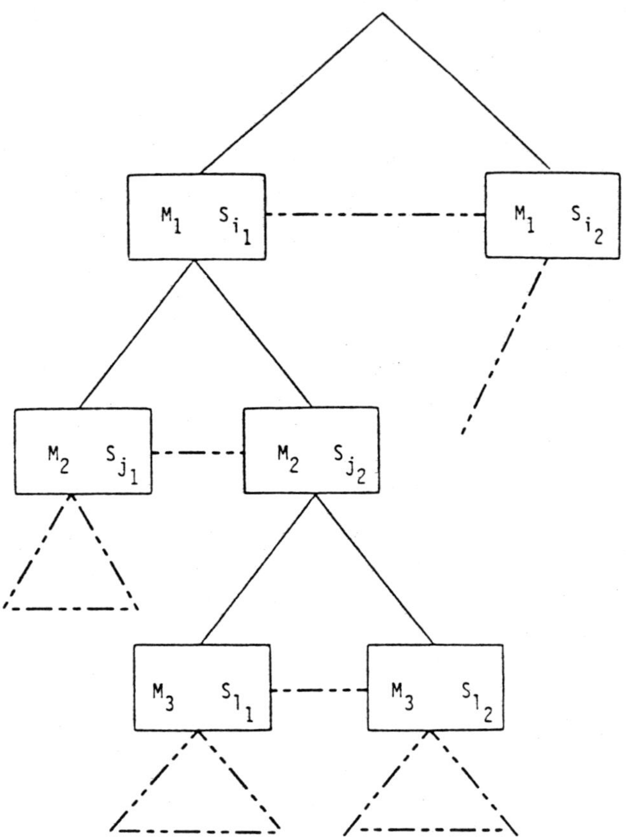

Figure 2.50
Interpretation tree.

itself very nicely to the implementation of the rigidity constraint. The basic paradigm that we use is that of recognizing while localizing. This can be achieved as follows. For every path from the root in the interpretation tree corresponding to a partial recognition $\mathscr{R}_k = ((M_1, S_{i1}), \ldots, (M_k, S_{ik}))$ $(k < n)$, we compute, using methods to be described later, the best rigid transformation $\mathbf{T}_k$ from model to scene for those primitives. Since the measurements are noisy, the residual error ε_k is nonzero. Also, k minus the number of model elements assigned to NIL is a measure g_k of how much of the object has been found in the scene so far. A combination of ε_k and g_k is the cost associated with the partial recognition $\mathscr{R}_k$. We can then apply $\mathbf{T}_k$ to the next unmatched model primitive M_{k+1} and consider only as possible candidates for M_{k+1} those unmatched scene primitives that are sufficiently close to $\mathbf{T}_k(M_{k+1})$, thus considerably reducing the breadth of the interpretation tree.

2.4.3 Localization of 3-D Objects from 3-D Measurements

We consider here the localization of 3-D objects from 3-D data [GL83, BH86, FH86]. Every three-dimensional rigid transformation can be considered in an infinite number of fashions as the product of a translation and a rotation. If we constrain the axis of the rotation to go through a given point and assume that it is applied first, then this decomposition is unique. We denote by $\mathbf{R}$ the orthogonal matrix representing the rotation and by $\mathbf{t}$ the translation vector.

We assume that the geometric primitives that define models and scenes are points, lines, and planes. Points are represented by three-dimensional vectors; planes are represented by their normal $\mathbf{n}$ and their distance to the origin d. Lines are usually defined by two points, and we represent them by their Plücker coordinates $(\mathbf{m}, \mathbf{m}')$ as follows. Let $\mathbf{p}_1$ and $\mathbf{p}_2$ be the two points defining the line; then its Plücker coordinates are given by

$$\mathbf{m} = \mathbf{p}_2 - \mathbf{p}_1,$$

$$\mathbf{m}' = \mathbf{p}_1 \wedge \mathbf{p}_2.$$

Notice that $\mathbf{m}.\mathbf{m}' = 0$. We have to determine how these representations are transformed through a rigid 3-D transformation.

A point M represented by $\mathbf{p} = \mathbf{OM}$ is transformed to $\mathbf{p}' = \mathbf{Rp} + \mathbf{t}$. A line L represented by $(\mathbf{m}_1, \mathbf{m}_2)$ is transformed to $(\mathbf{Rm}_1, \mathbf{Rm}_2 + \mathbf{t} \wedge \mathbf{Rm}_1)$. A plane P represented by $(\mathbf{n}, d)$ is transformed to $(\mathbf{Rn}, d + \mathbf{t}.\mathbf{Rn})$. Therefore, if we match a point $\mathbf{p}'$ in the scene to a point $\mathbf{p}$ in the model, the following relationship must hold:

$$\mathbf{p}' = \mathbf{Rp} + \mathbf{t}. \tag{13}$$

If we match a line $(\mathbf{m}'_1, \mathbf{m}'_2)$ in the scene to a line $(\mathbf{m}_1, \mathbf{m}_2)$ in the model, the following relationships must hold:

$$\|\mathbf{m}_1\|\mathbf{m}'_1 = \|\mathbf{m}'_1\|\mathbf{Rm}_1, \tag{14}$$

$$\mathbf{m}'_2 \wedge (\mathbf{Rm}_2 + \mathbf{t} \wedge \mathbf{Rm}_1) = \mathbf{0}. \tag{15}$$

Using equation (14), we rewrite equation (15) as

$$\mathbf{m}'_2 \wedge (\mathbf{Rm}_2 + r\mathbf{t} \wedge \mathbf{m}'_1) = \mathbf{0}, \tag{16}$$

where $r = \|\mathbf{m}_1\|/\|\mathbf{m}'_1\|$. Using the fact that $\mathbf{m}'_1$ and $\mathbf{m}'_2$ are orthogonal, (16) can be simplified as

$$\mathbf{m}'_2 \wedge \mathbf{Rm}_2 + r(\mathbf{m}'_2.\mathbf{t})\mathbf{m}'_1 = \mathbf{0}.$$

Finally, if we match a plane $(\mathbf{n}', d')$ in the scene to a plane $(\mathbf{n}, d)$ in the model, the following relationships must hold:

$$\mathbf{n}' = \mathbf{Rn}, \tag{17}$$

$$d' = d + \mathbf{t}.\mathbf{Rn}. \tag{18}$$

Using equation (17), equation (18) can be rewritten as

$$d' = d + \mathbf{t}.\mathbf{n}'.$$

In practice, due to measurement errors, these relations are not exactly satisfied and we are led to a mean-square solution. Specifically, given k point matches, we have to solve

$$\min_{\mathbf{R},\mathbf{t}} \sum_i \|\mathbf{p}'_i - \mathbf{Rp}_i - \mathbf{t}\|^2. \tag{19}$$

Given k line matches,

$$\min_{\mathbf{R},\mathbf{t}} \sum_i \alpha\|\mathbf{v}'_i - \mathbf{Rv}_i\|^2 + (1-\alpha)\|\mathbf{m}'_{2i} \wedge \mathbf{Rm}_{2i} + r_i(\mathbf{t}.\mathbf{m}'_{2i})\mathbf{m}'_{1i}\|^2, \tag{20}$$

where $\mathbf{v}_i = \mathbf{m}_{1i}/\|\mathbf{m}_{1i}\|$, $\mathbf{v}'_i = \mathbf{m}'_{1i}/\|\mathbf{m}'_{1i}\|$. The choice of α is left to the decision of the user. Finally, given k plane matches, we have to solve

$$\min_{\mathbf{R}} \sum_i \|\mathbf{n}'_i - \mathbf{Rn}_i\|^2, \tag{21}$$

$$\min_{\mathbf{t}} \sum_i (d_i' - d_i + \mathbf{t}.\mathbf{n}_i')^2. \tag{22}$$

Closed form solutions for these problems are not easy to obtain because of the constraints on the rotation matrix **R**. We now make good use of the various representations for rotations presented in appendix 2B to come up with either closed form or iterative solutions.

Let us start with the problem of matching planes and consider problem (21). Using the quaternion representation of rotations described in appendix 2B, we can rewrite this as

$$\min_{\mathbf{R}} \sum |\mathbf{n}_i' - \mathbf{q} * \mathbf{n}_i * \overline{\mathbf{q}}|, \tag{23}$$

where vectors and quaternions are identified. Because **q** represents a rotation $|\mathbf{q}|^2 = 1$. Therefore we can multiply (23) by $|\mathbf{q}|^2$:

$$\min_{\mathbf{R}} \sum (|\mathbf{n}_i' - \mathbf{q} * \mathbf{n}_i * \overline{\mathbf{q}}||\mathbf{q}|)^2,$$

and since the quaternion magnitude is compatible with the quaternion product, this is the same as

$$\min_{\mathbf{R}} \sum |\mathbf{n}_i' * \mathbf{q} - \mathbf{q} * \mathbf{n}_i * \overline{\mathbf{q}} * \mathbf{q}|^2$$

and since $\overline{\mathbf{q}} * \mathbf{q} = |\mathbf{q}|^2 = 1$ we finally have

$$\min_{\mathbf{R}} \sum |\mathbf{n}_i' * \mathbf{q} - \mathbf{q} * \mathbf{n}_i|^2. \tag{24}$$

It follows from the definition of the quaternion product given in appendix 2B that $\mathbf{n}_i' * \mathbf{q} - \mathbf{q} * \mathbf{n}_i$ is a linear function of the 4 coordinates of **q**. Therefore, there exists a 4×1 vector $\mathbf{a}_i$ such that

$$|\mathbf{n}_i' * \mathbf{q} - \mathbf{q} * \mathbf{n}_i|^2 = \mathbf{q}^T \mathbf{a}_i \mathbf{a}_i^T \mathbf{q},$$

where **q** denotes here the 4×1 vector attached to the quaternion **q**. Therefore, equation (24) can be rewritten as

$$\min_{\mathbf{q}} \mathbf{q}^T \mathbf{A} \mathbf{q}, \tag{25}$$

with

$$\mathbf{A} = \sum \mathbf{a}_i \mathbf{a}_i^T$$

and the constraint on the vector **q** is $\|\mathbf{q}\|^2 = 1$. The solution to equation

(25) is the eigenvector of unit length of matrix $\mathbf{A}$ corresponding to the smallest eigenvalue. The best translation in equation (22) can then be found using standard least-squares techniques.

Let us look at the case of points now. Introducing the centroids $\mathbf{p}'$ and $\mathbf{p}$ of the points $\mathbf{p}'_i$ and $\mathbf{p}_i$ and letting $\mathbf{n}'_i = \mathbf{p}'_i - \mathbf{p}'$ and $\mathbf{n}_i = \mathbf{p}_i - \mathbf{p}'$, we rewrite equation (19) as

$$\min_{\mathbf{R},\mathbf{t}} \sum_i \|\mathbf{n}'_i - \mathbf{R}\mathbf{n}_i + \mathbf{p}' - \mathbf{R}\mathbf{p} - \mathbf{t}\|^2. \tag{26}$$

Expanding each term in (26) and noticing that $\sum \mathbf{n}'_i = \sum \mathbf{n}_i = \mathbf{0}$, we are left with

$$\min_{\mathbf{R},\mathbf{t}} (k\|\mathbf{p}' - \mathbf{R}\mathbf{p} - \mathbf{t}\|^2 + \sum \|\mathbf{n}'_i - \mathbf{R}\mathbf{n}_i\|^2). \tag{27}$$

The next thing to do is to observe that for any rotation matrix, the first term in the criterion, $\|\mathbf{p}' - \mathbf{R}\mathbf{p} - \mathbf{t}\|^2$, can be made equal to zero by choosing $\mathbf{t} = \mathbf{p}' - \mathbf{R}\mathbf{p}$. Therefore, equation (27) is minimized by finding the best rotation that minimizes the second term, which then determines the translation. The problem for points is thus solved by first computing $\mathbf{R}^*$ minimizing $\sum \|\mathbf{n}'_i - \mathbf{R}\mathbf{r}_i\|^2$, and that can be done using the technique developed for planes, and then setting $\mathbf{t}^* = \mathbf{p}' - \mathbf{R}^*\mathbf{p}$.

We do not have an analytical solution to problem (20) for the lines, but we propose two reasonable solutions. The first solution consists in solving for the rotation $\mathbf{R}^*$ by the previous method while using only the first part of the criterion:

$$\sum \|\mathbf{v}'_i - \mathbf{R}\mathbf{v}_i\|^2,$$

and then use the second part of the criterion,

$$\sum \|\mathbf{m}'_{2i} \wedge \mathbf{R}^*\mathbf{m}_{2i} + r_i(\mathbf{t}.\mathbf{m}'_{2i})\mathbf{m}'_{1i}\|^2,$$

to solve for the translation (this is a standard least-squares problem for $\mathbf{t}$). This corresponds to a choice of $\alpha = 1$ in criterion (20).

Another possibility, which has proved to be extremely fruitful [AF87], is to do a linearized recursive least-squares estimation of the rotation, using its exponential representation (see appendix 2B), and a recursive least-squares estimation of $\mathbf{t}$. Recursive least-squares estimation techniques are related to Kalman filtering techniques [JAZ70, MAY79]. Here we simply want to stress the fact that the exponential representation of rotations allows us very simply to linearize the problem since there are no constraints on the 3×1 vector $\mathbf{r}$ representing a rotation. Therefore we have to compute

$$d(\mathbf{R}\mathbf{v}_i)/d\,\mathbf{r} \qquad \text{and} \qquad d(\mathbf{m}'_{2i} \wedge \mathbf{R}\mathbf{m}_{2i})/d\,\mathbf{r},$$

which is easily done using the results of appendix 2B.

2.4.4 Controlling the Search

There are two basic steps in the search process that is very similar to the one used in section 2.2.4 to find stereo matches. The first is the hypothesis formation, i.e., the search for sets of consistent pairings that provide enough information for the computation of the rigid transformation. The second step is the prediction of new matches using the estimated transformation and verification of the correctness of the initial hypothesis.

2.4.4.1 Hypothesis Formation This step is crucial because it determines the number of paths that are going to be explored in the interpretation tree before finding an acceptable solution. The main problem is that we need at least two pairings to estimate the transformation or part of it. The number of primitives required is summarized below:

	Translation	Rotation
Points	3	3
Lines	2	2
Planes	3	2

The two lines/planes should not be parallel and the three planes should not meet along a line—otherwise the transformation is underconstrained and cannot be estimated completely.

The hypothesis formation proceeds in three steps:

1. **Selection of a first pairing:** For each primitive in the model, the compatible primitives of the scene are considered. The choice of these primitives cannot make use of the rigidity constraint; only the position invariant features, such as the length of the segments and the area of the surface patches, can be used. All these features are highly sensitive to occlusion and therefore should be used carefully and with large tolerances.
2. **Selection of a second pairing:** Given a first pairing (M_1, S_1) and a second model primitive M_2, the candidates for the matching must satisfy the rigidity constraint. For points, this choice is quite simple, the only constraint being that $d(S_1, S_2) = d(M_1, M_2)$.

For planes, the only constraint is on the angle between the normals; i.e., S_2 must be chosen such that $(\mathbf{n}'_2, \mathbf{n}'_1) = (\mathbf{n}_2, \mathbf{n}_1)$. When two pairs of planes have been matched, the rotation is fixed and the translation is constrained to be parallel to the intersection of the planes M_1 and M_2.

Let us consider two lines M_1 and M_2 in the model and S_1 and S_2 in the scene represented by their Plücker coordinates $(\mathbf{m}_{11}, \mathbf{m}_{21})$, $(\mathbf{m}_{12}, \mathbf{m}_{22})$, $(\mathbf{m}'_{11}, \mathbf{m}'_{21})$, and $(\mathbf{m}'_{21}, \mathbf{m}'_{22})$. Just as in the case of planes, we have an angle constraint, namely, $(\mathbf{m}_{11}, \mathbf{m}_{12}) = (\mathbf{m}'_{11}, \mathbf{m}'_{12})$, but we also have a metric constraint—for example, that their shortest distances $d(M_1, M_2)$ and $d(S_1, S_2)$ be equal.

These constraints are general in the sense that they do not make any hypothesis about the size of the primitives being matched. In practice, lines are line segments and planes are polygonal faces. In particular, they will not discriminate between two coplanar faces or two aligned segments. Grimson and Lozano-Pérez [GL83] argue about using constraints obtained from bounds on the components of a vector joining two points on two faces or two line segments in the model. Indeed, if $\mathbf{n}_1$ and $\mathbf{n}_2$ are the two unit normals to the faces and $\mathbf{d}$ the separation vector between two points on each face, then when these two points vary in the faces, the components of $\mathbf{d}$ in the coordinate frame defined by $\mathbf{n}_1$, $\mathbf{n}_2$, and $\mathbf{n}_1 \wedge \mathbf{n}_2$ vary in some intervals. These intervals can be computed once and for all when building the model.

When making a hypothesis, these ranges can be used to filter further the list of potential candidates for the second match by choosing a few points on S_1 and S_2, computing the corresponding vectors $\mathbf{d}$, and verifying that their components in the coordinate frame defined by $\mathbf{n}'_1$, $\mathbf{n}'_2$, and $\mathbf{n}'_1 \wedge \mathbf{n}'_2$ fall in the ranges.

This idea can be extended to line segments as well by considering the coordinate frame defined by $\mathbf{m}_{11}, \mathbf{m}_{12}$, and $\mathbf{m}_{11} \wedge \mathbf{m}_{12}$. Our experience has shown that in general, these constraints do not add very much to the power of the original rigidity constraint.

3. **Estimation of the transformation**: The transformation is estimated (or partially estimated in the case of planar patches) using the techniques described in the previous section. We mentioned the fact that some primitives may not have a canonic orientation; therefore the transformation estimated from an initial hypothesis is not unique and several equivalent transformations are generated (two for lines and planes).

Another important item is the order in which the model primitives are considered for matching. Noninteresting branches of the interpretation tree might be explored if the order is not carefully determined. Consider, for example, the case in which the first two primitives are parallel planes. The estimated rotation is arbitrary and the rotation error vanishes. A large subtree may be explored based on a wrong estimation of the transformation. A number of rules can be applied in the ordering of primitives:

- Small primitives (in terms of length or area) should be avoided because of their probable sensitivity to measurement noise.
- The first two or three primitives must be linearly independent in order to avoid indetermination.
- If local symmetries exist in the object, the primitives that best discriminate between almost equivalent positions of the object should be among the first ones considered for matching. Notice that in some cases this might contradict the first two rules.

2.4.4.2 Prediction and Verification In this step, given an initial hypothesis and the associated transformation **T**, we want to *predict* the set of candidate scene primitives that can be matched with model primitives in order to *verify* the validity of the initial hypothesis.

The basic way of doing this is by applying the transformation to every unmatched model primitive M and find the scene primitives that are close enough to $\mathbf{T}(M)$ (see figure 2.51). An important issue here is that we want to avoid a sequential exploration of the scene description for each model primitive because the rigidity constraint is used precisely to reduce the number of potential candidates for a match.

One solution to this problem, which we used several times in this chapter, is the grouping of scene primitives into buckets corresponding to discretized values of the parameters defining the transformation. More precisely, the scene is divided into a number of cubic buckets. To each scene primitive is attached the list of buckets it intersects, and to each bucket is attached the list of scene primitives that intersect it. Moreover, the primives can also be grouped by orientation buckets as follows. The Gaussian sphere is used to represent orientations of lines or planes and can tessellated in a number of ways. Each tessella is a bucket, and all scene primitives whose orientations fall into that bucket are attached to it (figure 2.52).

Returning to our unmatched model primitive M, we compute $\mathbf{T}(M)$, and from its orientation determine which bucket of the Gaussian sphere it falls into. This in turn determines which scene primitives are candidates for the

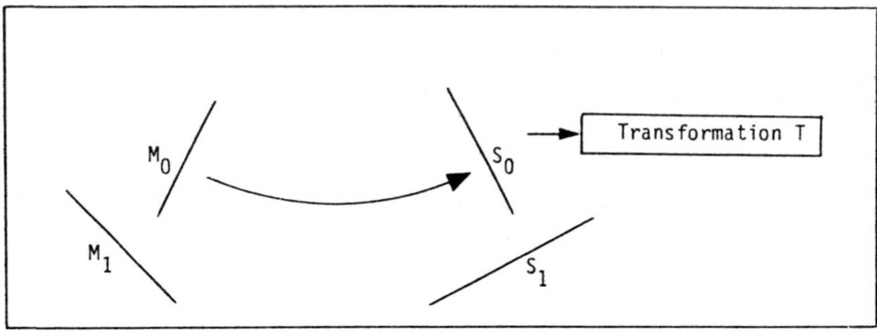

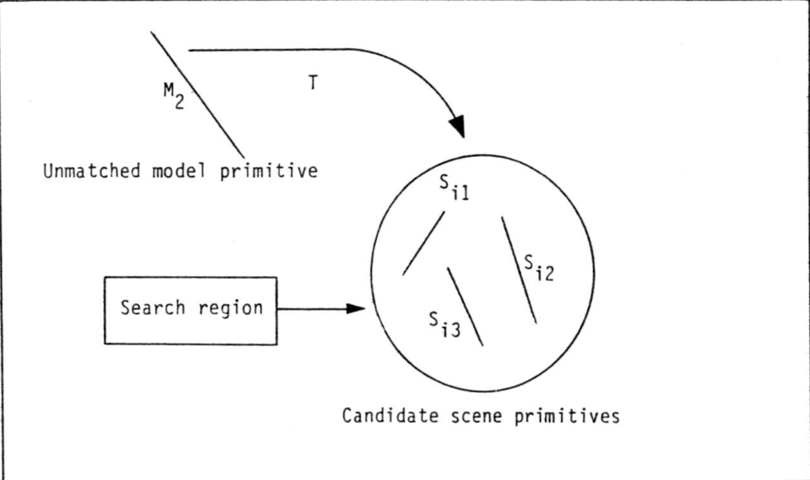

Figure 2.51
Finding potential matches.

match. If the translation is known, the corresponding buckets in the scene determine another set of candidates for the match. The final candidates are the intersection of these two lists. This basically controls the breadth of the exploration of the interpretation tree and allows us to find all possible positions of the model in the scene satisfying the constraint that the error is less than some threshold and also the number of model faces assigned to *NIL* below another threshold.

This second threshold can be used also to control the depth of the exploration by noticing that if at some point in the exploration we reach a point where, even if all the remaining model primitives are matched to scene primitives, the threshold cannot be reached, then the exploration might as well stop.

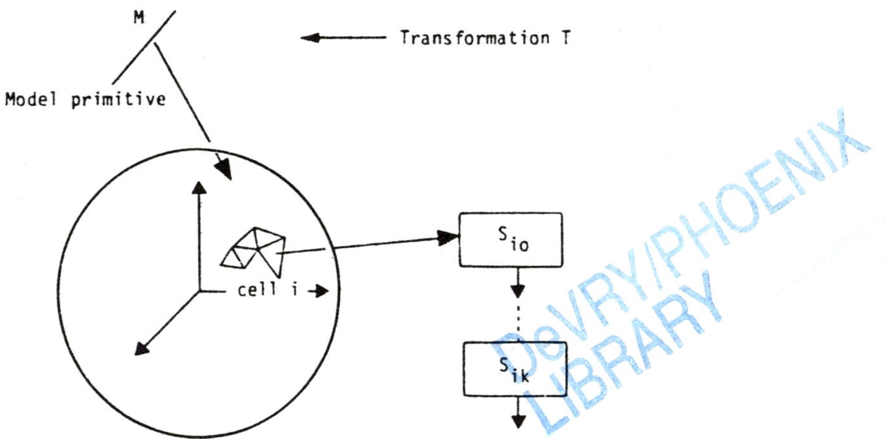

Figure 2.52
Orientation buckets.

Results obtained by a system that uses some of these ideas are shown in figures 2.53–2.55, which are taken from [FH86].

2.4.5 What Needs to Be Worked On

There are an enormous number of things that need to be worked on in this area; some of them we referred to briefly in section 2.3.4. Problems are of course difficult to isolate from the problems in shape representation, but we can identify a number of "hot" areas that need special attention:

1. There is the need to take into account explicitly in the recognition and localization process the uncertainty on the primitives that are matched. Some work has begun to appear [LC85, CRO86, DUR86, FAF86, SC86], but much remains to be done. In particular, nothing has yet been proposed to use results in probabilistic geometry, where we believe many potentially useful results can be found.
2. With respect to geometry, we believe that the community badly needs systems capable of manipulating symbolic geometric information. Again, some work is going on in this area [CHO84, MUN86, WU78], but more effort needs to be put into it. We need a geometric equivalent to the MACSYMA system.
3. It is also important to demonstrate systems that have the capability of recognizing and locating rigid objects from data coming from stereo and

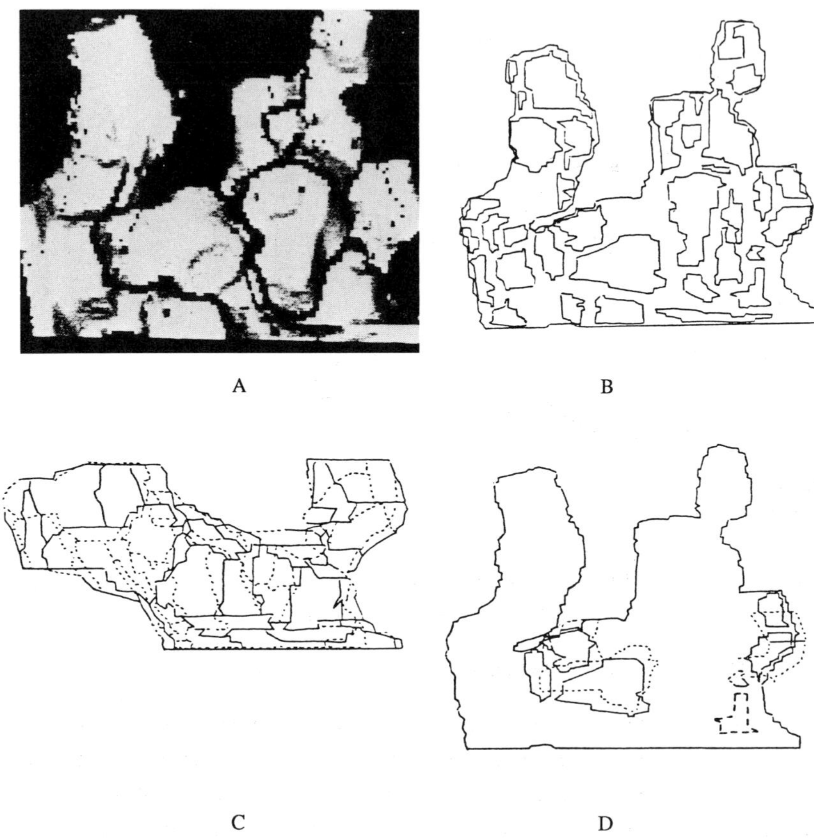

Figure 2.53
Results of the recognition and locating algorithm: (A) image of the normals to the scene; (B) scene segmentation in planar patches; (C) first identified model after rotation with the estimated rotation matrix; (D) superimposition of identified scene and model primitives.

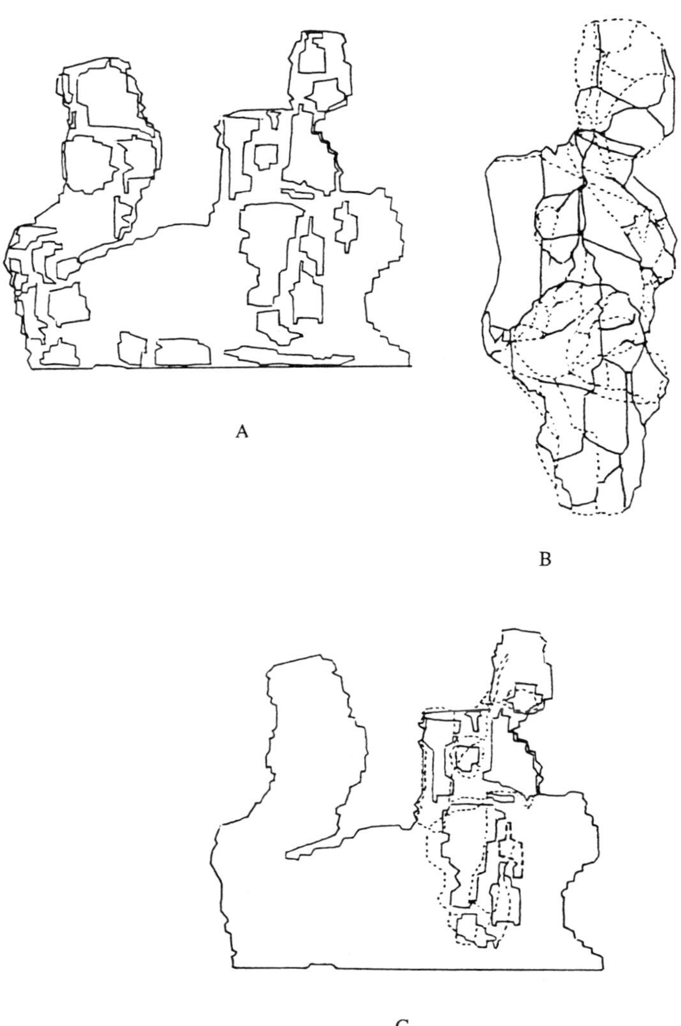

Figure 2.54
Results of the recognition and locating algorithm (continued): (A) scene segmentation in planar patches; (B) second identified model after rotation with the estimated rotation matrix; (C) superimposition of identified scene and model primitives.

Figure 2.55
Results of the recognition and locating algorithm (continued): (A) scene segmentation in planar patches; (B) third identified model after rotation with the estimated rotation matrix; (C) superimposition of identified scene and model primitives.

structure from motion where noise, matching errors, and sparseness of 3-D data make the problem a lot harder than in the case of dense range data. We believe that this is a somewhat easy problem, but one that should be clearly solved in order to make passive 3-D vision a reality.

4. In parallel, more complicated problems, such as articulated rigid objects, should also be investigated [GRI87]. These problems have obvious practical applications and are probably a good way of attacking an even more difficult problem, which we believe can now be tackled with some hopes of success:

5. We refer to the problem of recognizing classes of objects as opposed to specific instances of one object. Previously, we gave the example of the class of chairs; there are obviously many other examples. This problem raises a number of fascinating questions, such as the relation of shape and function and the relation of the language and our perception of the 3-D world. We believe that vision theory and practice have now reached a stage where these questions can be asked without coming up, as was the case so often in the past, with solipsistic systems.

2A Computing the Epipolar Geometry

We describe briefly the epipolar geometry of a stereo pair. Let us make a simple geometric remark, helping ourselves with figure 2.3. Indeed, in this figure we can see that given m_1 in retina plane 1, all possible physical points M that may have produced m_1 are on the infinite half-line $C_1 m_1$. As a direct consequence, all possible correspondents m_2 of m_1 in plane 2 are located on the image, through the second imaging system, of this infinite half-line. This image is also an infinite half-line (*ep*2) starting at point E_2, the intersection of the line going through C_1 and C_2 and the plane 2. E_2 is called the epipole of plane 2 with respect to plane 1, and the line (*ep*2) is called the epipolar line of point m_1 in the plane 2. The corresponding constraint is that for a given point m_1 in plane 1, its possible correspondents in plane 2 all lie on a half-line in plane 2 (see figure 2.13).

Therefore we have reduced the dimension of our search space from 2 to 1. The epipolar constraint is of course symmetric, and for a given point m_2 in plane 2, its possible correspondents in plane 1 all lie on a half-line (*ep*1) starting at the epipole E_1, the intersection of the line $C_1 C_2$ with plane 1. (*ep*1) and (*ep*2) are the intersections of the plane $C_1 M C_2$, called the epipolar plane, with planes 1 and 2, respectively.

Of course, when plane 1 or plane 2 (or both) is parallel to this line, one (or both) epipole goes to infinity and epipolar lines in one plane (or both) become parallel. The situation where both planes are parallel to the line C_1C_2 is often assumed because of its simplicity. But, in practice, it is difficult to align precisely and in a stable way the two optical systems, and we show next that the problem of computing the epipolar lines is just as simple in the general case.

The coordinates of C_i ($i = 1, 2$), the two optical centers in the world reference frame, are obtained by solving the following systems of linear equations:

$$\mathbf{M}_i\mathbf{C}_i = \mathbf{0}, \qquad i = 1, 2$$

where $\mathbf{C}_i = [x_i, y_i, z_i, 1]^t$ is the coordinate vector of C_i, $i = 1, 2$. Since each epipole E_i is the image by the ith camera of the other camera's optical center C_j $(j \neq i)$, the image coordinates of the epipoles E_i are obtained by applying matrices $\mathbf{M}_i$ to the vectors $\mathbf{C}_j$ $(i, j = 1, 2, i \neq j)$.

Let us now show how, for a given point m_1 in plane 1, its corresponding epipolar line can be easily computed from its coordinates. Using again equations 1 for camera 1, slightly modified to deal only with 3×1 column vectors, we have

$$\mathbf{l}_1^t\mathbf{X} - u_1\mathbf{l}_3^t\mathbf{X} + l_{14} - u_1 l_{34} = 0,$$

$$\mathbf{l}_2^t\mathbf{X} - v_1\mathbf{l}_3^t\mathbf{X} + l_{24} - v_1 l_{34} = 0.$$

These are the equations of two planes whose intersection is the line C_1m_1. A vector $\mathbf{n}$ parallel to the line is given by the cross product of the two normal vectors $\mathbf{l}_1 - u_1\mathbf{l}_3$ and $\mathbf{l}_2 - v_1\mathbf{l}_3$:

$$\mathbf{n} = u_1\mathbf{l}_2 \wedge \mathbf{l}_3 + v_1\mathbf{l}_3 \wedge \mathbf{l}_1 + \mathbf{l}_1 \wedge \mathbf{l}_2 = u_1\mathbf{g} + v_1\mathbf{h} + \mathbf{k}.$$

The line C_1m_1 is described by the vectors $\mathbf{c}_1 + \lambda\mathbf{n}$, where λ varies from $-\infty$ to $+\infty$. The epipolar line of m_1 is the image of that line through camera 2; therefore, from equation 1, the projective coordinates $\mathbf{u}_2 = [U_2, V_2, S_2]^t$ of a point on that epipolar line are given by

$$\mathbf{u}_2 = \mathbf{M}_2\begin{bmatrix}\mathbf{c}_1\\1\end{bmatrix} + \lambda_{\mathbf{n}} = \mathbf{e}_2 + \lambda\mathbf{M}_2'\mathbf{n},$$

where $\mathbf{M}_2'$ is the (3×3) matrix obtained by dropping the last column of $\mathbf{M}_2$. Letting $\mathbf{a} = [a_1, a_2, a_3]^t = \mathbf{M}_2'\mathbf{n}$, we see that both the epipole E_2

represented by $\mathbf{e}_2$ and the point represented by $\mathbf{a}$ belong to the epipolar line.

Letting $\mathbf{G} = \mathbf{M}'_2\mathbf{g}$, $\mathbf{H} = \mathbf{M}'_2\mathbf{h}$, and $\mathbf{K} = \mathbf{M}'_2\mathbf{k}$, we have

$$\mathbf{a} = u_1\mathbf{G} + v_1\mathbf{H} + \mathbf{K}$$

and therefore

$$\mathbf{a} = \mathbf{F}\mathbf{u},$$

where $\mathbf{F}$ is the 3×3 matrix $[\mathbf{G}, \mathbf{H}, \mathbf{K}]$ and $\mathbf{u} = [u_1, v_1, 1]^t$.

The projective equation of the epipolar line attached to the point m_1 is given by the determinant

$$\begin{bmatrix} U & u_{2e} & a_1 \\ V & V_{2e} & a_2 \\ S & S_{2e} & a_3 \end{bmatrix} = 0.$$

Therefore, the equation of the line can be written as

$$(V_{2e}a_3 - S_{2e}a_2)U + (S_{2e}a_1 - U_{2e}a_3)V + (U_{2e}a_2 - V_{2e}a_3)S = 0.$$

The cartesian equation is of course

$$(V_{2e}a_3 - S_{2e}a_2)u + (S_{2e}a_1 - U_{2e}a_3)v + (U_{2e}a_2 - V_{2e}a_3) = 0.$$

Notice that this encompasses the case where the epipole E_2 is at infinity ($S_{2e} = 0$) or at a finite distance ($S_{2e} \neq 0$). The coefficients of the equation of the line are affine functions of the coordinates u_1, v_1 of point m_1, with coefficients depending only on matrixes $\mathbf{M}_1$ and $\mathbf{M}_2$, and are therefore easily computed.

2B Representing 3-D Rotations

2B.1 Orthogonal Matrixes

An orthogonal matrix is a real square matrix satisfying:

$$\mathbf{R}\mathbf{R}^T = \mathbf{I}. \tag{28}$$

This implies that $\det(\mathbf{R})^2 = 1$ and therefore that $\det(\mathbf{R}) = \pm 1$. It can be proved that every 3×3 orthogonal matrix of determinant equal to 1 is the matrix of a rotation and inversely that every rotation can be represented as such a matrix.

Equation (28) implies that some very strict constraints must be satisfied by the row and column vectors of **R**. More precisely, it implies that the row (column) vectors are mutually orthogonal and of unit lengths; i.e., if we denote by $\mathbf{r}_i$ ($i = 1, 2, 3$) the column vectors, they satisfy

$$\mathbf{r}_i^T \mathbf{r}_j = \delta_{ij}, \qquad i, j = 1, 2, 3, \quad i \leq j, \tag{29}$$

where δ_{ij} is the usual Kronecker symbol:

$$\delta_{ij} = \begin{cases} 1 & \text{if } i = j \\ 0 & \text{if } i \neq j. \end{cases}$$

The relations (29) can be considered as a set of 6 independent quadratic constraints that are satisfied by the elements of an orthogonal matrix **R**.

2B.2 Rodrigues Formula

Given a rotation of angle θ with respect to an axis **u** (a vector of unit length), there is a simple relationship between the orthogonal matrix **R** representing the rotation and θ and **u**. This relationship is known as the Rodrigues formula from the nineteenth-century French mathematician who published it first [ROD40]. If we denote by **U** the matrix associated with $\mathbf{u} = [u_1, u_2, u_3]^T$,

$$\begin{bmatrix} 0 & -u_3 & u_2 \\ u_3 & 0 & -u_1 \\ -u_2 & u_1 & 0 \end{bmatrix},$$

then it can be easily verified that for any vector **x** we have $\mathbf{u} \wedge \mathbf{x} = \mathbf{Ux}$. Rodrigues formula then says that

$$\mathbf{R} = \mathbf{I} + \sin(\theta)\mathbf{U} + (1 - \cos(\theta))^2 \mathbf{U}^2. \tag{30}$$

2B.3 Quaternions

Quaternions have been found useful in robotics and vision [PW83, FH86, HOR86]. They can be understood very simple by looking at them as 4×1 real numbers that form a vector space in the usual sense and on which we define a multiplication to turn that vector space into a noncommutative field.

We therefore consider a quaternion **q** as being either a 4×1 vector $[q_1, q_2, q_3, q_4]^T$ or a pair $(s, \mathbf{v})$ where s is a real number equal to q_1, and **v** is the vector $[q_2, q_3, q_4]^T$. In this notation, a quaternion is very similar to a complex number, s being like the real part, and $\mathbf{v}$ like the imaginary part. A real number x is readily identified with the quaternion $(x, 0)$, the product of two real numbers is the real part of the product of the corresponding quaternions,

$$(x, \mathbf{0}) * (x', \mathbf{0}) = (xx', \mathbf{0}),$$

and a 3×1 real vector $\mathbf{v}$ is readily identified with the quaternion $(0, v)$.

We now define the product $*$ of two quaternions $\mathbf{q}$ and $\mathbf{q}'$ as follows:

$$\mathbf{q} * \mathbf{q}' = (ss' - \mathbf{v}.\mathbf{v}', s\mathbf{v}' + s'\mathbf{v} + \mathbf{v} \wedge \mathbf{v}'). \tag{31}$$

The definitions of the conjugate and the magnitude of a quaternion are very similar to the ones for the complex numbers:

$$\overline{\mathbf{q}} = (s, -\mathbf{v}) \qquad \text{and} \qquad |\overline{\mathbf{q}}|^2 = \overline{\mathbf{q}} * \mathbf{q} = \mathbf{q} * \overline{\mathbf{q}} = (s^2 + \|\mathbf{v}\|^2, \mathbf{0}) = (\|\mathbf{q}\|^2, \mathbf{0}).$$

In this formula, we have used $|$ for the quaternion magnitude and $\|$ for the usual Euclidean norm. It can be easily verified that the magnitude is compatible with the product in the sense that

$$|\mathbf{q} * \mathbf{q}'| = |\mathbf{q}||\mathbf{q}'|.$$

This is just about all we need to know about quaternions in order to use them to represent 3-D rotations. Indeed, a rotation of angle θ with respect to an axis $\mathbf{u}$ (a vector of length 1), can be represented by the two quaternions $\mathbf{q} = (s, \mathbf{v})$ and $-\mathbf{q}$, where

$$s = \cos(\theta/2), \tag{32}$$

$$\mathbf{v} = \sin(\theta/2)\mathbf{u}. \tag{33}$$

Notice that $\|\mathbf{q}\| = 1$.

It should not be surprising that there are two quaternions for one rotation, since a rotation of angle θ with respect to an axis $\mathbf{u}$ is the same as a rotation of angle $2\pi - \theta$ with respect to the axis $-\mathbf{u}$. Looking at the previous formula, the two rotations correspond precisely to the two quaternions $\mathbf{q}$ and $-\mathbf{q}$. Inversely, for a quaternion $\mathbf{q}$ of magnitude 1, it is clear that there exists a unique rotation defined by the two formulas (33).

The correspondence between rotations and quaternions of unit magnitude is even deeper, since it preserves the group operation. Indeed, given two rotations 1 and 2 and the associated quaternions $\mathbf{q}_1$ and $\mathbf{q}_2$ (for each rotation choose any one of the two possible quaternions—it does not matter which one you choose), the product of rotation 1 with rotation 2 (that is the rotation obtained by first applying 2 and then 1) corresponds to the products $\mathbf{q}_1 * \mathbf{q}_2$ and $-\mathbf{q}_1 * \mathbf{q}_2$.

Since the product of rotations is *not* commutative unless they have the same axis, this shows [as also does formula (31)] that the product of two quaternions is not commutative.

This correspondence between rotations and quaternions allows us to derive a very useful formula. Let $\mathbf{R}$ be the orthogonal matrix representing a rotation of angle θ with respect to an axis $\mathbf{u}$ (a vector of length 1), $\mathbf{q}$ one of the two corresponding quaternions, and $\mathbf{w}$ a 3×1 vector. Then we can write

$$\mathbf{Rw} = \mathbf{q} * \mathbf{w} * \bar{\mathbf{q}}. \tag{34}$$

In this formula, we have identified 3×1 vectors and the corresponding quaternion; i.e., the formula should be read

$$(0, \mathbf{Rw}) = \mathbf{q} * (0, \mathbf{w}) * \bar{\mathbf{q}}.$$

Formula (34) allows us to derive the relationship between the coefficients of the rotation matrix $\mathbf{R}$ and the coordinates of the quaternions $\mathbf{q}$ and $-\mathbf{q}$ representing it. If we write $\mathbf{q} = [s, l, m, n]^T$, where $\mathbf{v} = [l, m, n]^T$, it is easy to derive, using formula (31) that

$$\mathbf{q} * \mathbf{w} * \bar{\mathbf{q}} = (0, (\mathbf{v}.\mathbf{w})\mathbf{v} + (s^2 - \|\mathbf{v}\|^2)\mathbf{w} + 2s\mathbf{v} \wedge \mathbf{w}). \tag{35}$$

We can also write

$$\mathbf{v} \wedge \mathbf{w} = \tilde{\mathbf{v}}\mathbf{w},$$

with

$$\tilde{\mathbf{v}} = \begin{bmatrix} 0 & -n & m \\ n & 0 & -l \\ -m & l & 0 \end{bmatrix}$$

and

$$(\mathbf{v}.\mathbf{w})\mathbf{v} = \mathbf{Aw},$$

where $\mathbf{A}$ is a 3×3 matrix given by

$$A = [l\mathbf{v}, m\mathbf{v}, n\mathbf{v}] = \begin{bmatrix} l^2 & lm & ln \\ lm & m^2 & mn \\ ln & mn & n^2 \end{bmatrix}.$$

Writing that the imaginary part of (35) is equal to **Rw**, we find that

$$R = \begin{bmatrix} s^2 + l^2 - m^2 - n^2 & 2(lm - sn) & 2(ln + sm) \\ 2(mn + sn) & s^2 - l^2 + m^2 - n^2 & 2(mn - sl) \\ 2(ln - sn) & 2(mn + sl) & s^2 - l^2 - m^2 - n^2 \end{bmatrix}. \quad (36)$$

From (36), it appears that the coefficients of **R** are polynomial functions of the coordinates of **q**.

2B.4 Antisymmetric Matrices

For every orthogonal matrix **R**, these exists a unique antisymmetric matrix **H** such that

$$\mathbf{R} = e^{\mathbf{H}},$$

where matrix exponentials are defined as usual. Matrix **H** can therefore be written as

$$\mathbf{H} = \begin{bmatrix} 0 & -c & b \\ c & 0 & -a \\ -b & a & 0 \end{bmatrix}.$$

The three-dimensional vector $\mathbf{r} = [a, b, c]^T$ has some useful properties. Its direction is that of the axis of rotation and its squared norm is equal to the rotation angle squared. Moreover, matrix **H** represents the cross-product with vector **r**, which we denote by $c(\mathbf{r})$. By this we mean

$$\mathbf{Hx} = c(\mathbf{r})\mathbf{x} = \mathbf{r} \wedge \mathbf{x}.$$

A simple justification of the fact that every matrix $e^{\mathbf{H}}$, with **H** an antisymmetric matrix, is orthogonal is the following:

$$(e^{\mathbf{H}})^T = e^{\mathbf{H}^T} = e^{-\mathbf{H}} = (e^{\mathbf{H}})^{-1}.$$

We also know that the eigenvalues a and b of **R** and **H** are related by

$$\mathbf{a} = \mathbf{e}^b.$$

Since it is well-known that the eigenvalues of **R** are 1, $e^{-i\theta}$, and $e^{-i\theta}$, where θ is the rotation angle, the eigenvalues of matrix **H** are easily found to be equal to 0, $i(a_2 + b_2 + b_2)^{1/2}$, and $-i(a^2 + b^2 + b^2)^{1/2}$. Therefore we have

$$\theta = \pm(a^2 + b^2 + b^2)^{1/2}.$$

Let us now consider an eigenvector of matrix $\mathbf{H}$ associated with eigenvalue 0. Since matrix $\mathbf{H}$ represents the vector product with vector $\mathbf{v} = [a, b, c]^T$, $\mathbf{v}$ is such a vector:

$$\mathbf{Hv} = \mathbf{0}.$$

$\mathbf{v}$ is also an eigenvector of matrix $\mathbf{R}$ associated with eigenvalue 1, as can be easily verified by using the formula

$$\mathbf{R} = e^{\mathbf{H}} = \mathbf{I} + \mathbf{H}/1! + \mathbf{H}^2/2! + \cdots .$$

Therefore

$$\mathbf{Rv} = \mathbf{v}$$

and $\mathbf{v}$ gives the direction of the axis of rotation. $\mathbf{H}$ is thus a very compact representation of the rotation in terms of its axis and angle. Another property of this representation can be deduced from a theorem in [GAN77] that states that if we compute the Lagrange-Sylvester polynomial p of the exponential function for the eigenvalues of $\mathbf{H}$, i.e., the polynomial such that

$$p(0) = e^0 = 1,$$

$$p(i\theta) = e^{i\theta},$$

$$p(-i\theta) = e^{-i\theta},$$

then we have the nice relationship

$$\mathbf{e}^{\mathbf{H}} = p(\mathbf{H})$$

so that it can be easily verified that

$$p(\mathbf{H}) = \mathbf{I} + (\sin\theta/\theta)\mathbf{H} + ((1 - \cos\theta)/\theta^2)\mathbf{H}^2,$$

which is precisely the well-known Rodrigues formula [ROD40].

2B.5 Deriving Functions of a Rotation Matrix

2B.5.1 Key Remark As we saw in section 2B.1, a rotation matrix is orthogonal:

$$\mathbf{RR}^T = \mathbf{I}. \tag{37}$$

Let $(x_1, x_2, x_3) \to \mathbf{R}$ be a parametrization of class C^1. Deriving (37) with respect to x_i, we obtain

$$\frac{\partial \mathbf{R}}{\partial x_i}\mathbf{R}^T + \mathbf{R}\frac{\partial \mathbf{R}^T}{\partial x_i} = 0. \tag{38}$$

Equation (38) shows that matrix $\mathbf{B}_i = (\partial\mathbf{R}/\partial x_i)\mathbf{R}^T$ is antisymmetric:

$$\mathbf{B}_i + \mathbf{B}_i^T = \mathbf{0}.$$

Since clearly $\partial\mathbf{R}/\partial x_i = \mathbf{B}_i\mathbf{R}$, we can always write

$$\frac{\partial \mathbf{ROM}}{\partial x_i} = \mathbf{b}_i \wedge \mathbf{ROM}.$$

Computing $d\mathbf{ROM}/d\,\mathbf{x}$, where $\mathbf{x} = [x_1, x_2, x_3]^T$,

$$\frac{d\mathbf{ROM}}{d\mathbf{x}} = \begin{bmatrix} \dfrac{\partial \mathbf{ROM}}{\partial x_1} & \dfrac{\partial \mathbf{ROM}}{\partial x_2} & \dfrac{\partial \mathbf{ROM}}{\partial x_3} \end{bmatrix}$$
$$= [\mathbf{b}_1 \wedge \mathbf{ROM} \quad \mathbf{b}_2 \wedge \mathbf{ROM} \quad \mathbf{b}_3 \wedge \mathbf{ROM}].$$

Therefore

$$\frac{d\mathbf{ROM}}{d\mathbf{x}}\mathbf{v} = (v_1\mathbf{b}_1 + v_2\mathbf{b}_2 + v_3\mathbf{b}_3) \wedge \mathbf{ROM}, \tag{39}$$

where $\mathbf{v} = [v_1, v_2, v_3]^T$.

Now, the problem is to compute the matrixes $\mathbf{B}_i$ for the previous exponential representation.

2B.5.2 Exponential Representation We use the representation $\mathbf{R} = e^{\mathbf{H}}$, where $\mathbf{H}$ is antisymmetric:

$$\mathbf{H} = \begin{bmatrix} 0 & -c & b \\ c & 0 & -a \\ -b & a & 0 \end{bmatrix}.$$

The vector $\mathbf{r} = [a, b, c]^T$ is parallel to the axis of rotation and $\|\mathbf{r}\|^2 = \theta^2$, where θ is the angle of rotation. Let us derive a number of properties of matrix $\mathbf{H}$ that will be used in what follows. If $\mathbf{e}_1$, $\mathbf{e}_2$, $\mathbf{e}_3$ are the three unit vectors defining the standard coordinate system, we shall also use the antisymmetric matrixes $\tilde{\mathbf{e}}_1$, $\tilde{\mathbf{e}}_2$, $\tilde{\mathbf{e}}_3$. For example, we have

$$\tilde{\mathbf{e}}_1 = \begin{bmatrix} 0 & 0 & 0 \\ 0 & 0 & -1 \\ 0 & 1 & 0 \end{bmatrix}.$$

It is easy to show that

$$\frac{\partial \theta}{\partial a} = \frac{a}{\theta} \qquad \text{and} \qquad \frac{\partial \mathbf{H}}{\partial a} = \tilde{\mathbf{e}}_1.$$

Deriving with respect to b and c yields similar relationships.

Next, we have

$$\mathbf{H}^2 = \begin{bmatrix} -(b^2 + c^2) & ab & ac \\ ab & -(c^2 + a^2) & bc \\ ac & bc & -(a^2 + b^2) \end{bmatrix}.$$

From this we can deduce that

$$\mathbf{H}^3 = -\theta \mathbf{H} \tag{40}$$

and

$$\mathbf{H}^2 \tilde{\mathbf{e}}_1 \mathbf{H} = -a\mathbf{H}^2. \tag{41}$$

Notice that $\mathbf{H}^2 \tilde{\mathbf{e}}_1 \mathbf{H}$ is an even matrix.

Similarly, we have

$$\mathbf{H}\tilde{\mathbf{e}}_1 = \begin{bmatrix} 0 & b & c \\ 0 & -a & 0 \\ 0 & 0 & -a \end{bmatrix}$$

and therefore

$$\mathbf{H}\tilde{\mathbf{e}}_1 \mathbf{H} = \begin{bmatrix} 0 & ac & -ab \\ -ac & 0 & a^2 \\ ab & -a^2 & 0 \end{bmatrix} = -a\mathbf{H}.$$

Let us rewrite Rodrigues formula [equation (30)],

$$\mathbf{R} = \mathbf{I} + f(\theta)\mathbf{H} + g(\theta)\mathbf{H}^2, \tag{42}$$

where

$$f(\theta) = \frac{\sin \theta}{\theta}, \qquad g(\theta) = \frac{1 - \cos \theta}{\theta^2}.$$

Verifying that $\mathbf{R}\mathbf{R}^T = \mathbf{I}$ in this formula yields a useful relationship between f and g:

$$\mathbf{R}^T = \mathbf{I} - f(\theta)\mathbf{H} + g(\theta)\mathbf{H}^2.$$

Therefore [we use equation (40)]

$$\mathbf{R}\mathbf{R}^T = \mathbf{I} + (2g(\theta) - f^2(\theta) - \theta^2 g^2(\theta))\mathbf{H}^2.$$

This yields

$$2g(\theta) - f^2(\theta) - \theta^2 g^2(\theta) = 0. \tag{43}$$

Let us now compute $\mathbf{R}(\partial\mathbf{R}^T/\partial a)$ using equation (42):

$$\frac{\partial\mathbf{R}}{\partial a} = a\frac{f'(\theta)}{\theta}\mathbf{H} + a\frac{g'(\theta)}{\theta}\mathbf{H}^2 + f(\theta)\tilde{\mathbf{e}}_1 + g(\theta)(\mathbf{H}\tilde{\mathbf{e}}_1 + \tilde{\mathbf{e}}_1\mathbf{H}).$$

We notice that $\mathbf{H}^2$ and $\mathbf{H}\tilde{\mathbf{e}}_1 + \tilde{\mathbf{e}}_1\mathbf{H}$ are symmetric matrices, and write

$$\frac{\partial\mathbf{R}^T}{\partial a} = -a\frac{f'(\theta)}{\theta}\mathbf{H} - f(\theta)\tilde{\mathbf{e}}_1 + a\frac{g'(\theta)}{\theta}\mathbf{H}^2 + g(\theta)(\mathbf{H}\tilde{\mathbf{e}}_1 + \tilde{\mathbf{e}}_1\mathbf{H}).$$

From this, we compute

$$\begin{aligned}\mathbf{R}\frac{\partial\mathbf{R}^T}{\partial a} = &-\theta a\left(\frac{f'(\theta)}{\theta a} + g'(\theta)f(\theta) - g(\theta)f'(\theta)\right)\mathbf{H}\\ &- f(\theta)\tilde{\mathbf{e}}_1 + f(\theta)g(\theta)\mathbf{H}\tilde{\mathbf{e}}_1\mathbf{H} + (a/\theta)(g'(\theta)\\ &- f(\theta)f'(\theta) - \theta^2 g'(\theta)g(\theta))\mathbf{H}^2 + g(\theta)(\mathbf{H}\tilde{\mathbf{e}}_1 + \tilde{\mathbf{e}}_1\mathbf{H})\\ &- f^2(\theta)\mathbf{H}\tilde{\mathbf{e}}_1 + g^2(\theta)\mathbf{H}^2(\mathbf{H}\tilde{\mathbf{e}}_1 + \tilde{\mathbf{e}}_1\mathbf{H}).\end{aligned} \tag{44}$$

Using relation (41), we can rewrite the previous equation as

$$\begin{aligned}\mathbf{R}\frac{\partial\mathbf{R}^T}{\partial a} = &-\theta a\left(\frac{f'(\theta)}{\theta^2} + g'(\theta)f(\theta) - g(\theta)f'(\theta)\right)\mathbf{H}\\ &- f(\theta)\tilde{\mathbf{e}}_1 + f(\theta)g(\theta)\mathbf{H}\tilde{\mathbf{e}}_1\mathbf{H} + (a/\theta)(g'(\theta) - f(\theta)f'(\theta)\\ &- \theta^2 g'(\theta)g(\theta) - \theta g^2(\theta))\mathbf{H}^2 + g(\theta)(\mathbf{H}\tilde{\mathbf{e}}_1 + \tilde{\mathbf{e}}_1\mathbf{H})\\ &- (f^2(\theta) + \theta^2 g^2(\theta))\mathbf{H}\tilde{\mathbf{e}}_1.\end{aligned} \tag{45}$$

Rewriting $\mathbf{H}\tilde{\mathbf{e}}_1$ as the sum of an odd and even component,

$$\mathbf{H}\tilde{\mathbf{e}}_1 = \frac{\mathbf{H}\tilde{\mathbf{e}}_1 + \tilde{\mathbf{e}}_1\mathbf{H}}{2} + \frac{\mathbf{H}\tilde{\mathbf{e}}_1 - \tilde{\mathbf{e}}_1\mathbf{H}}{2},$$

the coefficient of $\mathbf{H}\tilde{\mathbf{e}}_1 + \tilde{\mathbf{e}}_1\mathbf{H}$ in equation (45) is

$$g(\theta) - \frac{f^2(\theta)}{2} - \frac{\theta^2 g^2(\theta)}{2},$$

which is equal to zero, thanks to equation (43). If we derive (43) with respect to θ, we obtain

$$g'(\theta) - f(\theta)f'(\theta) - \theta^2 g(\theta)g'(\theta) - \theta g^2(\theta) = 0,$$

which shows that the coefficient of $\mathbf{H}^2$ in (45) is zero.

We are therefore left with the sum of four odd terms, in $\mathbf{H}$, $\tilde{\mathbf{e}}_1$, $\mathbf{H}\tilde{\mathbf{e}}_1\mathbf{H}$, and $\mathbf{H}\tilde{\mathbf{e}}_1 - \tilde{\mathbf{e}}_1\mathbf{H}$. Finally we have

$$\mathbf{R}\frac{\partial \mathbf{R}^T}{\partial a} = -\theta a\left(\frac{f'(\theta)}{\theta^2} + g'(\theta)f(\theta) - g(\theta)f'(\theta) + \frac{f(\theta)g(\theta)}{\theta}\right)\mathbf{H}$$
$$- f(\theta)\tilde{\mathbf{e}}_1 + (1/2)(f^2(\theta) + \theta^2 g^2(\theta))(\mathbf{H}\tilde{\mathbf{e}}_1 - \tilde{\mathbf{e}}_1\mathbf{H}).$$

Using equation (43), the coefficient of $\mathbf{H}\tilde{\mathbf{e}}_1 - \tilde{\mathbf{e}}_1\mathbf{H}$ is $g(\theta)$. Using the expressions for f and g, the coefficient of $\mathbf{H}$ is

$$a\frac{\sin\theta - \theta}{\theta^3}.$$

Finally,

$$\mathbf{R}\frac{\partial \mathbf{R}^T}{\partial a} = a\frac{\sin\theta - \theta}{\theta^3}\mathbf{H} - f(\theta)\tilde{\mathbf{e}}_1 + g(\theta)(\mathbf{H}\tilde{\mathbf{e}}_1 - \tilde{\mathbf{e}}_1\mathbf{H}).$$

Matrix $\mathbf{H}\tilde{\mathbf{e}}_1 - \tilde{\mathbf{e}}_1\mathbf{H}$ is given by

$$\mathbf{H}\tilde{\mathbf{e}}_1 - \tilde{\mathbf{e}}_1\mathbf{H} = \begin{bmatrix} 0 & b & c \\ -b & 0 & 0 \\ -c & 0 & 0 \end{bmatrix}.$$

If we compute the partials with respect to b and c, we get

$$\mathbf{R}\frac{\partial \mathbf{R}^T}{\partial b} = b\frac{\sin\theta - \theta}{\theta^3}\mathbf{H} - f(\theta)\tilde{\mathbf{e}}_2 + g(\theta)(\mathbf{H}\tilde{\mathbf{e}}_2 - \tilde{\mathbf{e}}_2\mathbf{H}),$$

where

$$\mathbf{H}\tilde{\mathbf{e}}_2 - \tilde{\mathbf{e}}_2\mathbf{H} = \begin{bmatrix} 0 & -a & 0 \\ a & 0 & c \\ 0 & -c & 0 \end{bmatrix},$$

and also

$$\mathbf{R}\frac{\partial \mathbf{R}^T}{\partial c} = c\frac{\sin\theta - \theta}{\theta^3}\mathbf{H} - f(\theta)\tilde{\mathbf{e}}_3 + g(\theta)(\mathbf{H}\tilde{\mathbf{e}}_3 - \tilde{\mathbf{e}}_3\mathbf{H}),$$

where

$$\mathbf{H}\tilde{\mathbf{e}}_3 - \tilde{\mathbf{e}}_3\mathbf{H} = \begin{bmatrix} 0 & 0 & -a \\ 0 & 0 & -b \\ a & b & 0 \end{bmatrix}.$$

Plugging all this into equation (39), we obtain

$$\frac{d\mathbf{ROM}}{d\mathbf{x}}\mathbf{v} = -\left(\mathbf{v}.\mathbf{r}\frac{\sin\theta - \theta}{\theta^3}\mathbf{H} - f(\theta)\tilde{\mathbf{v}} + g(\theta)\mathbf{v} \barwedge \mathbf{r}\right)\mathbf{ROM} = \mathbf{b} \wedge \mathbf{ROM},$$

with

$$\mathbf{b} = \frac{1 - f(\theta)}{\theta^2}(\mathbf{r}.\mathbf{v})\mathbf{r} + f(\theta)\mathbf{v} + g(\theta)\mathbf{r} \wedge \mathbf{v}.$$

Acknowledgments

I want to thank a number of people without whom this chapter could not have been written: Nicholas Ayache, Jean Daniel Boissonnat, Michael Brady, Rodney Brooks, Eric Grimson, Martial Hebert, Ellen Hildreth, Thomas Huang, Elizabeth Le Bras, Francis Lustman, and Giorgio Toscani. I also want to thank Chantal Chazelas and Nathalie Rocher for their superb job at preparing and editing the manuscript and figures. Last but not least, this work was partially supported by ESPRIT Project P940.

References

[AF85] N. Ayache and B. Faverjon. A fast stereo vision matcher based on prediction and recursive verification of hypotheses. In *Proceedings of the Third Workshop on Computer Vision: Representation and Control*, pages 27–37, Bellaire, October 13–16, 1985.

[AF87] N. Ayache and O. D. Faugeras. Building, registrating and fusing noisy visual maps. In *Proc. 1st ICCV*, June 1987.

[AHU74] V. Aho, J. E. Hopcroft, and J. D. Ullman. *The Design and Analysis of Computer Algorithms*. 1974.

[AK71] Y. I. Abdel-Aziz and H. M. Karara. Direct linear transformation from comparator coordinates into object space coordinates in close-range photogrammetry. In *Symposium on Close-Range Photogrammetry*, pages 1–18, University of Illinois at Urbana, Illinois, January 26–29, 1971.

[AK74] Y. I. Abdel-Aziz and H. M. Karara. *Photogrammetric Potentials of Non-Metric Cameras.* Photogrammetry Series 36, University of Illinois at Urbana, Illinois, 1974. Civil Engineering Studies.

[AL87] N. Ayache and F. Lustman. Fast and reliable passive stereovision using three cameras. In *International Workshop on Industrial Applications of Machine Vision and Machine Intelligence*, Tokyo, February 1987.

[ARN86] V. I. Arnold. *Catastrophe Theory.* Springer-Verlag, Heidelberg, 1984–1986.

[BA84] M. Brady and H. Asada. Smoothed local symmetries and their implementation. *Int. J. Robotics Research*, 3(3), 1984.

[BB81] H. Baker and T. O. Binford. Depth from edge and intensity based stereo. In *Proceedings 7th Joint Conference on Artificial Intelligence*, pages 631–636, Vancouver, August 1981.

[BFLB87] J. D. Boissonnat, O. D. Faugeras, and E. Le Bras. *Representing Stereo Data with the Delaunay Triangulation.* Technical Report, INRIA.

[BH86] R. C. Bolles and P. Horaud. 3DPO: a three-dimensional part orientation system. *Int. J. Robotics Research*, 5(3), 1986.

[BIN] T. O. Binford. *Stereo Vision: Complexity and Constraints*, pages 475–487. First International Symposium of *Robotics Research*, MIT Press, Michael Brady and Richard Paul editors, 1984.

[BOI84] J. D. Boissonnat. Geometric structures for 3-dimensional shape representation. *ACM Transactions on Graphics*, 3(4), October 1984.

[BOI85] J. D. Boissonnat. Reconstruction of solids. In *First ACM Symposium on Computational Geometry*, Baltimore, June 1985.

[BOI86] J. D. Boissonnat. *An Automatic Solid Modeler for Robotics Applications*, pages 65–72. Third International Symposium of *Robotics Research*, MIT Press, O. D. Faugeras and G. Giralt editors, 1986.

[BOW81] A. Bowyer. Computing Dirichlet tesselations. *Computer Journal*, 24:162–166, 1981.

[BPYA85] M. Brady, J. Ponce, A. Yuille, and H. Asada. *Describing Surfaces.* Technical Report, MIT AI Memo 822, January 1985.

[BRO66] D. C. Brown. Decentering distortion of lenses. *Photogrammetric Engineering*, 32(3), May 1966.

[BRO71] D. C. Brown. Close-range camera calibration. *Photogrammetric Engineering*, 37(8): 855–866, 1971.

[BRO84] P. Brou. Finding the orientation of objects in vector maps. *Int. J. Rob. Res.*, 3(4), 1984.

[CB87] J. H. Connell and M. Brady. Generating and generalizing models of visual objects. *Artificial Intelligence*, 31:159–183, 1987.

[CG87] J. P. Cocquerez and A. Gagalowicz. Mise en correspondence de régions dans une paire d'images stéréo. In *MARI 87*, Paris, May 1987.

[CHO84] S. C. Chou. Proving elementary geometry theorems using Wu's algorithm. *Contemporary Mathematics*, 29:243, 1984.

[CRO86] J. L. Crowley. Representation and maintenance of a composite surface model. In *IEEE Conference on Robotics and Automation*, pages 1455–1462, San Francisco, April 7–10, 1986.

[DH73] O. Duda and P. E. Hart. *Pattern Classification and Scene Analysis.* Wiley-Interscience, New York, 1973.

[DUR86] H. F. Durrant-Whyte. Consistent integration and propagation of disparate sensor observations. In *Proceedings 1986 IEEE Conference on Robotics and Automation*, pages 1464–1469, San Francisco, April 7–10, 1986.

[FAF86] O. D. Faugeras, N. Ayache, and B. Faverjon. Building visual maps by combining noisy stereo measurements. In *Proceedings 1986 IEEE Conference on Robotics and Automation*, pages 1433–1438, San Francisco, April 7–10, 1986.

[FAI75] W. Faig. Calibration of close-range photogrammetric systems: mathematical formulation. *Photogrammetric Engineering and Remote Sensing*, 41(12): 1479–1486, 1975.

[FGK*83] O. D. Faugeras, F. Germain, G. Kryze, J. D. Boissonnat, M. Hebert, J. Ponce, E. Pauchon, and N. Ayache. *Toward a flexible Vision System*, chapter 3, pages 129–142. *Robot Vision*, IFS, 1983.

[FH86] O. D. Faugeras and M. Hebert. The representation, recognition, and locating of 3D shapes from range data. *Int. J. Robotics Research*, 5(3), 1986.

[FL87] O. D. Faugeras and F. Lustman. *Let Us Suppose That the World Is Piecewise Planar*, chapter 1, pages 33–40. Third International Symposium of *Robotics Research*, MIT Press, Olivier D. Faugeras and Georges Giralt editors, 1987.

[FLT87] O. D. Faugeras, F. Lustman, and G. Toscani. Motion and structure from motion from point and line matches. In *1st ICVV*, pages 25–34, 1987.

[FT86] O. D. Faugeras and G. Toscani. The calibration problem for stereo. In *Proceedings of the IEEE Conference on Computer Vision and Pattern Recognition CVPR-86*, pages 15–20, Miami Beach, Florida, June 22–26, 1986.

[GAN77] F. R. Gantmacher. *Matrix Theory*. Chelsea, New York, 1977.

[GAN84] S. Ganapathy. Decomposition of transformation matrices for robot vision. In *Proceedings of Int. Conf. on Robotics and Automation*, pages 130–139, 1984.

[GDS86] E. Gurewitz, I. Dinstein, and B. Sarusi. More on the benefit of a third eye for machine stereo perception. In *Proc. of the Eighth ICPR*, pages 966–968, Paris, 1986.

[GEN79] D. B. Gennery. Stereo camera calibration. In *Proceedings Image Understanding Workshop*, pages 101–108, November 1979.

[GEN80] D. B. Gennery. *Modeling the Environment of an Exploring Vehicle by Means of Stereo Vision*. Ph.D. Thesis, Stanford Artificial Intelligence Laboratory, 1980. also Artificial Intelligence Laboratory Memo 339.

[GL83] W. E. L. Grimson and T. Lozano-Pérez. Model-based recognition and localization from sparse three-dimensional data. *Int. J. Robotics Research*, 3(3): 35, 1983.

[GPSL86] A. Gerhard, H. Platzer, J. Steurer, and R. Lenz. Depth extraction by stereo triples and a fast correspondence estimation algorithm. In *Proc. of the 8th ICPR*, pages 512–515, Paris, 1986.

[GRI81] W. E. L. Grimson. A computer implementation of a theory of human stereo vision. *Phil. Trans. Roy. Soc. London*, B292: 217–253, 1981.

[GRI87] W. E. L. Grimson. Recognition of object families using parametrized models. In *Proc. 1st ICCV*, pages 93–101, 1987.

[Har87] C. G. Harris. Determination of ego-motion from matched points. In *Proc. Alvey Conference*, pages 189–192, University of Cambridge, 1987.

[HIL84] E. C. Hildreth. *The Measurement of Visual Motion*. MIT Press, Cambridge, MA, 1984.

[HOR74] B. K. P. Horn. Determining lightness from an image. *Computer Graphics and Image Processing*, 3(1): 277–299, 1974.

[HOR75] B. K. P. Horn. *Obtaining Shape from Shading Information*, chapter 4 in *The Psychology of Computer Vision*. McGraw-Hill, New York, P. H. Winston editor, 1975. 115–155.

[HOR77] B. K. P. Horn. Image intensity understanding. *Artificial Intelligence*, 8(2): 201–231, 1977.

[HOR86] B. K. P. Horn. *Robot Vision*. MIT Press, Cambridge, MA, 1986.

[HTMS82] E. L. Hall, M. B. K. Tio, C. A. McPherson, and F. A. Sadjadi. Curved surface measurements et recognition for robot vision. In *IEEE Workshop on Industrial Applications of Machine Vision*, Conference Record, 1982.

[HUA86] T. S. Huang. *Determining Three-Dimensional Motion and Structure from Two Perspective Views*, chapter 14. In *Handbook of Pattern Recognition and Image Processing*, Academic Press, 1986.

[II86a] M. Ito and A. Ishii. Range and shape measurement using three-view stereo analysis. In *Proc. CVPR86*, pages 9–14, Miami Beach, Florida, 1986.

[II86b] M. Ito and A. Ishii. Three view stereo analysis. *IEEE Trans. on PAMI*, PAMI-8(4): 524–531, 1986.

[IPS85] A. Izaguirre, P. Pu, and J. Summers. A new development in camera calibration: calibrating a pair of mobile cameras. In *Proceedings of Int. Conf. on Robotics and Automation*, pages 74–79, 1985.

[JAZ70] A. H. Jazwinski. *Stochastic Processing and Filtering Theory*. Academic Press, Orlando, 1970.

[KAR79] H. M. Karara. Handbook of non-topographic photogrammetry. *American Society of Photogrammetry*, 1979.

[KMM77] R. E. Kelly, P. R. H. McConnel, and S. J. Mildenberger. The gestalt photomapper. *Photogramm. Eng. Rem. Sens.* 43, 1407–1417, 1977.

[KOE84] J. J. Koenderink. *What Tells Us the Contour about Solid Shape?* Technical Report, Dept. Medical and Physiol. Physics, Univ. Utrecht, Netherlands, 1984.

[KOL74] O. Kolbl. Tangential and asymmetric lens distorsion. In *Proceedings of Symposium of Commission III: Determined by Self Calibration*, I. S. P., Stuttgart, 1974.

[LC85] J. P. Laumond and R. Chatila. Position referencing and consistent world modeling for mobile robots. In *Proceedings 1985 IEEE Conference on Robotics and Automation*, pages 138–145, Saint Louis, Missouri, 1985.

[LH86a] Y. Liu and T. S. Huang. Estimation of rigid body motion using straight line correspondences. In *Proceedings Workshop on Motion: Representation and Analysis*, pages 47–51, IEEE Computer Society, Charleston, South Carolina, May 1986.

[LH86b] Y. Liu and T. S. Huang. Estimation of rigid body motion using straight line correspondences, further results. In *Proceedings ICPR 1986*, pages 306–307, Paris, October 27–31, 1986.

[LIN72] K. Linkwitz. Some remarks on present investigations on calibration of close-range cameras. *International Archives of Photogrammetry*, 1972. Commission V.

[LL86] D. T. Lee and A. K. Lin. Generalized delaunay triangulation for planar graphs. *Discrete Comput. Geom.* 1, 201–217, 1986.

[LON81] H. C. Longuet-Higgins. A computer algorithm for reconstructing a scene from two projection. *Nature* 293, 133–135, 1981.

[LON84] H. C. Longuet-Higgins. The reconstruction of a scene from two projections—configurations that defeat the 8-point algorithm. In *Proceedings, First Conference on Artifical Intelligence Applications*, pages 395–397, Denver, Colorado, December 5–7, 1984.

[MAL71] K. Malhotra. A computer program for the calibration of close-range cameras. In *Proceedings of Symposium on Close-Range Photogrammetry*, University of Illinois at Urbana, Illinois, 1971.

[MAR82] D. Marr. *Vision*. Freeman, 1982.

[MAY79] P. S. Maybeck. *Stochastic Models, Estimation and Control*. Academic Press, 3 volumes, 1979.

[MBK81] H. A. Martins, J. R. Birk, and R. B. Kelley. Camera models based on data from two calibration planes. *Computer Graphics and Image Processing*, 17:173–180, 1981.

[MOR80] H. P. Moravec. *Obstacle Avoidance and Navigation in the Real World by a Seeing Robot Rover*. Technical Report Memo 340., Stanford Artificial Intelligence Laboratory, Stanford Artificial Intelligence, 1980. Ph.D. Thesis.

[MP76] D. Marr and T. Poggio. Cooperative computation of stereo disparity. *Science* 194, 283–287, 1976.

[MP79] D. Marr and T. Poggio. A computational theory of stereo vision. *Proc. R. Soc. Lond.* B204, 301–328, 1979.

[MSA86a] Amar Mitiche, Steven Seida, and J. K. Aggarwal. Interpretation of structure and motion using straight line correspondences. In *Proceedings ICPR 1986*, pages 1110–1112, Paris, October 27–31, 1986.

[MSA86b] Amar Mitiche, Steven Seida, and J. K. Aggarwal. Line based computation of structure and motion using angular invariance. In *Proceedings Workshop on Motion: Representation and Analysis*, pages 175–180, IEEE Computer Society, Charleston, South Carolina, May 1986.

[MUN86] J. L. Mundy. *Reasoning about 3-D Space with Algebraic Deduction*, pages 117–124. Third International Symposium of *Robotics Research*, MIT Press, Olivier D. Faugeras and Georges Giralt editors, 1986.

[NAG86] Hans-Hellmut Nagel. Image sequences—ten (octal) years—from phenomenology towards a theoretical foundation. In *Proceedings ICPR 1986*, pages 1174–1185, Paris, October 27–31, 1986.

[NIS84] H. K. Nishihara. *PRISM: A practical real-time imaging stereo matcher*. Technical Report Memo 780, MIT Artificial Intelligence Lab., 1984.

[NP84] H. K. Nishihara and T. Poggio. *Stereo Vision for Robotics*, pages 489–505. First International Symposium of *Robotics Research*, MIT Press, Michael Brady and Richard Paul editors, 1984.

[OK85] Y. Ohta and T. Kanade. Stereo by intra- and inter-scanline search. *IEEE Transactions on Pattern Analysis and Machine Intelligence*, PAMI-7(2):139–154, 1985.

[OKA81] A. Okamoto. Orientation and construction of models. *Photogrammetric Engineering and Remote Sensing*, 47(10):1437–1454, 1981. Part I: the orientation problem in close-range photogrammetry.

[OKA84] A. Okamoto. The model construction problem using the collinearity condition. *Photogrammetric Engineering and Remote Sensing*, L(6):705–711, 1984.

[OWI86] Y. Ohta, M. Watanabe, and K. Ikeda. Improving depth map by right angles trinocular stereo. In *Proc. of the Eighth ICPR*, pages 519–521, Paris, 1986.

[PAV78] T. Pavlidis. *Structured Pattern Recognition*. Springer-Verlag, New York, 1978.

[PB85] J. Ponce and M. Brady. *Toward a Surface Primal Sketch*. Technical Report MIT AI Memo 824, MIT AI April 1985.

[PH86] M. Pietikainen and D. Harwood. Depth from three camera stereo. In *Proc. CVPR86*, pages 2–8, Miami Beach, Florida, 1986.

[PS85] F. Preparata and M. Shamos. *Computational Geometry*. Springer-Verlag, 1985.

[PW83] E. Pervin and J. A. Webb. Quaternions in computer vision. In *Proc. CVPR*, Washington, 1983.

[ROD40] O. Rodrigues. Des lois géométriques qui régissent les déplacements d'un système solide dans l'espace, et de la variation des coordonnées provenant de ces déplacements considérés indépendamment des causes qui peuvent les produire. *Journal De Mathématiques Pures et Appliquées*, 1st Series(5): 380–440, 1840.

[SC86] R. C. Smith and P. Cheeseman. On the representation and estimation of spatial uncertainty. *International Journal of Robotics Research*, 5(4): 56–68, 1986.

[STR84] T. Strat. Recovering the camera parameters for a transformation matrix. In *Proceedings: DARPA Image Understanding Workshop*, pages 264–271, October 1984.

[SUT74] I. Sutherland. Three-dimensional data input by tablet. *Proc. IEEE*, 62(4): 453–461, 1974.

[TF87] G. Toscani and O. D. Faugeras. Structure and motion from two noisy perspective views. In *Proc. International Conference on Robotics and Automation*, Raleigh, North Carolina, 1987.

[TH82] R. Y. Tsai and T. S. Huang. Estimating three-dimensional motion parameters of a rigid planar patch, ii: singular value decomposition. *IEEE Transactions on Acoustics, Speech, and Signal Processing*, ASSP-30(4), 1982.

[THO61] D'Arcy Thompson. *On Growth and Form*. Cambridge University Press, 1961.

[THO72] R. Thom. *Stabilité Structurelle et Morphogénèse*. W. A. Benjamin, Inc., 1972.

[TSA85a] R. Y. Tsai. *Accuracy Analysis and Prediction for 3D Robotics Vision Metrology*. Technical Report, IBM Research Report RC 11348, 1985.

[TSA85b] R. Y. Tsai. A *Versatile Camera Calibration Technique for High Accuracy 3D Machine Vision Metrology using Off-the-Shelf TV Cameras and Lenses*. Technical Report, IBM Research Report RC 11413, 1985.

[TSA86] R. Y. Tsai. An efficient and accurate camera calibration technique for 3d machine vision. In *Proceedings of the IEEE Conference on Computer Vision and Pattern Recognition CVPR-86*, pages 364–374, Miami Beach, Florida, June 22–26, 1986.

[TWK87] D. Terzopoulos, A. Witkin, and M. Kass. Symmetry-seeking models for 3D object reconstruction. In *Proc. 1st ICCV*, 1987.

[VP87] A. Verri and T. Poggio. Against quantitative optical flow. In *1st ICCV*, 1987.

[WAT81] P. P. Watson. Computing the *n*-dimensional Delaunay triangulation with application to voronoi polytopes. *Computer Journal*, 24: 167–172, 1981.

[WBKL84] P. H. Winston, T. O. Binford, B. Katz, and M. Lowry. *Learning Physical Descriptions from Functional Definitions, Examples, and Precedents*, pages 117–135. First International Symposium of *Robotics Research*, MIT Press, Michael Brady and Richard Paul editors, 1984.

[WON75] Wong. Mathematical formulation and digital analysis in close-range photogrammetry. *Photogrammetric Engineering and Remote Sensing*, 41(11): 1355–1373, 1975.

[WU78] W. T. Wu. On the decision problem and mechanization of theorem proving in elementary geometry. *Scientia Sinica* 21, 159–172, 1978.

[YAC86] M. Yachida. *3-D Data Acquisition by Multiple Views*, pages 11–18. Third International Symposium of *Robotics Research*, MIT Press, Cambridge, Ma., O. D. Faugeras and G. Giralt editors, 1986.

[YC78] Y. Yakimovsky and R. Cunningham. A system for extracting three-dimensional measurements from a stereo pair of tv cameras. *Computer Graphics and Image Processing*, 7:195–210, 1978.

[YKK86] M. Yachida, Y. Kitamura, and M. Kimachi. Trinocular vision: new approach for correspondance problem. In *Proc. of the Eighth ICPR*, pages 1041–1044, Paris, 1986.

[YP83a] A. L. Yuille and T. Poggio. *Fingerprints Theorems for Zero-Crossings.* Technical Report, MIT Artificial Intelligence Laboratory AIM-730, 1983.

[YP83b] A. L. Yuille and T. Poggio. *Scaling Theorems for Zero-Crossings.* Technical Report, MIT Artificial Intelligence Laboratory AIM-722, 1983.

[ZH85] X. Zhuang and R. M. Haralick. Two view motion analysis. In *Proceedings IEEE Conference on Computer Vision and Pattern Recognition CVPR-85*, pages 686–690, San Francisco, June 19–23, 1985.

3 Contact Sensing for Robot Active Touch

Paolo Dario

3.1 Introduction

Although the most extensive effort in machine perception to date has been in machine vision, robotic active tactile perception, or "**robot haptics**," is now being recognized as important and is receiving increasing research attention [1–4]. As pointed out by many investigators [5, 6], not only is touch complementary to vision, but it offers powerful sensing capabilities on its own. For instance, touch is capable of measuring directly many physical properties of an object (such as shape, surface details, hardness), while vision can only infer these properties indirectly, by deducing them from optical properties.

In addition to this "practical" interest, tactile sensing poses intellectual problems that, like, and for many aspects even more than, vision, focus directly on the very essence of robotics, intended as the intelligent connection of perception to action [7]. Like vision, in fact, tactile sensing involves the two fundamental aspects of machine perception (i.e., the acquisition and the processing of information); more than vision, touch emphasizes those aspects related to data acquisition.

It is well-known that the acquisition of visual data in humans is an active process: however, most computer vision studies have been oriented so far to investigate methods of processing the visual information, rather than active strategies of image acquisition. Tactile sensing does not leave any room to ambiguity: the very same concept of machine touch is inextricably related to the notion of motion and activity, and the study of artificial tactile sensing requires the same high degree of sophistication both in the acquisition and in the processing of tactile data.

A number of investigators, especially among machine vision experts, have insisted that tactile sensing is analogous to visual sensing in many respects. This statement is certainly true to some extent; however, there are some differences between visual and tactile images. For instance, "tactile images" may have different physical meanings, and hence require different interpretations. For instance, a tactile image can be a representation of the distribution of the normal contact forces between the sensor and a touched object, or of the strain inside the compliant covering of the sensor, or, also, of the pattern of thermal, magnetic, or chemical parameters resulting from contact. On the other hand, like visual images the tactile images are strongly affected by the procedure and by the parameters actually used for their acquisition (for instance, by the contact force, by the orientation of the sensor, and by its velocity during the scanning process). For these reasons, it is reasonable to expect that a tactile sensor really suitable for the study of robotic perception will result only from a system design approach, and, also, that its sensing capabilities will be fully exploitable only within the specific robot system for which the sensor has been conceived.

The critical importance of system design for robot haptics is still often underestimated. For example, while the importance of a good tactile sensor is unanimously stressed by the robotic community, the need for compliant motion control [8], that is at least as fundamental as a good tactile sensor for accomplishing tactile exploratory tasks, is not so widely recognized (and, unfortunately, compliant motion is a feature not available, and very difficult to add, to most current robot systems). Also, the key role of a machine equivalent to the human kinesthetic response in haptic perception is not always pointed out clearly.

The main research interest of our laboratory in the field of robotics is on the study of robot systems ultimately able to replicate human haptic capabilities. For the reasons we have mentioned above, our approach to this problem attributes a fundamental importance to system design and, in such a context, to the synergism of the following components: contact sensors, kinesthetic sensors, end effector and controller. This approach will be illustrated in this chapter by referring to the simple robot system we have implemented in our laboratory, but with emphasis on the possible general applicability of the informing concepts we have elaborated.

In the present initial phase of our project, we have elected neither to use an existing industrial robot nor to develop a complete multifingered hand (as would be actually necessary for investigating artificial tactile perception exhaustively). We did not use a commercial robot equipped with a conventional gripper because an ordinary controller cannot usually be adapted for hybrid (force/position) control techniques, and because the type of humanlike tactile perception we intend to investigate requires fine manipulation, and hence a degree of dexterity that is impossible to obtain with a conventional robot gripper. On the other hand, we decided not to develop a multifingered hand because the design and control difficulties inherent to this task would have probably distracted too much of our limited resources from the main objective of our study. As have others [1, 2], we thought that a single robot "finger," provided that it possesses the same motion capabilities and incorporates about the same type of sensors as a human finger, would have been sufficient in order to investigate the basic paradigms of artificial tactile perception, without forcing us to face potentially overwhelming technical difficulties.

This chapter describes the conceptual fundamentals and some details of the one-finger robot system we have actually implemented. The chapter includes a first section that provides an extensive general discussion on the system design approach we followed, a second section that deals with the design philosophy of the skinlike tactile sensor that represents a distinctive feature of the robot system, a third section describing the robot finger, a fourth section reporting on system control, and a conclusion in which open research topics are discussed.

3.2 Considerations on System Design

A sensible approach to the study of robot haptics requires substantial emphasis on intelligent sensory-data gathering techniques.

Human perception seems to be the only adequate model for analyzing robot haptics (viewed as an aspect of machine perception). Unfortunately, while the understanding of the neurophysiological bases of human visual perception is quite accurate, analogous models of human haptic perception are not so well established [9, 10]. For this reason, both our observations on human haptics and the analogies of our robot system with the human model will tend to be only qualitative.

Human tactile perception results from **cutaneous** and **kinesthetic** responses. The cutaneous response conveys a wide range of tactile information relative to the local contact conditions between the skin and the touched object. The kinesthetic response provides the nervous central system with information on limb and joint positions. In order to explore actively the tactile environment, humans rely on the integration of cutaneous and kinesthetic responses.

A simple example of integration of kinesthetic and cutaneous data for tactile perception derives from our experience. When we manipulate an object, in fact, the kinesthetic response on the position of the finger joints and on the relative position of the fingers (that, in the case of unconstrained finger movements, is interpreted by the brain as a **proprioceptive** information on the internal hand environment) is perceived, if associated with cutaneous data relative to the location and condition of the contact between the fingers and the manipulated object, as an estimation of the shape of the object (**exteroceptive** sensing). Thus, a very general indication of the control architecture of a robot end effector intended for replicating this type of haptic sensing is that it should be hierarchical, so that the joint position signals detected by the kinesthetic sensors are processed not only at low level (for motion control) but also at high level (for perceptual purposes).

As far as the use of contact-related sensory data is concerned, the anthropomorphic analogy does not help very much, since the specific functions of the different force and tactile receptors of the human manipulative system, and, above all, the details of the peripheral and central neural mechanisms that underlie the processing of tactile information, have not been elucidated yet [9]. For this reason, although carefully considering the indications available from psychophysical and psychological research on active touch, we have elected to analyze contact sensing for robot tactile exploration without adhering strictly to any anthropomorphic model.

3.2.1 Force and Tactile Sensing

By the term **contact sensing** we imply the generality of effects occurring when the robot end effector touches an object. (It is interesing to note here the observation proposed by Brown [11], according to whom tactile sensing could be viewed as degenerate range sensing at or near zero range. As we shall discuss later, the conceptual association of contact and noncontact sensing may have some interesting implications for the design of robot tactile sensors.) Contact sensing includes **force sensing** and **tactile sensing**. The distinction between these two sensing modes—which we define, respectively, as the measurement of the global mechanical effects of contact and as the detection of the wide range of local parameters (physical and chemical) affected by contact—is sometimes rather subtle. Consider the effects originated by the contact between the robot end effector and an object. At a macroscopic level, a range of forces and moments is transmitted through the contact. At a microscopic level, those forces and moments result from a distribution of stresses produced by the interpenetration of the two contacting bodies. However, this interaction may also produce in the two contacting objects other physical effects (for instance, alterations of the thermal and/or magnetic equilibrium) or chemical modifications. An additional distinction between force and tactile sensing, perhaps more subtle, but coherent with a system approach, refers to the level of the system control hierarchy at which the related signal is processed. According to this distinction, force sensing is essentially related to low-level compliant motion control, while true tactile sensing implies high-level (perceptual) processing.

Having pointed out that true tactile sensing should also include the detection of a number of parameters other than just local contact forces, let us analyze more in depth, although still qualitatively, the mechanical effects of contact. Consider, for the sake of simplicity, the case of a rigid object indenting the compliant covering of the fingertip of an articulated end effector (hand). As observed by Fearing and Hollerbach [12], who discussed a theoretical approach to solid mechanics for tactile sensing, a finger must have a compliant covering to take advantage of the increased prehension stability possible at corners and also to facilitate, at a higher control level, distinguishing between object features. In fact, a compliant covering will provide contact areas larger than a hard skin.

The same physical effect originated by contact, i.e., the indentation of the compliant skin, should be analyzed and interpreted, for control purposes, in two conceptually different ways: in fact, measuring the actual **contact stresses inside the skin** is important for controlling fine manipulation tasks, while measuring the **deflection profile of the skin** would be useful for recognizing directly geometrical object features. It is important to recognize the distinct meanings that can be assigned to the same type of tactile information, even if the two effects (displacement profile and contact stress distribution) are obviously related.

There are examples of both deflection sensors and contact stress (more commonly strain) sensors in the growing literature on tactile sensor technology [13, 14]. For almost the totality of those sensing devices, however, performances are said to be unacceptably poor, especially in terms of linear response and hysteresis. We propose to reconsider this statement based on the above observations on the role of displacement and contact stress sensing, and on the following additional considerations.

3.2.2 Sensory-Motor Exploratory Paradigms

The very same concept of haptic exploration is inherently related to that of **dexterity**, a capability that may result, in an articulated end effector, from a combination of improvements in robot mechanisms and in the development of high-level control strategies [15]. Sensing and controlling forces and moments generated at the end effector (at the fingertip, in the case of a multifingered hand or of a single exploratory finger) is necessary in order to control compliant motion, a fundamental aspect of dexterous behavior. Forces and moments can be sensed at different locations of the robot manipulator. In general, the finer is the manipulation task, the closer to the contact area the force sensing device should be placed. Obviously, there is a trade-off between sensor sensitivity and ruggedness: a sensitive sensor capable of detecting locally small distributed contact forces is likely to be very delicate. The same holds for humans: in fine manipulation tasks (for instance, in the manipulation of delicate electronic components or small and flexible wires) force control is obtained using mostly the force-related sensations provided by cutaneous receptors, while the gross manipulation of heavy objects is controlled almost completely through the force feedback from deep (muscle and tendon tension, or joint torque) sensors, and the hand is often protected by gloves that reduce severily tactile sensations.

We can consider three different types of manipulation, which may require different force sensing modalities:

a. gross manipulation,
b. fine manipulation of regular objects, and
c. fine manipulation of delicate objects.

In the case of a multifingered end effector connected to a robot arm, some operations classifiable as type (a) (for instance, those in which the fingers grasp firmly the object, without truly manipulating it) could be controlled just by using torque sensors located in the robot joints, and/or wrist sensors [16]. For other type (a) tasks requiring some degree of dexterous manipulation, the measure of finger joint torques and fingertip contact forces and moments is probably more appropriate. Finger joint torques are usually measured in cable-actuated hands by strain gauge-based tendon tension sensors [17].

The dexterous manipulation of objects of regular shape and weight (i.e., easily and securely graspable) can usually be managed by sensing and controlling finger joint torques and fingertip contact forces and moments. The concept of fingertip force sensing has been proposed by Salisbury [18] and implemented in a semiconductor strain gauge-based device incorporated in the fingertips of the Stanford/JPL hand by Brock and Chiu [19]. This sensing technique has two attractive features: first, being equivalent to a miniature 6-axis force-sensing wrist located at the base of each fingertip, a fingertip sensor is capable of accurately resolving all the three components of the applied force and all the three components of the applied moment; second, if some assumptions on contact conditions are verified, a fingertip sensor is capable of measuring the magnitude and the direction of the resultant contact force as well as contact location. As the overall performances of this type of device can be excellent in terms of sensitivity, accuracy, linearity, and negligible hysteresis, it is reasonable to argue that, for the class of manipulation tasks mentioned above (which probably include most of the cases of practical interest), the function of force sensing could be conferred **entirely to the fingertip sensor** and that, therefore, there might even be no need for sensing force at the cutaneous level. A consequence of this hypothesis is that a tactile sensor should not be necessarily designed only for sensing distributed contact forces.

Accepting this conclusion means eliminating at the origin many ambiguous and potentially misleading interpretations of the role of cutaneous sensors during ordinary manipulation and some type of tactile exploration: in fact, if the resultant contact force can be controlled through contact-resolving fingertip sensors, cutaneous sensors would have the only function of sensing locally all the other parameters affected by contact. Being free from the constraints imposed on their performances by the necessity of accurately controlling contact force, tactile sensors could possibly be designed without worrying too much about linearity, hysteresis, and static response (all "drawbacks" that, incidentally, most receptors of the human skin possess to some extent).

The third case we are considering, i.e., fine manipulation of delicate objects, is the most challenging. In this situation, in fact, contact forces can be so small that even the sensitivity of a fingertip force sensor may not suffice. In addition, the accurate execution of the desired manipulative or exploratory task could require not only the control of the resultant force but also the fine control of local contact forces. For example, incipient slip conditions are different when the finger is touching a corner or an edge or when it is manipulating a tiny wire that indents deeply the fingertip covering, because the local distribution of contact forces varies, even if the resultant contact force may be the same. For this type of "extreme" applications, therefore, monitoring and controlling the distribution of contact stresses at the compliant covering of the fingertip, including both normal and tangential stresses, is important.

It is worth noting that, in the case of fine manipulation, the same information on the distribution of contact stresses that is necessary for controlling fine exploratory movements contains also data that can be interpreted at a superior level of the system control architecture as object geometrical features. Conversely the fingertip force sensor, although primarily intended for the control of low-level compliant motion tasks, is also usable by the high-level controller for perceiving some features of the environment [18, 19].

3.2.3 Tactile Subroutines

Consider now true tactile sensing. If the explored object has a regular shape, and thus controlling compliant exploratory motion does not require extreme care in force sensing, a convenient sensor configuration would consist of a tactile sensor array located on the external surface of a fingertip that incorporates also a 6-component force vector sensor. As already observed, since the resultant force vector sensor provides the data necessary for a reasonably accurate control of contact force and of fingertip orientation, the tactile sensor array would have a purely exteroceptive function. In order to define the sensing capabilities of this ideal tactile sensor, we shall analyze the possible operation of a robot exploratory system.

A convenient approach could involve identifying individual exploratory acts, each aimed at extracting one single property of the explored object, and on analyzing the type of cutaneous information that (associated with kinesthetic response) would allow us to infer that specific object property. As the hypothetical robot exploratory system we are considering should possess humanlike sensory-motor capabilities, it is reasonable to refer to the indications available from psychophysical and psychological research in the field of haptic perception, integrated by some intuitions derived from our everyday experience on tactile exploratory strategies, in order to define sequences of sensory-motor acts reproducible by the robot system. We have followed this approach in the design of a tactile sensor, whose skinlike sensing capabilities result from the attempt of imitating the overall perceptual properties of the human tactile sensing system, rather than that of truly mimicking the morphological disposition and/or the physiology of the skin receptors. The specifications for this sensor were derived from the analysis of some fundamental sensory-motor exploratory sequences ("**tactile subroutines**") carried out by a single articulated robot finger operating in a sensorized scenario [20]. For simplicity, we referred to the case of an object placed on a (sensorized) table, or grasped by a hand, and being explored by the robot finger.

A typical, "general purpose" sequence of exploratory tasks might include the following tactile subroutines (whose characteristics and temporal succession have been somewhat arbitrarily selected as representative of the many possible exploratory strategies that the human brain can decide to execute):

a. **APPROACH.** The sensorized fingertip approaches the object. Finger motion can be blind, or it can be guided by data provided by a vision system, or by the sensorized table [20], or by active and/or passive proximity sensors incorporated in the fingertip structure. Active range sensing could be based on various physical effects, such as acoustical, optical, or inductive. Ultrasonic (US) range sensing is particularly attractive for many reasons [11], and, as some animal species demonstrate, it can be usefully exploited (if the transducers are moved adaptively) to infer some geometrical features of a surface. Thus, and compatibly with the spatial resolution ultimately achievable by the range-sensing system, it would be possible to reconstruct a 3-D image of the explored object. Such an image, which could be provided by an array of US transducers appropriately disposed on the fingertip, would be obtained without any significant physical interaction between the object and the exploratory system; hence this image could even be regarded as the "displacement profile" of the object that we indicated as the ultimate goal of an ideal tactile sensor dedicated to haptic exploration. Note that this concept correlates very well with the already mentioned observation that tactile sensing may be viewed as range sensing at or near zero range [11].

A passive proximity-sensing system capable of detecting the infrared radiation emitted (or absorbed) by an object could mimic the thermal sensitivity of some receptors of the human skin. During the approaching motion, therefore, the high-level control system could be warned of a potentially dangerous object temperature.

In this subroutine we could include some instants immediately before and some immediately after the first contact between the fingertip and the object. These instants are particularly delicate because (a) if the surface of the object is sharp, the contact should be very soft, and (b) there is a transition between almost purely position controlled finger motion and the compliant motion necessary for true tactile exploration. In order to detect accurately these conditions, therefore, the tactile sensor could include receptors very sensitive to local indentation.

An interesting issue related to this first contact concerns the possibility of determining in advance the part of the fingertip sensor that will contact the object. This problem has intriguing implications. In fact, the controller could exploit range sensing to orient the fingertip conveniently before the actual contact occurs. Alternatively, the fingertip could be oriented during tactile exploration. If the spatial density of the tactile receptors were uniform over the sensorized fingertip area and the fingertip shape were reasonably regular, there would be no need for orienting the fingertip. However, if one follows the bionic criteria of taking advantage of active motion capabilities in order to optimize the use of available sensory resources, a convenient solution could be to assign privileged sensing functions to a selected, relatively small fingertip area (equivalent to the foveal region of the retina, and thus defined as the **tactile fovea**), in which most sensing receptors are concentrated. A tactile fovea, however, should possess not only the highest spatial density of the sensing sites but also a peculiar concentration of all the different types of sensory receptors. A consequence of this preferential disposition of sensing capabilities is that all the exploratory strategies should include, as a subtask, a procedure of fingertip orientation in order to **place the tactile fovea just on the desired object feature**. Another consequence of the foveal concept is that the attention of the high-level controller should be primarily focused on the sensors located in the tactile fovea.

b. **SHAPE.** After having touched the object, the fingertip explores the object surface, while pressing slightly against it. A reasonable exploratory strategy could involve first a coarse assessment of the shape of the object, carried out by moving the finger rapidly, without caring too much about local features, and, then, a second phase consisting of a slow and accurate exploration of the geometrical features of interest, whose existence and location were roughly detected during the first exploratory phase. An analysis of this strategy suggests that the coarse exploration could be performed either using the feedback signal provided by a fingertip force sensor or by exploiting a simpler method based on a tactile array sensor.

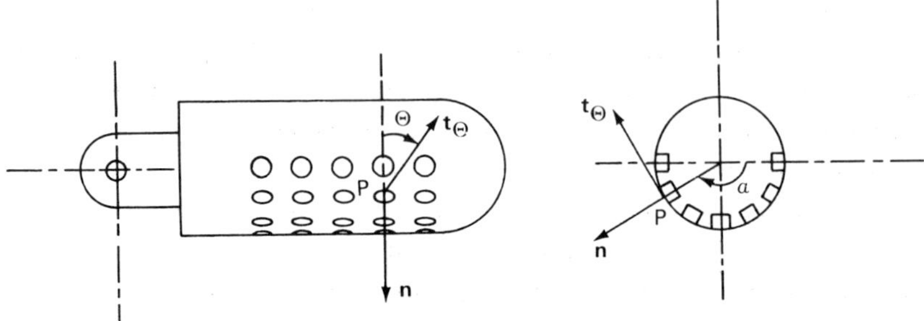

Figure 3.1
Fingertip sensor with coordinate frame. **n** is the normal vector to the fingertip surface in the contact point (in the fingertip space); **t** is the tangent vector to the fingertip surface in the contact point (along a direction θ specified by the high level controller); **P** is the contact point.

In both cases, a fundamental contribution to shape reconstruction would come from kinesthetic data.

The fingertip force sensor described in [19] allows for an almost complete characterization of the contact conditions: in fact, it provides a measure of the magnitude and direction of the contact force as well as of contact location, from which the surface contour can be inferred. Moreover, this calculation is virtually free from the errors due to the limited spatial resolution of an arrayed tactile sensor. The second method, described in detail in [25], is based on the concept illustrated in Figure 3.1, for the case of the first configuration of the sensorized fingertip we designed in our laboratory.

According to this method, a relatively low resolution array of uniformly spaced tactile sensors allows us to calculate approximately the point of contact as the sensing site at which the contact force is higher. As the geometry of the fingertip is known, it is also possible to calculate the common normal to the contacting bodies, along which the finger should exert the predetermined contact force, and the common tangent plane, along which the finger can move for the next exploratory step. Also in this case it is relatively easy to reconstruct the gross shape of the object from the succession of contact locations and orientations.

Unlike the first exploratory phase, which could be executed without necessarily orienting the fingertip, fine exploration requires the attention of the controller to perceive the spatial details of the indenting feature, a task that can be performed better if the feature of interest is probed with the tactile fovea. As already observed, the high **tactile acuity** of the tactile fovea should result not only from a higher concentration of tactile receptors but also from the amount of attention that the high-level controller would devote to the processing of foveal signals; such attention priority in a robot system has an anthropomorphic equivalent in the comparatively large portions of the somatosensory cortex that projects to the fingertip sensory receptors [9]. If the sensitivity of the fingertip force sensor is sufficient to sense and control the contact force appropriate for the extraction of object features, the tactile sensor has the only function of sensing (directly or indirectly) indentations. Otherwise, local contact forces should be sensed, and the information relative to indentation extracted from this measure.

It is important to remark that in the case of fine feature extraction processes a close interaction is necessary between force sensing and tactile sensing: in fact, the extraction of local features (edges, corners, holes) could be improved, depending on the adopted tactile sensing mechanism and on sensor configuration, by varying the intensity of the contact force, its orientation, and even its frequency of application. Thus, the high-level controller should evaluate the "quality" of the actual indentation profile, and use this parameter as a feedback for adjusting contact conditions. The proposed separation between force sensing and tactile (in this case, indentation) sensing might facilitate, when possible, the practical implementation of this subroutine.

c. **TEXTURE.** When the overall object shape has been reconstructed, the finger can explore the roughness of some (usually planar) parts of the object surface. Two different exploratory strategies can be followed, depending on the performance of the tactile sensor. A very-high-resolution sensor capable of resolving tiny features would allow detecting textures only by static exploration [22]. In this case, the exploratory strategy would consist simply of pressing with the fingertip on the textured surface and analyzing the pattern detected by the sensor. Alternatively, the robot finger could follow a strategy similar to that used by humans, that is, to explore dynamically the desired surface, while applying a very small contact force: thus, friction effects would be negligible. If sensible strain receptors are incorporated in the skin surface, they detect signals that can be analyzed and associated with the surface features that elicit them. Ridges can be incorporated in the tactile sensor in order to enhance strain sensitivity [12].

d. **HARDNESS.** A figure of merit of the elastic properties of the material of which an object is made can be obtained by measuring the ratio between the force that the finger exerts on the object and the corresponding displacement. A possible strategy could involve pressing and releasing the fingertip on the object, while monitoring simultaneously contact force and joint rotations. A disadvantage of this technique is that fingertip displacement must be calculated from joint rotation measures, and this may introduce errors due to friction, backlash, and elasticities in the chain of serially connected links forming the mechanical structure of the finger. A different approach is possible using a tactile sensor that incorporates both strain and stress sensing capabilities. Measuring directly at the fingertip fovea the stress/strain ratio (as is possible if the touched surface is reasonably regular) avoids measurement errors.

e. **THERMAL.** This subroutine is aimed at identifying different materials based on their different heat transmission properties. A further development of this approach, involving a matrix of temperature sensitive elements, allows us to obtain "thermal images" of the touched feature [23]. The execution of this subroutine implies that the tactile sensor incorporates both a heat source and temperature sensors. The fingertip is heated up to a temperature higher than that of the sourroundings. Then the fingertip touches a convenient (i.e., planar) part of the object surface, and presses it softly (once more, this implies low-level, compliant motion control capabilities). The amount of heat flowing from the fingertip to the object varies depending on the thermal diffusivity (during the temperature transient) or thermal conductivity (at the equilibrium) of the material of which the object is made. The temperature sensors located in the tactile fovea at the interface between the fingertip and the object detect temperature variations. The foveal signals are processed at the high-control level, and correlated with different classes of materials. As heat transfer phenomena are relatively slow, the fingertip should keep pressing the object statically until the high-level controller has fully "perceived" the signal detected by the temperature receptors, and distinguished it as clearly belonging to a class of materials.

"Pain" sensors distinct from the "regular" temperature sensors and having the same function as the skin "nociceptors" could be also incorporated in the tactile sensor. Alternatively, the high-level controller could exploit proximity information on radiant heat, or a threshold signal detected by the "regular" temperature sensors, to prevent damaging the tactile sensor.

A similar approach could be followed for other sensing modalities that also require a relatively long interaction between the sensor and the object, such as the detection of some chemical species. In this case, the exploratory strategy would not involve finger motion, but only the adjustment of fingertip position and contact force so as to optimize the detection of the parameter of interest (e.g., humidity, pH) by means of chemical sensor(s) incorporated in the tactile fovea.

These subroutines exemplify concretely a type of sensory-motor approach that one could follow for reproducing humanlike active touch strategies in a robot system. The analysis of the role and functions of the sensory receptors that are required to carry out the proposed exploratory sequences provides guidelines for the design of an appropriate tactile sensor, and helps in defining the characteristics of the other components of the robot system.

The scheme of an overall control architecture capable of managing the exploratory subroutines previously described, and also additional ones, is given in figure 3.2.

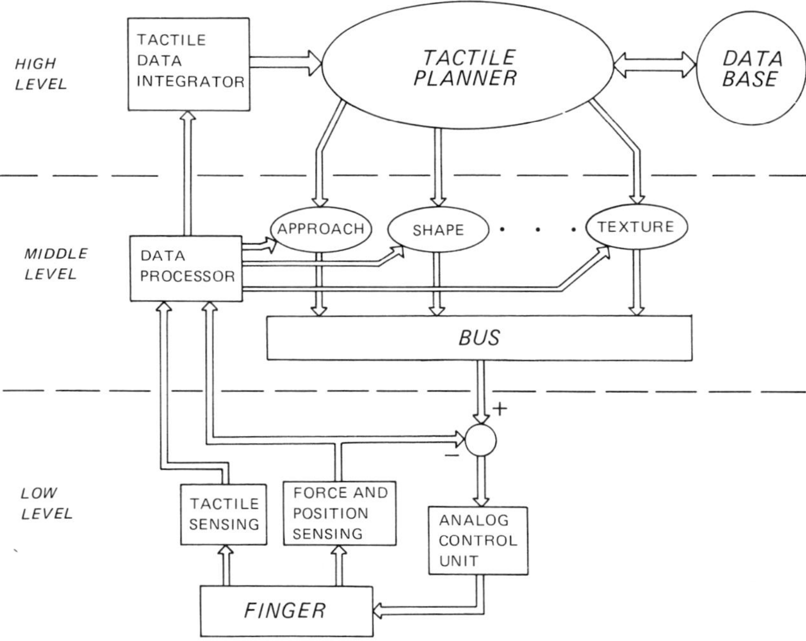

Figure 3.2
Three-level control architecture for the robot system. According to the concepts outlined in this chapter, joint position and fingertip force data are processed both at low level (for compliant motion control) and at high level (for perceptual purposes), while true tactile sensations are processed only at high level.

This three-level control architecture comprises a low-level control system designed primarily to execute hybrid commands sent by the middle-level controller. A peculiar feature of such low control level is that, implementing a concept we have already pointed out, contact force and joint position signals are used both for motion control and, after they have been delivered to the middle-level controller, to obtain kinesthetic response. The middle control level coordinates the execution of the tactile subroutines selected by the high-level planner. Note also that the true tactile signals are sent directly to the middle-level controller, where they are integrated with kinesthetic data, and then delivered to the top, perceptual level.

3.2.4 Tactile Sensor Design

Let us try to summarize now the various indications obtained from the above discussion on the design of a tactile sensor to be incorporated in the ideal robot system we have outlined. In the following paragraphs we shall discuss the main features of a tactile sensor we have actually fabricated following the above indications, and of the robot exploratory system we have assembled "around" this sensor for investigating robot haptics.

An ideal tactile sensing system suitable for the intended exploratory tasks (although limited to the single-finger configuration) should be designed according to the following general specifications:

a. The exploratory finger must be equipped with joint position and velocity sensors for servoing its actuators, as well as for providing kinesthetic response. Those sensors should if possible, be located directly at the finger joints in order to increase the accuracy of their measurement.
b. The finger should include sensor devices to measure joint torque. This is necessary in order to control separately the stiffness of each joint, as required in order to improve compliant motion control. As joints are usually actuated through tendons, it is convenient to measure torque directly at the joints or to measure tendon tension after the major sources of friction.
c. The fingertip should essentially possess two types of sensing capabilities: force and tactile. Force sensing for general exploratory tasks can be obtained through a contact-resolving fingertip sensor. Only for particularly fine exploratory tasks may local sensing of distributed contact forces be required.

The tactile sensor should have the shape of a fingertip, or of a part of it. A hemispherical fingertip may simplify the analysis of contact conditions.

The contact-resolving sensor placed in the fingertip base detects not only the resultant contact force vector but also contact position and orientation. The basic function of the tactile sensor is to detect, with the highest possible spatial resolution, the indentation profile originated by the explored object in the compliant layer that should cover the fingertip. Although measuring strain or stresses may be easier than measuring deflections, the displacement profile is the information ultimately necessary for feature recognition purposes. The sensing elements should be disposed according to the concept of a tactile fovea, i.e., concentrating the available sensory resources in a small area in which the contact is, either casually or deliberately (using voluntary movements), more likely to occur.

The sophisticated tactile sensing capabilities that we have identified as necessary for haptic perception can result from incorporating in the same tactile sensor several different sensing elements, appropriately arranged in order not to mutually interfere, but rather to cooperate in sensing tasks. Possible redundancy of sensory data can also be useful to disambiguate uncertain situations, provided that suitable decision strategies are implemented at high control level. For instance, the sensor could incorporate active transducing elements for range sensing, passive elements for proximity sensing, vibration sensitive elements, temperature sensitive receptors, and chemical sensors.

Of course, a sensor including all the above capabilities would have some value only for research purposes; furthermore, it would be difficult to fabricate using state-of-the-art sensor technology. Despite these considerations, we have attempted to design an "unconventional" tactile sensor aimed specifically at investigating the validity of the concepts we have discussed. The most relevant features of the tactile sensor, of the anthropomorphic finger, and of the control architecture that compose the robot system are summarized in the next sections.

3.3 The Tactile Sensor

As some characteristics of the tactile sensor we have designed have been already discussed in previous papers [24, 14], we shall only summarize here the most significative features of the sensor, and point out some recent evolutions of the original configuration that were suggested, in part, by experiments on the exploratory robot system.

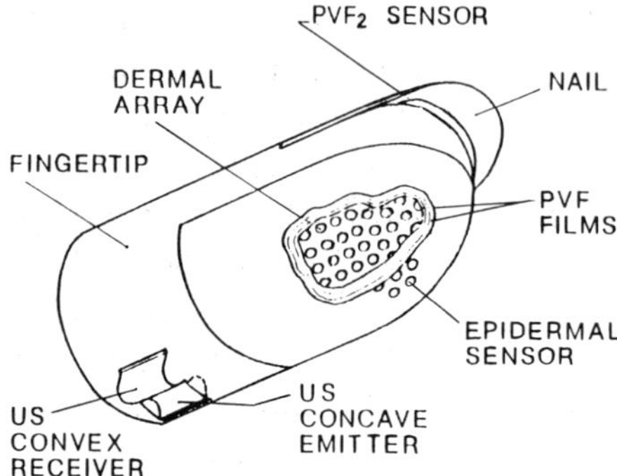

Figure 3.3
Schematic view of the fingertip tactile sensor, showing also the "nail" sensor for detecting surface roughness, and the curved PVF_2 transducers for US range sensing.

The sensor is based on the technology of the ferroelectric polymer polyvinylidene fluoride (PVF_2), a material that, like all ferroelectrics, is both piezoelectric and pyroelectric. Furthermore, PVF_2 has excellent chemical and mechanical properties and it can be prepared in thin, flexible films. As the bandwidth of the piezoelectric response of PVF_2 extends from almost dc to the MHz region, we can anticipate that the choice of this material favors some factors, such as easy conformability, high strain sensitivity, high temperature sensitivity, and dynamic response, over other factors, such as static response. A scheme of the fingertip tactile sensor is shown in figure 3.3.

The sensor comprises a superficial ("epidermal") sensing layer, an intermediate compliant layer, and a deep ("dermal") sensing layer. Both sensing layers are made out of PVF_2, while the compliant layer is made out of natural rubber.

A ferroelectric polymer sensor generates electric charge when stressed mechanically or heated, exhibiting a behavior that is phenomenologically analogous to the mechanoelectric and thermoelectric sensitivity found both in living and in dead skin. This analogy, along with the attractive mechanical and technological features of ferroelectric polymer materials, provided additional significance to the adoption of PVF_2 for our tactile sensor. The transduction of a mechanical (or thermal) signal into electric charge depends on the crystalline structure of the ferroelectric material. For a PVF_2 film electroded on both sides, thickness mode and extensional mode operations are possible. In practice, the electric charge detected between the two electrodes of the film results from strains both across the thickness and along the length of the sensor, as well from temperature variations. The sign of the generated charge also depends on the type of strain (compressive or tensile) and on the thermal variation (heating or cooling).

Conceptually, the dermal sensor, which comprises 128 sensing sites, is intended to mimic the role of the slowly adapting (SA) skin receptors, very sensitive to fine spatial details, while the epidermal sensor, comprising, in the present implementation, only 7 sensing sites, simulates the function of the quickly adapting (QA) receptors, those which preserve only the coarsest spatial details of an indenting profile.

The pattern of dermal sensing sites is defined by electrodes on the surface of a 21-mm-diameter cylindrical Plexiglas frame and of a hemispherical tip on top of it. A 110-μm-thick PVF_2 film is bonded to the Plexiglas frame. The charge generated by pressure or temperature variations is transferred capacitively to the pattern of electrodes. The intermediate rubber layer has the twofold function of providing some compliance to the sensor and of shielding the dermal sensors from rapid temperature transients (a 1.5-mm-thick rubber layer introduces a time delay of about 1 sec between the detection of the thermal signals in the epidermal and dermal layers). The epidermal sensor is made out of 80-μm-thick PVF_2 film, on which a pattern of 7 circular sensing sites (diameter 1.5 mm, center-to-center spacing 2.5 mm) is obtained. Also the 128 dermal sensing sites have a diameter of 1.5 mm and center-to-center spacing of 2.5 mm. The epidermal sensor comprises a bottom, graphite ink-based resistive layer (100 μm thick, 500-Ω resistance), and it is located in a central area above the dermal array. This central area, comprising the epidermal sensors and those dermal sensors directly underneath, represents the tactile fovea.

The epidermal and dermal sensors have distinct, but closely correlated, sensing roles. Being made of the same transducer material, they are both sensitive to strain and temperature: however, they have different mechanical and thermal boundary conditions. Owing to its mechanical constraints, the dermal sensor works predominantly in the thickness mode: thus, it measures essentially the normal component of the contact force acting on each sensing site. As the amount of cross-talk between the dermal elements is very small, tactile images (representing the spatiotemporal distribution of normal contact forces) having negligible blur can be obtained. The main characteristic of the dermal sensor is that it provides reliable and accurate tactile signals, with good sensitivity (about .01 N resolvable force, about 40 N maximum allowable load) and linearity (within about 1%, between .01 and 2.56 N peak to peak), small hysteresis (essentially only that determined by the elastomer layer), large bandwidth, and limited response to temperature variations.

Opposite are the features of the thin epidermal sensor, which is extremely sensitive to any variation of the contact (and even of the proximity) conditions. The epidermal sensor feels the proximity of hot objects through the pyroelectric effect, and delivers very large signals when it touches sharp profiles (which produce deep indentations and, consequently, large membrane strain in the PVF_2 film). Moreover, the epidermal sensors detect roughness, when rubbed gently on the object surface, and assess material thermal properties when heated by the resistive layer and pressed slightly against the object.

As already pointed out, the epidermal sensor responds primarily to indentation (i.e., strain), while the dermal sensor measures primarily stress. This effect is exploited to measure object material "hardness." In fact, when the tactile sensor presses a flat part of the object surface, the ratio between the epidermal and dermal signals provides a measurement of material elastic properties.

Scanning a matrix of piezo- and pyroelectric sensors is a complex task. We have devised a method that combines the advantages of charge amplification with the possibility of reducing the encumbrance of the electronic circuitry, typical of multiplexing techniques [4]. This method implies scanning the sensors by analog multiplexers, delivering the output to a charge amplifier, and reconstructing by computer the true force signal from the voltage increments corresponding to charge variations.

A distinctive feature of the sensorized fingertip (whose shape closely resembles that of the human fingertip) is its capability of exploring complex surfaces and cavities [25]. Two additional sensing features have been added to the present version of the finger: the first is an elastic structure (which mimics a nail) located on the fingertip that incorporates a PVF_2 film sensor working as a sensitive strain gauge. When the nail is rubbed against a surface, tiny, irregular grooves can be detected by the "nail" sensor, which vibrates like the pickup of a gramophone. The second feature is represented by an array of curved US transducers, also made out of thin PVF_2 film. At present, this array comprises only three adjacent elements (a central, concave focusing emitter and two convex, wide acceptance angle receivers).

In a future version of the tactile sensor, a denser array of US transducers will be incorporated in the same structure of the epidermal sensor and will have the function of both a range-sensing imager and a true tactile sensor.

A limitation of the present version of the tactile sensor is that no contact-resolving sensor is mounted yet in the fingertip base. At present, therefore, compliant motion must be controlled by using the relatively inaccurate technique described in section 3.2.3.

3.4 The Anthropomorphic Robot Finger

The finger we have designed and fabricated has an anthropomorphic configuration, and is composed of four rigid links connected by hinge joints that provide a total of four degrees of freedom. The two-degrees-of-freedom articulation of the proximal phalanx of the human finger is reproduced by two separate joints with perpendicular axes. Intended to be eventually connected to a many-degrees-of-freedom manipulator in order to become capable of following complex object surfaces, the finger is presently mounted on a rigid fixture that limits quite severely its exploratory capabilities. Despite this limitation, finger dexterity is sufficient to investigate the fundamental problems associated with the simple exploratory procedures we wish to perform.

Each articulation is driven via plastic coated stainless steel tendons routed through flexible and incompressible sheaths, and actuated by remotely located dc servomotors. We have reduced friction effects associated with this configuration by including these effects in the force control loop, and by superimposing a "dither" vibration to the motor signal. One dc motor actuates each finger joint through a pair of opposed tendons, which are pretensioned to half the maximum required tension to avoid slackering of one of the tendons during high force exertion. The mechanical proportioning of the finger was based on the assumption that it should be capable of exerting a maximum contact force of about 10 N and moving at a maximum speed of .1 m/s. Joint position and velocity are monitored by incremental encoders located coaxially with the driving motors (the accuracy of those measures could be largely improved by utilizing miniature encoders located in the finger joints). Joint torques are measured by tendon tension sensors consisting of strain gauge-instrumented cantilevers located at the outlet of the conduits guiding the drive cables.

3.5 System Control

The general features of the control architecture devised for controlling the robot system have already been described [21, 4] and summarized in this chapter (section 3.2.3). A detailed discussion of the implementation of most of the above tactile exploratory subroutines has been given in [4]. Here we present a brief description of a subroutine (TEXTURE), which illustrates adequately the way in which the different components of the exploratory system are coordinated by the control architecture.

The aim of TEXTURE, i.e., to estimate the roughness of parts of the object surface, is pursued by rubbing the tactile sensor against the surface, while exerting a predetermined, adjustable force on it, and by analyzing the signal elicited by this procedure in the sensitive epidermal sensors. TEXTURE is similar to SHAPE in the strategy adopted to track object surfaces, but it differs in that the high-level controller is "attentive" to perceiving specifically the signal detected by the PVF_2 epidermal sensors. A scheme of TEXTURE is depicted in figure 3.4.

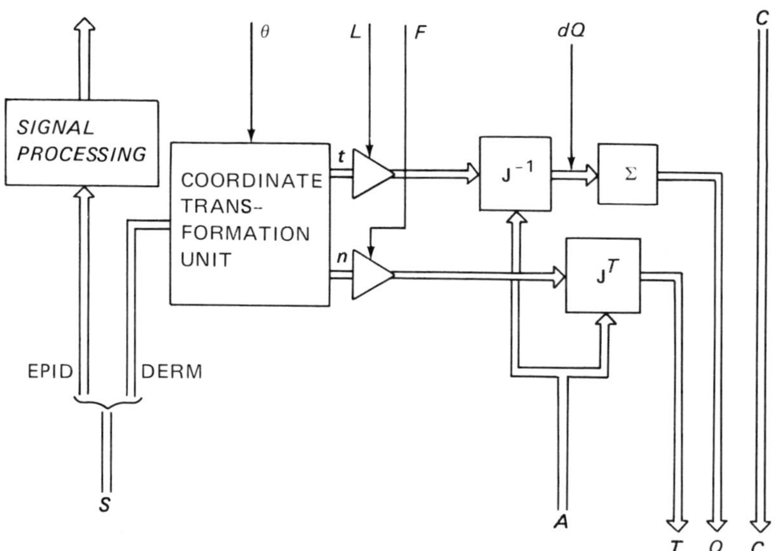

Figure 3.4
Block diagram of the exploratory subroutine TEXTURE, which we have implemented on a DEC MicroPDP-11/73 computer. *S* indicates the composite tactile sensor, which produces the two different outputs EPID and DERM; *A* is the actual position vector. The meanings of the other symbols are given in the text.

The high-level controller sets the value of θ (the direction of motion in the plane tangent to the contact point), L (step length), F (contact force intensity), and $\boldsymbol{C}$ (a "compliance" matrix that controls motion stiffness). Fingertip sensor signals are processed in a coordinate transformation unit that considers only the dermal sensors and calculates from them the coordinates of the contact point (P) in the base frame. The transformation unit also calculates normal ($\mathbf{n}$) and tangent ($\mathbf{t}$) directions in the fingertip frame. The tangent unit vector is multiplied by L and the result applied to the Jacobian inverse matrix $\mathbf{J}$, which provides the angular increments dQ of each joint. The sum of those increments is sent to the low level control as the nominal position command (Q). The normal unit vector is multiplied by F, and the result applied to the Jacobian transpose matrix, providing the nominal torque command ($\boldsymbol{T}$). Details of the computations required for the case of our 4 d.o.f. (degrees of freedom) finger have been given in [4].

While in SHAPE the "attention" of the high-level controller is focused on P and $\mathbf{t}$ (which it uses to reconstruct the shape of the object), during TEXTURE the dermal sensor signal is processed only at low level, without any "conscious" involvement of the upper hierarchies of the controller. The high-level controller "feels" and analyzes the epidermal sensor signal. As far as this aspect is concerned, a method to detect surface features involves first the high-pass filtering of the epidermal sensor signal to avoid the influence of possible slow variations of contact force and object temperature. Then the spectrum of the resulting signal can be analyzed in order to extract features useful to the high-level controller to classify and recognize different vibration patterns [26].

3.6 Discussion and Conclusions

In this chapter we have discussed a number of basic issues related to the design of robot systems capable of executing the sensory-motor exploratory sequences that constitute the basic paradigms of machine haptic perception. We have also described an approach aimed at preliminarily assessing the practical feasibility of some of the concepts outlined in the first part of this chapter. Although we have only touched on the formidable problems associated with the study of robot haptics, our approach served to identify critical areas and to suggest subjects for future research. In the following we discuss some of these indications, referring to each class of the components of the robot system as we have classified them (i.e., sensors, end effector, controller).

We have attempted to elucidate the role of different sensors in the context of the robot system. The idea we proposed—i.e., that for many (perhaps most) fine manipulation tasks the tactile sensor should not necessarily measure distributed forces for reliably controlling compliant motion, but just deflection patterns—might lead, if supported by additional experimental evidence, to evaluating more favorably many of the already available tactile sensors, and to paying more attention to the design of the contact sensors required for true contact force control. Accordingly, research emphasis should perhaps be shifted, in one direction, toward investigating more efficient arrays of deflectometers, and, in another direction, toward the formulation of a more general theory for force-based compliant motion control.

Measuring the local distribution of contact forces might become important for very fine tactile exploration. Thus, the design of tactile sensors specifically aimed at detecting local contact force components could also deserve a large part of future research efforts in the field of tactile sensor technology.

Another observation relates to the increasingly important role that other sensing modalities could have in tactile sensing. Sensors that associate multiple sensing capabilities, and even some ability to process sensor signals locally, would broaden significantly the effectiveness of dexterous manipulation for robot exploration. For instance, range (in particular, transducer arrays for US imaging), proximity, temperature, and chemical sensing are features that we intend to incorporate in the next generation of tactile sensors we are currently investigating in our laboratory. We would also like to use the same transducer technology for all the sensors (we have already demonstrated that it is possible to use PVF_2-based transducers for stress, strain, range, proximity, and temperature sensing; we are now investigating the possibility of integrating all these sensors in the same transducer structure, and are also pursuing a similar approach by using optical fiber sensor technology) [27]. Adopting the same transducer technology might be convenient not only from a conceptual viewpoint (in fact, interpreting sensor signals may be easier) but also for practical reasons (signal processing can be performed using the same methods and even the same components). In this context, it is interesting to note once more the importance of a system approach in optimizing component design. In fact, as we have illustrated in this chapter, different sensing functions can be performed by the same sensor during different exploratory subroutines, depending on the type of sensory-motor act that is executed and on the control level at which the signal is processed.

An additional field for further research is the control of dexterous manipulation, a fundamental aspect of robot haptics. A contribution to the study of this complex problem, which is already being addressed by a number of investigators, could be represented by the extension to the case of a multi-fingered hand of some of the considerations we have discussed for the single exploratory finger.

An important aspect of dexterity is related to the actuation system. As none of the presently available actuators are fully satisfactory for fine manipulation tasks, we are working on the development of new actuators, which have the potential of being incorporated in the finger links and, eventually, even in each finger joint. A possible "direct drive finger" would in fact eliminate most of the problems (friction, backlash, and related inaccuracies) associated with present tendon-actuated hands [28].

Finally, the definition of the control architecture of a robot system capable of haptic perception is an open research field for control theory and artificial intelligence. The considerations we have proposed in this chapter had the sole purpose of pointing out both the number and the significance of the topics that should be investigated in this area. Here we remark that the state of the art in the study of human and robot haptics (at an early stage in both fields) suggests the opportunity of combined efforts. As already occurred in the more mature field of vision, the synergism of robotics and neuroscience might lead to the elucidation of many still obscure aspects of machine and human active touch.

Acknowledgments

Some of the work described in this chapter has been supported by the Italian Government (MPI 40%) and by the NATO Scientific Affairs Division (Grant C.R.G. 85/224).

References

1. Hillis, W. D., 1981. Active touch sensing. MIT AI Memo AIM-629, Cambridge, MA.

2. Bajcsy, R., 1984. What can we learn from one finger experiments? In *Robotics Research*, Brady, M., and Paul, R. eds. Cambridge, MA: MIT Press, pp. 509–527.

3. Stansfield, S. A., 1986. Primitives, features and exploratory procedures: building a robot tactile perception system. *Proc. of IEEE Int. Conf. on Robotics and Automation*, San Francisco, pp. 1274–1279.

4. Dario, P., and Buttazzo, G., 1987. An anthropomorphic robot finger for investigating artificial tactile perception. Int. *J. Robotics Res.* (in press).

5. Harmon, L. D., 1982. Automated tactile sensing. *Int. J. Robotics Res.* 1(2):3–32.

6. Bajcsy, R., and Allen, P. 1984. Converging disparate sensory data. *Proc. of Int. Symp. on Robotics*, Tokyo, pp. 17–22.

7. Brady, M., Gerhardt, L. A., and Davidson, H. F., eds., 1984. *Robotics and Artificial Intelligence.* NATO ASI Series, Springer-Verlag, Heidelberg.

8. Mason, M. T., 1982. Compliant motion. In *Robot Motion: Planning and Control.* Brady, M., Hollerbach, J. M., Johnson, T. L., Lozano-Perez, T., and Mason, M. T., eds. Cambridge, MA: MIT Press.

9. Gordon, G., ed., 1978. *Active Touch.* Oxford: Pergamon Press.

10. Shiff, W., and Foulke, E., 1982. *Tactual Perception.* Cambridge: Cambridge University Press.

11. Brown, M. K., 1986. The extraction of curved surface features with generic range sensors. *Int. J. Robotics Res.* 5(1):3–18.

12. Fearing, R. S., and Hollerbach, J. M., 1984. Basic solid mechanics for tactile sensing. *Proc. of IEEE Int. Conf. on Robotics and Automation*, Atlanta, pp. 266–273.

13. Ogorek, M., 1985. Tactile sensors. *Manufacturing Engineering* 94(2):69–77.

14. Dario, P., and De Rossi, D., 1985. Tactile sensors and the gripping challenge. *IEEE Spectrum* 22(8):46–52.

15. Salisbury, J. K., 1984. Design and control of an articulated hand. Int. Symp. on Design and Synthesis, Tokyo.

16. Dario, P., Bergamasco, M., and Fiorillo, A. Force and tactile sensing for robots. In *Sensors and Sensory Systems for Advanced Robots*, P. Dario, ed. NATO ASI Series, Springer-Verlag, Heidelberg (in press).

17. Salisbury, J. K., and Craig, J. J., 1982. Articulated hands: force control and kinematic issues. *Int. J. Robotics Res.* 1(1):4–17.

18. Salisbury, J. K., 1984. Interpretation of contact geometries from force measurements. In *Robotics Research*, Brady, M., and Paul, R. eds. Cambridge, MA: MIT Press, pp. 565–577.

19. Brock, D., and Chiu, S., 1985. Environment perception of an articulated robot hand using contact sensors. Proc. of ASME Winter Annual Meeting, Miami.

20. Dario, P., Bicchi, A., Fiorillo, A. Buttazzo, G., and Francesconi, R., 1986. A sensorized scenario for basic investigation on active touch. In *Robot Sensors*, Vol. 2: *Tactile and Non-Vision*, Pugh, A., ed. Kempston, U.K.: IFS (Publications) Ltd. and Berlin, Heidelberg, New York, Tokyo: Springer-Verlag, pp. 237–245.

21. Bicchi, A., Dario, P., and Pinotti, P. C., 1985. On the control of a sensorized artificial finger for tactile exploration of objects. Proc. of '85 IFAC Symp. on Robot Control, Barcelona.

22. Ellis, R. E., 1986. A multiple-scale measure of static tactile texture. *Proc. IEEE Int. Conf. on Robotics and Automation*, San Francisco, pp. 1280–1285.

23. Siegel, D. M., Garabieta, I., and Hollerbach, J. M., 1986. An integrated tactile and thermal sensor. *Proc. IEEE Int. Conf. on Robotics and Automation*, San Francisco, pp. 1286–1291.

24. Dario, P., De Rossi, D., Domenici, C., and Francesconi, R., 1984. Ferroelectric polymer tactile sensors with anthropomorphic features. *Proc. 1st IEEE Int. Conf. on Robotics*, Atlanta, pp. 332–340.

25. Buttazzo, G., Dario, P., and Bajcsy, R., 1986. Finger based explorations. Proc. SPIE Conference on Advances in Intelligent Robotic Systems, Cambridge, MA.

26. Patterson, R. W., and Nevill, G. E., 1986. Performance of an induced vibration touch-sensor. In *Robot Sensors*, Vol. 2: *Tactile and Non-Vision*, Pugh, A., ed. IFS (Publications) Ltd, U.K., Springer-Verlag, pp. 219–228.

27. Dario, P., Bicchi, A., Femi, D., and Fiorillo, A., 1987. Multiple sensing fingertip for robot active touch. *Proc. of 3rd Int. Conf. on Advanced Robotics*, Versailles, France, pp. 347–358.

28. Dario, P., Bergamasco, M., Bernardi, L., and Bicchi, L., 1987. A shape memory alloy actuating module for fine manipulation. Proc. of IEEE Workshop on Microrobots and Teleoperators, Hyannis, MA.

4 Learning and Recognition in Natural Environments

Alex Pentland and Robert Bolles

4.1 Introduction

DARPA's (Defense Advanced Research Projects Agency) Autonomous Land Vehicle (ALV) project is intended to develop and demonstrate vision systems that can navigate in outdoor, natural environments. As we see it, the major challenges faced by this project are to develop (1) a general-purpose vocabulary of models that is sufficient to describe most of the important landmarks that the vehicle will encounter, (2) recognition techniques that will allow us to locate these landmarks from sensor data in a directed, top-down manner, and (3) learning techniques that will allow us to compute stable, accurate descriptions of interesting landmarks in an unguided bottom-up fashion.

If a machine vision system possessed these capabilities, it could then identify potentially useful landmarks, enter their descriptions into a database, and then, during a subsequent traversal of the same region, search for and recognize previously seen landmarks. Thus the capacity to describe real-world scenes using a vocabulary of models that is recognizable from the data would make it possible to accumulate a comprehensive catalog of landmarks that can be recognized and then used to guide the vehicle. This, then, is the goal of the work described herein.

The reason for exploring these structure-learning and structure-matching techniques at this time is the availability of a dense range sensor, which produces data that are commensurate with the task—the world through which the ALV travels is three-dimensional and the range sensor produces three-dimensional data. Without such a sensor one is forced to estimate three-dimensional properties of the world from two-dimensional data. Although the "shape-from-x" techniques, which make these estimates, are steadily improving, they are still computationally expensive and are generally not capable of producing dense results.

There has been a common belief for some time that the availability of dense range data would make the perception problem trivial. This prediction has not come true. Image segmentation is still a problem, although somewhat more manageable with range data. The structure-learning technique presented in this chapter is in effect a fancy segmentation technique. It builds descriptions of range data by identifying patterns in the data that can be concisely represented in terms of higher-level primitives. The structure-matching technique is also specifically designed for range data. It makes data-level predictions, which it then correlates with the sensed data. This approach is much more successful for range data than intensity data because it is significantly easier to make high-quality predictions of range data.

4.1.1 Approaches to Machine Vision

Most machine vision systems can be divided into a *prediction* phase and a *description* phase, as is illustrated by figure 4.1. In addition, they can be categorized by the type of representation used for recognition—that is, by the type of descriptions they match when establishing a correspondence between the predicted object appearance and the image description.

We identify four major types of representation, each corresponding to a certain level of abstraction. The lowest level of abstraction is the *image*: an array of numbers that represent either the sensor data or a continuous, topology-preserving transform of the sensor data. This level of representation retains the analog, nonsymbolic nature of the sensor data. Slightly more abstract is the *feature* level: a representation of image appearance in terms of such discrete elements as edges, corners, and circles. This level of representation is symbolic rather than analog.

At a higher level of abstraction is the representation of the viewed objects in terms of their component *parts*, by means of constructive solid geometry (CSG) [1], generalized cylinders [2], or the superquadrics-and-deformations representation employed in [14]. This is the first level of representation that refers primarily to the three-dimensional world being viewed, rather than

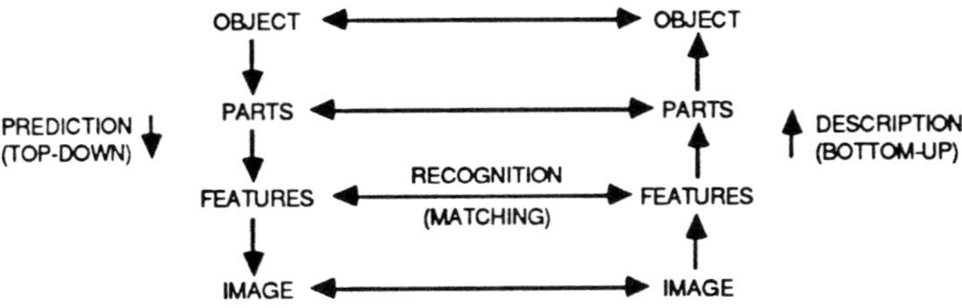

Figure 4.1
Possible levels of analysis for the learning-recognition process.

being tightly tied to the image. Finally, there is the representation of the scene in terms of *objects*, such as people, cars, and airplanes. At this level it is the *goals* of the vision system that become important, because what constitutes "an object" depends upon the viewer's purpose.

4.1.1.1 Recognizing Objects The recognition task to be performed dictates the level of abstraction that is most appropriate. For instance, most industrial vision systems use the *feature* level of abstraction, skipping the part and image levels almost entirely. In such a system, the object to be recognized is first described in terms of the stable, easily found features, such as edges. This results in a wire-frame model, in which the "wires" of the model correspond to edges of the object. The appearance of these edges can be predicted, given a particular viewing perspective. This prediction process is normally accomplished beforehand, if possible, as a precompilation step.

To recognize the modeled object we then form a *description* of the image in terms of our chosen feature, as is depicted in the right-hand side of figure 4.1. Typically, this means using an edge finder in order to locate long, straight edges. The actual step of object recognition, therefore, is reduced to establishing a correspondence between the features predicted relative to some viewpoint and the image description, as is shown by the arrows running between the prediction and description sides of figure 4.1. This is accomplished by searching for a good match between some viewpoint's predicted edges and the edge description recovered from the image itself.

4.1.1.2 Learning Object Descriptions The most important limitation of the foregoing industrial vision approach is that there is no way to learn the object description used for recognition *automatically*. Although relatively less important in most industrial applications, this ability is crucial in the ALV domain.

As summarized on the right-hand side of figure 4.1, the process of learning object descriptions consists of computing a description—in terms of image, feature, or part representations—of interesting objects. For such a description to be useful, we must be able to compute this description in a totally bottom-up manner. This is why the task of learning object descriptions is the greatest hurdle faced by the ALV, since it requires that the program *automatically* segment the data, select the appropriate modeling primitives, and finally instantiate the model's parameters.

4.1.2 Learning and Recognition in the ALV Domain

The levels of representation we choose depend upon the task to be performed. We must start, therefore, by listing the important tasks confronting our ALV. We believe that the following requirements are both necessary and sufficient to ensure that potentially useful landmarks will be identified, their descriptions learned, and these descriptions then used for recognition during subsequent traversals of the terrain.

- Obstacle detection—Deviations from "smooth" terrain that are large enough to impede the progress of the vehicle, or which are distinctive enough to be useful in navigation, must be detected.
- Description—Stable and accurate description of these obstacles must be constructed.
- Prediction—The appearance of such obstacles, when seen from other viewpoints or when other sensors are used, must be predicted.
- Recognition—These predictions must be used to recognize and locate the known obstacles in new imagery.
- Refinement—Information based on the new imagery is used to update and refine object descriptions.

In the context of figure 4.1, these tasks translate into the following procedures:

- Separate figure from ground in the image data to obtain an image-level description of the second obstacle.
- Compute an object description from this image description (proceed up the description side of figure 4.1).
- Use this object description to predict either part, feature or image level descriptions (i.e., proceed down the prediction side of figure 4.1).
- Establish a correspondence between the predicted appearance from some viewpoint and the (bottom-up) computed image description (i.e., connect between the description and prediction sides of figure 4.1).
- Use this correspondence to update the object model (i.e., use the connection established between the prediction and description sides of figure 4.1 to compute an improved object model).

With regard to figure 4.1, then, there are only two remaining questions: (1) What level of representation is best for recognition? and (2) What level is best for learning descriptions?

The answer to the recognition question, we believe, is that *each* level—image, feature, and part—should contribute to the recognition process. Exactly how each will be utilized depends upon the precise nature of the context, sensor, and object description. We envision a flexible, opportunistic vision system built within this framework, one that can draw upon whatever information is available. For instance, if we know exactly what a landmark looks like and we have a good estimate of our position, then it is most efficient to match at the image level. On the other hand, if we know only that there is "a pole" out there, we might better match at the part level—i.e., to find "a thing" that is approximately vertical, and is much longer than it is wide.

Our response to the learning problem, however, is quite different. The fact that people can recognize objects and learn their descriptions strongly supports the view that there is some "natural," stable method for structuring object descriptions and that, moreover, people somehow recover this "natural" structure from imagery and use it to support object learning. There is strong psychological evidence [10–16] that this natural structuring of object descriptions is closely related to people's naive perceptual notion that objects have "parts." We argue, consequently, that most natural objects have a *part structure* that is recoverable from image data, and that the recovery of this part structure is a necessary precondition for learning object descriptions. The specifics of the part-level representation employed in this paper were originally suggested by psychological evidence about people's notions of part structure [13, 14].

This is not to say that feature and image levels do not both contribute to learning object descriptions; it is clear, after all, that the presence of T-junctions, parallel lines, and the like strongly constrain part structure [2-4, 10-12]. Nevertheless, feature-level or image-level descriptions *alone* do not seem sufficient for generating descriptions that support general activities. There is, for instance, a medical literature concerning patients whose spatial and feature-recognition abilities remain intact, but who are unable to recognize objects except in very special, highly constraining situations [34].

4.1.3 Outline of This Chapter

In this chapter we first describe a part representation that we believe facilitates the learning and recognition processes by providing the requiste descriptive power without involving a large number of parameters. Second, we describe a method for learning object descriptions, starting from ALV range data, that are based upon this part representation. Finally, we describe a method for recognizing previously learned objects by matching between predicted image appearance and measured image data.

With regard to the diagram in figure 4.1, the techniques described herein represent two extremes among the possible approaches. The learning algorithm builds object models from the data in a completely unguided, bottom-up fashion. Moreover, it does this by direct image-to-part search, skipping the feature level entirely. The recognition algorithm predicts an object's image appearance and matches at the image level, again skipping the feature level.

Although we have chosen to investigate initially only part-level representation and image-level matching, we anticipate that in a fully developed system *all* levels of representation would be present and that, especially in the recognition phase, there would be interactions at each of these levels. Our motivation for first exploring image-level matching derives from the similarity of range data to Marr's $2\frac{1}{2}$-D sketch: both are relatively dense measurements of the scene's intrinsic geometry. We believe that our range data techniques for learning descriptions and recognizing objects are transferable to situations where shape-from-x and depth-from-x method, rather than a laser rangefinger, have provided us with image-like descriptions of a scene's intrinsic geometry.

Further, it seems that a good strategy for developing a robust, flexible vision system is to first learn to deal directly with the image. For instance, it seems necessary to have the capacity to recover [reasonably] canonical part descriptions directly from image data before we can have reliable part-level matching. Similarly, if we can recognize without using feature-level information then it seems likely that our system's performance can only improve when that information becomes available.

4.2 A Part Representation

Many modern psychologists [10–12], as well as the psychologists of the Gestalt school, have argued that we conceive of the world in terms of *parts*, and that the first stages of human perception are primarily concerned with decomposing an image into these parts. This part-structure is seen as forming the building blocks for the rest of our perceptual interpretation.

Following in this tradition, we have examined [13] the manner in which people describe shape by using both the tools of protocol analysis and the psychophysical method devised by Triesman. One concise characterization of our results is that we observed our subjects describing 3-D shapes *procedurally*—namely, by describing how one would make the shape using a malleable material such as clay, using a few generic forming actions [14]. As an example, our subjects might have described the back of a chair as a rounded, flattened cube that has been slightly cupped or bent to accommodate the human form. The bottom of the chair might be described as a similar object, but rotated 90°. By "oring" these two parts together with elongated rectangular primitives describing the chair legs, they would obtain a complete description of the chair. This description is illustrated in figure 4.2 (bottom left).

It is important to note, however, that in our experiments it did *not* seem that people tried to describe the exact surface shape. Rather, they appeared to be trying to describe the general, "global" structure of the form, with the detailed surface shape being described (where necessary) as variations from the overall shape. For instance, a person's forearm would be described in two parts: *globally* as a tapered cylinder, and *locally* by small deviations from the global form that describe the detailed musculature. Thus this sort of parts-level description should be thought of as being in addition to, for example, thin-plate [32] or fractal [27, 28] surface models. It is the frame upon which hangs detailed descriptions of surface shape.

One reason people may favor such a part-and-process method of description is that it offers considerable potential for reasoning tasks. It seems, for instance, that people employ such descriptions in learning as well as in commonsense and analogical reasoning [15–17, 31, 35, 36]. This may be because such descriptions refer to the world in something like "natural kind" terms: they speak qualitatively of whole forms and relations among the parts of objects, rather than of local surface patches or of particular instances of objects.

Moreover, recent research in graphics, biology, and physics has given us good reason to believe that it may be possible to describe *objectively* our world by means of a few, commonly occurring types of formative processes [18–21]. In essence, we think that our world can be modeled as a relatively small set of generic processes—such as bending, twisting, and interpenetration—that occur again and again, with the apparent complexity of our environment being produced from this limited vocabulary by compounding these basic forms in myriad different combinations.

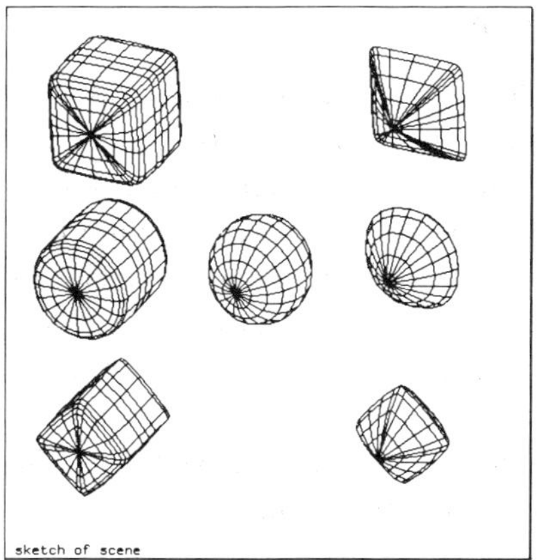

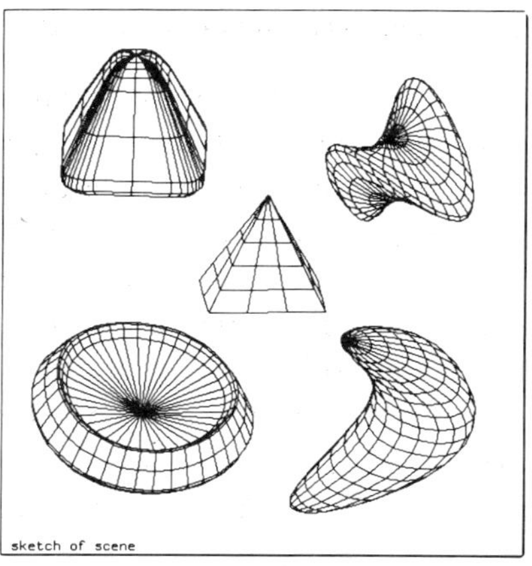

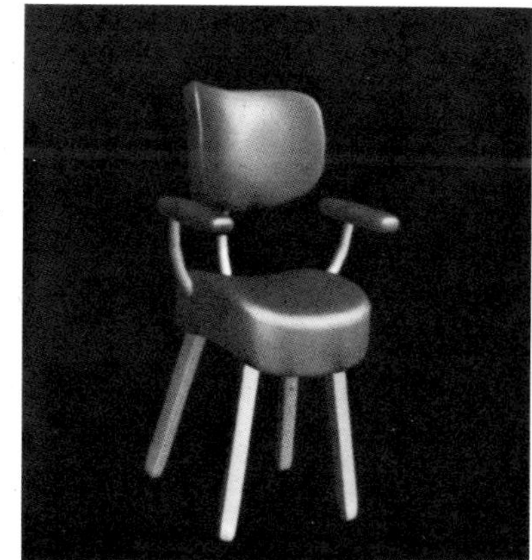

Figure 4.2
(Top left) A sampling of the basic forms allowed, (top right) deformations of these forms, and (bottom left) a chair formed from Boolean combinations of appropriately deformed superquadrics.

4.2.1 Our Representation of Part Structure

Inspired by the way people seem to describe shape, we have adopted a representation that describes shape in a similar manner—that is, by explaining how one would create a particular shape by forming and combining lumps of clay. The most primitive notion in this representation is analogous to a "lump of clay," a modeling primitive that may be shaped and deformed, but that is intended to correspond roughly to our naive perceptual notion of "a part." For this basic modeling element we use a parameterized family of shapes known as a *superquadrics*, invented by Danish designer Peit Hein [22, 23]. These are described (adopting the notion $C_\eta = \cos\eta$, $S_\omega = \sin\omega$) by the following equation:

$$\mathbf{X}(\eta, \omega) = \begin{pmatrix} C_\eta^{\varepsilon_1} C_\omega^{\varepsilon_2} \\ C_\eta^{\varepsilon_1} S_\omega^{\varepsilon_2} \\ S_\eta^{\varepsilon_1} \end{pmatrix}, \tag{1}$$

where $\mathbf{X}(\eta, \omega)$ is a three-dimensional vector that sweeps out a surface parameterized in latitude η and longitude ω, with the surface's shape controlled by the parameters ε_1 and ε_2. This family of functions includes cubes, cylinders, spheres, diamonds, and pyramidal shapes as well as the round-edged shapes intermediate between these standard shapes. Some of these shapes are illustrated in figure 4.2 (top left). Superquadrics are, therefore, a superset of the CSG modeling primitives that are currently in common use.

These basic "lumps of clay" are used as prototypes that are then deformed by linear stretching and tapering, or by quadratic bending, and finally combined through Boolean operations to form new, complex shapes. Our representation, therefore, is both a generalization of the CSG approach and a modification of the generalized cylinders approach; we are combining a restricted class of generalized cylinders using Boolean operations. We have found that this representational system has a surprising generative power that makes it possible to create a wide variety of form, as illustrated in figures 4.2 and 4.3.

We have constructed a 3-D modeling system, called "SuperSketch," that employs this shape representation. This real-time, interactive modeling system, implemented on the Symbolics 3600, allows users to create "lumps" interactively, to change their squareness/roundness, to stretch, bend, and taper them, and finally to combine them through Boolean operations. We have found that, with SuperSketch, users can quickly model a wide range

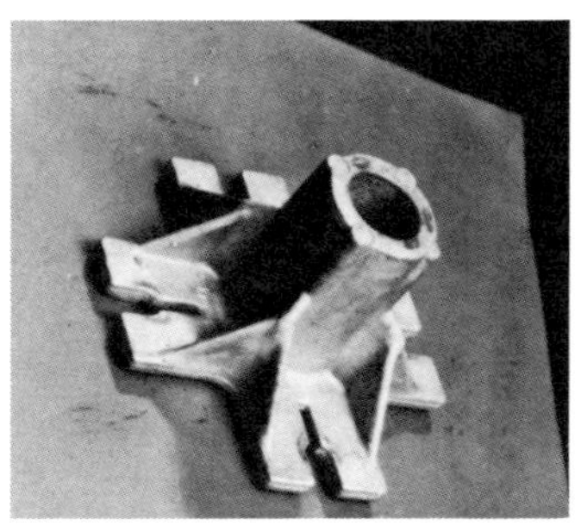

A

B

C

Figure 4.3
(A) An industrial casting, (B) its SuperSketch model [approximately 300 parameters; total elapsed construction time (user's time): 23 minutes], and (C) A SuperSketch model of two people (approximately 1,000 parameters; total elapsed construction time (user's time): about 4 hours)

of man-made and natural shapes in a way that correctly captures the intuitive part structure of the object. This system was used to make the images in this chapter.[1]

It is clear, however, that any such parameterized shape vocabulary is not capable of describing the surface shape in full detail. Rather, it provides a coarse-grain description of the *general* structure of the form. Having captured a first-order approximation to the overall, 3-D form by use of our shape language, it is now easy to use a 2-D surface description language (such as the thin-plate or fractal surface models) to account for the detailed surface shape in a viewpoint-invariant manner.

There are several advantages to describing shape by use of a combination of a global parameterized model together with a point-by-point surface model. First, separating our description into global and local representations nicely supports the differing requirements of various reasoning tasks. For instance, in calculating an object's dynamics or kinematics we need only know that the object is approximately cylindrical, while determining a gripping point requires knowing about local deviations from the overall cylindrical shape. Second, such a global-local representation allows a much more concise encoding of the detailed surface shape, because the larger-scale variations in the form are succinctly encoded by our simple parameterized shape vocabulary. And finally, having a global description is necessary to integrate positional information correctly when measurement errors are normal to the surface rather than parallel to the viewer.

4.2.2 Using Part Descriptions for Machine Vision

Perhaps the most important fact about this representation is that it can describe a surprisingly general class of 3-D shapes by use of only 14 parameters per part model.[2] This compares favorably with the 9 parameters needed to describe intensity, first, and second derivatives at a single point; the 9 parameters needed merely to describe the position, orientation, and size of a rectangular solid; or the hundreds of parameters that might be needed to describe an equally detailed generalized cylinder.

As a consequence we have found that even complex objects and scenes can usually be modeled quite concisely; for example, the models shown in figure 4.3 have only a few hundred parameters each. These modeling results have led us to believe that such "part" representations provide a concise, natural level of description, and thus may provide sufficient descriptive constraint that we can search for optimal shape descriptions without a fatal combinatorial explosion.

Given such a part representation, we must now develop techniques for recovering these "parts" from image data. There are two cases to consider: (1) the directed, top-down case in which we expect a certain object to occur near some particular location; (2) the bottom-up case in which we must compute a new description solely on the basis of raw image data. The following discussion will be with reference to range data only, as the primary sensor now being used by the ALV is a laser rangefinder.

4.3 Learning New Object Descriptions

Reliable, bottom-up learning of object descriptions is perhaps the most difficult task faced by a vision system. Certain characteristics of range data, however, make the problem much simpler. The primary simplifying characteristic is the fact that range imagery allows simple separation of figure from ground: one locates the ground plane, marks that data as "ground," and the remaining data as "figure." The technique we are currently using to separate figures from ground is described in [25]; briefly, is a matter of fitting a "rubber sheet" model to the surface observed in the range data and then finding those places where the surface changes elevation rapidly enough to "tear" a hole in the sheet.

Having identified patches of image data as obstacles that need to be described and having adopted a particular representation, we can now pose the problem of learning an object description as the process of using this shape vocabulary to find the "best" account of the data. That is, we can define the learning process as one of optimizing our description over our shape vocabulary relative to some goodness-of-fit criterion.

In this chapter we employ encoding length as our means of evaluating an explanation of the image data; in other words, our goodness-of-fit criterion is the cost (in bits) of encoding the image data by use of our 3-D shape vocabulary and a simple noise model. The total cost of encoding some image data will be calculated as being the sum of several subterms; these are the cost of specifying (a) the parts description itself, (b) the "noise" in the data (small deviations from the value predicted by the parts description), (c) errors of commission (points predicted that don't occur in the data), and (d) errors of omission (image points missed by the parts description). Thus the "best" explanation of some image data we shall take to be that encoding (in terms of part models and noise) that has the minimum sum of costs (a) through (d).

There are two motivations for adopting the minimal-length encoding approach. First, finding the minimal-length encoding of a scene is equivalent to finding the maximum a posteriori (MAP) explanation of the scene, where the assumed encoding costs are inversely proportional to the log of the prior probabilities. Thus by finding a minimal-length encoding of image data using our vocabulary, we actually recover the MAP estimate of both the large-scale image structure, which we shall model by use of our parts vocabulary, and the small-scale image structure, which we shall model as pointwise independent shot noise in the range sensor's distance measurement.

Second, such application of minimal-length encoding is the formal version of "Occam's Razor": the principle that the simplest explanation is the best explanation.[3] For instance, if we know the causally important "atoms" that make up our scene, then the minimal-length encoding of the scene in terms of those atoms is the one that provides the best available explanation of the scene's history. This technique is exactly what geologists or paleontologists use when reconstructing the folding of rock strata or development of a family of animal species.[4]

Both the MAP and Occam's Razor motivations for the minimal-length encoding approach require that we employ a shape vocabulary that describes the scene naturally, i.e., in a way that captures the structural distinctions needed by subsequent reasoning processes. If we choose our vocabulary correctly, therefore, then a minimal-length encoding using that vocabulary will provide us with a meaningful segmentation of the data. This fact provided the motivation for our choice of a shape vocabulary based upon psychological evidence concerning people's notions about the intrinsic structure of 3-D objects [10–18].

The primary difficulty in computing a minimal-length encoding is that it requires global optimization of a fitting function and, unfortunately, there are no efficient, general-purpose global optimization techniques for such nonlinear problems. We can, however, take advantage of the special properties of this particular problem in order to achieve an adequate solution.

The first property we can take advantage of is that we can decompose our search for the best explanation into two phases: a local phase and a subsequent global phase. We do this because the part models in our shape vocabulary are compact, with surfaces that are opaque to the sensor. Thus, changes that are far enough from a particular image point do not affect the description at that point. For instance, if the largest element in our shape vocabulary has a projected radius of 50 pixels, we do not have to look more than 60 or 70 pixels in any direction in order to find the part model that provides the best fit in the immediately surrounding image region.

The second property we can take advantage of is that the search for a "best fit" within a particular image region seems to be convex for parameter values near the optimal solution. Figure 4.4 (top) illustrates rendering a range image of a bananalike SuperSketch model. Figure 4.4 (bottom left) shows the comparison of this predicted image appearance to the measured range data; by histogramming the difference between predicted and measured range we compute a local estimate of our encoding length "goodness-of-fit" functional (explained in more detail below). Figure 4.4 (bottom right) shows how this functional changes as the parameters of the SuperSketch model is varied. At the center of each graph in figure 4.4 (bottom right) is the exactly matching parameter value; it can be seen that our "goodness-of-fit" functional varies slowly as we move away from the correct parameter value (note that these graphs do not show the functional's value over the entire parameter range).

Figure 4.4, therefore, illustrates that the problem of finding the best fit is convex, at least at this point in the parameter space. By repeating the experiment illustrated by figure 4.4 at points scattered throughout the entire parameter space we have been able to show that this locally convex behavior obtains uniformly. The convex region surrounding the correct parameter setting is often quite broad: For instance, we can usually vary length, width, or depth by more than 50% before we leave the convex region of the parameter space. In contrast, the "goodness-of-fit" functional is relatively sensitive to orientation: We sometimes can vary rotation angles by only 10 or 15° before leaving the convex region.

These two properties, together with the small number of parameters in our modeling primitives, mean that we can use a coarse-grained search over the entire parameter space as our optimization procedure. That is, by comparing each combination of parameter settings with the image data surrounding an image point, we can find a part model that provides the best *regional* fit to the image data. We have found that, if the (x, y, z) position parameters are excluded, about 2^{16} "goodness-of-fit" evaluations are required to search the entire parameter space adequately, sampling most parameters at three different values (e.g., object widths of 10, 20, and 40

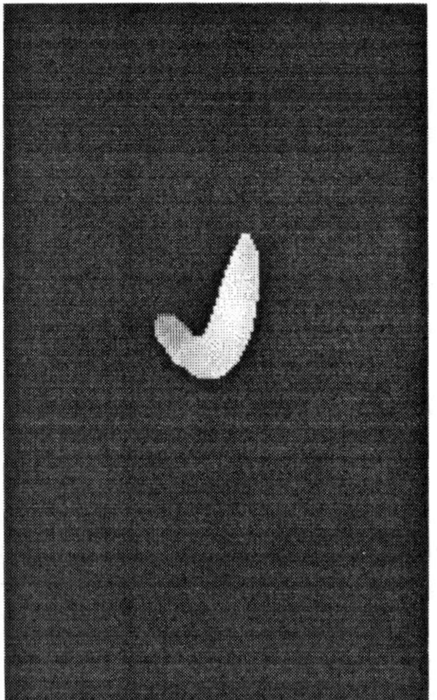

Figure 4.4
Finding the best fitting part descriptor. (Top) We take a 3-D SuperSketch part with the hypothesized parameters (e.g., orientation, length, squareness), and render it using known sensor characteristics to produce a predicted image $R^*(x, y)$. (Bottom left) We then histogram point-by-point differences [i.e., $R^*(x, y) - R(x, y)$] between the R^* and the range data $R(x, y)$; the size of the largest peak is used to calculate our goodness of fit measure. (Bottom right) The fit between these range data and a 3-D part model as the parameters of the 3-D part are varied; the correct fit occurs at the center of each graph.

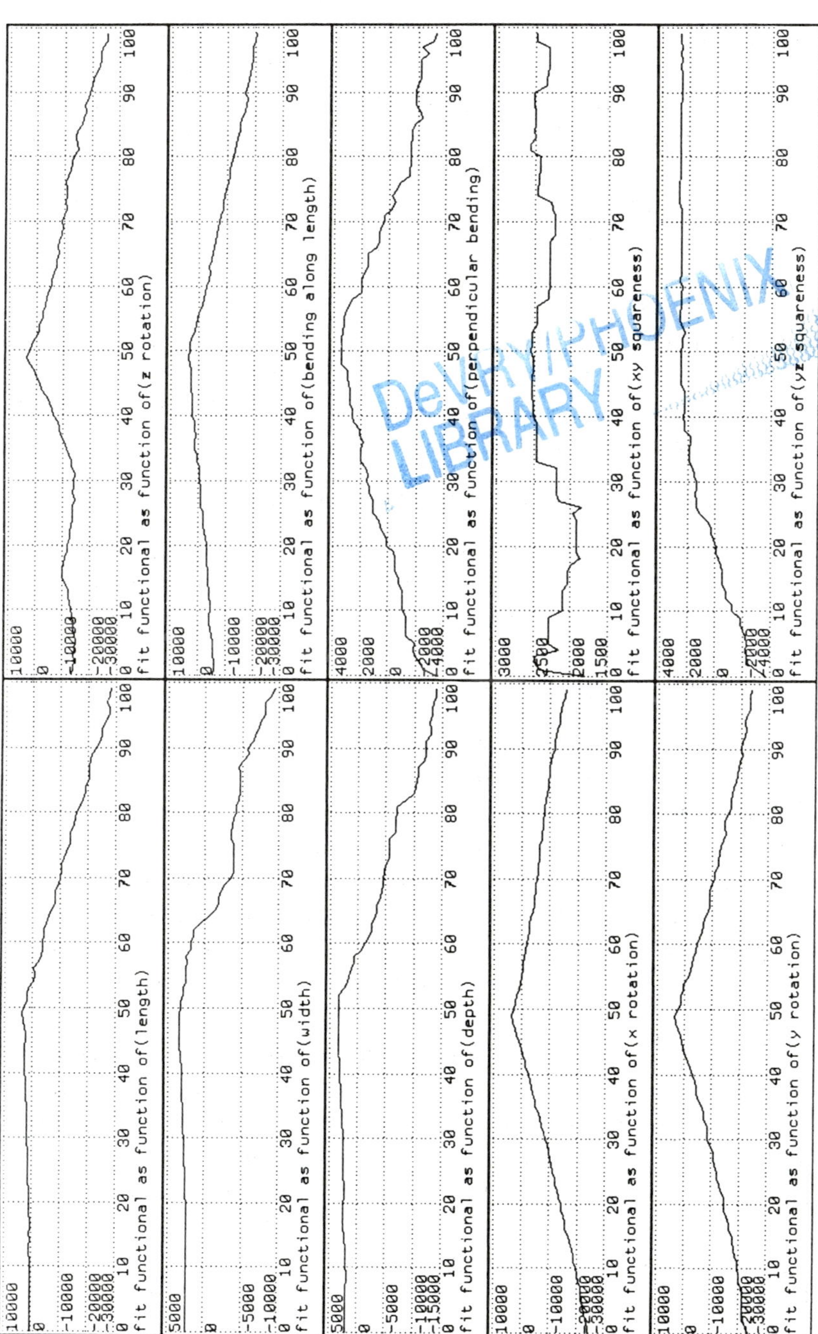
fit functional as function of(length)
fit functional as function of(width)
fit functional as function of(depth)
fit functional as function of(x rotation)
fit functional as function of(y rotation)
fit functional as function of(z rotation)
fit functional as function of(bending along length)
fit functional as function of(perpendicular bending)
fit functional as function of(xy squareness)
fit functional as function of(yz squareness)

inches) and some critical parameters, such as orientation, more frequently (e.g., every 22.5°).[5] Note that all of these evaluations can be carried out in parallel, so that given massively parallel hardware (such as the Connection Machine, which has 2^{16} processors) we can use this method to find the best-fitting SuperSketch model in a very short period of time.

By picking the parameter setting that provides the best regional encoding of the data for *each* [coarsely quantized] image position, we obtain a relatively small set of part models that together closely describe all the image data. We then choose from among this set of regional fits a subset of part models that provides the best explanation (i.e., the shortest encoding) of the image data, and, finally, conduct a gradient descent on this final ensemble of part models to obtain a good approximation to the minimal length encoding of the image data.

4.3.1 Evaluation of the Goodness-of-Fit Functional

One of the key elements assuring the success of this approach is evaluation of the localized goodness-of-fit functional in a manner that is insensitive to occlusions and that, in addition, takes into account all the information we have about edge placement, surface shape, perspective effects, and sensor characteristics. The procedure we use is to

1. Construct a 3-D SuperSketch part with the hypothesized parameters (e.g., orientation, length, squareness).
2. Render that part by using known sensor characteristics, thereby constructing a range image that correctly accounts for the effects of perspective, surface shape, and other known variables that affect image appearance. By using a rendering technique that includes a full camera and sensor model, we make *explicit* all of the edge, surface, perspective, and other relations that are normally *implicit* in our model parameters. Steps (1) and (2) are illustrated by figure 4.4 (top).
3. Histogram all of the point-by-point differences in depth between the rendered part and the image data. This produces a histogram with "buckets" at each possible offset between the hypothesized part's depth and the actual depth, so that the value in each bucket is the number of pixels at that particular offset. While doing this, we also keep track of the number of pixels ε that fall off the figure entirely (when the "figure" of interest can be separated from the surrounding "ground," as is generally possible with range data). By using a RANSAC-style [24] histogramming procedure, we

allow large portions of the figure to be occluded without disturbing the matching process. This is illustrated by figure 4.4 (bottom left).

4. From this histogram we estimate the position p of the largest peak. This peak is the most frequently occuring distance between the hypothesized and the measured surfaces; the number of counts in this peak is the area (in pixels) of the hypothesized part's surface that would match the image data if the part were moved in depth by a distance p. By using buckets of width σ, therefore, we can employ this histogramming technique to determine the number of pixels μ that would match to within $\pm\sigma$ at the optimal depth positioning of the hypothesized part.

5. Compute the value of the goodness-of-fit functional Γ for this set of parameters:

$$\Gamma = \mu - \lambda\varepsilon. \tag{2}$$

In our examples we have used $\sigma = \lambda = 4$; the procedure seems to be relatively insensitive to the value of these parameters.

The measure μ gives us the number of pixels accounted for by this hypothesized part to within $\pm 1/2\sigma$, and the measure ε gives us the number of pixels that are errors of commission. As a consequence the quantity

$$s = -\kappa_0 + (8 - \kappa_1)\Gamma \tag{3}$$

estimates the amount of savings (in bits) gained by encoding 8-bit range data using the hypothesized part parameters, where κ_0 is the bit cost for describing the hypothesized part (assumed constant for all parameter settings), κ_1 is $\log(\sigma)$, the average cost of encoding the difference between the range data and the hypothesized surface (data noise errors), and λ is set to the cost of errors of commission divided by $(8 - \kappa_1)$.

Thus we can determine the *localized* minimial-length encoding by maximizing Γ over all of our model parameters. Having accomplished this at each image point, we can then use these localized encodings to compute a global minimal-length encoding.

4.3.2 Finding the Global Minimal-Length Encoding

To determine the global minimal-length encoding, we choose from among these localized minimal encodings a subset that best describes the image. That is, we must find the set of part models that minimizes the combined cost of part description, noise errors, and errors of both commission and omission. This task is difficult because the localized part models are not independent; they overlap and obscure each other, so that encoding cost for the set is not closely related to the sum of the individual encoding costs.

In general, then, we must consider n-ary relationships in order to find the global minimal-length encoding. Binary relationships, however, constitute almost all of the part-to-part interactions in calculating the minimal-length encoding, because it is relatively rare for three or more parts to overlap across a significant area. Thus if we can correctly account for binary relationships we can produce an encoding whose length is close to that of the global minimal-length encoding. Luckily, accounting for binary relationships turns out to be relatively easy.

We have four types of cost to consider in computing the minimal-length encoding. The first is the part description cost, which is a constant number of bits for each hypothesized part descriptor. Thus there are no part-to-part interactions for this portion of the encoding cost. The second is the noise cost, the cost of encoding a pixel value relative to the part's surface. Here we have binary interactions, because the same pixel can be covered by more than one part. Similarly, in calculating the cost of errors of commission we must take binary relationships into account, because two parts can overlap the same pixel.

To account for these binary relationships we form what we call area *matrices*. We let $\mathbf{M}_a$ be a matrix whose ith diagonal element is the number of pixels covered by the ith part. The (i, j)th, $i \neq j$, elements of this matrix are minus one-half the number of pixels covered by *both* the ith and jth parts. A similar matrix, $\mathbf{M}_c$ is formed for the pixels that are errors of commission.

Let us describe a particular set of models by the vector $\mathbf{X}$, which has a one in the ith slot if model i is a member of the set, and zero otherwise. To calculate the total cost of encoding image data by use of the set of part models $\mathbf{X}$ we evaluate

$$E(\mathbf{X}) = \kappa_0 \mathbf{X}\mathbf{X}^T + \kappa_1 \mathbf{X}\mathbf{M}_a\mathbf{X}^T + \kappa_2 \mathbf{X}\mathbf{M}_c\mathbf{X}^T + \kappa_3(N - \mathbf{X}\mathbf{M}_a\mathbf{X}^T), \quad (4)$$

where κ_0 is the cost of describing one part model, κ_1 is the cost of encoding a pixel to within $\pm 1/2\sigma$, κ_2 is the cost of encoding an error of commission, κ_3 is the cost of encoding an error of omission, and N is the number of nonzero pixels in the image.

By minimizing equation (4) we can thus approximate the minimal-length encoding of the image data using our shape vocabulary.

Equation (4) is a quadratic form in $\mathbf{X}$; the minimum value of this equation can be found by gradient ascent as long as the combined matrix

$$\mathbf{M} = \kappa_0 \mathbf{I} + (\kappa_1 - \kappa_3)\mathbf{M}_a + \kappa_2 \mathbf{M}_c \quad (5)$$

is negative definite. This condition obtains whenever the part models do not overlap each other greatly, i.e., whenever our initial local encodings are reasonable.

The major difficulty in minimizing this equation is that the values of $\mathbf{X}$ must be either zero or one. In the following examples we have used a simple iterative, gradient descent algorithm. This equation can, alternatively, be solved by such diverse methods as relaxation, Hopfield-style neural nets, or integer programming techniques.

By minimizing equation (4) we obtain a set of part models that furnishes a reasonable approximation to the minimal-length encoding of the image data in terms of our part representation. We have found, however, that it is useful to perform a final step of optimization on all of the parameters of this final set of parts. We accomplish this final optimization by means of a numerical gradient descent algorithm that renders all the hypothesized part models together (thus completely accounting for occlusion relations) and computes the encoding cost; the algorithm then changes one of the part's parameters and ascertains whether the cost is improved. If improvement does indeed occur, the change is accepted.

4.3.3 Practical Considerations

Straightforward implementation of the above operations requires about 10^9 operations per image region. The number of operations required can be reduced considerably by pruning the search space during the search, in a manner similar to that used by Bolles [5], Goad [7], Faugeras et al. [8], or Grimson and Lozano-Pérez [9] in their model-based, global-search-and-match vision systems.

In our current implementation, this pruning is accomplished by keeping track of the current n best fits (largest Γ values) within an image region and by using their Γ values to (1) abort evaluations of Γ as soon as it becomes clear that the eventual value will be smaller than any of the current n largest Γ values (e.g., when, even if all of the remaining pixels match exactly, Γ will still be smaller than the current n best Γ values), (2) discard any parameter setting that a priori cannot generate a value of Γ that is larger than one of the current n best Γ values, and (3) order the search so that the above pruning techniques will be maximally effective (e.g., search over the parameter settings with the largest potential Γ values before searching over other parameter settings).

By taking advantage of these and other efficiency expedients, the examples shown here have required an average of about 10^{10} operations each, roughly $2\frac{1}{2}$ hours of CPU time on a Symbolics 3600. For industrial applications, in which bending and tapering are not typical, the search space is smaller and therefore the required computation time can be as much as 1/30th that of the full algorithm. Because of the inherent parallelism of the technique (thousands of identical searches within each region) a full global search is expected to take approximately 1 sec per image with today's large, parallel computers.

Perhaps more important, however, a fully developed system would also make use of such image features as edges and corners to prune the search space. Additionally, hierarchical coarse-to-fine techniques could be used to guide the search, thus improving efficiency and perhaps eliminating the need for gradient-descent improvement at the end of the global parameter-space search. We have not as yet had time to explore these possibilities.

4.4 Learning Results

One of the major advantages of the above global search technique is the certainty of convergence to a good answer (in terms of accounting for the metric properties of the data) within a fixed period of time. It may be more important, however, that we obtain a stable account of the object's *part structure*, for this is what we shall use to determine the object's class identity (e.g., "gate," "car") for purposes of reasoning. Whether a model that accounts accurately for metric properties also accounts for part structure depends upon whether our shape vocabulary actually expresses a robust aspect of the object's true 3-D structure, as well as whether that structure is conserved under projection. This issue is one we shall address in this section.

The following examples involve range imagery; for purposes of evaluation we present one synthetic example, followed by three examples using the ALV's laser rangefinder. One characteristic of range data is that it is generally easy to obtain a rough "figure/ground" separation [25]; we have taken advantage of that ability. Some simple preprocessing of the laser rangefinder data was used to remove mixed-range pixels (i.e., pixels for which the laser beam, at the time a measurement was being made, illuminated 2 or more objects at different ranges) as well as the inherent ambiguity-interval problems of those data [25, 26].

4.4.1 A Synthetic Data Example

The first example uses simulated range data, with approximately six-bit resolution. The purpose of this example is to demonstrate (1) the performance of the algorithm independent of special data characteristics and (2) the ability of the technique to recover part structure despite the problems of scale and configuration.

Figure 4.5A shows a SuperSketch model (here and in succeeding figures we shall show side views of SuperSketch models as insets placed in the lower-right-hand corner of the frame surrounding the model); Figure 4.5B shows a range image generated from this model. This is a fairly accurate model of the articulated human form; as such, it illustrates the necessity for a part-structure representation of the overall shape: without such a representation, we would have to store descriptions of every possible positioning of the figure in order to recognize it as it moves about. Figure 4.5C shows the initial explanation of the image data, found by our local, parallel search technique applied to points sampled along the figure's 2-D skeleton,[6] followed by our procedure to select an optimal subset from among the regionally best-fitting part models.

The main thing to note about this initial shape description is that, although not perfect, it seems sufficiently similar to the original generating description that we can use it to index into a database of known forms and to recognize the figure as human.

Figure 4.5D depicts the final learned model, the result of gradient descent from figure 4.5C. Figure 4.5E shows "blow-up" views of both the original and the learned model. Note the similarity of part structure and of part-by-part parameters.

Perhaps the major question to be asked about this procedure for recovering part structure is whether or not the recovery is *stable*, since stability in structuring the image data—i.e., in producing a *segmentation*—is the primary property required for reliable higher-level reasoning functions such as class formation and generalization [16, 35, 36]. Some of the particulars of this example are illustrative in this regard.

Note, for instance, that the feet are described by bent primitives that account for both ankle and foot, even though in the final model the "ankle"

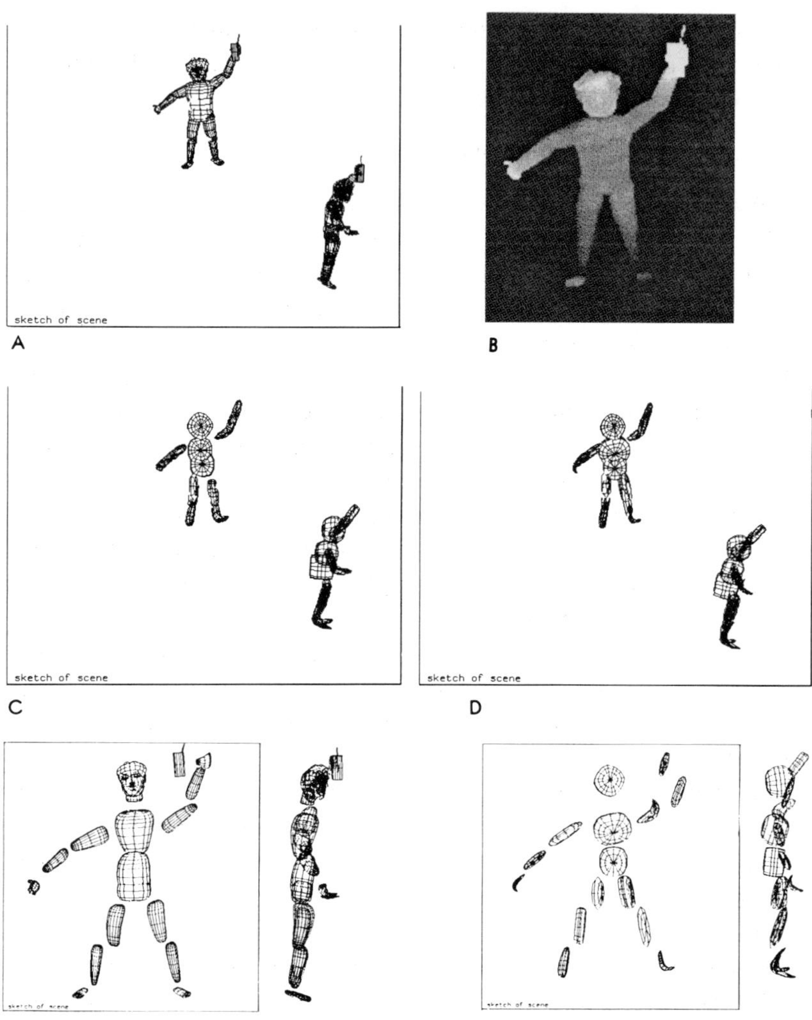

Figure 4.5
(A) A SuperSketch model, (B) a range image generated from this model, (C) initial explanation of the image data, (D) final learned model, and (E) "blow-up" views of the original model and the learned model. Note the similarity of part structure and part-by-part parameters.

part is not visible, having been occluded by the "calf." Such fitting of a bent primitive to two unbent parts is also observed in the right arm—here too, the upper part is occluded, by the "forearm." Although these assignments of part structure are perhaps not perfect, they are entirely plausible segmentations when only one view is given. Such occasional merging of connected primitives seems unavoidable with only one view; thus, we must allow for *both* possible descriptions—as one bent primitive and as two straight but connected primitives—when forming object classes or searching for similar stored models.

A more interesting case occurs in the recovery of descriptions for the hands and head, for although both head and hands are actually quite complex shapes, they are recovered as being a single, undifferentiated part. These examples show the effect of *scale*; when the image features become smaller than the range of scales searched, there is a sort of "summarizing" effect as a fit is attempted to the overall composite form. Marr and Nishihara [37] pointed out the need for this type of "summarizing"; they proposed that we must employ a *multiscale* representation in comparing the learned model with stored models. In this example we can see how a multiscale representation, with descriptions for each distinct scale of part structure, might be combined with this recovery procedure to resolve some of the difficult problems associated with scale.

4.4.2 Learning Descriptions Outdoors

The remaining examples make use of data from the ALV's ERIM time-of-flight laser rangefinder. This rangefinder, which collects a 256×64 pixel image in 0.4 sec, has a useful range of about 128 feet and an advertised accuracy of about 5%. Its unusual imaging geometry is similar to that of a very-wide-angle lens.

Figure 4.6A shows a range image of the upper part of a person, taken with this sensor. This example is interesting, especially in comparison with the synthetic data example above, in that the amount of depth information within the figure is negligible; from a practical point of view, this is merely a silhouette. Figure 4.5B shows the initial explanation of the image data, while figure 4.6C displays the result of gradient descent from figure 4.6B.

Perhaps the most salient point of this example is that a reasonable 3-D part structure can be learned even from what is essentially only silhouette data; the left (upraised) arm, head, and right arm are clearly present in the learned description. Thus our learned description provides a plausible

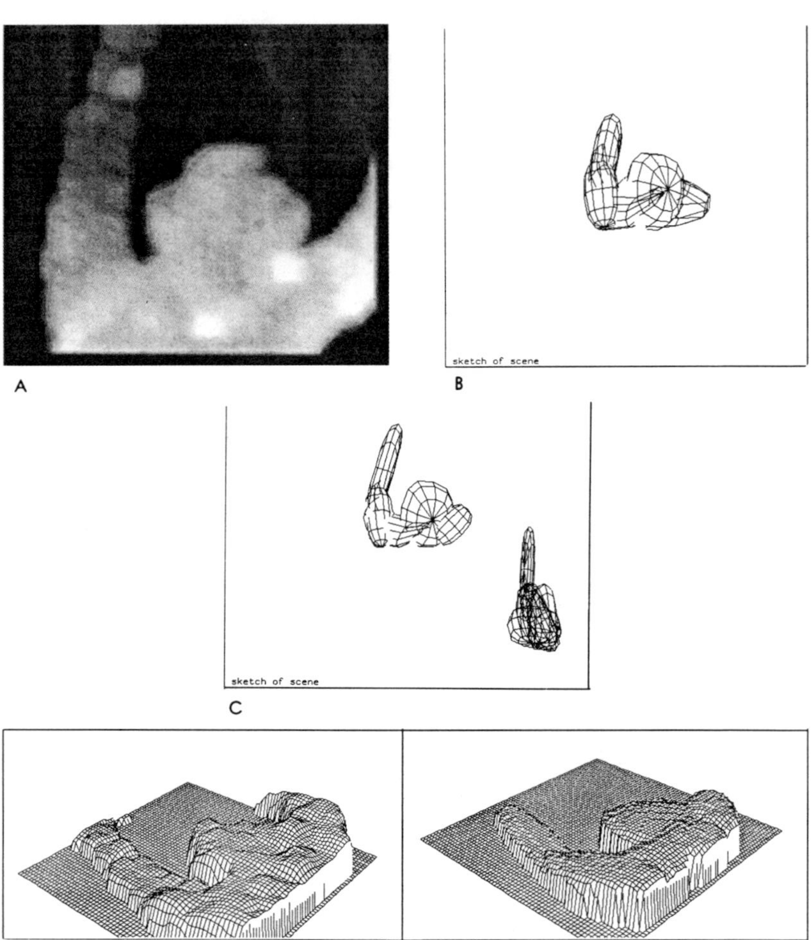

Figure 4.6
(A) A range image of the upper part of a person (for practical purposes this is merely a silhouette), (B) the initial explanation of the image data [note that the correct part structure of his left (upraised) arm, head, and right arm are clearly present], (C) the final learned model, and (D) comparison of the original range data with a range image produced by the final learned model.

segmentation of the measured range data. Figure 4.6D shows a comparison of the original range data with a range image produced by the learned model of figure 4.6C. This comparison shows that the simple (70-parameter) learned description provides an accurate encoding of the data's original metric information, demonstrating descriptive adequacy.

Figure 4.7A shows a range image of a gate by the side of the road. Again, the data contain little more information than a silhouette; the linear elements of this figure average two pixels across. Note that our figure/ground procedure has inadvertently included a bush near the left-most gatepost as part of the figure. The most interesting aspect of this example is the small size of the imaged features; these data thus provide a severe test of noise sensitivity.

Figure 4.7B shows the initial explanation of the image data. Once again a reasonable part-structure description is learned, although the left gatepost is seen as a block reflecting the fact that the data include a bush as well as the gatepost. Figure 4.7C contains a comparison of the original range data with a range image produced by the learned model of figure 4.7B; here too, the learned description provides a good encoding of the data's metric information.

One of the advantages of having a high-level description like this part language is that it provides good "hooks" for applying domain-specific knowledge. This is illustrated by figure 4.7D, which shows the result of "prettifying" figure 4.7B by making use of the domain knowledge that nearly horizontal or vertical parts are likely to be horizontal or vertical. Figure 4.7E contains a comparison of the learned model of figure 4.7D with a manually constructed SuperSketch model of the gate. The thickening of the posts and bars in the learned gate can be largely attributed to preprocessing intended to remove mixed-range pixels.

Figure 4.8A shows a range image of a few roadside bushes. The most important aspect of this example is that it is *not* an example in which there is an obvious part structure; nor is it an example with smooth surfaces. These data, therefore, allow us to examine some of the limits of our technique's descriptive adequacy, as well as its ability to produce stable segmentations. Figure 4.8B shows the initial explanation of the image data, as found by iterative, best-first search among the best fits at points in a 10×10 grid covering the image data.

Figure 4.8C shows a range image produced by the learned model of figure 4.8B; the outlines of figure 4.8A and 4.8C can be seen to be similar,

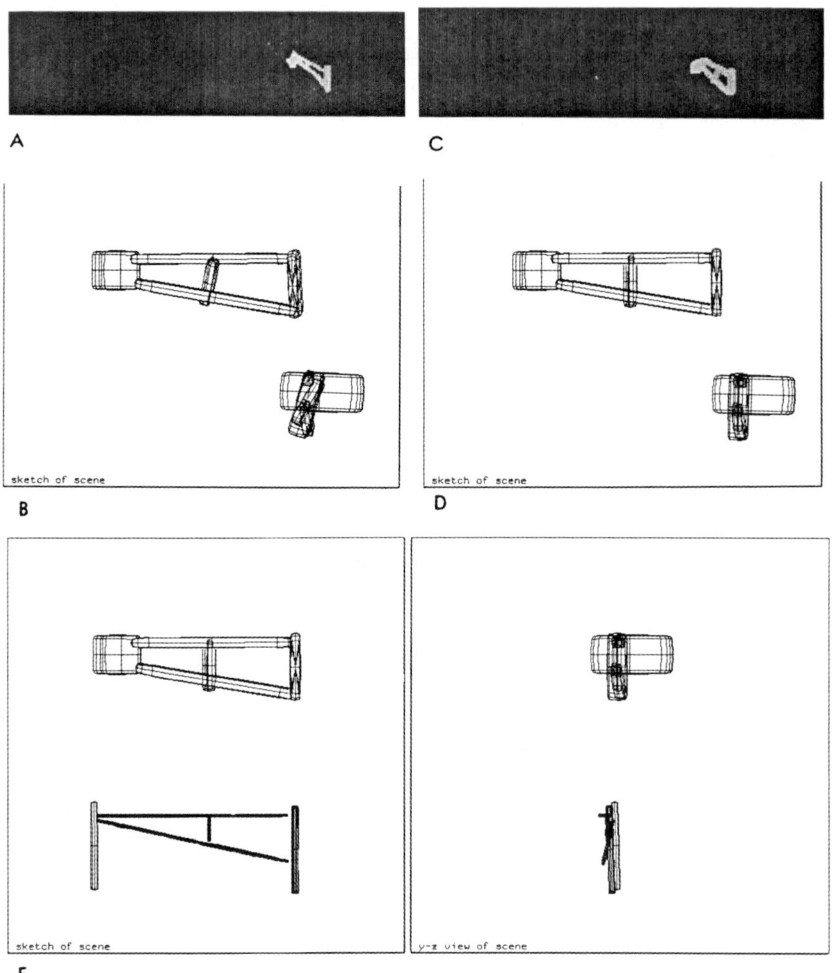

Figure 4.7
(A) A range image of a gate by the side of the road (our figure/ground procedure has unintentionally included a bush near the leftmost gatepost as part of the figure), (B) the final learned model, (C) a range image produced by the learned model, (D) the result of "prettifying" (B) using the domain knowledge to the effect that nearly horizontal/vertical parts are likely to be horizontal/vertical, and (E) comparison of the learned model of (D) with a SuperSketch model of the gate that was constructed by hand.

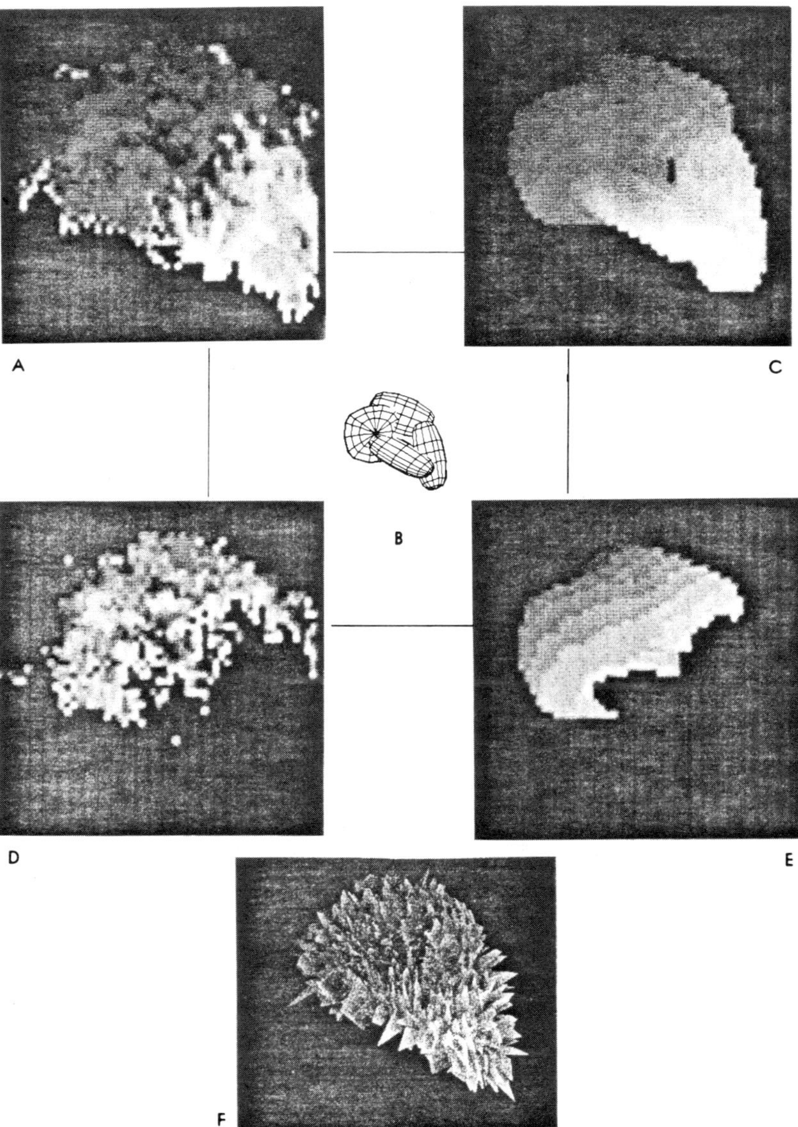

Figure 4.8
(A) A range image of a few roadside bushes, (B) final learned model, (C) a range image produced by the learned model, (D) the original range data, with points closer than the median distance removed, (E) a range image produced by the learned model again with points closer than the median distance removed, showing the model's close match with the range data's internal structure, and (F) range image produced by adding a fractal surface model to the part representation.

thus confirming that the learned description (in this example a total of 56 parameters) retains most of the data's original metric information. Figure 4.8D shows a comparison of the original range data, with points closer than the median distance removed and a range image produced by the learned model of figure 4.8B, again with points closer than the median distance removed. This comparison shows that the learned model does in fact provide a good encoding of the measured range data.

This example shows that even very complex shapes can be usefully "summarized" by our shape vocabulary, thus supporting our dual claims of descriptive adequacy and stable part structure recovery. We can improve the accuracy of the learned model somewhat by modeling the discrepancies between the recovered part structure and the range data by using a fractal surface model [27, 28]. The result of adding a fractal surface model to the recovered part model is shown in figure 4.8E. A considerably better description of the differences between part model and image data could be obtained by using a particle model [33] of the bush's branches and leaves.

4.4.3 Summary

We have described a representation that is fairly general purpose, despite having only a few parameters.[7] Capturing this expressive power with a small number of parameters has allowed us to approach the problem of learning object descriptions using the model-based vision technique of global search-and-match. On the basis of our experiments we believe that our method is robust; our examples, for instance, confirm that the recovery procedure is reasonably stable with respect to noise and scale.

This system does not now make use of feature-level cues either to determine part structure or to estimate parameter values. Nor does it make use of hierarchical, coarse-to-fine strategies to guide the search. It seems clear that adding such capabilities would only enhance the efficiency and robustness of the algorithm.

4.5 Recognizing Objects

The previous section showed that, by using a part representation, one could compute an accurate description of interesting objects from ALV range imagery in a bottom-up way. Having accomplished the task of learning the position and description of various landmarks, our vehicle is now faced with the very different problem of using these learned descriptions to navigate.

In the navigational problem the flow of information is largely top-down, in contrast to the learning task in which the information flow was almost exclusively bottom-up. This is because in the navigational task, we already know the landmark's approximate position, and have a description of it in our database. Thus, we can simply "monitor" the appearance of the landmark—that is, we can update our knowledge of vehicle position and landmark structure through a simple gradient descent procedure that compares predicted appearance with measured appearance, adjusting position and structure to minimize any discrepancy.

The main requirements for our recognition procedure, therefore, are that it be able to (1) recognize the landmark from any vehicle viewpoint and (2) handle small errors in predicted orientation (e.g., $\pm 20°$), predicted position (e.g., ± 15 feet), and our object's description (e.g., ± 2 feet in height or relative position). If our recognition algorithm fails, it should be capable of notifying us, so that we can invoke the more expensive bottom-up learning procedure, or, alternatively, a hybrid procedure that searches for parts that are roughly similar to those that were expected.

Such a monitoring function has the advantage that it can be quite efficient: the algorithm described here, for instance, could be executed in a few dozen seconds on a Symbolics 3600. The more computational expensive task of learning objects would only be required in unknown terrain.

4.5.1 Recognition by Image-Level Prediction

The technique we are using at present for landmark recognition is based on predicting the appearance of each of the landmark's component parts in the range data. In most machine vision applications data-level matching is risky because it is difficult to predict sensor values. However, since this sensor produces range (instead of reflected intensity), it is plausible to match directly to the raw image values.

Even though range prediction is more straightforward than intensity prediction, it is still very important to use a detailed model of the sensor in the prediction process. We currently take the following variables into account:

- the range sensor's spherical geometry,
- the motion of sensor (as measured by an on-board internal navigation system),
- "mixed pixels" along object boundaries, which we account for by predicting image appearance at twice the normal resolution and then subsampling appropriately, and
- sensor ringing and blur, again accounted for by appropriate subsampling of a super-resolution predicted image.

By taking these effects into account we can make range predictions that are sufficient for correlation-type matching.

Using this sensor model, we construct two "masks" that give (1) the expected pixel-by-pixel range values for each *part* in our model of the object and (2) the pixel-by-pixel expected variance for the predicted range values. We then perform individual searches for each *part* in our model of the object, using the first of these masks to compute the sum-of-squares difference between the expected range values and the actual range values, with the second mask serving as a set of weights for combining these differences to form a weighted sum-of-squares fit for each part at every point in the image. We then factor in our knowledge of the vehicle's position and the object's interpart geometry to obtain least-mean-square-error estimates for object's location and geometric structure.

By modeling the error in position, sensor sampling, quantization, and noise, we are able to account for the primary known sources of error in our model of the landmark's appearance. Perhaps even more important, however, is the fact that by performing the search on a part-by-part basis enables us to minimize the effects of unknown *global* distortions, such as those that arise from small deviations in viewing angle or from errors in the model's geometry. This is because parts are smaller and have less structure than the entire object model, and so their projected shape tends to be relatively unaffected by global distortions. Thus, by performing the search on a part-by-part basis, we attain much of the robust, qualitative descriptive capabilities of, for example, Konderink's catalog of characteristic views [38].

4.5.2 Detailed Recognition Strategy

In the learning examples reported above, there was no need to incorporate a priori information; it was all done by strictly bottom-up processing. However, in the case at hand, the primary concern is how to integrate top-down, a priori knowledge of vehicle position, sensor characteristics, and object structure with sensor data to obtain the best possible estimates of object location and structure.

The approach we have taken is to model uncertainties concerning our a priori knowledge by using the first and second moments of the distribution of errors. That is, we characterize our knowledge by use of mean, variance, and covariance statistics. Our choice of this representation for uncertainty is motivated by the availability of clear, simple combination rules, as well as by the small amount of information required to characterize the uncertainty. (See references [29, 30] for more details regarding this approach to modeling uncertainty.)

Adopting this representation for uncertainty, our overall recognition strategy can be stated as follows:

4.5.2.1 Prediction of Model Position The first step is to use the ALV's internal navigation system and our knowledge of object position, together with a model of the uncertainties in these data sources, to predict the mean, variance, and covariance for the location of each of the object's parts in the range imagery. We use this information in three ways: (1) to select an image region within which we shall search for the part, (2) to predict the most likely location of each part *i*, and (3) to predict the variance of that prediction.

Figure 4.9 illustrates the prediction process. It shows an ALV image with an ellipsoid superimposed upon the range data. The ellipsoid represents the expected range of positional uncertainty associated with one of the gate's components. See reference [26] for additional details.

4.5.2.2 Prediction of Range Values We use our object and sensor models to predict both the pixel-by-pixel range values for each part and the

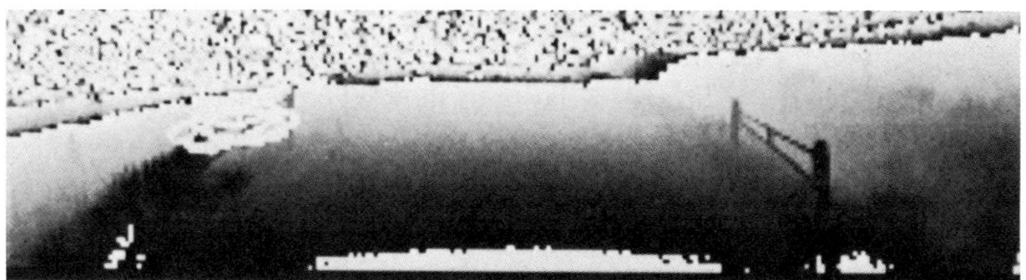

Figure 4.9
Prediction of gate position with error ellipses projected onto actual ALV data.

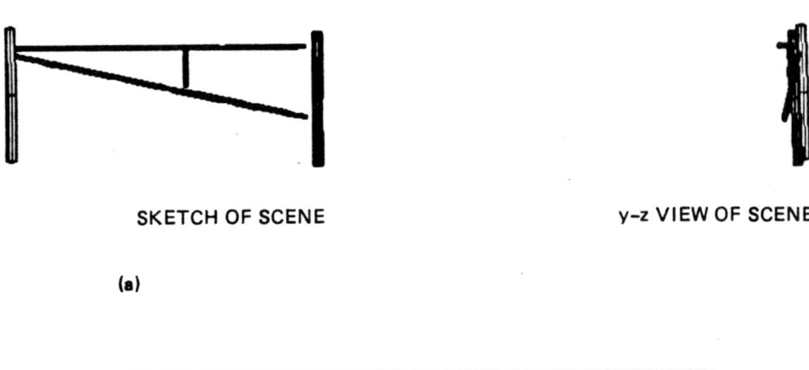

Figure 4.10
(a) A model of a gate constructed using SuperSketch and (b) an intensity image illustrating the gate model placed on a digital map (DTM).

variances associated with these values. Figure 4.10 shows a SuperSketch model of the gate shown in figure 4.9, together with an intensity image of this gate model placed on a digital terrain map (DTM).

By combining the SuperSketch model, the DTM, and the range sensor model we can predict the range values the ALV would measure at a specific point in space. This is shown in figure 4.11b, along with the actual ALV range data in figure 4.11a; it can be seen that the prediction is reasonably accurate.

Finally, figure 4.11c depicts the variance "masks" produced; these are arrays that hold the expected variance of the range predictions shown in figure 4.11b. In this illustration, light areas are predicted to have relatively low variance, while dark areas are predicted to have relatively high variance. It can be seen that pixels falling entirely on the object are predicted to have small variances, pixels falling on the surrounding ground are predicted to have medium-sized variances, and the mixed pixels along the part's edge are predicted to have large variances.

4.5.2.3 Computing Sums of Squares Our objective is to find the part centers (x_i, y_i, z_i) that minimize (1) the sum of the squared errors between the predicted range values and the measured range values, (2) the squared error between the final interpart center-to-center distances and the predicted interpart center-to-center distances, and (3) the squared error between the predicted part locations and the final estimated part locations.

As we change the hypothesized location of the part, our range predictions change. For a single part, however, the change is almost entirely translational; this is because single parts tend to be compact and smoothly featureless. We can thus save an enormous amount of computation at very little cost in accuracy by making the assumption that the appearance of a single part is invariant under small changes in viewpoint. This allows us to predict range values and variances once for each part, and then simply displace our predictions in x, y, and z so as to evaluate the fit between our range predictions and the measured data for each possible part location in the volume shown in figure 4.9.

Let z_i be the predicted range at the center of part i and $\Delta z_{i,dx,dy}$ the predicted difference in range between the center of part i and the pixel displaced $(\Delta x, \Delta y)$ from (x_i, y_i, z_i), the part's imaged center. Then our predicted range values $z_{i,x,y}$ are simply

$$z_{i,x,y} = z_i + \Delta z_{i,\Delta x,\Delta y}, \tag{6}$$

where $x = x_i + \Delta x$, $y = y_i + \Delta y$.

The first stage of our algorithm, therefore, is to calculate the squared error $s^2_{x_i,y_i,z_i}$ between the predicted range $z_{x,y}$ and the measured range $z^*_{x,y}$ when part i is centered at location (x_i, y_i, z_i):

$$s^2_{x_i,y_i,z_i} = \frac{1}{n^2} \sum_{\Delta x} \sum_{\Delta y} \left(\frac{z^*_{x_i+\Delta x, y_i+\Delta y} - z_i - \Delta z_{i,\Delta x,\Delta y}}{\sigma_{i,\Delta x,\Delta y}} \right)^2, \tag{7}$$

where $\sigma_{i,\Delta x,\Delta y}$ is our estimate of the variance of our range prediction at the pixel displaced $(\Delta x, \Delta y)$ from part i's center. Note, however, that instead of searching over all x_i's, y_i's, and z_i's we can "factor out" z_i because there is a unique z_i^{opt} that produces the minimum squared error for each image location (x_i, y_i). Thus we can simplify the above computation by instead finding

$$s^2_{x_i,y_i,z_i^{opt}} = \min_{z_i} \frac{1}{n^2} \sum_{dx} \sum_{dy} \left(\frac{z^*_{x_i+\Delta x, y_i+\Delta y} - z_i - \Delta z_{i,\Delta x,\Delta y}}{\sigma_{i,\Delta x,\Delta y}} \right)^2. \tag{8}$$

(b)

(a)

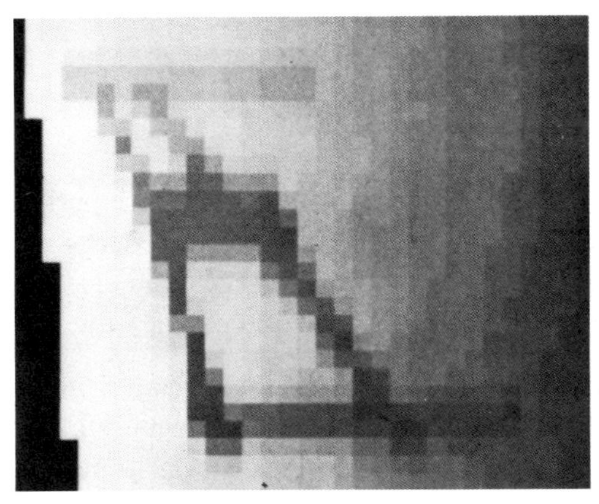

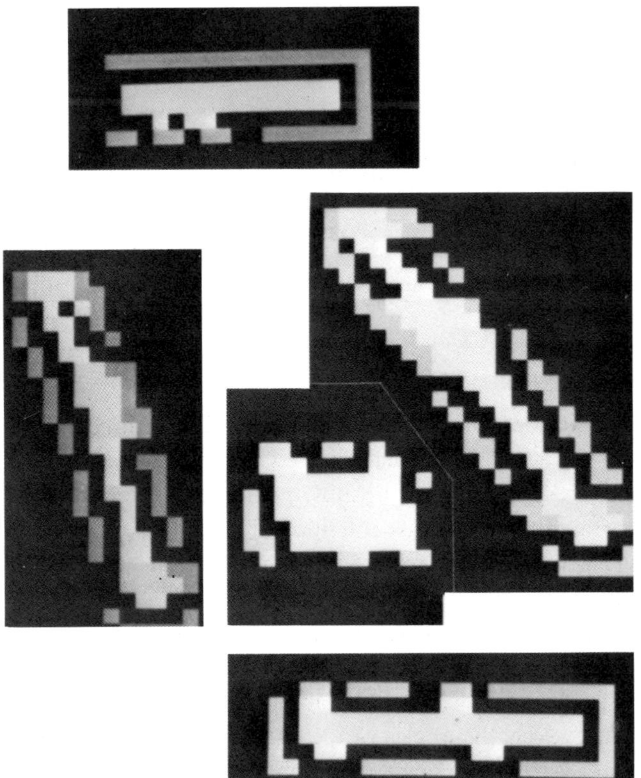

Figure 4.11
(a) ALV data of gate, (b) predicted image appearance of model, and (c) predicted range data together with "masks" showing predicted variance for pixels on and near each part; light areas have low variance, dark area have high variance.

We can calculate $s^2_{x_i, y_i, z_i}$ from $s^2_{x_i, y_i, z_i^{opt}}$, where $z_i = z_i^* + k$, as follows:

$$s^2_{x_i, y_i, z_i} = s^2_{x_i, y_i, z_i^{opt}} + \frac{1}{n^2} \sum_{\Delta x} \sum_{\Delta y} (k/\sigma_{i, \Delta x, \Delta y})^2. \tag{9}$$

That is, we search the measured range data for each part, computing the sum of squared errors between the predicted range values and actual values for each possible location of each part, at each point assuming the part is centered at distance z_i^{opt}, which is the best possible value for z_i.

4.5.2.4 Combining to Form Minimum-Mean-Square-Error Estimates The final stage of our algorithm is to combine the mean square error between predicted and measured data values with mean square errors for part position and interpart distances to obtain a final estimate of object location and geometry.

To do this, we first calculate the interpart distance errors and the part position errors. Let $d_{i,j}$ be the predicted distance between the center of part i and the center of part j, and $d^*_{i,j}$ be the distance between the current estimates of center positions for parts i and j (i.e., the "measured" distance). Then we find part centers (x_i, y_i, z_i) such that

$$\varepsilon = \min_{\text{all part centers}} \left[\sum_i s^2_{x_i, y_i, z_i} + \sum_i \sum_j \left(\frac{d_{i,j} - d^*_{i,j}}{\sigma_{d_{i,j}}} \right)^2 + \sum_i \left(\frac{z_i - z^*_i}{\sigma_{z_i}} \right)^2 \right]. \tag{10}$$

Currently we perform this final optimization by a pruned combinatorial search; for a typical examples, this search takes only a few dozen seconds on a Symbolics 3600.

4.5.3 Results

One method of testing the robustness of our recognition technique is to introduce errors in the a priori information used for prediction, and then to observe the effect of these errors on our subsequent recognition accuracy. Thus, in the following example we have intentionally included errors in the gate's position and geometry that are near the limit of what we expect to occur operationally.

Figure 4.12a shows the consequences of these errors for the predicted location and orientation of each of the gate's component parts. In this example, each part's center is displaced by about 5 feet horizontally and 1 foot vertically; moreover, the angles of the gate's cross-bars are off by about 15°, so that the further gatepost is positioned several feet lower than the nearer one. Because of the low viewing angle, the predicted range values for the ground are even more seriously wrong, averaging almost 20 feet in error. Note that because of the wide-angle-lens character of the sensor, each of these errors can result in large (15–20 pixel) changes in imaged position.

Figure 4.12b shows the results of applying our recognition algorithm to the ALV range data shown in figure 4.11a using the (in error) predictions of figure 4.12a. In this figure the final estimated position for each of the object's parts is overlayed on the original range data. The positions of each part in figure 4.12b should be compared to the initial estimate of each part's position shown in figure 4.12a. It can be seen that, despite substantial errors in the predicted part position and orientation, as well as severe errors in the predicted ground and road ranges, the algorithm has nicely located the gate and at the same time appropriately corrected the geometry of the gate model.

4.5.4 Summary

We have implemented a system that performs top-down recognition of an object by searching for its component parts, a procedure that minimizes the effects of global distortions caused by unknown relative orientation or poorly known overall object geometry. This procedure has the advantage of being very efficient, requiring only a few dozen seconds on a Symbolics 3600.

We accomplish object recognition by modeling the image appearance of its component "parts" sufficiently to obtain a useful estimate of each of their positions; we can then match the *ensemble* of these parts to estimate the object's position and orientation, as well as to update our model's interpart geometry. Because this algorithm employs only image-level matching, we believe that—as with the learning algorithm—we shall eventually be able to apply this technique directly to the outputs of shape-from-x and depth-from-x vision modules.

This system does not now make use of matching at the level of image features, or of matching bottom-up-derived part descriptions to stored object descriptions. Both of these other matching strategies can be useful in many situations; adding such capabilities will only enhance the robustness of our algorithm.

It is clear that this approach to recognition will not work when the object's structure or position is very poorly known. For such cases, we

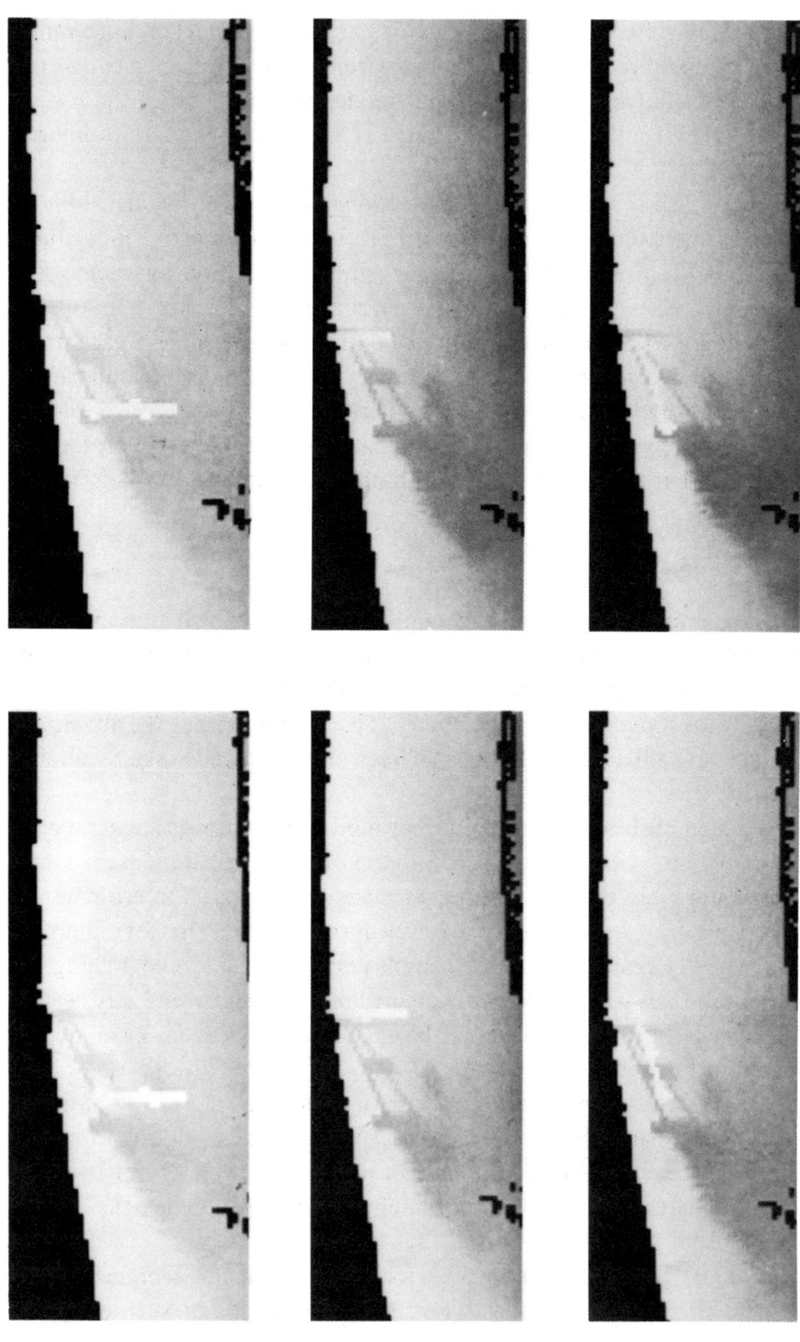

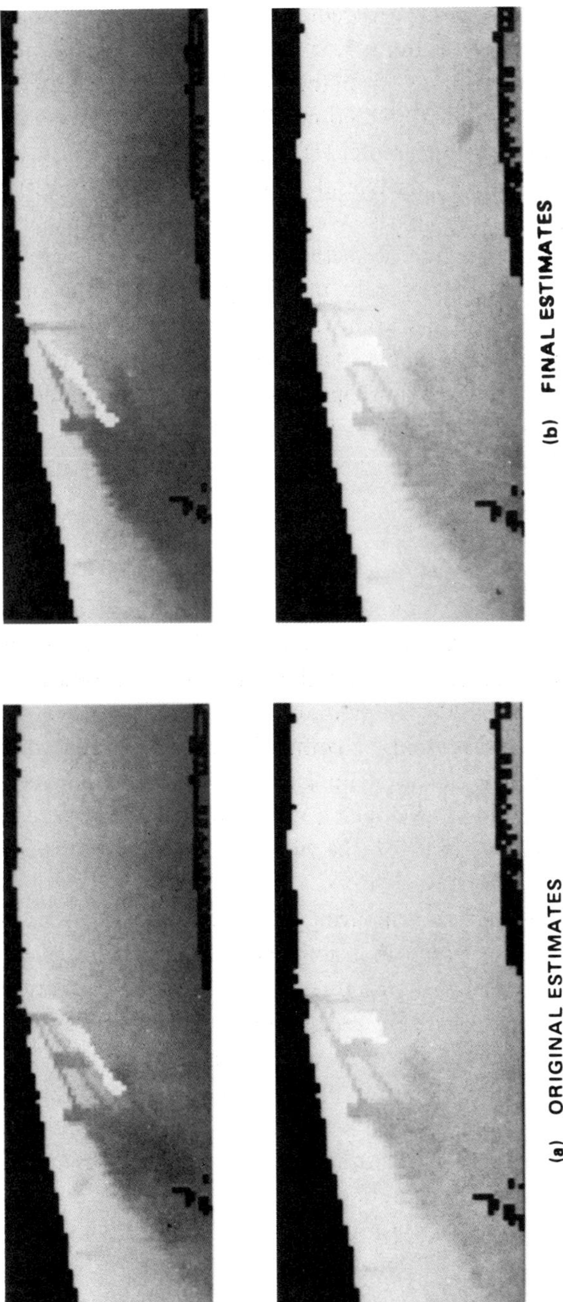

Figure 4.12
(a) Original estimates of part location and (b) final estimates of part location.

envision a hybrid of this top-down predictive approach and the bottom-up approach of the previous sections. For instance, if we are told only that "there is a gate on the left-hand side of the road," we do not possess enough information to make detailed predictions. We can, however, use a stored model of a "typical" gate to predict (1) that there are likely to be long, thin horizontal and vertical parts and (2) the likely image position and orientation of each of these parts. We can then use this information to restrict our bottom-up search algorithm to a reasonable range of part dimensions and orientations, thereby massively decreasing the amount of search required. This limited search will result in a set of horizontal and vertical parts that are reasonable constituents of a gate. Then, as with the above image-level recognition algorithm, we can search among these constituents for a combination of horizontal and verticals that best match our "typical gate" model.

4.6 Summary

The goal of our DARPA Autonomous Land Vehicle (ALV) project is to identify potentially useful landmarks, enter their descriptions into a database, and then, when traveling through the same region at some later time, search for and recognize these landmarks. This capability will make it possible to accumulate eventually a comprehensive catalog of landmarks that can be recognized and then utilized for accurate guidance of the vehicle. We have therefore developed a system that accomplished each of these tasks in a manner that, on the basis of these initial experiments, appears to be robust.

This system relies on the structuring of objects into their component "parts," using a representation that accords with psychological evidence about people's notions of part structure [13, 14]. It appears that this representation facilitates the learning and recognition processes by providing the requisite descriptive power without using a large number of parameters. This property of concise description is especially important when the minimum-length-encoding approach to learning object descriptions is employed. In fact, we believe that, for such an approach to be successful, it *must* utilize a representation roughly similar to this one.

At present, our system does not make use of feature-level descriptions. The learning algorithm, for instance, builds object models from the data by means of direct image-to-part search. Similarly, our recognition algorithm predicts an object's image appearance and matches directly at the image level. Thus, although we anticipate that in a fully developed system *all* levels of representation would be present, the particular algorithms we have investigated represent only the outside information paths in the system structure depicted in figure 4.1.

Our reason for choosing to investigate image-level algorithms first is that the capability to describe and recognize structures in raw image data is a prerequisite for robust part-level matching, and, further, it seems possible to generalize these image-level algorithms to feature-level tasks. The other reason for beginning to investigate image-level matching of range data is the future extension of such techniques to the output of shape-from-x and depth-from-x vision modules.

Acknowledgments

This research was made possible by National Science Foundation, Grant No. DCR-83-12766, by Defense Advanced Research Projects Agency Contract No. DACA76-85-C-0004, and by a grant from the Systems Development Foundation. We wish especially to thank Marty Fischler for his help, comments, and insight.

Notes

1. SuperSketch is available free to colleges and universities by writing to the first author.

2. Three for position, three for orientation, three for size, two for squareness/roundness along the various axes, two for bending, and one for tapering. We restrict bending to being along at most two axes, one of which is the longest axis, and tapering to only the longest axis. These restrictions stem from our finding that additional degrees of freedom were almost never used when people constructed models with SuperSketch.

3. That is, one can prove that if a body of data is generated by a vocabulary V with parameter settings P_i, then the minimal-length encoding of that data (using V) will recover the P_i—given sufficient resolution, noise-free data, and modulo ambiguities in the vocabulary. Thus one can formally define the intuition that the "simplest explanation" is in fact the correct one.

4. Our protocol analysis research indicates that people tend to think of the world in terms of formative processes that are analogous to sculpting in clay. If the world *really is* formed of processes analogous to our vocabulary of construction and deformation, then by use of minimal-length encoding we can use the image data to infer a description that can be related to the scene's causal history, and thus to its functional significance. It is important to understand that the formative processes in the scene do not actually have to *be* clay sculpting in order to determine causal structure: They only have to be *isomorphic* to the clay-sculpting operations.

5. In our sampling we restrict bending to occur along at most two axes, one of which must be the longest axis, and we have typically not included tapering (it seems to make a difference only on large forms), except when using a reduced sampling in orientation.

6. Originally the skeleton was found by hand simulation of a grassfire technique; later we confirmed that the points we used could have been found automatically.

7. We note that any part structure representation must have at *least* nine parameters: three each for position, orientation, and size. Our representation adds only five more parameters in order to achieve its reasonably general-purpose descriptive power. We note the fact that we could use our modeling vocabulary to obtain accurate fits to the data in these examples lends support for its descriptive adequacy.

References

[1] Badler, N., and Bajcsy, R., (1978). Three-dimensional representations for computer graphics and computer vision, *Computer Graphics, 12*, 153–160.

[2] Binford, T. O. (1971). Visual perception by computer, *Proceeding of the IEEE Conference on Systems and Control*, Miami, December.

[3] Agin, G. A., and Binford, T. O., (1973). Computer description of curved objects, *Proc. Am. Asso. for Artificial Intelligence '73*, pp. 629–635, Stanford, CA, August.

[4] Nevatia, R., and Binford, T. O. (1977). Description and recognition of curved objects *Artificial Intelligence, 8*, 1, 77–98.

[5] Bolles, B., and Horaud, R. (1986). Edge-chain analysis for object verification, *IEEE Conf. on Robotics and Automation*, San Francisco, CA, April 7–10.

[6] Brooks, R. (1985). Model based 3-D interpretation of 2-D images, in *From Pixels to Predicates*, Pentland, A. (Ed.), Norwood N.J.: Ablex Publishing Co.

[7] Goad, C. (1985), A fast model-based vision system, in *From Pixels to Predicates*, Pentland, A. (Ed.), Norwood N.J.: Ablex Publishing Co.

[8] Faugeras, O. D., Hebert, M., Pauchon, E., and Ponce, J., (1984). Object representation, identification, and positioning from range data, *Robotics Research: The First Symposium* (Brady, M., and Paul, R., eds., MIT Press).

[9] Grimson, W. E. L., and Lozano-Perez, T. (1985). Recognition and localization of overlapping parts from sparse data in two and three dimensions, *Proc. IEEE Robotics Conference*, pp. 140–150, St. Louis, MO.

[10] Hoffman, D., and Richards, W. (1985) Parts of recognition, in *From Pixels to Predicates*, Pentland, A. (Ed.), Norwood, N.J.: Ablex Publishing Co.

[11] Leyton, M. (1984). Perceptual organization as nested control, *Biological Cybernetics 51*, 141–153.

[12] Beiderman, I. (1985). Human image understanding: recent research and a theory, *Computer Vision, Graphics and Image Processing, 32*, 1, 29–73.

[13] Pentland, A. (1987). Towards an ideal 3-D CAD system, *SPIE Conf. on Machine Vision and the Man-Machine Interface*, Jan. 12–16, San Deigo, CA. Order No. 758–20.

[14] Pentland, A. (1986). Perceptual organization and the representation of natural form, *Artificial Intelligence Journal, 28*, 2, 1–38.

[15] Teversky, B., and Hemenway, K. (1984). Objects, parts and categories, *J. Exp. Psychol. Gen., 113*, 169–193.

[16] Rosch, E. (1973). On the internal structure of perceptual and semantic categories, in *Cognitive Development and the Acquisition of Language*. Moore, T. E. (Ed.), New York: Academic Press.

[17] Hayes, P. (1985). The second naive physics manifesto, in *Formal Theories of the Commonsense World*, Hobbes, J. and Moore, R. (Eds.), Norwood, N.J.: Ablex.

[18] Thompson, D'Arcy (1942). *On Growth and Form*, 2d Ed., Cambridge: The University Press.

[19] Stevens, Peter S. (1974). *Patterns in Nature*, Boston: Atlantic-Little, Brown Books.

[20] Smith, A. R. (1984). Plants, fractals and formal languages, in *Computer Graphics 18*, 3, 1–11.

[20] Mandelbrot, B. B. (1982). *The Fractal Geometry of Nature*, San Francisco: Freeman.

[22] Gardiner, M. (1965). The superellipse: a curve that lies between the ellipse and the rectangle, *Scientific American*, September 1965.

[23] Barr, A. (1981). Superquadrics and angle-preserving transformations, *IEEE Computer Graphics and Application, 1*, 1–20.

[24] Fischler, M., and Bolles, R. (1981). Random sample consensus: a paradigm for model fitting with applications to image analysis and automated cartography, *Communications of the ACM, 24*, 6, 381–395.

[25] Fischler, M., et al. (1986). Knowledge-based vision techniques for the autonomous land vehicle program, SRI Project Report 8388.

[26] Barnard, S., Bolles, R., Marimont, D., and Pentland, A. (1986). Multiple representations for mobile robot vision, *SPIE Cambridge Symposium on Optical and Optoelectronic Engineering*, October 26–31, 1986, Cambridge, MA. Available SPIE Proceedings, Vol. 727.

[27] Pentland, A. (1985). On describing complex surfaces, *Image and Vision Computing, 3*, 4, 153–162.

[28] Pentland, A. (1984). Fractal-based description of natural scenes, *IEEE Pattern Analysis and Machine Intelligence, 6*, 6, 661–674.

[29] Morrison, D. F. (1967). *Multivariate Statistical Methods*, McGraw-Hill Inc., New York, N.Y.

[30] Papoulis, A. (1965). *Probability, Random Variables, and Stochastic Processes*, McGraw-Hill Inc., New York, N.Y.

[31] Bajcsy, R., and Solina, F. (1987). Three-dimensional object representation revisited, *First International Conf. on Computer Vision, '87*, June 8–11, London, England.

[32] Terzopoulos, D. (1987). Regularization of inverse visual problems involving discontinuities, *IEEE Pattern Analysis and Machine Intelligence, 8*, 6, 413–424.

[33] Reeves, W. T. (1983). Particle systems—a technique for modeling a class of fuzzy objects, *ACM Transactions on Graphics 2*, 2, 91–108.

[34] Saches, O., *The Man Who Mistook His Wife for A Hat*, Boston.

[35] Winston, P. H. (1975). Learning structural descriptions from examples, Ph.D. Thesis, in *The Psychology of Computer Vision*, Winston, P. H. (Ed.), New York: McGraw-Hill.

[36] Winston, P., Binford, T., Katz, B., and Lowry, M. (1983). Learning structural descriptions from examples, *Proceedings of the National Conference on Artificial Intelligence (AAAI-83)*, pp. 433–439, Washington, D.C., August 22–26.

[37] Marr, D., and Nishihara, K. (1978). Representation and recognition of the spatial organization of three-dimensional shapes, *Proceedings of the Royal Society—London B200*, 269–296.

[38] Koenderink, Jan J., and van Doorn, Andrea J. (1979). The internal representation of solid shape with respect to vision, *Biological Cybernetics, 32*, 211–216.

5 3-D Vision for Outdoor Navigation by an Autonomous Vehicle

Martial Hebert and Takeo Kanade

5.1 Introduction

Research in robotics has recently focused on the field of autonomous vehicles, that is, mobile units that can navigate under computer control based upon sensory information. Several components are involved in the design of such a system. A high-level cognitive module is in charge of making decisions based on the perceived environment and the mission to be carried out. Sensory modules, such as video image analysis and range data analysis, transform the sensors' output into a compact description that can be used by the decision-making modules. Low-level control software converts decisions into actions performed by the hardware.

In this chapter, we focus on one type of sensory component, the analysis of range data for an autonomous vehicle navigating in an environment with features such as trees, uneven terrain, and man-made objects. In particular, we study the use of the Environmental Research Institute of Michigan (ERIM) laser range finder to perform three tasks: obstacle detection, surface description, and object recognition (figure 5.1). Obstacle detection is the minimum capability required by an autonomous vehicle in order to navigate safely. Surface description is needed when the obstacle detection is not sufficient for safe navigation—in the case of uneven terrain, for example, or when a more accurate description of the environment is needed, e.g., for object recognition. Object recognition capabilities are required when the vehicle must recognize and locate known landmarks, e.g., a traffic sign, in order to carry out its mission.

5.2 Intermediate Representations of Range Data

5.2.1 Sensor Data

The ERIM sensor derives the range at each point by measuring the difference of phase between a modulated laser beam and a reflection from the scene. A two-mirror scanning mechanism directs the beam onto the scene so that an image of the scene is produced. In the ERIM-ALV version, the

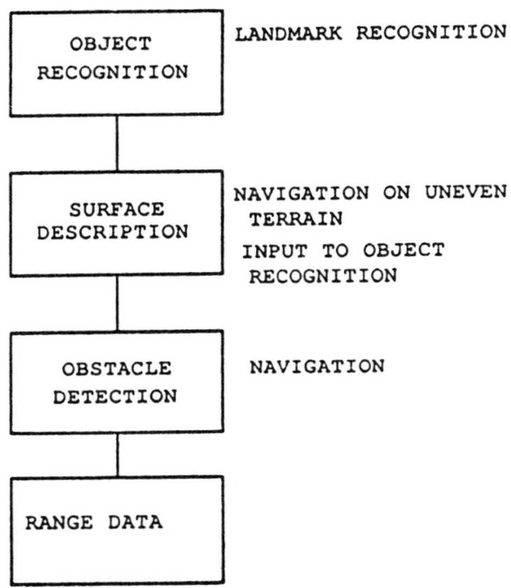

Figure 5.1
Range data processing.

field of view is $\pm 40^\circ$ in the horizontal plane and 30° in the vertical plane, from 15° to 45°. The resulting range image is a 64×256 8-bit image. The frame rate is currently two images per second. The nominal range noise is 0.4 feet at 50 feet. The sensor also produces reflectance images in which the value of each pixel is the amount of light reflected by the target. Figure 5.2 shows an example of a range image and the corresponding reflectance image.

The ERIM sensor presents some limitations: Since only the phase difference is measured, the range values are known only *modulo* 64 feet. This causes ambiguity in the range data. The resolution degrades rapidly as the range increases. This is due to the divergence of the beam, which produces larger footprints as the distance increases, and to the scanning mechanism. Since the scanning angles are discretized using constant increments, the density of points decreases as the range increases. Points may be incorrectly measured at the edges of objects due to multiple reflections of the beam. This effect is common to all active scanning techniques and is known as the "mixed points" problem. We have found that applying a median filter to the original image eliminates most mixed points.

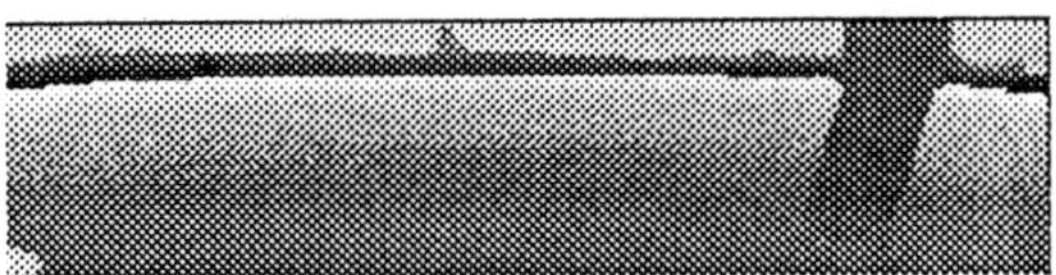

(a)

(b)

Figure 5.2
An example of range and reflectance images: (a) range image (the darker pixels are closest); (b) reflectance image.

5.2.2 Vehicle-Centered Coordinates

The raw data from the ERIM scanner represent range as a function of angles θ and ϕ of the two mirrors.

As shown in figure 5.3, we assume that a coordinate frame **u**, **v**, **w** is attached to the scanner. We use another coordinate frame, the "vehicle" frame, **i**, **j**, **k** to express the measured points so that the resulting values are vehicle-centered and are therefore independent of the sensor configuration. We can thus derive the coordinates (x, y, z) of the point measured at pixel (row, col) with range D. If ϕ is the angle between $(\mathbf{u}, \mathbf{v})$ and the direction of the measured beam **a** at pixel i, j, θ is the angle between $(\mathbf{u}, \mathbf{w})$ and the direction of the measured beam **a** at pixel i, j, ϕ_s is the starting vertical scanning angle, $\Delta\theta$ and $\Delta\phi$ are the angular increments, and τ is the tilt angle of the scanner, that is the angle between the planes $(\mathbf{t}, \mathbf{j})$ and $(\mathbf{u}, \mathbf{v})$ as shown in figure 5.3, then the conversion is

$$\theta = (j - 128) \times \Delta\theta,$$

$$\phi = \phi_s - i \times \Delta\theta,$$

$$x = D\,(\cos\theta(\cos\tau\cos\phi - \sin\tau\sin\phi)),$$

$$y = D\sin\theta,$$

$$z = D\,(\cos\theta(\sin\tau\cos\phi + \cos\tau\sin\phi)).$$

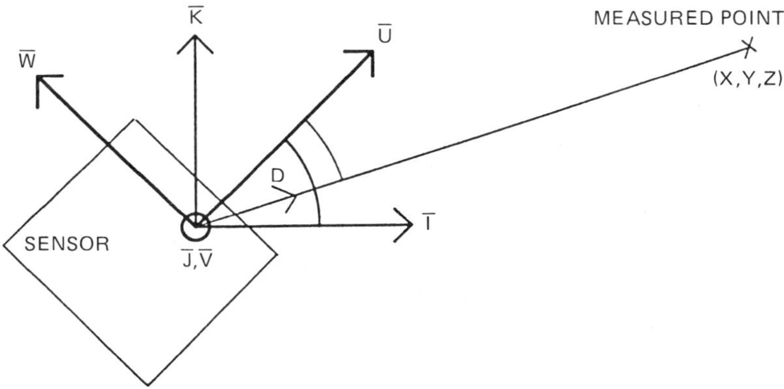

Figure 5.3
Conversion to vehicle coordinates. (Note that letters are capitalized and that vectors are indicated by overbars.)

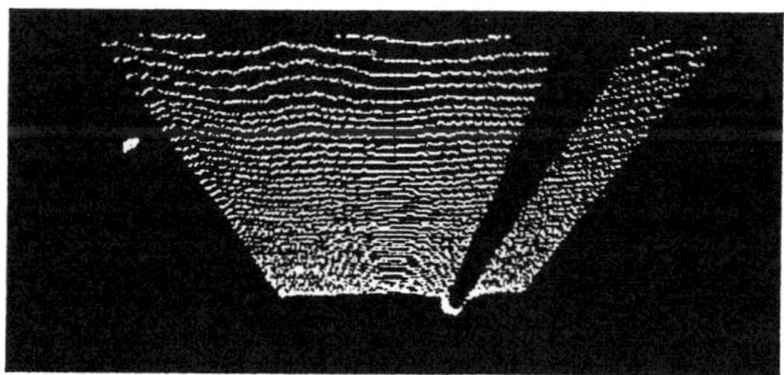

Figure 5.4
Overhead view of the data of figure 5.2.

Figure 5.4 shows the data of the image of figure 5.2 after conversion to vehicle coordinates as viewed in the direction of the **w**.

5.2.3 Bucket Map

In outdoor environments, the ground plane (**i**, **j**) has a privileged role: the terrain can be modeled as a function $z = f(x, y)$, x and y being the coordinates on the ground plane. In order to take advantage of the ground plane, we used an intermediate representation, the bucket map.

A bucket map is defined by a regular grid on a reference horizontal plane. Each cell of the grid, or bucket, contains a set of measured points as shown in figure 5.5. The points within a bucket may all be from the same image or from several consecutive images. The size of a bucket is typically 30 cm × 30 cm.

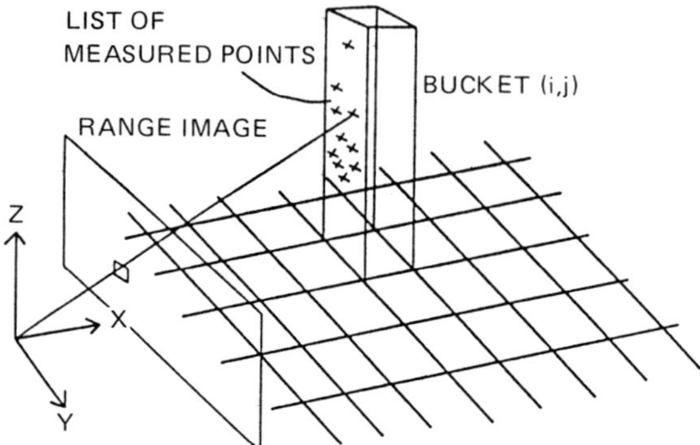

Figure 5.5
The bucket map structure.

5.3 Obstacle Detection

The first task of range data analysis for navigation is to report the portions of the environment that are potentially hazardous. We must identify individual objects in the environment that the vehicle must avoid. Most obstacle detection algorithms combine surface discontinuities and surface slope [6, 10] to extract intraversable regions in the image using a vehicle model [3]. Faster algorithms use a priori knowledge of the terrain, e.g., flat ground assumption, by computing the difference between the range image and the expected ideal image [4]. Since we want to be able to navigate in a variety of environments, we chose the first approach, which, although computationally expensive, allows us to handle uneven terrains. Specifically, we identify points in the bucket map at which the elevation exhibits a large discontinuity and points at which the surface slope is above a given threshold. The first set of points corresponds to the edges of the objects; the second lies within the surface of an object facing the sensor. The obstacle detection algorithm is divided into four steps:

1. Detect elevation discontinuities on the bucket map. The discontinuities are computed in 2×2 windows around each point.
2. Compute the surface normal for each bucket.
3. Detect the buckets for which the surface normal makes an angle with the vertical direction greater than a given threshold.
4. Extract connected regions from the set of buckets detected at step 3 so that each region corresponds to an object. Two buckets are connected if they do not cross the line of elevation discontinuity.

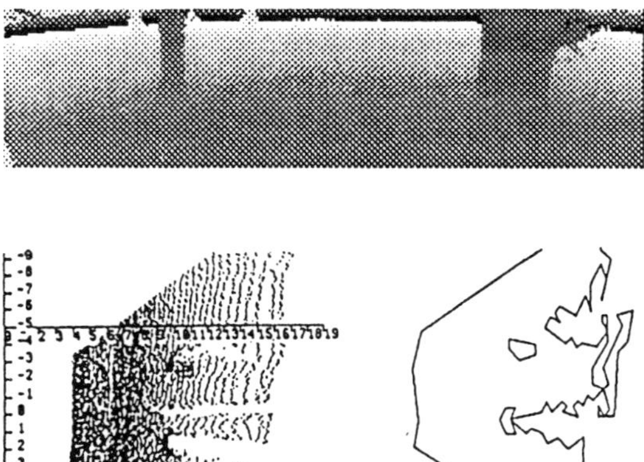

Figure 5.6
Obstacle detection.

The rationale for using two criteria, elevation and surface normals, is that the surface normals are meaningless at the edge of an object, and the elevation cannot be used alone without some knowledge of a ground-plane on which objects are known to rest. One possible undesirable result of the detection algorithm is that a small portion of the terrain might be reported as an obstacle due to the resolution of the bucket map. This error can be corrected only when a more complete object description is created. It does not, however, significantly affect the behavior of the vehicle since only small regions are involved.

For the purpose of obstacle avoidance, the detected objects are represented by polygons on the ground surface in order to be used by a path planner. Figure 5.6 shows a range image, the location of the buckets classified as part of objects, and the polygonal representation of the obstacle map. The squares indicate the buckets in which the objects have been found. The large polygon enclosing the map is the boundary of the portion of the environment seen by the sensor.

5.4 Surface Description

The obstacle detection algorithm is sufficient for vehicle navigation in a simple environment that includes only a smooth terrain and discrete obstacles. A typical example of such an environment occurs in road-following applications. We need a more sophisticated representation in two cases:

- The surrounding terrain is uneven. In that case, part of the environment that may be hazardous or costly to navigate cannot be described as discrete objects.
- The mission requires the recognition of specific objects given a priori models. In that case, the mere knowledge of the existence and position of an object in the world is not sufficient; we need a more detailed description of its shape.

We describe surfaces by a set of connected surface patches. Each patch corresponds to a smooth portion of the surface and is approximated by a parameterized surface. In addition to the parameters and the neighbors, each region has two uncertainty factors: σ_a and σ_d. σ_a is the variance of the angle between the measured surface normal and the surface normal of the approximating surface at each point. σ_d is the variance of the distance between the measured points and the approximating surface. Those two attributes are used in the object recognition algorithm.

The surface description is obtained by segmenting the range image into regions. Several schemes for range image segmentation have been proposed in previous work [1]. These techniques are based either on clustering in some parameter space, or on region growing using smoothness criteria of the surface. We chose to combine both approaches into a single segmentation algorithm. The algorithm first attempts to find groups of points that belong to the same surface, and then uses these groups as seeds for region growing, so that each group is expanded into a smooth connected surface patch. The smoothness of a patch is evaluated by fitting a surface, plane or quadric, in the least-squares sense.

The strategy for expanding a region is to merge the best point at the boundary at each step. This strategy guarantees a near optimal segmentation. It has, however, two major drawbacks: it may be computationally expensive, and it may lead to errors due to sensor errors on isolated points, such as mixed points. To alleviate these problems, we use a multiresolution approach. We first apply the segmentation to a reduced image in which each pixel corresponds to an $n \times n$ window in the original image, n being the reduction factor. This first, low-resolution, step produces a conservative description of the image (figure 5.7c). The low-resolution regions are then expanded using the full-resolution image (figure 5.7d). No new regions are

(a)

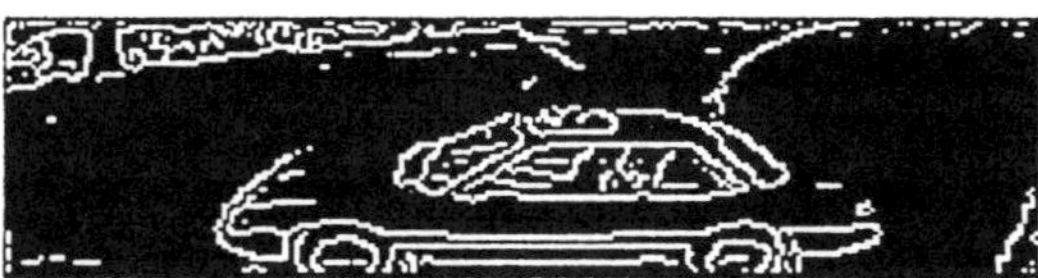

(b)

(c)

(d)

Figure 5.7
Range image segmentation: (a) range image; (b) edges from reflectance image; (c) low-resolution segmentation ($n = 2$); (d) final segmentation.

created at full resolution. Figure 5.8 shows the segmentation of an image of uneven terrain.

In addition to the pure region segmentation, we use the edges extracted from the reflectance image to improve the description. In the low-resolution segmentation step, pixels that correspond to a window that contains at least one edge pixel are removed. In the full-resolution step, regions are expanded so that they do not cross an edge. As a result, edge pixels are all part of the regions boundaries. Explicitly including edges improves the segmentation in two ways. First, edges that correspond to low-amplitude occluding edges separates regions that may be merged in the range image segmentation. Second, reflectance edges can delineate surface markings that are not visible in the range image. Figure 5.7b shows an edge image obtained by applying a 10×10 Canny edge detector.

5.5 Object Recognition

5.5.1 Recognition Strategy

The goal of an object recognition algorithm is to find the most consistent interpretation of a scene given a stored list of primitives $(M_{i1}, \ldots, M_{in})$, the model, and a segmentation of the observed scene $(S_{j1}, \ldots, S_{jn})$. The algorithm must therefore search all the possible matchings $((M_{i1}, S_{j1}), \ldots, (M_{in}, S_{jn}))$. This search being a combinatorial problem, the main issue is to prune the search space in order to be able to process complex scenes. Many strategies have been proposed for solving the object recognition problem [1]. The most common approach is to use geometric constraints to constrain the search, assuming the objects are rigid. This approach may require an accurate geometric model, which is usually not available in our application. Another approach is to generate beforehand the possible appearances of the object to be recognized in order to reduce the search space by precompiling the constraints in the model [8, 5]. We use a combination of both approaches in which geometric constraints are precompiled in the model. The model has two components: a set of surface patches, M, and a set of constraints, C. The constraints encapsulate knowledge about the object's shape, such as "surfaces M_i and M_j are orthogonal". A constraint, c, associated with a set of regions $(M_{i1}, \ldots, M_{in})$ can be viewed as a function that decides whether a partial matching $((M_{i1}, S_{j1}), \ldots, (M_{in}, S_{jn}))$ is acceptable. The number n of regions involved may be different depending

(a)

(b)

(c)

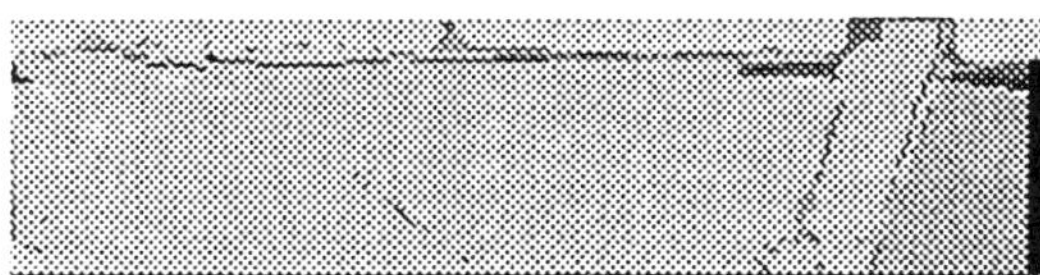

(d)

Figure 5.8
Range image segmentation: (a) range image; (b) edge from reflectance image; (c) low-resolution segmentation ($n = 2$); (d) final segmentation.

on the constraint. For example, the constraint on the area of a region is a unary constraint, while the orthogonality constraint is a binary constraint. The list of constraints and their implementation are discussed in the next two sections.

The search algorithm first constructs a list of candidates for each model region, M_i, by applying the unary constraints associated with M_i to every scene region, S_j. This provides a first reduction of the search space according to unary constraints. The algorithm then explores the remaining search space, discarding the partial solutions that do not satisfy the remaining constraints. In other words, each time a new pairing, (M_i, S_j), is added to a partial solution, $((M_{il}, S_{j1}), \ldots, (M_{in}, S_{jn}))$, the constraints associated with M_i are evaluated over the new set of pairings. The partial solution is not explored further if one of the constraints is not satisfied. The result of all the constraint evaluations are stored in tables, so that a given constraint is never evaluated twice on the same set of pairings.

The result of the search algorithm is a small set of solutions. The last step of the recognition algorithm is to compute a score that reflects the quality of each solution. This last step is necessary since there is no way of forcing the search to produce only one solution because of near-symmetries in the model, segmentation errors, or even the presence of several instances of the object in the scene. The actual computation of the score is discussed in detail in section 5.5.3. The next sections describe the details of the algorithm. We use a car as an example of an object model throughout the discussion.

5.5.2 Constraints

The constraints we currently use are

- *NX*, *NY*, *NZ*: constrains the components of the surface normal of a region—this constraint is used to implement natural limitations on the orientation of an object, such as "the roof of a car cannot be vertical."
- *Z*: constrains the vertical position of a region.
- *AREA*: constrains the area of one region.
- *ANGLE*: constrains the angle between two regions.
- *NEIGHBOR*: constrains two regions to be neighbors by computing the distance between the boundaries of two regions.
- *EXCLUDE*: forbids two regions to be visible at the same time. This constraint is based on the notion of aspects [8, 7]. An aspect is a set of

regions that can be observed from a given viewpoint. Instead of explicitly enumerating the possible aspects of an object, the *EXCLUDE* constraint describes them implicitly.

- *DISTANCE*: constrains the distance between two surfaces.

The constraints are precomputed and stored in the model. Each constraint is described by the following structure:

- **number of arguments**, N: An example is the constraint *ANGLE*, which constrains the angle between two regions has two arguments. The maximum number of arguments is currently three.
- **evaluation function**, F: This is a function that returns an interval given N regions. For example, the constraint *ANGLE* computes an interval centered around the angle between two input regions. The size of the interval is determined at run-time by the uncertainty on the parameters of the regions. In the case *ANGLE*, the interval width is given by the angular variance σ_a within the two input regions.
- **interval**, I: An interval, or set of intervals, must intersect the computed interval to satisfy the constraint.

This representation of constraints is flexible: A new type of constraint can be easily added to a model by simply defining the appropriate evaluation function. Building a new model is easier since the constraints are not hardcoded in the recognition program.

5.5.3 Evaluation of the Solutions

One would like to have a recognition program that generates only one solution that is reported as the recognized object in the scene. Unfortunately, the search algorithm generates many solutions that have to be evaluated in order to determine the best one. There are three reasons why the search generates multiple solutions: First, the constraints we use are very liberal so that the same model can be used for a wide range of scenes; consequently false solutions are difficult to avoid. Second, the object may have near-symmetries that lead to several equally valid interpretations. Third, the image segmentation being imperfect, an object region may be broken into several pieces in the image, thus producing several equivalent solutions.

Our approach to evaluating solutions is first to compute the position and orientation, or pose, of the object for each solution, to then generate a synthesized, or *predicted*, range image using the estimated pose, and finally to correlate the predicted image and the original range image. We derive a score from the correlation measure between the two images that is used to discard erroneous solutions, and to sort the other solutions.

The pose is calculated for each solution by minimizing the two sums

$$\sum_{(i,j)} area_i \| R\mathbf{n}_i^{model} - \mathbf{n}_j^{scene} \|^2$$

and

$$\sum_{(i,j)} area_i \| R\mathbf{c}_i^{model} + \mathbf{t} - \mathbf{c}_j^{scene} \|^2,$$

where R and $\mathbf{t}$ are the estimated rotation and translation and $\mathbf{n}_i^{model}$ and $\mathbf{c}_i^{model}$ (respectively, $\mathbf{n}_j^{scene}$ and $\mathbf{c}_j^{scene}$) are the surface normal and center of region i (respectively, j) of the model (respectively, scene). The summation is over all the pairs (scene region j, model region i).

In order to compute the predicted image, we have to compute the position of each point of the model in image coordinates, as well as the predicted range. The position in the predicted range image of a point $\mathbf{p}$ on the surface of the model is given by

$$\phi^{predicted} = \operatorname{atan}(x/z),$$

$$\theta^{predicted} = \operatorname{atan}(y \times \cos(\phi))/x),$$

$$row = (\phi_s - \phi^{predicted})/\Delta\phi,$$

$$col = (\theta^{predicted} - \theta_s)/\Delta\theta,$$

$$d^{predicted} = \sqrt{x \times x + y \times y + z \times z}.$$

In this equation x, y, z are the coordinates of the transformed point $R\mathbf{p} + \mathbf{t}$, (row, col) is the predicted location in the image, and $d^{predicted}$ is the predicted range value at that location.

The correlation between predicted and actual range images is given by

$$C = \sum_{(i,j) \in P \cap O} a_{ij} \times |d_{ij}^{predicted} - d_{ij}^{original}|,$$

where a_{ij} is the area intercepted by pixel i, j, and the summation is made over the intersection $P \cap O$ of the object in the predicted image, P, and the object in the observed range image, O.

The correlation must be normalized to obtain a score S that is between 0 and 1:

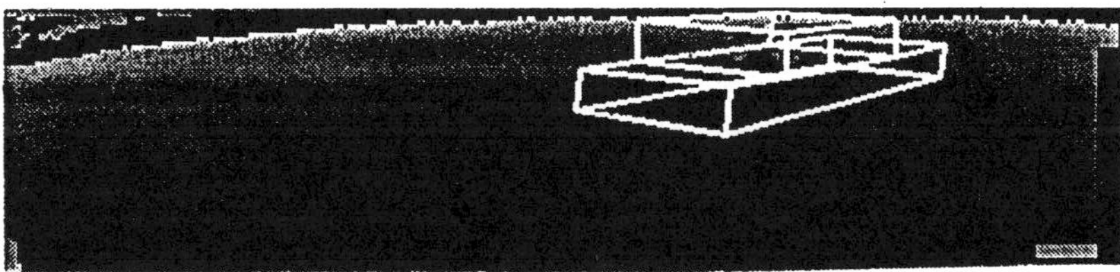

(a)

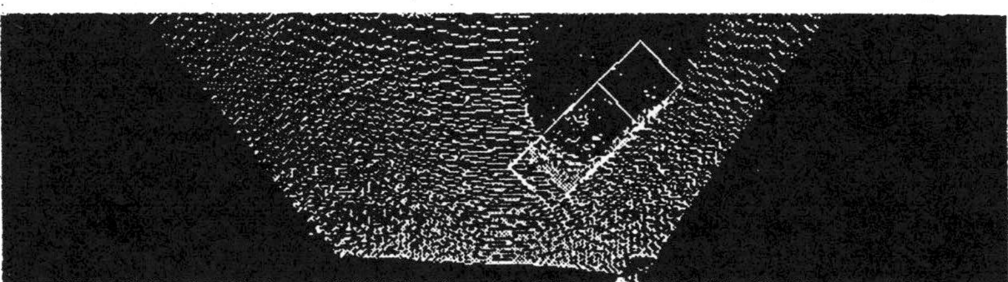

(b)

Figure 5.9
Object recognition: (a) best solution; (b) overhead view.

$$S = area(P \cap O)/area(P \cup O) \times (1 - C/(K \times area(P \cap O))).$$

In this equation, K is the maximum range difference allowed between predicted and observed images (independent of the image). S is normalized by $area(P \cap O)/area(P \cup O)$ to avoid problems when $P \cap O$ is very small, in which case we would give a high score to a very poor solution. We use the score S to eliminate false solution (typically $S < 0.5$), and to sort the solutions by decreasing score. Figure 5.9 shows the solution of highest score found on one image: the top image shows the superimposition of the recognized model and the range image; the bottom image is the overhead view of the superimposition.

5.5.4 Results

We have tested the object recognition program on 23 images of two different objects, each image corresponds to a different aspect of the objects, or to a different distance between the object and the sensor. Figure 5.10 shows a sample of the set of images. The failure modes of the program are as follows:

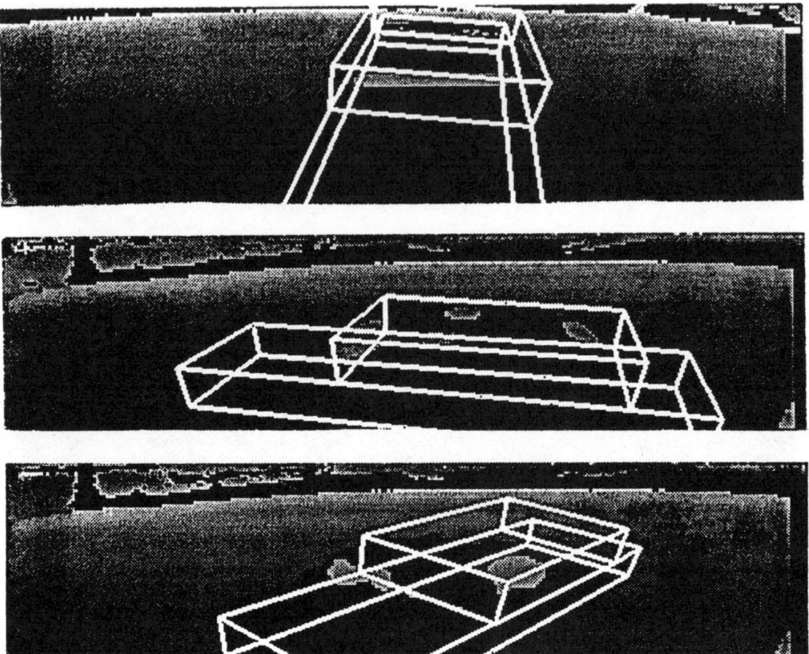

Figure 5.10
Object recognition on a sample set of images.

• In two cases, the program failed to produce any interpretation. These cases occur when not enough regions have been extracted to perform the recognition.
• In two cases, the program produced a set of interpretations, but the correct solution was not in the top-score set of solutions.
• In three cases, the program produced the correct interpretation as part of the top-score set of solutions but not as the best solution. These cases occur when a near symmetry in the image cannot be disambiguated based on the shape information only; as a result, both the correct solution and its symmetrical have the same score.

The object recognition algorithm can be improved in several ways: The recognition of one object requires an average of 1,500 constraint evaluations, most of which are rarely used for rejecting a partial solution. One optimization is to order the constraints in the model so that the constraints most likely to prune the search are evaluated first.

Figure 5.11
Symmetrical solutions.

The scoring procedure leads to very close scores for symmetric or ambiguous solutions. As an example, figure 5.11 shows two symmetric solutions for which the difference between the two scores is below 1%. This can be improved by including other sources of information, such as video or reflectance images, in the scoring procedure.

5.6 Conclusion

We have presented a set of techniques for producing 3-D environment descriptions for an autonomous vehicle. Those techniques and their output descriptions have been designed to fulfill the requirements of the major tasks of an autonomous vehicle, such as obstacle avoidance, landmark recognition, and map building. We have conducted numerous demonstrations in the CMU autonomous vehicle system, the NAVLAB, for navigation and map building applications.

We are currently pursuing further research along three lines: implementation on faster hardware, handling uncertainty, and developing a more general scheme for object recognition.

The primary bottleneck in the development of an autonomous vehicle is the computation time required by the sensory modules, such as the range image analysis module. We are in the process of porting the algorithms to a fast systolic machine, the WARP [11]. The obstacle detection module has been already successfully demonstrated with a cycle on the order of one second on the WARP. The second line of work is the representation of uncertainty. Sensor measurements, vehicle position estimates, and segmentation process induce errors in the final 3-D description. These errors can be quantified and taken into account in all the range analysis algorithms. We plan to use an explicit representation of the uncertainty building surface description [9, 2]. Second, we plan to develop a more general scheme for object recognition. Third, we plan to apply the object recognition algorithm to a larger class of objects and algorithms.

References

[1] P. J. Besl and R. C. Jain. Three-Dimensional Object Recognition. *ACM Comp. Surveys* 17(1), March 1985.

[2] R. M. Bolle and D. B. Cooper. On Optimally Combining Pieces of Information, with Application to Estimating 3-D Complex-Object Position from Range Data. *PAMI* 8(5), September 1986.

[3] M. J. Daily, J. G. Harris, and K. Reiser. Detecting Obstacles in Range Imagery. In *Image Understanding Workshop*. Los Angeles, 1987.

[4] R. T. Dunlay and D. G. Morgenthaler. Obstacle Detection and Avoidance from Range Data. In *Proc. SPIE Mobile Robots Conference*. Cambridge, MA, 1986.

[5] C. Goad. Special Purpose Automatic Programming for 3D Model-Based Vision. In *Proc. Image Understanding Workshop*. 1983.

[6] M. Hebert and T. Kanade. First Results on Outdoor Scene Analysis. In *Proc. IEEE Robotics and Automation*. San Francisco, 1985.

[7] M. Hebert and T. Kanade. The 3D Profile Method for Object Recognition. In *Proc. Computer Vision and Pattern Recognition*. San Francisco, 1985.

[8] K. Ikeuchi. Precompiling a Geometrical Model into an Interpretation Tree for Object Recognition in Bin-Picking Tasks. In *Image Understanding Workshop*. Los Angeles, 1987.

[9] L. Matthies and S. A. Shafer. *Error Modelling in Stereo Navigation*. Technical Report CMU-RI-TR-86-140, Carnegie-Mellon University, the Robotics Institute, 1986.

[10] D. Y. Tseng, M. J. Daily, K. E. Olin, K. Reiser, and F. M. Vilnrotter. *Knowledge-Based Vision Techniques Annual Technical Report*. Technical Report ETL-0431, U.S. Army ETL, Fort Belvoir, VA, 1986.

[11] H. Webb and T. Kanade. Vision on a Systolic Array Machine. *Computing Structures and Image Processing*. Academic Press, 1985.

II PLANNING

6 Geometric Issues in Planning Robot Tasks

Tomás Lozano-Pérez and Russell H. Taylor

6.1 Introduction

The goal of research in robot planning is to develop algorithms for specifying the robot commands required to achieve high-level goals such as "*Grasp cylinder A and insert it into the hole in part B.*" For example, a planner will have to choose where the robot should grasp part *A* so that it is stable in the gripper and so that the gripper will not interfere in the assembly. Also, all the paths of the robot must be planned to avoid all known obstacles. Finally, an assembly strategy must be chosen to insert the part *A* reliably into part *B*. Today, this level of planning is carried out by a human programmer who constructs programs to carry out the task. Such a programmer must specify the *motions* by which a task is to be accomplished. The input to an automatic robot planner, by contrast, is merely a *description* of the task to be accomplished. For this reason, these planners are known as *task-level* planners.

Geometric constraints on the operation of the robot and on its interaction with its environment are an essential part of task planning. Of course, there are a host of other constraints, notably the dynamics of the interaction, which are discussed in chapter 7 of this book by Mason. In this chapter, we focus primarily on geometry, but we cannot ignore physics. We shall see that planning grasping and parts-mating strategies must incorporate physical constraints to be meaningful.

Before we begin a detailed discussion of robot planning, we must answer a crucial question: Why should we study planning at all? The answer is not self-evident. There are a number of plausible alternatives. For example, we could continue to improve tools for programming robots, using more modern programming languages, or we could determine that there is a finite repertoire of tasks that need to be automated and then could build a program library for these tasks. Although both of these approaches are sometimes viable in the structured environments of large-scale manufacturing, they are inadequate in relatively unstructured environments such as those in repair tasks or very-small-batch manufacturing.

In unstructured environments, modern programming languages and program libraries are not sufficient. Fundamentally, the problem is that by changing the relative positions and orientations of even a few parts we can create a tremendous variety of environments, each requiring a different robot program. It would take a programmer too long to construct a program to handle all the variations. Even in very structured environments, programming time is often a bottleneck in the application of robots. It may take a human programmer months to work all the bugs out of a program to perform an assembly involving ten parts. Such a program will have hundreds of individually specified motions and hundreds of error detection and recovery segments. In unstructured environments, the situation is much worse; each instance of picking up a part from an arbitrary orientation in the presence of obstacles presents a formidable planning challenge.

Although this argument for the importance of planners turns on the application of robots in unstructured environments, we believe that once task-level planners are available, they will be helpful in reducing the programming bottleneck even in structured environments.

Fundamentally, we need to understand the process of generating commands to achieve tasks. A planner is one embodiment of that understanding. Once we achieve that understanding, the resulting planners will be helpful not only in planning robot motions but in other tasks as well. For example, an automatic planner could be used to evaluate the suitability of a particular part design for assembly.

6.2 Robot Planning Systems

The essence of any robot planning problem is the *prior* specification of actions that manipulate the physical environment. In other words, the goal is to construct a program *now* that, when executed *later*, will achieve a desired task. One may ask whether such prior planning is really necessary since a feedback control system can position the manipulator at any specified point in the workspace without prior planning. Difficulties arise, however, when the robot must interact with (or avoid) other objects in its work environment. Typically, the planner must construct a sequence of position commands that will reach the goal without causing a collision. This command sequence is a *program*, albeit a simple one.

Robot programs embody assumptions about the geometry of the physical environment (e.g., the locations of parts), the semantics of the robot commands (e.g., the robot is position controlled), and the mechanics of the task (e.g., the masses and coefficients of friction of objects to be manipulated). When any of these assumptions are violated (e.g., one of the parts is displaced from its expected position), then the program may fail to achieve the intended task.

All the information necessary to achieve a task is seldom available at planning time. Indeed, much of the information, whether known a priori or obtained through sensing at plan execution time, will never be available exactly, but only within uncertainty bounds. For example, the general position of a part on a conveyor, together with a set of possible stable orientations, may be known at planning time. However, the *exact* position and orientation usually must be determined by sensing (usually, vision) at plan execution time. The sensing is seldom perfect, and one must carefully analyze whether the information returned is accurate enough to assure that subsequent plan steps that rely on it will achieve the desired result.

Planning problems can be categorized on the basis of the amount of uncertainty in the planner's knowledge of the task environment:

1. When there is little or no uncertainty, the planner can determine the details of all the necessary actions at planning time, ignoring physical interactions.
2. When there is substantial, but bounded, uncertainty, the planner must plan a class of actions that will reduce uncertainty (possibly including sensing) and achieve the task in spite of remaining uncertainty. In the presence of uncertainty, physical interactions will typically be necessary.
3. When there is nearly total uncertainty, little prior planning can be done. In rare circumstances, there may be some actions that are guaranteed to achieve the task even in the absence of nearly all information. Generally, the planner must command the appropriate sensing actions and wait until more information is available to do planning.

We will separate our discussion of planning systems into these three cases.

6.3 Planning with Complete Prior Knowledge

The most prevalent method for programming industrial robots is by "teaching." In this type of programming, the robot and part feeders are bolted into fixed positions at the workstation. A human programmer uses joysticks or other means of teleoperator control to guide the robot through a fixed sequence of positions, which are recorded and played back at execution time.

The planning assumption embodied in such programs is that any variations in layout, part locations and shape, and manipulator behavior are negligible. So long as this assumption is valid, such programs work very well, and teaching remains an effective means to exploit a human programmer's spatial planning abilities. But teaching has several important drawbacks even when the assumption is valid. One key drawback, as with all robot programming, is the time required to generate the programs and to edit them when necessary. The key potential advantage of automatic planning systems is the reduction of the time to generate and update robot programs.

Another drawback of teaching is that it requires direct access to the robot in its intended workspace, which might be difficult either because it entails stopping production or because all the elements of the workspace are not available. Also, teaching systems are limited by the user's intuitive understanding of what may be the most efficient way of carrying out the task. It is possible, in principle, for a computer program to choose a better path than a human programmer would choose. The first of these problems, the availability of the robot and workspace, can be alleviated through the use of *off-line* simulation systems [15]. Typically these systems provide graphical facilities for users to describe the workspace layout and robot motion sequences, and carry out graphical simulations of the task. Such systems permit easy experimentation with different layouts and motions, but they still place all the burden of specifying the motions on the human user.

In the next section, we consider the steps in planning such tasks. The same assumptions that make tasks with little or no uncertainty easier for humans also make them easier for automatic planners. The problems faced in planning such tasks are exclusively geometric. We use a prototypical assembly task to focus our discussion, which concentrates on the problem of achieving desired geometric relationships between parts.

6.3.1 Paradigm

Consider the simple two-dimensional assembly task illustrated in figure 6.1 Part A is to be mated to part B so that the notch in the bottom of A fits over the stud on the top of B. Part A is presented in a feeder tray F_A and part B is held in an assembly fixture F_B. The robot consists of a Cartesian

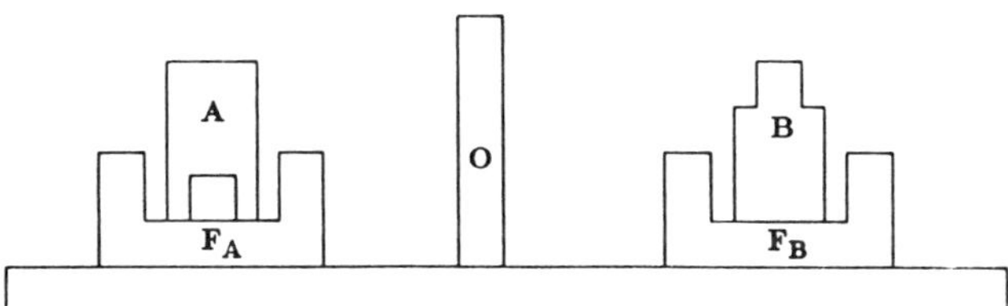

Figure 6.1
A two-dimensional assembly task.

gantry capable of moving in the *xy* plane and a two-fingered gripper whose opening can be set to any desired width. An obstacle, *O*, is also present in the work area.

The plan to be constructed consists of four steps:

1. *Approach* part *A* so that the gripper is positioned to grasp the part. This requires planning a path between the known initial position of the robot and the desired grasp location that does not cause a collision with any of the objects in the workspace. Note that the grasp position must be chosen by the planner. The choice of grasping position is constrained by a number of considerations. First, the robot must grasp the part stably, so that its position relative to the gripper is not easily disturbed by external forces applied to it. Second, the robot must be able to reach the grasp point on the part without colliding with any obstacles (including the part itself). Third, the grasp point must allow the robot to reach the intended destination without collisions.
2. *Grasp* part *A* by closing the fingers.
3. *Transport* part *A* to its final destination. Planning this step requires computing a collision-free path between the initial and final positions of the part.
4. *Release* part *A* by opening the fingers.

So long as the assumptions about the runtime environment remain valid, the plan can be executed (without replanning) as many times as desired.

The problem statement considered above ignored three important practical problems in the design of robot applications.

1. *Describing the task.* The shapes and initial locations of all the parts must be specified, as must the goal of the operation. The specification should be as natural and simple to provide as possible, while being specific enough to support geometric planning.

2. *Design feeders for the parts.* Parts A and B must be placed in the fixtures. This may be done by humans, robots, or automatic feeders. The problem of the design of feeders is a large topic in itself (see [4]).
3. *Layout the workspace* by selecting the positions for F_A and F_B. A layout should satisfy a number of criteria. First, and foremost, the robot must be able to reach all the required parts. Second, the parts should be as close as feasible to each other so as to minimize the execution time.

These steps are done once, before the task is ever attempted, whereas the four steps of the plan mentioned above will be executed for each iteration of the task.

6.3.2 Describing the Task

Ideally, the task could be described to a planner as informally as we did above. In practice, we require computationally effective means of representing shapes, initial positions, goal positions, and physical constraints. Part shapes are presumed to be represented by 3-D solid models generated by CAD systems. Initial and goal positions are typically specified by geometric relationships between parts, and many physical constraints may be specified by similar relationships that must be preserved or avoided. There have been three main approaches to specifying such relationships:

1. *Nominal transformation between the coordinate systems of the parts.* Thus for parts A and B,

$$A = {}^{A}T_{B} * B.$$

Here ${}^{A}T_{B}$ is a transformation that when applied to the position of B gives the desired position of A; therefore, it specifies the relative position between A and B. This transformation is easily represented in a computer as a 4×4 homogeneous transformation matrix [53] or as a quaternion and offset [58]. The part positions (and orientations) themselves may also be represented as the transformation between a fixed world coordinate frame and a coordinate frame fixed in the part. One problem with this type of specification is that it may overconstrain the task; for example, a round peg in a round hole has a degree of rotational freedom that is not captured in this specification. Another problem is that such coordinate transformations are notoriously difficult for people to specify.
2. *Constraint relationships between the location variables associated with the parts.* For example, the desired location of a cylindrical peg can be

specified as constraints on its position and orientation parameters relative to the hole, assuming fixed reference frames in each of the parts. The position of the peg's origin relative to the hole's origin is fixed, and so are the rotations about the x and y axes of the hole. But the rotation of the peg's coordinate frame around the hole frame's z axis is totally unconstrained. In general, these relations may involve both equalities and inequalities among the position and orientation parameters. This technique avoids overconstraint of task specification and is compatible mathematically with the representation of motion constraints used in motion planning. But these constraints are still hard to specify directly.

3. *Symbolic relationships between object features.* The relationships can be translated automatically to constraint equations and simplified [1, 59]. This method attempts to achieve the generality of constraint equations with the convenience of a simple user interface. The computational problem involves converting from user-specified relationships, such as *edge x AGAINST face y*, into algebraic equations on the position parameters. Symbolic algebra techniques for doing this conversion are limited to relatively simple cases. The general problem of symbolically solving the constraint equations is intractable. In general, even determining whether a given symbolic specification is feasible will require numerical solution.

6.3.3 Motion Planning

Whenever the robot makes a motion under position control, it must avoid unwanted collisions with other objects in the environments. Even necessary "collisions," such as part-gripper contact during grasping or part-workpiece contact during placement, must be planned so that the contact happens only at the correct place. Therefore, a robot planner must possess the ability to plan collision-free motion paths from an initial location X to a final location Y. This subject has received a great deal of attention in recent years.

There are two related, but distinct, lines of research in collision-free motion planning. One is focused on approximate numerical algorithms and the other is focused on exact combinatorial algorithms. Research on approximate algorithms has led recently to practical algorithms e.g., [41, 23, 24, 5]. Research on combinatorial algorithms has generated several new approaches to motion planning and refined our understanding of the inherent complexity of the problem. Our discussion below emphasizes approximate numerical methods. [63] provides an excellent discussion of results in combinatorial methods and complexity.

All approaches to motion planning can be characterized as search strategies in the space of the position parameters of the robot. This parameter space is known as the *configuration space.* For example, the position of the simple gantry robot in our example can be specified by the (x, y) position of some representative point on the robot. In general, we also need to specify the finger opening f. The triple (x, y, f) represents a *configuration* of the robot. These parameters specify the position of every point on the robot. If the orientation of the hand were controllable, we would also need to specify its angle. For a six-degree-of-freedom revolute manipulator with a parallel jaw gripper, one would need to specify the six joint angles and the gripper opening. *Configuration space* is the parameter space of positions. The obstacles in this space represent values of parameters that cause collisions.

Most path planning algorithms operate by building explicit representations of subsets of the robot's configuration space that are free of collisions (the *free space*). Path planning is then the problem of finding a path, within these subsets, that connects the initial and final configurations. Most methods differ primarily in the particular subsets of free-space that they use for planning and in the representation of these subsets.

The earliest motion planning methods based on characterizing free-space were proposed by Widdoes, in an unpublished paper at Stanford, by Udupa at Caltech [60], by Lozano-Pérez at MIT [39], and by Lozano-Pérez and Wesley at IBM [45].

[38] describes an algorithm for computing the free-space for a Cartesian manipulator. It is based on computing the free configurations for the manipulator. Figure 6.2 illustrates the method in two dimensions for finding collision-free paths for a polygon without rotation. The moving object and the fixed obstacles are decomposed into unions of convex polygons. A

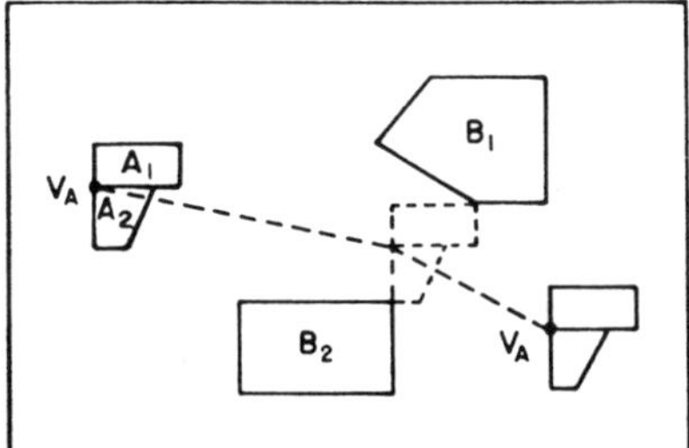

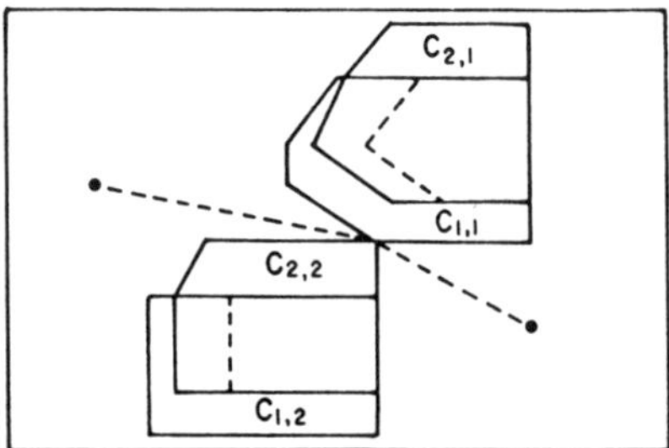

Figure 6.2
Configuration space path planning example.

reference point is chosen on the moving object (V_A in the figure). Then each obstacle is grown to compensate precisely for shrinking the moving polygons to the point V_A. A collision-free path for the point in the *configuration space* corresponds to a collision-free path for the original object in the original space. When rotation is allowed, a third dimension must be added to the configuration space for two-dimensional problems. Three-dimensional problems with rotation result in a six-dimensional configuration space.

[38, 42] describe a method for computing the exact configuration space obstacles for a Cartesian manipulator under translation. The free-space is represented as a tree of polyhedral cells at varying resolutions. The rotational motion of the manipulator is handled by defining several free-space representations, each using a manipulator model that represents the volume swept out by the rotational links over some range of joint angles (a generalization of Widdoes's use of three grids). Path searching is done by searching a graph whose nodes are cells in the free-space representation and whose links denote overlap between the cells. Extensions of this method can be used for computing the motion of realistic articulated manipulators [23, 24, 41] and solids moving with six-degrees-of-motion freedom [18]. An example of the configuration space with obstacles for a two-link manipulator is shown in figure 6.3. An example of a path for a six-degree-of-freedom articulated manipulator is shown in figure 6.4.

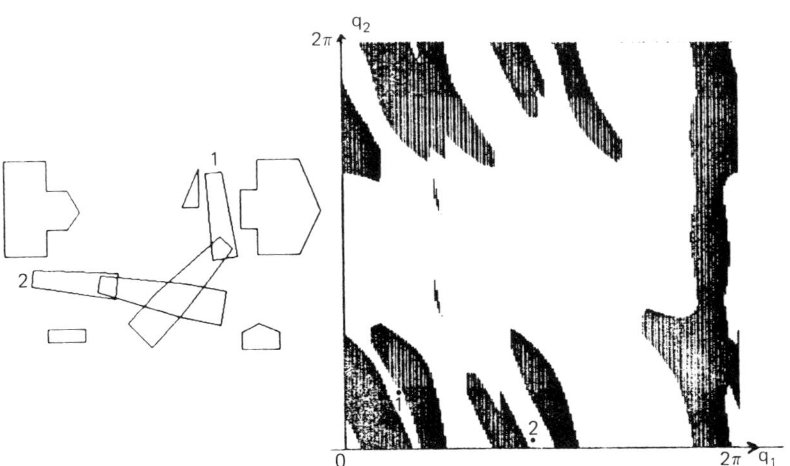

Figure 6.3
An example of the configuration space of a two-link manipulator (from [41]).

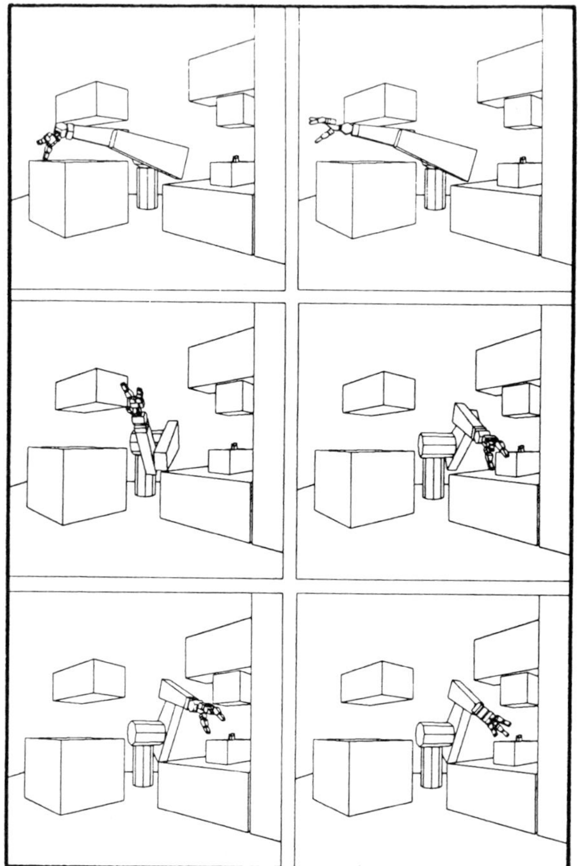

Figure 6.4
An example of a three-dimensional path found for an articulated manipulator (from [41]).

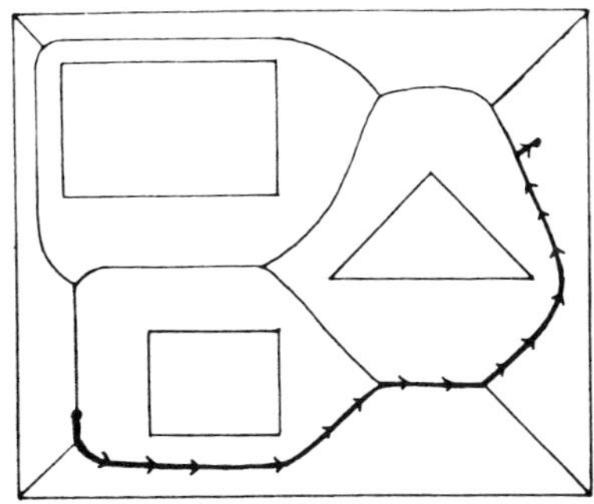

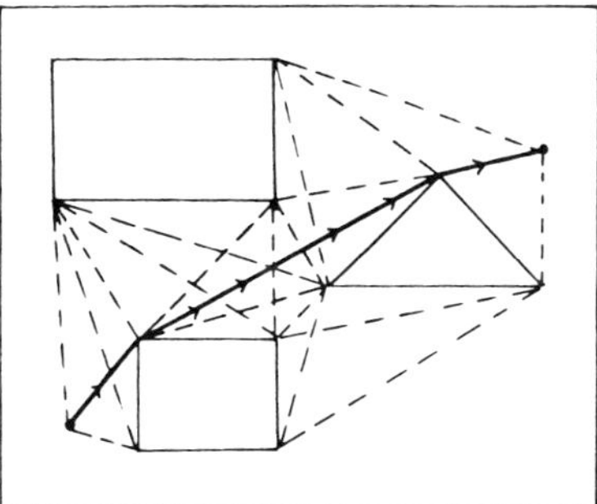

Figure 6.5
(Left) The Voronoi diagram of a set of obstacles. (Right) The visibility graph of the same obstacles.

[7] introduced a free-space method based on high-level descriptions, as generalized cones, of the empty corridors between obstacles. These descriptions can then be used to compute motion constraints for convex objects moving along the center of the corridors. Subsequent work [5] showed how to generalize this approach for a robot with six revolute joints engaged in pick-and-place motions with four Cartesian degrees of freedom (three translation and one rotation of the end-effector). Both of these algorithms have been implemented and are very efficient.

Other representations of free space are illustrated in figure 6.5. Figure 6.5a shows the *Voronoi diagram* [63] of a set of obstacles in the configuration space. The Voronoi diagram represents the locus of points that are equidistant from at least two of the obstacle boundaries. In other words, it is the locus of maximally distant points from the obstacles. Figure 6.5 (right) shows the *visibility graph* (*V-graph*) [45] of the same obstacles. The V-graph connects those vertices of obstacles that can "see" each other, that is, that can be connected by a straight line that does not penetrate any of the obstacles. Both of these representations are convenient because they define one-dimensional subsets of the configuration space that can be searched using traditional graph-search methods.

The advantage of free-space methods is that (by explicitly characterizing the free-space) they can guarantee to find a path if one exists within the free-space. Of course, this guarantee is forfeited if an algorithm only computes a subset of the free-space. Another advantage of having an explicitly represented free-space is that it is feasible to search for short paths, rather than simply finding the first path that is safe. The disadvantage is that the computation of the free-space may be expensive.

Many proposed path planning methods are fundamentally tied to the use of object approximations. The underlying assumption is that when avoiding collisions object details cannot possibly be important. This assumption does not hold in most applications where the goal is to bring objects into contact. If the path planning algorithm fails when the robot is near objects, how can one generate plans to grasp objects or to approach the site of an assembly operation? The configuration space algorithms, in contrast, allow motion near to obstacles and are therefore applicable both to planning transfer motions, to choosing grasp configurations, and to planning assembly motions.

6.3.4 On-Line Obstacle Avoidance

The most common proposals for obstacle avoidance are based on defining a penalty function on manipulator configurations that encodes the presence of objects [33, 32, 29, 34]. In general, the penalty is infinite for configurations that cause collisions and drops off sharply with distance from obstacles. The total penalty function is computed by adding the penalties from individual obstacles and, possibly, adding a penalty term for distance from the goal. At any configuration, we can compute the value of the penalty function and estimate its partial derivatives with respect to the configuration parameters. On the basis of this local information, the path search function must decide which sequence of configurations to follow. The decision can be made so as to follow local minima in the penalty function. These minima represent a compromise between increasing path length and approaching too close to obstacles.

The approach proposed in [32] is the best developed of these methods. The method uses a penalty function that satisfies the definition of a potential field; the gradient of this field at a point on the robot is interpreted as a repelling force acting on that point. The motion of the robot results from the addition of these repelling forces and an attractive force pulling the robot toward the goal, subject to kinematic constraints.

The penalty function methods are attractive because they seem to provide a simple way of combining the constraints from multiple objects. This simplicity, however, is achieved only for a circular or spherical robot, since the penalty function will be a simple transformation of the obstacle shape only in these cases. For articulated robots, penalty functions must act on each link and the forces must be combined [32]. Also, the computation of the penalty between nonspherical bodies will typically require computing minimum distances. In fact, the cleanest representation of the penalty function can be found in the configuration space of the moving body, where the moving object can be treated as a point.

The key drawback of using penalty functions to plan safe paths is the strictly local information that they provide for path searching. Pursuing the local minima of the penalty function can lead to situations where no further progress can be made (local minima). In these cases, the algorithm must choose a previous configuration where the search is to be resumed, but in a different direction from the previous time. These *back-up* points are difficult to identify from local information. This suggests that the penalty function method might be combined profitably with a more global method of hypothesizing paths. The free-space methods discussed earlier might serve this function. Penalty functions are more suitable for applications that require only small modifications to a known path. In these applications, search is not as central as it is in the synthesis of robot programs.

6.3.5 Grasping

A typical robot operation begins with the robot grasping an object; the rest of the operation is influenced by choices made during grasping. In this section, the *target object* is the object to be grasped. The surfaces on the robot used for grasping, such as the inside of the fingers, are *gripping surfaces*. The manipulator configuration that has it grasping the target object at that object's initial configuration is the *initial grasp configuration*. The manipulator configuration that places the target object at its destination is the *final grasp configuration*.

There are three principal considerations in choosing a grasp configuration for objects whose configuration is known: first, *safety*—the robot must be safe at the initial and final grasp configurations; second, *reachability*—the robot must be able to reach the initial grasp configuration and carry the object along a collision-free path to the final grasp configuration; third, *stability*—the grasp should be stable in the presence of forces exerted on the grasped object during transfer motions and parts mating operations.

If the initial configuration of the target object is subject to substantial uncertainty, an additional consideration in grasping is *certainty*—the grasp motion should reduce the uncertainty in the target object's configuration.

Choosing grasp configurations that are safe and reachable is related to path planning but has its own unique characteristics. Grasp planning methods must consider the detailed interaction of the manipulator's shape and that of the target object, since candidate grasp configurations are those having the gripping surfaces in contact with the target object while avoiding collisions between the manipulator and other objects. Grasp planning methods must also deal with interactions between the grasp configuration and the constraints imposed by subsequent operations involving the grasped object.

Most approaches to choosing safe grasps consist of three steps: choose a set of candidate grasp configurations, prune those that are not reachable by the robot or that lead to collisions, and then choose the best from those that remain.

Among the geometric constraints that have been considered in choosing grasps are

• grasps that are not within the manipulator's workspace [52, 59, 43];
• potential collisions of gripper and neighboring objects at initial grasp configuration [39, 38, 62, 11, 36, 54, 43, 35];
• existence of a collision-free path to initial grasp configuration [39, 38, 62, 43];
• potential collisions of any part of the manipulator and neighboring objects at initial grasp configuration, potential collisions of gripper and neighboring objects at final grasp configuration, potential collisions of any part of the manipulator and neighboring objects at final grasp configuration, and existence of a collision-free path from initial to final grasp configuration [39, 38, 43].

Having discarded infeasible grasps, a choice of optimal grasp must be made. A common criterion is choosing the configuration that leads to the most stable grasp. The stability condition used in early grasping planners [52, 11] amounts to checking that the center of mass of the target object is on or near the axis between the gripper jaws. This condition tends to minimize the torque on the grip surfaces, assuming that the finger contact is a point. Recently, a more thorough analysis of the torque generated by the contact between parallel jaws and an object has been done [2]. This analysis considers the whole area of contact, assuming a linear pressure distribution.

The development of multifingered hands has led to increased attention to automatic grasp selection. An early example of this work is [28]; it describes an elegant analysis of stable grasping in the absence of friction, ignoring reachability. More recently, there have been several studies of stable grasping using multifingered hands [16, 51, 31, 30]. A theory of stable multifingered grasping is emerging, but it has not yet connected with previous work on reachability considerations.

Most proposals for grasp planning have focused on only one aspect of the problem—typically, on finding safe grasp configurations or stability. Even within the aspect of safety, most proposed methods consider only a subset of the constraints needed to guarantee a safe and reachable grasp configuration. In particular, most methods consider only the constraints on gripper configuration in the initial configuration of the target object. No method adequately handles constraints on grasp configuration required by subsequent motions carrying the grasped object. Furthermore, all of the proposed methods assume a limited class of object models (usually combinations of polyhedra and cylinders) and only a simple type of hand (usually parallel-jaw grippers). These methods are seldom easy to generalize to more complex object models or hands.

Our discussion so far has considered grasping when the configuration of all objects in the environment is known exactly. Grasp planning in the presence of uncertainty is only beginning to be studied. Often the initial uncertainty must be reduced substantially for subsequent plan steps to succeed or (even) for the target object to be grasped at all. Vision or touch sensing may be used to identify the configuration of an object. Many tricks for grasping under uncertainty have been developed for particular applications, but a general theory for synthesizing grasping strategies does not exist. One common class of grasping strategies relies on touch sensing to achieve compliance of the robot gripper to the configuration of the target object [52]. [49, 10] have studied alternative grasping strategies where the target object complies to the gripper motion while under the influence of friction; the goal is to synthesize strategies that grasp the target object at a known configuration in the presence of initial uncertainty.

6.4 Planning with Incomplete Planning-Time Knowledge but Complete Execution-Time Knowledge

The assumption that all relevant information can be known accurately at plan generation time significantly limits the scope of plans. For example, it is often very difficult to feed parts in exactly the same position at every iteration of the task. Part feature locations may vary somewhat between iterations. Other relevant factors, such as the position of assembly fixtures, may also vary if the same plan is to be replicated across several workstations or as a result of routine maintenance.

One of the main reasons for the introduction of robot programming languages e.g., WAVE, AL, VAL, and AML (see [40]), was to provide a means to handle this variability in the task. Typically, sensors are used to update program variables representing part and feature positions, robot calibration offsets, etc., and the necessary position commands are computed from these variables at runtime. In some cases, conditional expressions may be used to vary the sequence or selection of program steps, depending on the sensed situation.

Planning actions that are functions of future measured states is considerably harder than planning with complete prior knowledge, whether the planning is done by a person or by a computer. For this reason, an alternative approach relying on on-line sensing and planning is often considered. If the on-line planner is a person, the result is essentially a teleoperator system. If an automatic planning system is used, the problem is essentially the same as that discussed in section 6.3. The main drawbacks of such an approach are slow execution speed, loss of an opportunity to ask a human planner for friendly advice where needed, and the necessity of planning the appropriate sensing operations, which must still deal with imperfect information.

An alternative approach is to produce plans whose primitive elements are guaranteed to work over the whole range of possible execution states. For example, compliant motions may be used to overcome small part misalignments, or the tendency of displaced parts to self-align in a gripper may be exploited. If such primitives are available, then the planning problem is again essentially the same as that of section 6.3. The additional concern is to verify that the assumption of robustness is correct, given bounds on the variability. The generation of such primitives will be discussed in later sections on planning with uncertainty.

6.4.1 Paradigms

Suppose that feeder F_A in our simple assembly task is a tray that only partially constrains the position of part A, but that a vision system is available to locate the part accurately at execution time. Similarly, suppose that the position of fixture F_B relative to the robot may vary as a result of routine maintenance procedures, but that the tips and inner surfaces of the gripper fingers are equipped with contact sensors that can be used for calibration.

Once again, there will be some steps that are done once and others that will be repeated for each execution of the task. The steps to be done once are

1. *Describe the task.* This is as before, except that a description of what is known at planning time and what will be known only at runtime must be provided.
2. *Design feeders for the parts.*
3. *Lay out the workspace* by selecting positions for feeder tray F_A and assembly fixture F_B. This step is similar to the perfect information case except that we must now assure that part A will be visible to the vision system and reachable by the robot for all possible positions in F_A. Also, we must guarantee that possible runtime variations in F_B will not affect the feasibility of subsequent plan steps.

The run-time plan now consists of six steps:

1. *Calibrate the workspace* by planning an appropriate sequence of motion and sensing steps to locate F_B relative to the robot. This step is executed as part of a setup procedure or whenever the workspace is significantly disturbed. Typically, it might consist of a sequence of "guarded" motions in which the robot touches known features on the fixture and a new position value for F_B is computed from the robot coordinates recorded when contact is detected. This value can then be used to compute an accurate position for part B, whose displacement relative to F_B is assumed to be known. Since the runtime value of F_B will not be known until after the calibration sequence is executed, planning such sequences inherently requires reasoning about uncertainty and will be discussed in subsequent sections. As a practical matter, however, "canned" calibration strategies that are guaranteed to work for reasonably bounded variations of F_B are often available, at least in skeleton form.
2. *Locate* part A in feeder tray F_A by using the vision sensor. As with calibration, the effect of this step is to remove the runtime uncertainty about the location of part A, although it is impossible to predict at planning time just where the part will be in the tray. Again, the generation of such visual strategies will generally require reasoning about runtime uncertainties, but we shall assume for the moment that a suitable canned function is available.

3. *Approach* part A so that the gripper is in the correct position for grasping the part. The approach plan will generally consist of a sequence of motion segments, with the runtime variable representing the position of part A appearing as a parameter for at least the last segment. The principal additional planning consideration is that the approach trajectory must be feasible for all possible displacements of the part in the tray. We must consider both the work envelope of the arm and the possibility of collisions between the gripper fingers and the sides of the tray. In the latter case it may be necessary to exclude some grasping positions from consideration or to plan additional motions to move the part into the center of the tray. Such *conditional* plans will include code to test the sensed position and selecting the appropriate motion sequence.
4. *Grasp* part A by closing the fingers to the specified opening.
5. *Transport* part A to its final destination relative to part B. As with the grasp approach, the transport plan will consist of a motion sequence parameterized by the updated position variable for part B. In principle, the same variability factors must be considered in planning the motion. As a practical matter, it may be convenient to *assume* that any variations in F_B will be small enough so that a geometrically feasible strategy planned for a nominal position of F_B and then parametrically modified to reflect the runtime value will also be feasible. However, it is important to check the validity of this assumption.
6. *Release* part A by opening the fingers.

From the discussion above, it is apparent that incomplete planning information introduces a number of significant issues, including the representation of geometric variability, the generation of sensor plans, and the generation of parameterized plans. We discuss each of these issues in more detail below.

6.4.2 Representing Variability

The planner requires a computationally effective means for representing *sets* of object configurations and for predicting the effects of actions over such sets. Typically, these sets are described in terms of free scalar parameters corresponding to geometric degrees of freedom (positions, shape parameters, etc.). For example, the initial position of part A relative to F_A may be described by parameters (x, y, θ). Constraint expressions are then introduced to bound the permissible variation in the parameters [59, 8].

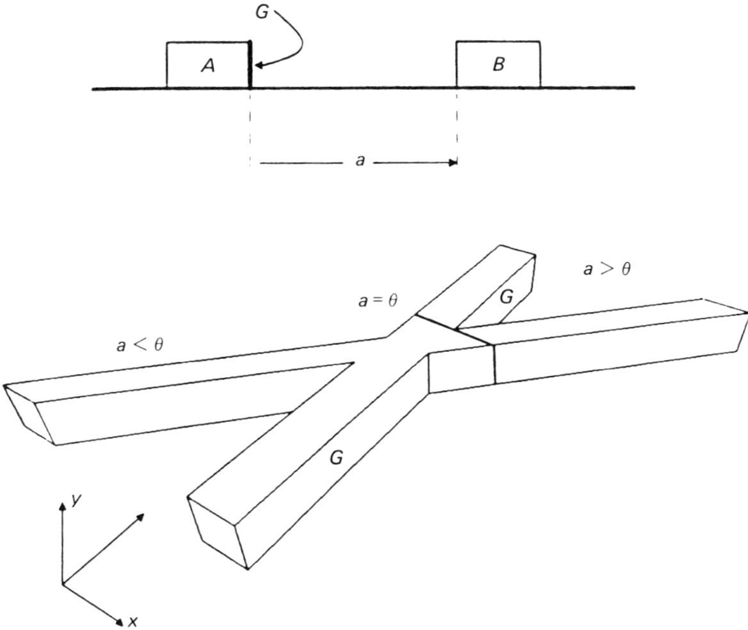

Figure 6.6
Representing the variability of a task by added dimensions in a configuration space (from [17]).

Task parameters may be viewed as new dimensions in the configuration space of the task. Figure 6.6 (from [17]) shows an example of a three-dimensional C-space in which one of the dimensions represents the relative position of two blocks and the other two dimensions represent the position of the robot hand. This type of modeling enables the planner to treat all the degrees of freedom in the task uniformly, albeit at a large computational cost.

6.4.3 Parameterized Plans

Parameterized plans are essentially subroutines whose inputs are variables representing those elements of the execution-time environment that cannot be known at plan generation time. Generally, these variables are chosen to correspond to the configuration parameters discussed in section 6.4.2. Often, the parameters simply modify an action whose character is more-or-less fixed. For example, the *Approach* motion to pick up part *A* in our example depends on the value returned by the *Locate* step. In other cases, the parameter is used to make a choice between alternative actions. For example, if part A is too close to the wall of F_A, then either a different grasping position on part A must be chosen or an additional action sequence to push part A out into the middle of F_A must be executed. Thus, plans can be thought of as decision trees where the branches have fixed sequences of actions and the nodes test the values of some parameters to choose between alternative sequences.

The simplest kind of parameterized plan is one in which all the motions are defined in Cartesian coordinates relative to some nominal coordinate frame, whose actual value will only be known at execution time. This process normally goes under the name of *calibration* and is widely used in industrial tasks. Although this type of parameterized plan seems straightforwardly correct, it is not foolproof, since small variations in the position commanded for the robot's gripper can correspond to very large variations in the kinematic configuration of the rest of the robot. Such variations can cause the robot's "elbow" or other part of its structure to collide unexpectedly with an obstacle in the workspace or can require that the joint travel limits of the robot be exceeded.

Since parameterized plans, even (or especially) those supplied by an expert human programmer, can fail, verification is an important planning function. There are two main approaches one might take:

1. *User-provided verification conditions:* Require that the user who provides a plan also provide a condition on the values of the task parameters that ensure reliable execution. The planner verifies that these conditions will be satisfied before including the plan in the execution sequence [8, 59, 55].
2. *Fully automatic verification:* The planner can also simulate the plan taking into account the range of variability in the environment and in the control system. [12] shows how to verify conditional plans made up of sequences of compliant motions in a known environment. The *forward projection* computation defined in [44, 21, 17] is also relevant here.

Option 2 is desirable since it reduces the burden on the user, but it is also extremely difficult. A great deal more work is required for both options.

6.4.4 Planning Sensing

In our discussion of parameterized plans, we have treated the sensing necessary to obtain the actual values of the task parameters as a primitive operation. This is unrealistic. The sensor values are complex functions of the task parameters. The sensor planning problem is twofold:

1. Produce a series of sensor readings that can be used to unambiguously solve for the task parameters themselves.
2. Predict the effect of the runtime sensing information on the rest of the plan.

The type of sensing and sensory interpretation required for a task will depend on the variability in the task. If a part is known to be very close to a nominal position, it may be that locating one edge of the part (by looking near its nominal position) is sufficient to locate the part unambiguously. The planning problem in this case is identifying which features of the predicted scene are most likely to be visible and determining how they should be used to solve for the task parameters [3].

If the variability in the part's position is large, quite a few features may have to be located before the part's location can be known with confidence. In this situation, it is not usually worthwhile to anticipate what the scene will be like. Two important approaches to this more general problem of interpreting sensory data are

1. Gather data for a relatively small number of features with unambiguous interpretation, e.g., a hole or a corner; each reading generates constraints on the task parameters. The model of the variability in the parameters can be used (on-line) to select sensing steps that guarantee either unambiguous interpretation or most information [26, 56, 20].
2. Gather a larger number of less distinctive features, e.g., edge fragments, and use the geometric constraints implicit in the model to prune inconsistent matches [27, 25, 22].

Which of these approaches is better depends on the avaiability of distinctive features, the variability in the position and shape of parts, and the extent of the interaction of the relevant features with the rest of the environment.

Once a match between the sensor data and the task model is found, the solution for the position parameters of an object is done by a fairly straightforward mechanical process, either numerical [3, 37, 22], symbolic [9], or hybrid [59].

In all cases, the planner will have to predict the impact of the on-line information on the rest of the plan. This problem has received little attention, but some relevant work exists [64, 19].

6.5 Planning with Uncertainty

Naive robot programs typically assume that the robot goes exactly where it is told to go, that the information reported by the sensors is accurate and complete, that the robot's internal model of the world is accurate, and that parts interact as expected. These assumptions are, of course, wrong (even dangerous), although they may be sufficiently good to allow at least some plans based on them to succeed. In this section, we consider the problem of planning when naive assumptions cannot be used.

We should distinguish between two types of discrepancies between idealized models and the actual state of the world. There are *errors*, such as an unexpected object along a trajectory or a missing feature on a part, and there is *uncertainty*, such as that arising from the limited precision of the robot. We shall postpone our discussion of errors until section 6.5.3 and concentrate initially on uncertainty, i.e., on relatively small parametric variations between model and reality.

The most common case where uncertainty affects the feasibility of a naive plan is assembly of parts whose clearances are on the same order as the precision of the robot and/or the sensors. In these circumstances, a program that simply measures the part locations and then commands motions to move parts to their nominal final destinations is almost bound to fail. Another important example of the effect of uncertainty arises in grasping an object with parallel-jaw grippers. Unless the object is exactly centered, one of the jaws will strike it first and (typically) cause it to rotate. Depending on the speed of the motion, the amount of initial misalignment, the mass distribution of the object, friction, and other factors, the grasping operation may or may not result in the object being firmly gripped and in a predictable orientation. Although sensing is often used to measure geometric variations, we shall concentrate on the residual uncertainty after all measurement operations that can be planned have been done. For example, the plan outline developed in section 6.4.1 used a vision sensor to locate part A in feeder tray F_A. Although we did not know at plan generation time just what value would be returned, we assumed that the runtime value would be sufficiently accurate. The measurement error, however, may be enough to prevent an accurate insertion of part A into a low-clearance hole.

The residual uncertainty can be handled by constructing *strategies* that combine compliant motions and additional sensing. These strategies must be planned so as to guarantee that the task can be successfully carried out in spite of the remaining uncertainty in task parameters.

6.5.1 Paradigms

Suppose that assembly fixture F_B in our example consists of a nest that holds the bottom of the part while leaving the top clear. Although a chamfer is provided to facilitate placing part B into the fixture, some clearance remains at the bottom, so that the part's exact position relative to the fixture is not known exactly. Similarly, manufacturing tolerances may introduce small variations in the location of the stud on part B and of the notch in part A. Suppose, further, that the robot's positional resolution, repeatability, and absolute accuracy are limited, but that sensors in the fingers are capable of measuring forces exerted in the X and Y directions and that the robot's control system permits it to use this information in suitable ways (discussed later).

The steps in the plan are

1. *Calibrate the workspace*, as before. The primary additional considerations are that we must now consider the limited accuracy with which the robot can locate (or return to) the calibration points, limited resolution of the calibration fixtures, and tolerance buildup between the calibration points and the functionally important point, i.e., the location of the stud on part B.
2. *Locate* parts A and B as precisely as possible, but uncertainty will remain.
3. *Approach* part A as closely as possible with a *gross motion* that is guaranteed to be safe, assuming worst case bounds on uncertainty. For example, this approach motion should open the fingers wide enough to allow for inaccuracies in the robot and for the residual uncertainty in part A's location.
4. *Grasp* part A so that its position relative to the hand is well defined. If this operation is improperly planned, complex interactions between the part and the fingers during the grasping motion can cause a "failure" result such as a cocked part. Although planning methods for grasping motions that lead to a predictable result in at least some situations have been developed [10, 48], there will still be some residual uncertainty.
5. *Transport* part A to a point near to part B. Once again the destination is chosen so that the uncertainties in the positions of the parts may be ignored. The planning then reduces to planning a path that avoids collisions.

6. *Assemble* the parts using compliant motions that are guaranteed to succeed in spite of the remaining uncertainty. The compliant motion planning requires finding commands that will bring part *A* into the desired relationship with part *B* in such a way that the executor can detect the termination of each motion. This problem is known as *fine motion planning.*
7. *Release* part *A* by opening the finers.

The introduction of uncertainty into our plan brings up several significant issues: estimating and propagating uncertainty, error detection and recovery, and planning fine motion strategies. We discuss these issues in more detail below.

6.5.2 Estimating and Propagating Uncertainty in Sensing and Action

A planner must be able to represent the expected uncertainty in the task parameters. One common method [59, 8] uses constraint expressions similar in form to those discussed in the section on uncertainty. Propagating worst case bounds to estimate the bounds on other parameters often leads to very large bounds on the parameters. Since the likelihood that an action will be robust over a range of parameters sharply decreases as the variation grows, it is desirable to find alternatives to the use of worst case bounds. When dealing with variability there was no choice, since all the values of the parameters were assumed to be equally likely. In the case of uncertainty it is often justified to assume that the parameters are random variables with a known distribution and to use methods for combining such distributions [57, 19]. In order to obtain computationally reasonable algorithms, however, these methods compute statistical measures whose interpretation is only clear if assumptions are made about the distributions (typically that the distributions are Gaussian). These assumptions need to be justified empirically, but when justifiable they lead to simple combination methods and much narrower estimates of the values of task parameters.

6.5.3 Error Detection and Recovery

It has been noted (e.g., [59, 40]) that in complex robot applications, more than half of the code is devoted to checking for unusual circumstances. Typically, instructions are inserted after every major plan step to verify that that step has proceeded as expected, usually by checking position and force readings against expected values. If an undesirable result is detected, additional instructions attempt to diagnose the problem, correct it, and resume the plan at an appropriate point. In simple cases, it may suffice to repeat the failing step. In extreme cases, the only recourse is to abandon the current task, clear the workspace, and begin anew. The combination of these activities is usually called "error detection and recovery," which we shall abbreviate as EDR [17].

It is important, however, that we distinguish between three very different types of situations that are normally called EDR.

1. A common technique for building robust manipulation strategies in the presence of uncertainty is to use a *logic-branching* strategy [61]. In such strategies the robot is commanded to perform a motion until some termination condition based on measured force and/or position is reached. Usually, the first motion commanded is the one that would achieve the task in the absence of uncertainty. Because of uncertainty, there may be several distinctly different outcomes to that command. The strategy attempts to determine which of these outcomes occurred and commands a subsequent motion to achieve the goal from that new state. It is common to call all the branches of such a strategy, except the nominal one, EDR actions. This is misleading. All the branches of such a strategy should be put on an equal footing. The fact that one branch did not achieve the goal directly is not an "error" but simply one of the expected results of performing an action in an uncertain world. We explore the synthesis of logic-branching strategies for assembly in the next section; they are not properly EDR strategies at all.
2. In some situations, the range of variability in a task is such that the task is impossible. The most common case where this happens is due to the interaction between manufacturing tolerances, e.g., a peg too wide for a hole. These situations are true errors. But we do not want to say that the task as a whole is not feasible because there are some values of the parameters for which the program is bound to fail. This is the domain of EDR strategies.
3. In other situations, the goal is unattainable because the environment has some unexpected surprise in it, e.g., an unanticipated obstacle. This is actually a stronger version of case 2 above, where we allow the environmental variability to include practically anything. Under these circumstances, it is nearly impossible to do any anticipatory planning. The robot will have to rely heavily on on-line sensing in building up its model of the world. This type of domain is more typical of mobile robots than industrial robots [6, 14].

Note that the possibility of failure has to do with the variability in the task parameters, e.g., the dimensions of the peg and the hole, and not with any uncertainty in the robot's knowledge of these parameters. When the variability in task parameters indicates that the goal may not be reachable, there is no *guaranteed* strategy for achieving the goal. The best we can hope for is a strategy that will succeed when the goal is reachable and indicate failure when it is not. We call such a strategy a *strong EDR strategy*. Note that the strong EDR condition is far from trivial in the presence of uncertainty. The difference between an avoidable failure due to sensing limitations and an unavoidable failure due to the absence of a path to the goal may be very difficult to detect. In practice, we have to settle for even weaker conditions. See [17] for a thorough discussion of these important issues.

6.6 Fine Motion Planning

In this section, we review in more detail the problem of planning sequences of compliant motions to achieve a goal in the presence of uncertainty but no task variability. The approach described here is based on that described in [44, 46, 21, 12, 17].

6.6.1 Problem Definition

The fine-motion planning problem is defined as follows:

1. The robot starts out in some point in the *initial region* I, a subset of the robot's configuration space.
2. The goal is to reach any point in a specified *goal region* G recognizably, that is, to *reach* the region and *recognize* that the region has been reached. We shall speak of the reachability and recognizability conditions on termination.
3. There are known *obstacle regions* O_i in the configuration space.

The goal is to find a sequence of robot commands that guarantee that the robot will reach *some* point in the goal region from *any* point in the initial region.

These conditions define the geometric nature of the problem. In fact, at this level the description of the problem is very similar to the gross-motion planning problems we have considered, except that the start and goal are regions instead of single configurations. The new factor in fine-motion planning is that we can use compliant motions to carry out tasks where the robot's precision is not sufficient or the uncertainty in the task is too large. The planner, therefore, must have a model of the compliant motions and their interaction with the environment.

There are many models of compliant motion [47, 61, 12]; we have picked one that is a good compromise of simplicity and realism. The choice of compliant motion model will affect the technical details of the planner, but will not change the basic approach.

The control system of the robot is assumed to implement a *generalized damper* described by the following equation:

$$\mathbf{f} = \mathbf{b}(\mathbf{v} - \mathbf{s}),$$

where $\mathbf{f}$ is the external force (and torque) vector acting on the robot, $\mathbf{v}$ is the robot's velocity vector (translational and rotational), and $\mathbf{s}$ is the commanded velocity vector. Note that when the external forces are zero, the generalized damper model is essentially a position control system; that is, the robot's actual velocity is constrained to be the commanded velocity. When external forces are acting on the robot, the actual velocity is adjusted away from the measured force.

Compliant motions are used in situations when the robot comes in contact with the environment. We assume that the objects are rigid and that the motions are slow enough that collision forces are negligible. When the robot is moving in contact with task surfaces, we model the contact forces using Coulomb friction.

There is friction on the surfaces of obstacles in the robot's workspace. We assume that the coefficients of static and dynamic friction are both equal to μ. This means that the reaction forces from a surface can be visualized as a cone, centered on the surface normal, whose half-angle is $\tan^{-1}\mu$. For two- and three-dimensional translation, if the commanded velocity $\mathbf{s}$ is within the *friction cone*, then the robot will stick. If the commanded velocity is outside of the friction cone, the robot will slide on the surface. The path on the surface is the projection of the commanded velocity vector on the surface.

The planner must also model the limitations in the robot's capabilities to sense the environment and control its motion.

The robot has bounded errors in control and sensing. The position of the robot can only be know to within a ball of radius ε_p and the actual velocity of the robot can only be controlled within a cone of half angle ε_v.

6.6.2 The Planner

The basic step in the proposed planner is computing the set of configurations, for a given commanded velocity vector **s** that are guaranteed to reach the goal region G. This new region is called the *backprojection* of the goal under the commanded velocity. The backprojection changes drastically as the command changes.

Figure 6.7 shows two backprojections of the goal in a two-dimensional peg and hole example. In configuration space, this example is transformed into a point in a transformed hole, where the diameter of the transformed hole is the clearance between the peg and the hole. In computing these backprojections, we have assumed that the half-angle of the friction cone is $\pm 15°$, but we have ignored control error.

In figure 6.7 (left), the commanded velocity points straight down. Because we are temporarily assuming that the robot has perfect control, the backprojection is an infinite band whose width is the diameter of the hole in configuration space, that is, the clearance between the peg and hole. Note that the commanded velocity is such that the robot will slide on the vertical surfaces but not the horizontal ones. Commanding a downward motion from any configuration outside of the backprojection will cause the robot to stick on one of the horizontal surfaces on either side of the hole. Therefore, the robot must be absolutely sure that it is inside the backprojection before issuing that command. But there is uncertainty as to the robot's actual position. The uncertainty in the position of the robot can be modeled as a ball in the configuration space. If the diameter of this ball is greater than the diameter of the backprojection, then there is no commanded position where the robot can be certain that it is within the backprojection.

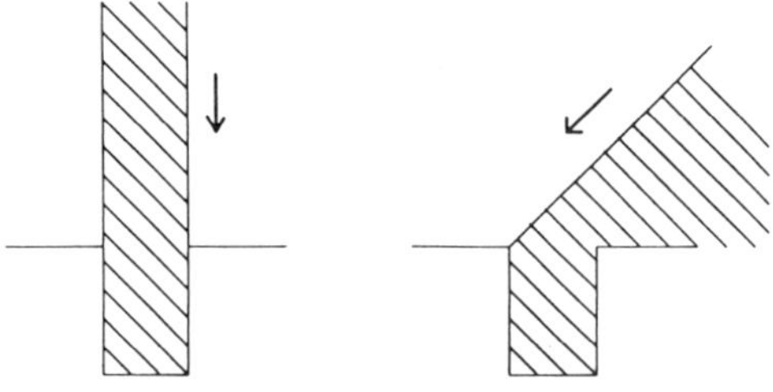

Figure 6.7
Backprojections of the goal in the absence of control uncertainty.

In figure 6.7 (right), the commanded velocity points down and to the left. The resulting backprojection is much larger than that in figure 6.7 (left). The reason is that with this commanded velocity, the robot will slide on both the horizontal and vertical surfaces. Because the backprojection grows as we move farther to the right, the robot can guarantee that the uncertainty ball is completely within the backprojection by commanding a position sufficiently to the right of the hole.

If all the initial robot positions start out within a backprojection of the goal, then the goal can be achieved with a single motion. Otherwise, the planner must identify a sequence of subgoals, with the last subgoal being the goal region G. From the discussion above, it is tempting to say that the subgoal sequence can be constructed by starting with G and repeatedly constructing the backprojections for the previous subgoal in the sequence (that is, by *backwards chaining*). The backwards chaining stops when the backprojection includes the starting region.

The idea of constructing subgoal sequences by chaining backprojections is approximately correct, but there are a number of subtle issues (see [44, 46, 21]). The most important issue was illustrated above, namely, it is not sufficient to know that the goal can be reached from some region. The robot must also know that it is within the region. Every commanded motion has a *termination predicate* that decides when the motion is to terminate. This predicate has access to the measured position and force at any point in time, but may also have access to the whole history of measured positions and forces. This predicate must be constructed by the planner. A backprojection is a suitable subgoal only if a termination predicate can be constructed for it. For example, the backprojection in figure 6.7 (left) is not a suitable subgoal when the uncertainty ball is greater than the clearance.

The interaction between the backprojection and termination predicates leads us to define a new construct, called the *pre-image* of the goal. The pre-image is made up of a set of configurations such that the robot can reach the goal and recognize that it has reached it, that is, such that a termination predicate can be constructed that will end the motion inside the goal. Therefore, it is the pre-image, and not the backprojection, that should be used to construct the subgoal sequence described earlier. But what is the relationship between the pre-image and the backprojection of the same goal? [21] has answered this question for termination predicates

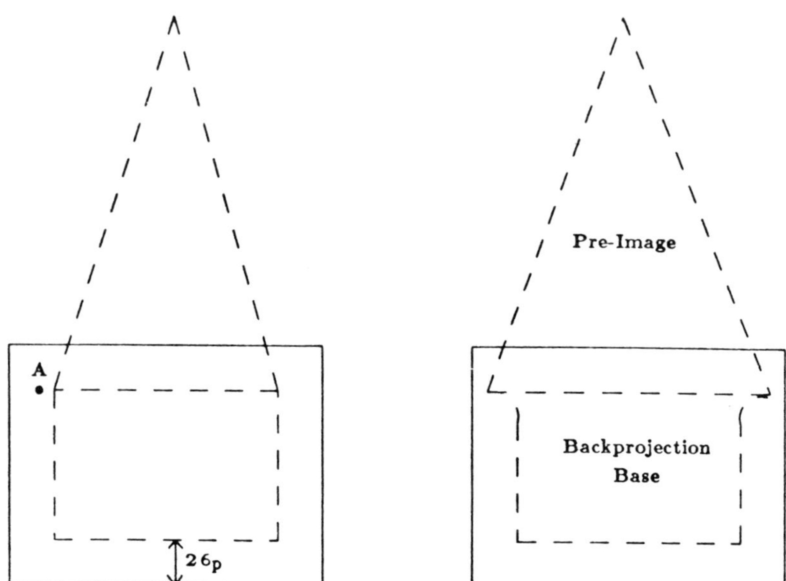

Figure 6.8
The definition of the backprojection base, due to Erdmann.

without history. To summarize, the pre-image is the backprojection of a subset of the goal, called the *backprojection base*.

Let us illustrate the definition of the backprojection base by approaching it in steps. Consider a goal that is simply a rectangular region of space. If this rectangular region in shrunk by ε_p, the uncertainty ball radius, then the robot can reliably detect entry into this region. The backprojection of this shrunken region is, at least, contained in the pre-image of the goal (figure 6.8). But this shrinking is overconservative; the point A in figure 6.8 is outside of the shrunken region, but it can also be reliably detected to be within the goal. The reason is that all nongoal configurations in the uncertainty ball at A could not have been reached by the particular commanded velocity we are considering. The actual backprojection base is shown in figure 6.8. In practice, the termination predicate may base its operation on force as well as position information. This significantly complicates the construction of pre-images for practical problems.

We have thus far ignored the effect of control error; figure 6.9 shows the effect (on the backprojection in figure 6.7) of introducing control error modeled by a cone, centered on the commanded velocity, whose half-angle

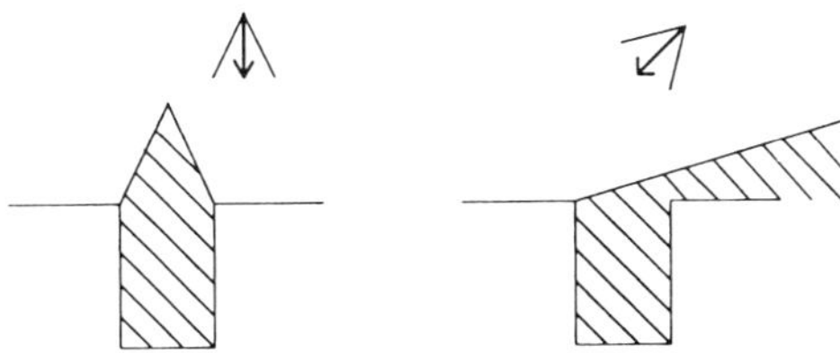

Figure 6.9
Backprojections of the goal in the presence of control error.

is also $\pm 15^\circ$. Essentially, the resulting backprojection in the presence of control error is the intersection of the backprojections (without control error) for all the velocities in the error cone. All the other considerations on the use of backprojections remain unchanged.

6.7 Conclusion

In this chapter we have attempted to characterize a number of problems and approaches in the automatic planning of robot tasks. In each case, we have stressed the importance of precise models of tasks, sensors, and controllers in clarifying the nature of the planning problem and the limitations of proposed methods. These models have a number of distinct uses. One use is as the theoretical basis for implementing prototype robot planning systems. But that is not their only function. Because the models are precise, they can serve as the basis for mathematical investigations.

One line of investigation in robot planning has explored the inherent computational complexity of planning under these models [63, 13, 50]. The results indicate that many of these problems, in their most general form and especially in the presence of uncertainty, are computationally intractable in the worst case. These results make it clear that either we must restrict automatic planning problems to computationally tractable subsets or that we must devise heuristic algorithms that give acceptable plans while offering good expected time performance.

Probably the most important uses of precise models of tasks, sensors, and controllers is to give precise answers to questions such as "How do we combine information from the sensors, control, and geometric model?" "How does uncertainty affect the construction of subgoals?" and "What is the value of information saved during execution of a manipulation program?" These obviously important questions cannot be given definite answers outside of a precise model of the planning and execution.

References

[1] A. P. Ambler and R. J. Popplestone. Inferring the positions of bodies from specified spatial relationships. *Artificial Intelligence*, 6(2): 157–174, 1975.

[2] J. Barber, R. Volz, R. Desai, R. Runinfeld, B. Schipper, and J. Wolter. Automatic two-fingered grip selection. In *IEEE International Conference on Robotics and Automation*, pages 890–896, San Francisco, April 1986.

[3] R. C. Bolles. *Verification Vision within a Programmable Assembly System.* Technical Report AIM-295, Stanford University, Artificial Intelligence Laboratory, December 1976.

[4] G. Boothroyd, C. Poli, and L. E. Murch. *Automatic Assembly.* Marcel Dekker, Inc., 1982.

[5] R. A. Brooks. Planning collision-free motions for pick-and-place operations. *International Journal of Robotics Research*, 2(4), 1983.

[6] R. A. Brooks. A robust layered control system for a mobile robot. *IEEE Journal of Robotics and Automation*, RA-2(1): 14–23, March 1986.

[7] R. A. Brooks. Solving the find-path problem by good representation of free space. *IEEE Transaction on Systems Man and Cybernetics*, SMC-13(3): 190–197, March/April 1983.

[8] R. A. Brooks. Symbolic error analysis and robot planning. *International Journal of Robotics Research*, 1(4), 1982.

[9] R. A. Brooks. Symbolic reasoning among 3d models and 2d images. *Artificial Intelligence*, 17, 1981.

[10] R. C. Brost. *Planning Robot Grasping Motions in the Presence of Uncertainty.* Technical Report CMU-RI-TR-85-12, Carnegie Mellon University, Robotics Institute, July 1985.

[11] P. Brou. *Implementation of High Level Commands for Robots.* Master's thesis, Massachusetts Institute of Technology, December 1980.

[12] S. J. Buckley. *Planning and Teaching Compliant Motion Strategies.* Technical Report TR-936, Massachusetts Institute of Technology, Artificial Intelligence Laboratory, 1987.

[13] J. F. Canny. *The Complexity of Robot Motion Planning.* MIT Press, 1987.

[14] R. Chatila and J. P. Laumond. Position referencing and consistent world modeling for mobile robots. In *IEEE International Conference on Robotics and Automation*, pages 138–144, St. Louis, 1985.

[15] J. J. Craig. Issues in the design of off-line programming systems. In *International Symposium of Robotics Research*, Santa Cruz, August 1987.

[16] M. R. Cutkosky. *Robotic Grasping and Fine Manipulation.* Kluewer Academic Press, 1985.

[17] B. R. Donald. *Error Detection and Recovery for Robot Motion Planning with Uncertainty.* Technical Report TR-982, Massachusetts Institute of Technology, Artificial Intelligence Laboratory, 1987.

[18] B. R. Donald. A search algorithm for motion planning with six degrees of freedom. *Artificial Intelligence*, 31(3): 295–354, March 1987.

[19] H. F. Durrant-Whyte. Consistent integration and propagation of disparate sensor observations. *International Journal of Robotics Research*, 6(3): 3–24, Fall 1987.

[20] R. E. Ellis, E. M. Riseman, and A. R. Hanson. Tactile recognition by probing: identifying a polygon on a plane. In *Fifth National Conference on Artificial Intelligence (AAAI)*, pages 632–637, Philadelphia, 1986.

[21] M. A. Erdmann. Using backprojections for fine-motion planning with uncertainty. *International Journal of Robotics Research*, 5(1): 19–45, Spring 1986.

[22] O. D. Faugeras and M. Hebert. The representation, recognition, and locating of 3-d objects. *International Journal of Robotics Research*, 5(3): 27–52, Fall 1986.

[23] B. Faverjon. Obstacle avoidance using an octree in the configuration space of a manipulator. In *IEEE International Conference on Robotics and Automation*, Atlanta, March 1984.

[24] B. Faverjon and P. Tournassoud. Object-level programming of industrial robots. In *IEEE International Conference on Robotics and Automation*, pages 1406–1411, San Francisco, April 1986.

[25] C. Goad. Special purpose automatic programming for 3d model-based vision. In *DARPA Image Understanding Workshop*, 1983.

[26] W. E. L. Grimson. Sensing strategies for disambiguating among multiple objects in known poses. *IEEE Journal of Robotics and Automation*, RA-2(4): 196–213, December 1986.

[27] W. E. L. Grimson and T. Lozano-Pérez. Localizing overlapping parts by searching the interpretation tree. *IEEE Transactions on Pattern Analysis and Machine Intelligence*, PAMI-9(4): 469–482, July 1987.

[28] H. Hanafusa and H. Asada. Stable prehension of objects by robot hand with elastic fingers. In *Seventh International Symposium on Industrial Robots*, pages 361–368, Tokyo, October 1977.

[29] N. Hogan. Impedance control: an approach to manipulation. In *American Control Conference*, June 1984.

[30] W. Holzmann and J. M. McCarthy. Computing the friction forces associated with a three-fingered grasp. In *IEEE International Conference on Robotics and Automation*, pages 594–600, San Francisco, April 1986.

[31] J. W. Jameson and L. J. Leifer. Quasi-static analysis: a method for predicting grasp stability. In *IEEE International Conference on Robotics and Automation*, pages 890–896, San Francisco, April 1986.

[32] O. Khatib. Real-time obstacle avoidance for robot manipulator and mobile robots. *The International Journal of Robotics Research*, 5(1): 90–98, Spring 1986.

[33] O. Khatib and J. F. Le Maitre. Dynamic control of manipulators operating in a complex environment. In *Third CISM-IFToMM*, Udine, Italy, September 1978.

[34] B. H. Krogh. Feedback obstacle avoidance control. In *21st Allerton Conference*, University of Illinois, October 1983.

[35] C. Laugier. A program for automatic grasping of objects with robot arm. In *Eleventh International Symposium on Industrial Robots*, Tokyo, October 1981.

[36] C. Laugier and J. Pertin. Automatic robot programming: the grasp planner. In *International Symposium on Advanced Software in Robotics*, Liège, Belgium, 1983.

[37] D. G. Lowe. Three-dimensional object recognition from single two-dimensional images. *Artificial Intelligence*, 31(3), 1987.

[38] T. Lozano-Pérez. Automatic planning of manipulator transfer movement. *IEEE Transaction on Systems Man and Cybernetics*, SMC-11(10): 681–698, October 1981.

[39] T. Lozano-Pérez. *The Design of a Mechanical Assembly System.* Technical Report A.I. T.R. 397, Massachusetts Institute of Technology, Artificial Intelligence Laboratory, December 1976.

[40] T. Lozano-Pérez. Robot programming. *Proceedings of the IEEE*, 71(7):821–841, July 1983.

[41] T. Lozano-Pérez. A simple motion planning algorithm for general robot manipulators. *IEEE Journal of Robotics and Automation*, RA-3(3):224–238, June 1987.

[42] T. Lozano-Pérez. Special planning: a configuration space approach. *IEEE Transaction on Computers*, C-32(2):108–120, February 1983.

[43] T. Lozano-Pérez, J. L. Jones, E. Mazer, A. Lanusse, W. E. L. Grimson, and P. Tournassoud. Handey: a robot system that recognizes, plans and manipulates. In *IEEE International Conference on Robotics and Automation*, Raleigh, 1987.

[44] T. Lozano-Pérez, M. T. Mason, and R. H. Taylor. Automatic synthesis of fine-motion strategies for robots. *International Journal of Robotics Research*, 3(1):3–24, Spring 1984.

[45] T. Lozano-Pérez and M. A. Wesley. An algorithm for planning collision-free paths among polyhedral obstacles. *Communications of the ACM*, 22(10):560–570, October 1979.

[46] M. T. Mason. Automatic planning of fine motions: correctness and completeness. In *IEEE International Conference on Robotics and Automation*, pages 492–503, Atlanta, 1984.

[47] M. T. Mason. Compliant motion. In J. M. Brady, J. M. Hollerbach, T. Johnson, T. Lozano-Pérez, and M. T. Mason, editors, *Robot Motion*, MIT Press, 1983.

[48] M. T. Mason. Mechanics and planning of manipulator pushing operations. *The International Journal of Robotics Research*, 5(3):53–71, Fall 1986.

[49] M. T. Mason and J. K. Salisbury. *Robot Hands and the Mechanics of Manipulation.* MIT Press, 1985.

[50] B. K. Natarajan. *On Moving and Orienting Objects.* Technical Report TR86-775, Cornell University, Department of Computer Science, 1986.

[51] V. D. Nguyen. *The Synthesis of Stable Force Closure Grasps.* Technical Report TR-905, Massachusetts Institute of Technology, Artificial Intelligence Laboratory, July 1986.

[52] R. P. Paul. *Modelling, Trajectory Calculation, and Servoing of a Computer Controlled Arm.* Technical Report AIM-177, Stanford University, Artificial Intelligence Laboratory, November 1972.

[53] R. P. Paul. *Robot Manipulators: Mathematics, Programming, and Control.* MIT Press, 1981.

[54] J. Pertin-Troccaz. On-line automatic robot programming: a case study in grasping. In *IEEE International Conference on Robotics and Automation*, Raleigh, 1987.

[55] J. Pertin-Troccaz and P. Puget. Dealing with uncertainty in robot planning using program-proving techniques. In *International Symposium of Robotics Research*, Santa Cruz, August 1987.

[56] J. L. Schneiter. An objective tactile sensing strategy for object recognition and localization. In *IEEE International Conference on Robotics and Automation*, pages 1262–1267, San Francisco, 1986.

[57] R. Smith and P. Cheeseman. On the representation and estimation of spatial uncertainty. *International Journal of Robotics Research*, 5(4):56–68, Winter 1986.

[58] R. H. Taylor. Planning and execution of straight line manipulator trajectories. *IBM Journal of Research and Development*, 23:424–436, 1979.

[59] R. H. Taylor. *The Synthesis of Manipulator Control Programs from Task-Level Specifications.* Technical Report AIM-282, Stanford University, Artificial Intelligence Laboratory, 1976.

[60] S. Udupa. Collision detection and avoidance in computer controller manipulators. In *Fifth International Joint Conference on Artificial Intelligence*, Cambridge, 1977.

[61] D. E. Whitney. Historical perspective and state of the art in robot force control. *International Journal of Robotics Research*, 6(1): 3–14, Spring 1987.

[62] M. Wingham. *Planning How to Grasp Objects in a Cluttered Environment*. Master's thesis, University of Edinburgh, 1977.

[63] C. K. Yap. Algorithmic motion planning. In J. T. Schwartz and C. K. Yap, editors, *Advances in Robotics*, Volume 1, Lawrence Erlbaum Associates, 1987.

[64] B. Yin. Using vision data in an object-level robot language. *International Journal of Robotics Research*, 6(1): 43–58, Spring 1987.

7 Robotic Manipulation: Mechanics and Planning

Matthew T. Mason

7.1 Introduction

This chapter describes research in robotic manipulation, in collaboration with colleagues at Carnegie-Mellon University, the Massachusetts Institute of Technology, and the IBM T. J. Watson Research Center. The long-range goal of this research is autonomous robotic manipulation in unpredictable environments. Our main approach is to analyze the behavior of specific, well-defined manipulation systems working in well-defined task domains. By focusing on the mechanics of primitive manipulation operations, we have tended toward a position that treats these key issues as inseparable:

- What are the fundamental mechanical processes of manipulation?
- How should we model physical actions?
- How can a robot construct a plan to achieve specified goals?
- What is the correspondence between manipulation task domains, manipulator operations, and automatic planning procedures?

Our primary motivation is to understand the relation between robots and tasks. For a given set of tasks, what manipulation techniques, and what planning capabilities will suffice? And, for a given robot, what class of tasks can be performed? Precise answers require precise models of tasks and robots.

The robots described in this chapter are very simple, being limited to a single application of a single operation, except in one case, where repeated application is allowed. The three operations we explore are

- Pushing an object to align it. When a ruler pushes a matchbox across a tabletop, the matchbox will rotate to a stable orientation, with one edge aligned with the ruler. More generally, a robot can exploit the mechanics of pushing to control the final orientation of the object with great accuracy.
- Squeezing an object with a parallel-jaw gripper. When an object is being grasped by a common parallel-jaw gripper, misalignments are corrected by the pushing and squeezing motions that occur in the grasping process. As with pushing, we can accurately control the object's orientation, and obtain a stable grasping configuration at the same time.
- Orienting an object with a tiltable tray. A flat object in a rectangular tray can be completely oriented by a sequence of tilting motions, without

sensing. The tray is tilted so that the object slides along the walls, and into and out of the corners, until the orientation (and position) of the object is completely determined. Although humans find these plans quite difficult to construct, we have implemented a system that constructs the plans automatically, for objects of specified polygonal shape.

For each of these operations, we shall begin with the mechanics of the task, and derive procedures for synthesizing plans. Although the set of tasks that can be solved by any one of these operations is quite small, the operation and the planning procedure are built on an analytical foundation. Our hope is that results of this nature will define a rational foundation with which we can systematically address a broad range of robotic manipulation problems.

Balancing the work on specific operations, we are also pursuing more abstract descriptions of action and planning. The model of planning first described by Lozano-Pérez, Mason, and Taylor (1984) in the context of fine-motions can also be applied to other domains, as long as the actions of the domain can be modeled as (possibly nondeterministic) mappings on some state-space. To some extent, this abstract work has guided our approach to planning specific operations. At the same time, our experience with specific operations may help us resolve some of the difficulties uncovered in the more abstract work.

7.1.1 Related Work

This chapter summarizes work that is more fully described in Mason (1986), Brost (1986), and Erdmann and Mason (1986). A number of others have contributed in the areas of mechanics and planning of manipulation tasks.

Contact, Impact, and Friction Many of the relevant studies of contact have been in the context of mechanical assembly. Simunovic (1975) played a seminal role with his work on the mechanics of peg-insertion, along with Whitney (1982), Drake (1977), McCallion and Wong (1975), and Ohwovoriole and Roth (1981). Friction and impact also figure prominently in the problem of grasping objects, either in a robot hand or in a fixture. Key works in grasping are Asada and By (1985), Salisbury (1982), Kerr and Roth (1986), Nguyen (1986), and Mishra, Schwartz, and Sharir (1986).

Contact occurs in almost all manipulation problems, and is the source of endless difficulties. For instance, it has been known for some time that Coulomb's model of sliding friction can lead to indeterminacies in the motion of rigid bodies. If we incorporate simple models of impact, the situation worsens. Lotstedt (1981) provides some interesting insights into the problems of contact. Erdmann (1984) and Rajan, Burridge, and Schwartz (1986) provide fairly general frameworks for analyzing frictional contact problems, and address the indeterminacies that arise with Coulomb friction. Some of the problems of impact and friction are treated by Featherstone (1984). Keller (1986) addresses frictional impact in three dimensions. Wang and Mason (1987) treat two-dimensional impact problems, based on a construction by Routh (1960).

Pushing and Squeezing Until recently, the mechanics of pushing has not received much attention, with the notable exceptions of MacMillan (1936) and Prescott (1923). Our work in the automatic planning of pushing and squeezing operations is similar to the work reported by Mani and Wilson (1985) and Peshkin (1986). Mani and Wilson independently developed planning tools for pushing very similar to our pushing space diagram. Peshkin refined and extended the mechanics of pushing, and developed a method to orient objects on a conveyor belt, using operations equivalent to pushing. Barber et al. (1986) analyze the frictional forces of a parallel-jaw gripper for the purpose of selecting a grip.

Sensorless Manipulation The use of mechanical means to orient objects, without the use of sensors, has been in widespread use for many years. Examples abound in automated manufacturing, where a variety of techniques have been developed to orient and feed parts. Our work was directly inspired by the work of Grossman and Blasgen (1975), who studied the use of a tilted tray, at a fixed orientation, that was vibrated to reduce friction. Objects dropped into the tray would assume one of a few possible orientations. A robot used a touch probe to determine which of the remaining possibilities had been attained.

7.1.2 Overview

The body of this chapter describes mechanics and planning of manipulator operations. Section 7.2 describes work on pushing and squeezing operations, while section 7.3 describes work on tray-tilting operations. Section 7.4 discusses the results, the methodology, and the potential for progress towards autonomous manipulation.

7.2 Pushing and Squeezing

This section describes the mechanics of pushing and automatic planning of a restricted class of pushing and squeezing operations.

Pushing is a key element of many manipulator operations. Pushing is a good way to reduce uncertainty in the locations of objects, to move many objects at once, and to move objects that are hard to grasp. Pushing, whether intentional or accidental, can readily be observed in almost any manipulation system:

• To clean a floor, we push the dirt around with a broom. The broom can move lots of particles at once, without worrying about the precise locations of the particles. To locate, grasp, and discard each dust particle one at a time would be tedious, if not impossible.
• Rather than pick a coin directly off a table, we might use one hand to push the coin off the table, into our other hand. This allows us to avoid a difficult grasp when the coin is thin.
• Rearranging furniture, or any other objects that cannot be readily grasped, may involve lots of pushing.
• When arranging a pack of playing cards, we usually sweep them into a pile, squeeze the edges of the pack, and tap the pack on a table top. The process is much faster, and much more accurate, than placing the cards one at a time onto the pack.
• When a robot gripper grasps an object, one finger will usually make contact before the other. From the time of the first contact until the grasp is complete, the object is being pushed or squeezed between the two fingers.

Whenever grasping is difficult, whenever several objects might be moved at once, whenever there are large uncertainties in object locations, pushing might be preferable to pick-and-place operations.

7.2.1 Mechanics of Pushing

Pushing is a very complex mechanical process—frictional forces play the dominant role, and the applied forces are usually unpredictable. As a result, the motion of an object being pushed is typically underdetermined. It is nonetheless possible to analyze and plan effective pushing operations. In this section we will follow the development reported in Mason (1986). The basic mechanical problem is formulated as follows:

1. A rigid object is in planar motion on a horizontal support. The object has a single point contact with a pusher. The pusher velocity is given.
2. The support forces and the pushing force obey Coulomb's law. The support forces are distributed over finite area, giving finite pressure. The only other force applied to the object is gravity. The object's center of mass is given.
3. Inertial forces are assumed to be dominated by frictional forces (quasi-static pushing).
4. The locations and magnitudes of the support forces are indeterminate.

Point 4 is an important one: the results would have little practical value if they required detailed knowledge of the support forces. In many practical situations the support forces are sensitive to small variations in the shape of the object, and cannot be predicted.

Two problems will be considered:

1. What is the motion of the object? The answer to this question is partially indeterminate. The object's motion depends on the distribution of support forces, which is undetermined. If we assume the support forces to be given, deleting assumption 4, then a numerical procedure, illustrated in figure 7.1, can readily determine the object's instaneous rotation center.
2. Does the object rotate? If so, is the rotation clockwise or counter-clockwise? The answer to this problem is determined, and in fact is quite easily found. We use the same three rays as in figure 7.1: the two edges of the friction cone, and a ray parallel to the pusher velocity. The rays vote on the direction of rotation, with ties resulting in a pure translation (see figure 7.2). For instance, if two rays pass to the left of the center of mass, then the object will rotate clockwise.

The simplicity of these results belies the difficulty of proving them. If the force applied at the pushing contact, or the motion imparted at the pushing contact, were known, then the problem would be simpler. The problem is that neither the applied force nor the motion of the contact point is generally apparent. A detailed solution appears in Mason (1986).

7.2.2 Planning a Pushing Operation

The mechanics of pushing presented above are incomplete, yet sufficient to plan some simple pushing operations. We proceed by constructing a specific class of pushing operations, for which the mechanics is virtually com-

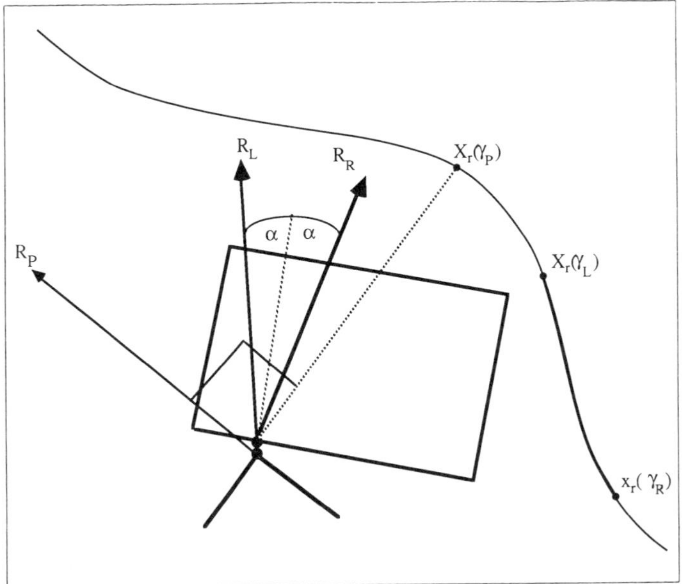

Figure 7.1
Finding the motion of the object, given the distribution of support forces. First a plot of rotation centers is constructed by numerical means, parameterized by the angle of force γ. We also construct three rays, comprising the velocity of the pusher R_P and the edges of the friction cone R_R and R_L, making angles γ_R and γ_L. The feasible rotation centers must lie in the segment of the plot of rotation centers delimited by $x_r(\gamma_R)$ and $x_r(\gamma_L)$. A line perpendicular to the ray of pushing is constructed. If this line strikes the plot within the delimited region, the intersection is the rotation center. Otherwise the nearest of $x_r(\gamma_R)$, $x_r(\gamma_L)$ is the rotation center.

plete. The restricted class of pushing operations is defined by the following assumptions:

- A planar finger is pushing a polygonal object.
- The finger moves in a straight line with constant velocity and constant orientation.
- The initial object orientation is known to within a given tolerance.

We also include the earlier assumptions of planar motion, frictional forces dominating inertial forces, known coefficient of friction, etc.

This operation has two variables that must be determined by a planning procedure: the finger orientation relative to the object, ϕ, and the direction of finger motion, δ. The planning problem is to find values for ϕ and δ such

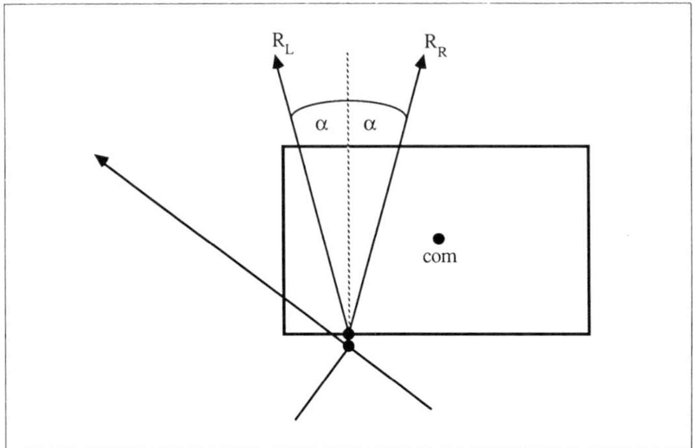

Figure 7.2
Finding the direction of rotation. $\alpha = \tan^{-1} \mu$. The three rays vote on the direction of the object's rotation. Ties result in pure translation. Each vote is determined by the relation of the ray to the object's center of mass. In this case the vote is unanimous for clockwise rotation.

that the object will reliably converge to a particular orientation. Here we follow the development of Brost (1986).

The construction in figure 7.3 simplifies the choice of values for ϕ and δ. In fact, *every* good choice is represented explicitly in the diagram 7.3e. This diagram is called a *pushing space* diagram, because it is defined by the two pushing parameters ϕ and δ. Each region shown contains exactly those pushing motions that will reliably converge to a predictable object orientation. To plan a pushing motion, we can choose any point from the appropriate region, and execute the corresponding pushing motion.

Construction of the pushing space diagram is simple. For any given pair of values ϕ and δ, we can construct the three rays, and poll them to determine a direction of object rotation. A rotation of the object changes the value of ϕ, as indicated by the arrows in 7.3b and 7.3c. For two of the rays, R_L and R_R, the vote decision boundaries are straight vertical lines,

Figure 7.3 ▶
Construction of a pushing diagram for a triangle. Diagram (a) shows the regions corresponding to contact with each vertex of the triangle; (b) shows the votes of the three rays; (c) shows the result of tallying the votes; (d) shows regions grouped according to rotation direction; and (e) shows regions of pushing motions that will have the same outcome [From Brost (1986).]

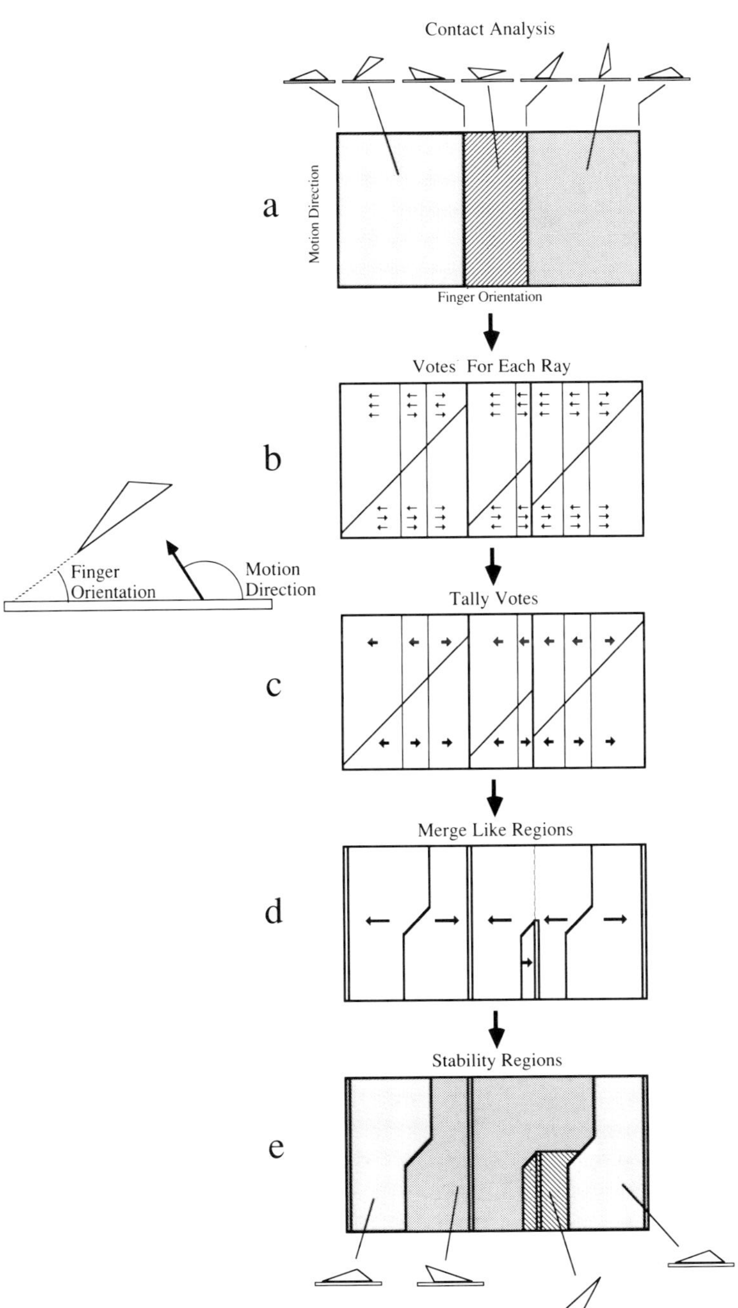
Contact Analysis
Motion Direction
Finger Orientation
a
Votes For Each Ray
b
Finger
Orientation
Motion
Direction
Tally Votes
c
Merge Like Regions
d
Stability Regions
e

since they depend on ϕ only. For the third ray R_P, the decision boundaries are diagonal lines, since the ray's vote depends on the difference between ϕ and δ. The pushing-space diagram results from tallying the votes in each region of the diagram, then grouping to form the convergence regions.

The last problem is to account for uncertainty in the object's orientation. The problem statement stipulates that the object's orientation is known to within some given tolerance. If, for instance, the object's orientation may be off by 15°, then we must not come closer than 15° to a boundary of the desired region. The easiest way to take care of this is to shrink each region by 15° horizontally. Uncertainty in the coefficient of friction and error in the control of ϕ and δ can be handled similarly.

An automatic planning procedure has been implemented that constructs pushing diagrams for any specified polygon, coefficient of friction, and uncertainty. It deals correctly with tolerances for the coefficient of friction, and error in controlling the parameters ϕ and δ. The resulting motions have been tested informally, and seem to bear out the analysis quite well.

We have neglected one important problem with pushing: to predict the distance required before convergence occurs. There are three ways to address this problem. Peshkin (1986) derives tighter bounds on the rotation center, that permit a worst-case prediction of the required pushing distance. Mani and Wilson (1985) empirically determine pushing distances. A third aapproach is to define a different operation that does not require an estimate of rotation rate. The squeeze-grasp, which we shall explore next, is such an operation.

7.2.3 Squeeze-Grasp

The *squeeze-grasp* is a restricted form of the most common parallel-jaw grasping operation. By applying the mechanics of pushing, we can predict the effect of simple squeeze-grasp operations, and plan the operation automatically. Again we are following the development of Brost (1986).

We define a special class of operations by the following assumptions:

- A polygonal object is squeezed between parallel planar fingers.
- The fingers have symmetric motions: they have equal velocities tangential to their faces, and equal but opposite velocities normal to their faces. Orientation is constant, and the translations are uniform.
- The operation continues until further squeezing is impossible, i.e., until continued squeezing would violate rigidity of the object.

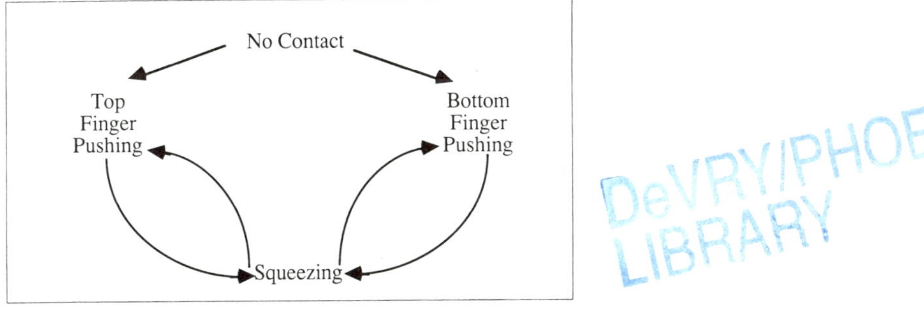

Figure 7.4
Squeeze-grasp state diagram. The operation starts in the no-contact state and ends in the squeezing state, but until the end the state can be upper-finger-pushing, lower-finger-pushing, or squeezing.

These assumptions, along with the earlier assumptions (known shape, known center of mass, uniform coefficient of friction, etc.), define the squeeze-grasp operation. As with pushing, there are two parameters: the orientation of the fingers relative to the object (ϕ), and the direction of motion relative to the finger faces (δ).

Our approach to squeeze-grasp planning is similar to push planning. The two parameters of a squeeze-grasp define a two-dimensional squeeze-grasp space. Each point in the squeeze-grasp space corresponds to a pair of values for the parameters of the operation, and hence determines a specific squeeze-grasp command. The problem is to associate the proper outcome with each point in the squeeze-grasp space.

To construct the squeeze-grasp diagram, we must consider the train of events in a squeeze-grasp, represented by the state diagram in figure 7.4. After a period of no contact, a pushing state is obtained when one finger or the other comes in contact with the object. When both fingers have made contact, a period of squeezing occurs, which usually persists until the operation is complete. However, a transition from squeezing back to pushing is possible. Generally, given the information available to the planer, the timing of these transitions are unpredictable.

Figure 7.5 shows the construction of a squeeze-grasp diagram for a triangle. Because of the symmetry between the two fingers' motions, we construct a diagram for the top finger, which is then used to derive a diagram for the bottom finger. The squeeze-grasp diagram is obtained by combining these two diagrams.

Brost has implemented an automatic planning procedure, based on

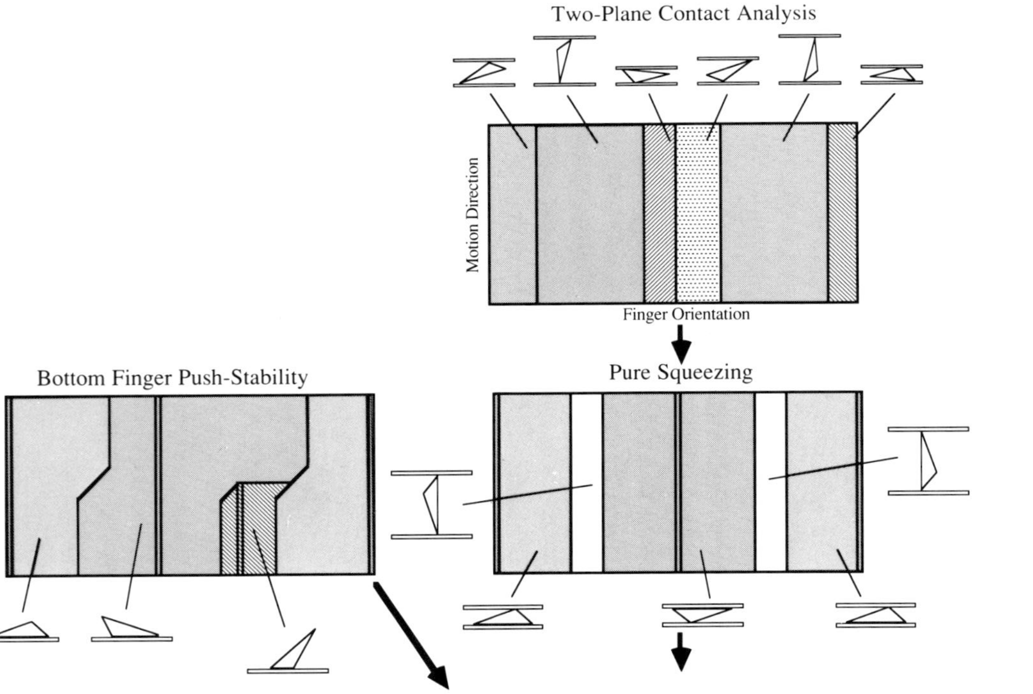

Figure 7.5
Construction of the squeeze-grasp diagram. A squeezing diagram is combined with the pushing diagram of Figure 7.3 to construct a "bottom finger pushing, or squeezing" diagram. The "top finger pushing, or squeezing" diagram is obtained by inverting and shifting the other. These two diagrams are combined to produce a squeeze-grasp diagram. Finally we plan for uncertainty in initial orientation, error in finger motion, and uncertainty in the coefficient of friction by shrinking the convergence region. [From Brost (1986).]

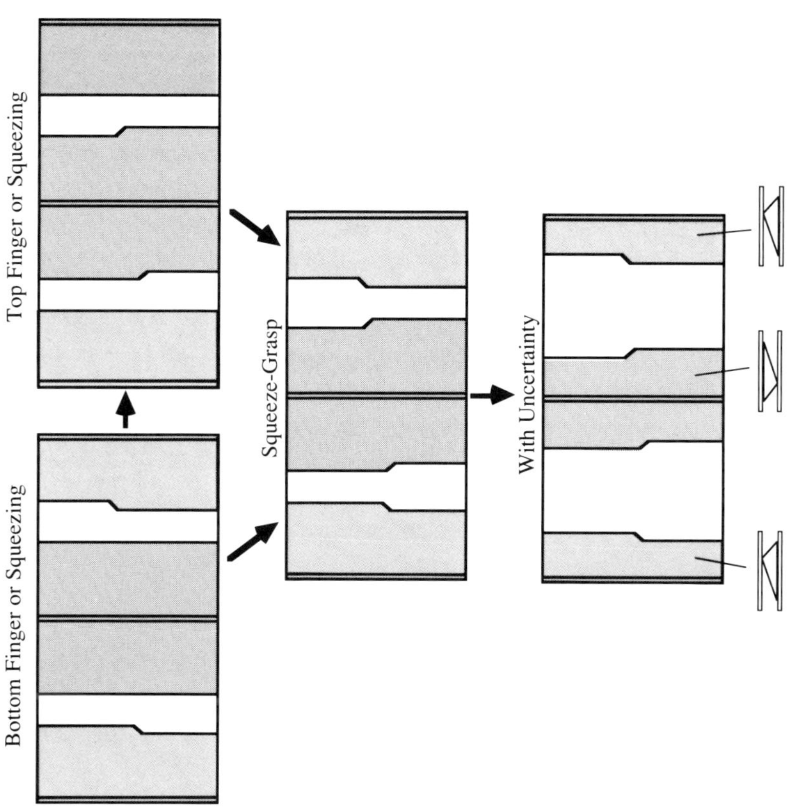
Top Finger or Squeezing
Bottom Finger or Squeezing
Squeeze-Grasp
With Uncertainty

the squeeze-grasp diagram. Uncertainty in the initial orientation, in the coefficients of friction, or in control of the finger motions can be handled by shrinking the diagrams as before. Informal testing of the operations bears out the analysis.

7.3 Parts-Orienting Tray

A randomly oriented object in a tray can sometimes be precisely oriented by a sequence of tilting motions. This section describes a planner that automatically constructs sequences of tray-tilting operations to orient and position an object of given shape. This section substantially follows the development of Erdmann and Mason (1986).

Define the *tilt* operation by the following set of assumptions:

- A planar object lies in a rectangular walled tray, which can be tilted at any angle desired.
- The forces acting on the object are gravity and friction with the walls. Friction with the bottom of the tray is neglected.

We also include the usual assumptions of known shape, known center of mass, negligible inertial forces, etc.

The configuration of the object in the tray is encoded by an ordered pair (e, p), where e is an edge of the object that is against a wall, and p is L, M, or R, if the object is in the left corner, middle, or right corner of the wall, respectively. (For simplicity's sake, we shall neglect a fourth possibility J, which is used to encode the underdetermined orientation of an object that results when the object has just slid across the middle of the tray.) Of all possible configurations of an object in a tray, only some can be encoded by a pair (e, p), but by choosing tilt angles judiciously we shall avoid all unrepresented configurations.

The mechanics of a tilt operation are conceptually straightforward, though quite complicated in application. A simple case is shown in figure 7.6, with an allen wrench in the middle of a wall. Every possible tilt angle falls into one of four equivalence classes. The wrench will either slide to the right, slide to the left, stay put, or slide into the middle of the tray. Sliding into the middle of the tray is avoided except in special circumstances, because the orientation can vary unpredictably, and lead to configurations not representable by our encoding. A more complicated case, with the wrench in a corner, is shown in figure 7.7.

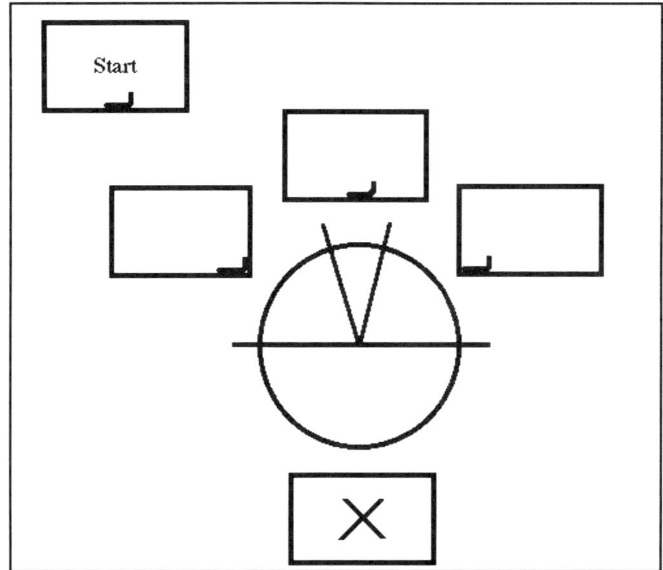

Figure 7.6
Contact analysis for the allen wrench. The pie chart groups tilt angles with the same effect, with tilt angles described by the direction of steepest ascent. Hence the upper-right sector corresponds to lifting the upper-right corner of the tray, with the effect of sliding the wrench to the lower-left corner, as shown. The X indicates a region of "illegal" tilt angles that would lose contact with the wall. [From Erdmann and Mason (1986).]

The diagrams of figures 7.6 and 7.7 represent the motions that are possible from the two initial configurations shown. We can repeat the analysis for every other configuration (e, p) of the allen wrench. The resulting collection of diagrams represents the repertoire of available actions to move the allen wrench in the tray, and is an effective means of representing the mechanics of tray-tilting for the allen wrench.

Planning of tray-tilting programs occurs in two steps:

1. Contact analysis: diagrams similar to figures 7.6 and 7.7 are constructed for every object location (e, p).
2. Search: the plan is constructed by searching a graph, where each node encodes a set of possible locations, and each edge corresponds to a set of equivalent tilt angles.

Figure 7.8 shows parts of a tree resulting from the search. We assume that the object is initially in the middle of a wall, but the orientation is un-

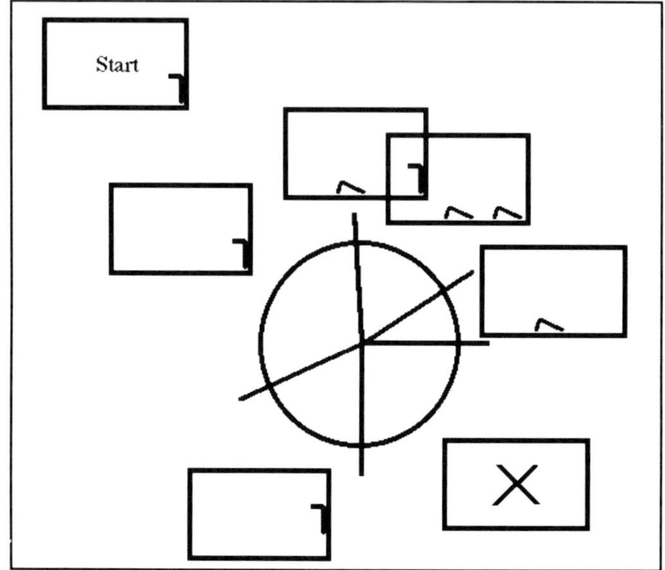

Figure 7.7
Contact analysis for the allen wrench—a more interesting tilt operation for the allen wrench. Multiple outcomes are the result of indeterminacy in rigid body mechanics using Coulomb's model of sliding friction. [From Erdmann and Mason (1986).]

known. So the root node of the tree is the set of configurations (e, M), for all edges e. Descending from the root node, we construct all nodes that can be attained by a single tilt operation. This process continues until a node is reached with just one possible configuration, or until the entire graph has been searched unsuccessfully.

The implementation uses a simple breadth-first search, although a heuristic search might be required for more complicated objects. The planner has been run on a few different objects:

- Allen wrench in *L* reflection. The search tree is shown in figure 7.9 and the resulting 8-step tray-tilting program is shown in figure 7.10. The program seems to orient correctly the wrench at least half the time. Failures are due primarily to excessive acceleration and impact forces, although a variety of other failure modes are discussed in Erdmann and Mason (1986).
- Allen wrench in *J* reflection. Obviously, the *L* wrench program can be "reflected" to orient a *J* wrench.

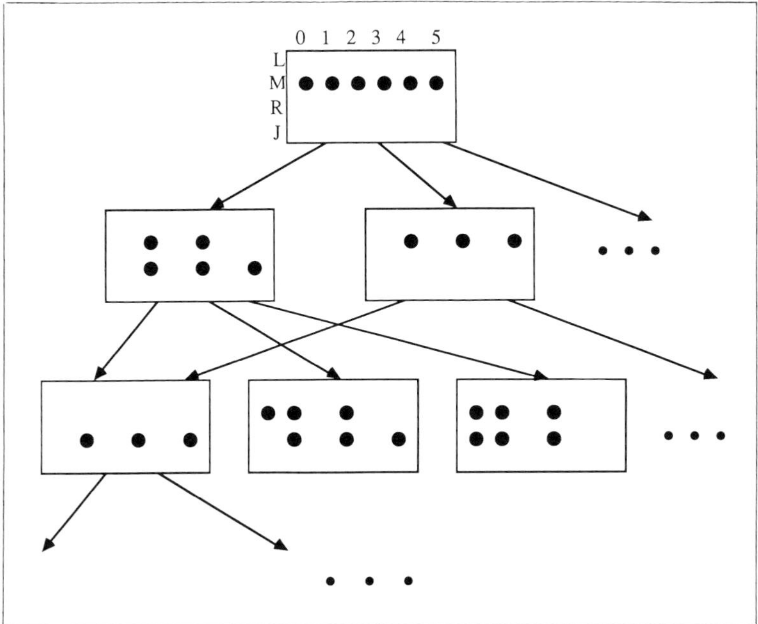

Figure 7.8
Part of the search tree for an allen wrench. Each node corresponds to a set of possible configurations (e, p), represented in tabular form. A configuration (e, p) is included if and only if a dot appears in column e and row p. [From Erdmann and Mason (1986).]

• Allen wrench in unknown reflection. Since the operations are assumed to be planar, it is impossible to orient completely a wrench whose reflection is uncertain. The best that can be done is to reach a state with two possible configurations, one for each reflection. The planner does find such a plan, which is quite a bit longer than the plans for the wrench in a known reflection.

• Binder clip. For a binder clip with the lever arms pulled together to the front, the planner finds a 14-step plan.

One problem that remains open is to characterize the class of objects that can be oriented by a tilting tray. The binder clip has left-right symmetry, and is oriented without difficulty. On the other hand, an object with rotational symmetry about the center of mass cannot be oriented. There are also asymmetrical objects that cannot be oriented. For example, an

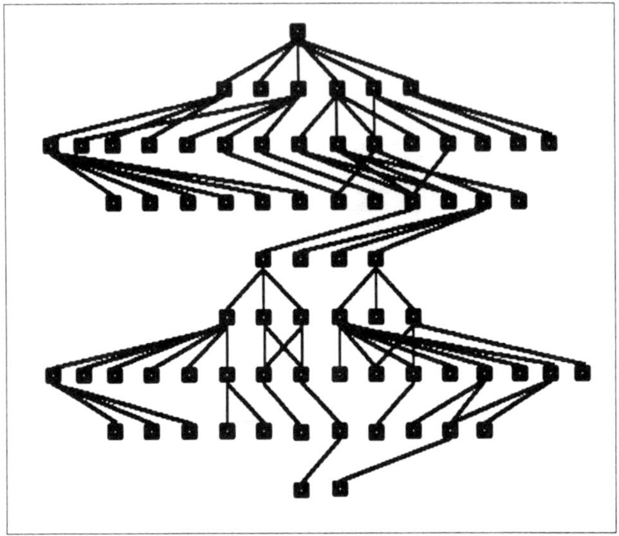

Figure 7.9
The full search tree for an allen wrench. For readability, edges pointing sideways or upwards have been suppressed. [From Erdmann and Mason (1986).]

acute triangular object cannot be oriented if the coefficient of friction is small enough. Sliding along the walls will cause no rotations, and passing through a corner will just rotate the object to an adjacent edge.

7.4 Discussion

This section discusses the philosophical background of our research, where the research is leading, and the key outstanding issues. Our primary goal is to develop a scientific foundation for robotic manipulation. Our approach is to explore the *competence relation* of robots to tasks, i.e., to characterize robots by the tasks they can solve. The methodology can be summarized in three steps:

1. *Define task domain D.* The task domain is a set of well-defined tasks, comprising models of physical objects, system dynamics, possible initial states, and goal states.
2. *Define robot R.* A well-defined robot must include models of sensation, control, a priori knowledge of the task state, and the planning procedure.

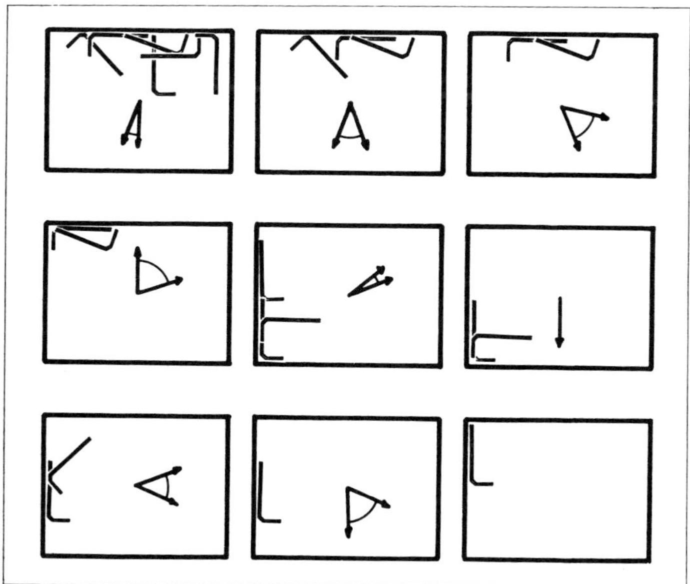

Figure 7.10
A tray-tilting program to orient an allen wrench. Reading from left to right, each frame shows the set of possible configurations, and the new tilt azimuth angle to be applied. [From Erdmann and Mason (1986).]

3. *Prove R solves D.* A proof of the robot's competence might be either formal or empirical, or both.

This methodology suggests that we build our theory around results of the form "Robot *R* is competent in task domain *D*," but there are many different ways of exploring the competence relation. For example, we can explore the relative competence of different robots or the relative complexity of two task domains. We can also expand the framework to include performance issues, specifically addressing the algorithmic complexity of the planner and the resultant plans, and time requirements for the plans.

How do our results compare with the paradigm proposed above? First, consider squeeze-grasp planning. A squeeze-grasp task is defined as follows:

• objects: two infinitely broad parallel fingers, a rigid polygonal object and center of gravity, all in the plane;
• dynamics: quasi-static motions, Coulomb friction (coefficient of friction given with tolerance), undetermined distribution of support forces;

• possible initial states: a nominal orientation for the object, and a bound on the possible deviation from nominal; and
• goal state: some specified orientation of the object, relative to the fingers.

The squeeze-grasp robot is modeled as follows:

• Sensing: none.
• Control: orientation, and translational velocity of fingers are controlled, with error subject to known tolerances.
• Planning: given shape, center of gravity, tolerances for initial orientation, coefficient of friction, and control error, use Brost's procedure to construct the squeeze-grasp diagram. If there is a convergence region for the goal orientation, choose control parameters from the convergence region, else report failure.

So squeeze-grasp tasks are parameterized by the object shape and center of gravity, the coefficient of friction, the goal orientation, and the various tolerances. We can describe any task in this class of tasks to the robot, and it will either find a solution or report failure. Various subsets of this task domain are of potential interest:

• The *scope of the operation*—the subset of tasks for which a squeeze-grasp motion exists.
• The *scope of the planner*—the subset of tasks for which the planner produces a squeeze-grasp. This is equivalent to the scope of the robot.
• The *scope of the decision procedure*—the subset of tasks for which the planner either produces a squeeze-grasp or reports failure.

In the case of squeeze-grasping, these sets bear a simple relation. The scope of the planner is equal to the scope of the operation, meaning that the planner is *complete*, and the scope of the decision procedure is the entire class of tasks.

Second, let us compare the tray-tilting work with the ideal. A tray-tilting tasks is defined as follows:

• objects: a rigid planar polygon in a rigid rectangular walled tray;
• dynamics: quasi-static motions, Coulomb friction with the tray walls, frictionless with the tray bottom;
• possible initial states: any orientation, in approximate center of tray; and

• goal states: any desired set of configurations, typically the goal is to achieve any completely determined configuration.

The tray-tilting robot is defined as follows:

• sensation: none;
• control: the direction of the gravity vector may be specified; and
• planner: for given object shape, center of gravity, coefficient of friction, and goal, apply the contact analysis and search the graph.

Again we can define the three different scopes: of the operation (a sequence of tray-tilts), of the planner, and of the decision procedure. For tray-tilting, the scope of the planner is smaller than the scope of the operation—some sequences are not considered because the planner cannot represent the intermediate sets of configurations. Subject to the same restriction, the scope of the decision procedure is equal to the entire class of tasks, but only because the search space is finite. (There are a finite number of stable configurations so there are a finite number of nodes in the graph. This provides a rather large upper bound on the size of the graph—$2^{24} - 1$ nodes for the allen wrench. The existence of better upper bounds is an open problem.)

So, for both the squeeze-grasp and the tray-tilting problems, the tasks and the robots are well-defined. Further, we can list lots of tasks for which the squeeze-grasp or the tray-tilting robot is competent. In fact, for any finite task domain, we have decision procedures that will tell us whether the robots are competent. Still, we do not have a simple characterization of the scope of these robots. Whether a simple characterization exists is an open problem.

Another troublesome aspect of this approach is that the task domains are terribly narrow. Mechanical analysis of a task requires a list of assumptions that severely limit the applicability of the results. One cannot avoid the feeling that the squeeze-grasp domain is a vanishingly small proportion of what a useful robot should be able to do. The problem of narrow scope applies to other manipulator operations, too. For example, consider the dominant operation in robotic manipulation: position control of a rigid object. A careful definition of the scope of position control would probably include the following conditions:

• The object must be rigid, of known shape. The surrounding environment shape must be known.

- The object must be rigidly attached to the hand, in a known relative location.
- The object mass must be within the manipulator specifications.
- The start and goal hand configurations, and a path connecting them, must lie in the manipulator workspace.
- The path must be free of collisions with other objects in the environment.

While perhaps less narrow than pushing, position control is far from universal. Falling outside of this scope, for instance, are the problems of placing ice cubes in a glass, where a dropping operation is required, and moving a couch, where a rigid grasp is impossible. The real power of one of these operations, such as positioning or squeeze-grasping, arises from its combination with other operations. When the squeeze-grasp is combined with a positioning operation and a placing operation, we have a system whose scope is considerably larger than the scopes of the isolated components. An important goal for future research should be to determine the scope of more complex robots that employ a variety of operations.

Perhaps the most notable aspect of this discussion is that we are forced to total speculation when comparing the scope of specific robots to a set of problems we would comfortably label "general-purpose manipulation." Apparently robotics research has contributed very little to our understanding of manipulation tasks, a situation that we believe should be rectified.

7.4.1 Models of Physical Action

We have reviewed the mechanics and automatic planning of three manipulation techniques. The stated goal is to learn something fundamental about manipulation that might be applied to the enormous variety of techniques that robots and people use. Our research on these three operations has one key common feature: we model action as a mapping from initial task configurations to result task configurations. Such a mapping can be one-to-many, representing a partially indeterminate action. The mapping is explicitly embodied in the pushing-space diagrams, the squeeze-grasp space diagrams, and in the tilt angle pie charts. This mapping proved to be a good way to support automatic planning, especially in unpredictable environments.

An important issue is whether the model of action as a mapping on task configurations is general. In its simplest form, the question is whether an action can be modeled as a change in the world. In general, action must be modeled as a change in the robot's *world model*, not just a change in the world itself (Lozano-Pérez, Mason, and Taylor, 1984; Mason, 1984). The reason is that a change in the world is usually not very useful, unless the robot *knows* that the change has occurred.

Changes in the robot's world model occur through two intermingled processes: sensation and prediction. Sensation allows the robot to adjust its world model on the basis of information arriving through its sensors. Prediction allows the robot to adjust its world model based on knowledge of its own actions, and its knowledge of common physics. Many actions involve both processes. For instance, we usually place ice cubes in a glass by dropping the cubes at the mouth of the glass. We might infer the success of the action by hearing it (sensation) or by our faith in gravity (prediction), but more likely by combining sensation *and* prediction.

For the operations described in this chapter, we can structure the planner's modeling of its actions using the following assumptions:

- The state of the robot's world model can be described as a set of possible task states.
- There is no sensory feedback; the world model will change based on prediction alone.
- Prediction uses the same mechanics as planning.

These assumptions define a special case, for which changes in the model correspond exactly to the expected changes in the world, as represented by the operation-space diagrams. The execution-time world model will make exactly the same predictions that the planner would. For this special case, then, it is quite proper to model actions by their expected effect on the world.

What about the general problem, where an action cannot be modeled as a mapping on the task state space? Strangely enough, we do not know whether the general problem needs to be solved. To simplify the discussion, the term *simple robot* will refer to a robot whose actions can be modeled as mappings on the task state, and we shall use the term *simple task* to refer to a task that can be performed by a simple robot. All of the robots explored in this chapter are simple robots, and the tasks they perform are simple tasks. We know that not all robots are simple, but we do not know whether all tasks are simple, or whether all of the tasks we care about are simple. Unless an important task falls outside the class of simple tasks, the general problem is nothing more than a mathematical curiosity.

7.5 Future Work

Our present and future efforts are meant to address specific goals: increasing the repertoire of our robots by defining and analyzing new operations; extending our models and planning procedures to include sensory operations; and demonstrating a robot that can construct or refine its models based on its own observations.

7.5.1 Increased Repertoire

From observations of robotic and human manipulation, it seems likely that effective manipulation in uncertain environments will require a repertoire of different manipulation techniques. An important part of our research is to increase the number of operations that can be effectively modeled and synthesized. A key focus is operations involving impact.

Impact is an important, neglected, process in manipulation. Yu Wang (Wang and Mason, 1987) has developed tools for predicting the effect of planar impact of an object with a flat surface subject to the following assumptions:

- Frictional forces obey Coulomb's law.
- The forces during restitution are proportional to the forces during compression (Poisson's hypothesis).
- Rolling friction and local deformations are negligible.
- The forces are impulsive—no displacement occurs during the collision.

Given an object's shape, coefficients of friction and restitution, and the position and velocity at impact, we can predict the net impulse, using a construction by Routh (1960). We can also determine the effect of uncertainty in the coefficients of friction and restitution.

There are many operations that use impact. The most obvious operations are throwing, dropping, and high-speed placing, which will require extension of our results to three dimensions, perhaps along the lines of Keller (1986). High-speed grasping involves planar impact problems. Another interesting application would be striking or tapping operations, where the robot would lightly tap the object to obtain incremental motions along a supporting surface. This operation is commonly used by machinists, who will clamp an object down and hit it with a hammer to adjust its position. By varying the clamping force, extremely fine adjustments are possible.

Another area of continuing research is the extension of our studies of pushing and squeezing. The pushing space and squeeze-grasp space diagrams require a plane-faced pusher. An extension to allow a point pusher, or multiple point pushers, would increase the power of the operations considerably. For example, for some objects it will be possible to eliminate all three degrees of uncertaintly in the plane in one operation, rather than the current maximum of two degrees of uncertainty. Point pushers can also model internal grasps.

7.5.2 Sensor-Based Planning

In collaboration with Russell H. Taylor at the IBM T. J. Watson Research Center, we are exploring the problem of planning sensor-based manipulator programs (Taylor, Mason, and Goldberg, 1987). We model sensory operations exactly the same as we model actions. We represent a priori and a posteriori states by the set of task configurations that are consistent with the robot's knowledge of the world. In the case of an action, every a priori configuration is mapped to an a posteriori configuration by the mechanics of the action. In the case of a sensory operation, the a posteriori set of configurations is a subset of the a priori set of configurations—we discard those that are inconsistent with the sensory datum. This approach works well where sensor noise is modeled using tolerances. Where useful tolerances are not available, as with Gaussian noise, the approach would require modification.

This model allows us to search a graph as we did with the tray-tilting planner. The difference is that, with an action, we can choose which edge to take from a node, but with sensing, the different edges correspond to different data, with the choice being made by nature. The result is a game tree, where an action represents the robot's move, and a sensor response represents nature's move. A plan is constructed by searching the game tree. We have implemented planners for two domains: the tray-tilting problem with simple optical interrupt type sensors in the tray; and sequences of squeeze-grasps, with sensing of finger separation.

7.5.3 Learning Robot

In collaboration with Tom Mitchell we are beginning a project to explore the process by which a robot can construct and refine models of action, through observation of the environment. The goal of this effort is to take a somewhat broader view of the purposes and uses of our models of robot operations. At present, our models are designed to support two activities: prediction and planning. In a more complete system, there should also be support for other activities: we would like the robot to recognize an operation that is being used by someone else (another robot, a human, a program written for the robot by a human, or the robot being driven by a human teleoperator); we would like the robot to adjust its models based on experience; and we would like the robot to discover new operations and construct the appropriate models.

7.6 Conclusion

We are exploring low-level models of robotic manipulation, hoping to develop a rational foundation for automatic planning of robotic manipulator programs in uncertain environments. This approach has already succeeded in providing an analytical basis for planning of pushing, squeeze-grasp, and tray-tilting operations. We are extending the approach to include impact, sensation, and learning.

Acknowledgments

Most of the work described above was performed in the Computer Science Department and the Robotics Institute, Carnegie-Mellon University. Besides Randy Brost, Yu Wang, and Mike Erdmann, whose works were cited, Tom Wood, Dan Christian, and Alan Christiansen have made substantial contributions. Some of the ideas arose during collaborative work with Russ Taylor and Tomás Lozano-Pérez. I am grateful for discussions with Marc Raibert, Tom Mitchell, Jon Doyle, and Hirochika Inoue. This research was supported by grants from the System Development Foundation and the National Science Foundation.

References

Asada, H., and By, A. B. Kinematics of Workpart Fixturing. *IEEE International Conference on Robotics and Automation*, St. Louis, March, 1985.

Barber, J., et al. Automatic Two-Fingered Grip Selection, *IEEE International Conference on Robotics and Automation*. St. Louis, March, 1985.

Brost, R. C. Automatic Grasp Planning in the Presence of Uncertainty. *IEEE International Conference on Robotics and Automation*, San Francisco, April, 1986.

Drake, S. H. Using Compliance in Lieu of Sensory Feedback for Automatic Assembly. PhD thesis, MIT Department of Mechanical Engineering, September 1977.

Erdmann, M. A. On Motion Planning with Uncertainty. AI-TR-810, MIT Artificial Intelligence Laboratory, August, 1984.

Erdmann, M., and Mason, M. T. An Exploration of Sensorless Manipulation. *IEEE International Conference on Robotics and Automation*. San Francisco, April, 1986.

Featherstone, R. *Robot Dynamics Algorithms*. Boston: Kluewer Academic Publishers, 1987.

Grossman, D. D., and Blasgen, M. W. Orienting Mechanical Parts by Computer-Controlled Manipulator. *IEEE Transactions on Systems, Man, and Cybernetics* SMC-5(5), September, 1975.

Keller, J. B. Impact with Friction. *Journal of Applied Mechanics* 53:1–4, March, 1986.

Kerr, J., and Roth, B. Analysis of Multifingered Hands. *International Journal of Robotics Research* 4(4):3–17, Winter, 1986.

Lotstedt, P. Coulomb Friction in Two-Dimensional Rigid Body Systems. *ZAMM* 61(12): 605–615, 1981.

Lozano-Pérez, T., Mason, M. T., and Taylor, R. H. Automatic Synthesis of Fine-Motion Strategies for Robots. *International Journal of Robotics Research* 3(1):3–24, Spring, 1984.

MacMillan, W. D. *Dynamics of Rigid Bodies*, New York: McGraw-Hill, 1936.

Mani, M., and Wilson, W. R. D. A Programmable Orienting System for Flat Parts. *Proceedings, NAMRI XIII*. University of California, Berkeley, May, 1985.

Mason, M. T. Automatic Planning of Fine Motions: Correctness and Completeness. *International Conference on Robotics*. IEEE Computer Society, Atlanta, GA, March, 1984.

Mason, M. T. Mechanics and Planning of Manipulator Pushing Operations. *International Journal of Robotics Research* 5(1), Spring, 1986.

McCallion, H., and Wong, P. C. Some Thoughts on the Automatic Assembly of a Peg and a Hole. *Industrial Robot*, December, 1975.

Mishra, Schwartz, and Sharir. On the Existence and Synthesis of Multifinger Positive Grips. TR 259. Courant Institute of Mathematical Sciences, New York University, November, 1986.

Nguyen, V. Constructing Force-Closure Grasps. *IEEE International Conference on Robotics and Automation*. San Francisco, April, 1986.

Ohwovoriole, M. S., and Roth, B. A Theory of Parts Mating for Assembly Automation. *Proceedings of Ro.Man.Sy-81*. September, 1981.

Peshkin, M. Planning Robotic Manipulation Strategies for Sliding Objects. PhD thesis, Carnegie-Mellon University, November, 1986.

Prescott, J. *Mechanics of Particles and Rigid Bodies*, London: Longmans, Green, and Co. 1923.

Rajan, V. T., Burridge, R., and Schwartz, J. T. Dynamics of a Rigid Body in Frictional Contact with Rigid Walls: Motions in Two Dimensions. *IEEE International Conference on Robotics and Automation*. Raleigh, NC, April, 1987.

Routh, E. J. *Dynamics of a System of Rigid Bodies*, New York: Dover, 1960.

Salisbury, J. K. Kinematic and Force Analysis of Articulated Hands. PhD thesis, Department of Mechanical Engineering, Stanford University, May, 1982.

Taylor, R. H., Mason, M. T., and Goldberg, K. H. Sensor-Based Manipulation Planning as a Game with Nature. *Fourth International Symposium on Robotics Research.* Santa Cruz, August 1987.

Wang, Y., and Mason, M. T. On Impact Dynamics of Robot Operations. *IEEE International Conference on Robotics and Automation.* IEEE, Raleigh, NC, April, 1987.

Whitney, D. E, Quasi-Static Assembly of Compliantly Supported Rigid Parts. *Journal of Dynamic Systems, Measurement, and Control* 104:65–77, March, 1982.

III CONTROL

8 A Survey of Manipulation and Assembly: Development of the Field and Open Research Issues

Daniel E. Whitney

8.1 Introduction

My colleagues and I, first at MIT and then at Draper, have been interested in manipulation and assembly for about 20 years, since the early days of robotics. One of the opportunities and responsibilities associated with working in a new field is that of giving names to things, of making up the terminology so that it is possible to talk and think about the problems.

This process involves a kind of evolution, since at first a problem area is only crudely perceived. There are only a few gross distinctions, and most of the problems are unrecognized, hence unnamed. As we make progress, we make distinctions, as if we are adding gray levels and pixels to an image. Where things used to look the same, we can now see differences, perhaps large ones. As we proceed to define and refine in this way, we continue to enrich our language and vocabulary about the problem. So, in one sense, this chapter is a small dictionary, intended to convey the meaning of some words.

Among these words are

1. gross and fine motion,
2. open loop and force-touch mediated motion,
3. geometrically determined motion, and
4. stochastically mediated or adaptive motion.

Gross motion is the most self-contained kind of motion. That is, of the types discussed here, gross motion can be formulated with the least information about circumstances outside the mover. The aim is simply to transfer an object or tool from one place to another while avoiding obstacles. Often there is little need to follow a path precisely, and the emphasis is on speed. Not much of task-related interest happens during gross motion. There is mainly the expectation that something will happen after the gross motion is finished. The size of gross motions is comparable to the size of the mover or larger.

Fine motions have the opposite characteristics in almost every respect. They are dominated by information or circumstances from outside the mover, and much of interest happens *during* the motion. Whereas the outcome of gross motion is rarely in doubt, both the outcome and the evolution toward the outcome of fine motion might be unpredictable. Speed is less dominant, and adjustment and reaction are the main features. Fine motions are usually small compared to the size of the mover.

Usually, gross motions bring objects to locations at which fine motions can accomplish tasks. An interesting counterexample is provided by spray painting, which in effect is a *large* fine motion: accuracy and outside circumstances are important, and something of interest does happen during the motion.

Open loop motion, like gross motion, depends little on outside information, and is generally preplanned. Open loop does not usually imply any particular size. The opposite of open loop is *closed loop* motion. Within this category we can include motion mediated by active sensory feedback, such as vision, force, or touch. In addition, I would like to include motions that are mediated by contact forces but without any active sensors or actuators.

Finally, I would like to distinguish *geometrically determined* motion from *stochastically mediated* or *adaptive* motion. Geometrically determined motions can be accomplished with information about the shapes and material properties of the manipulated objects. Most currently feasible robot motions are of this type. By contrast, adaptive motions require obtaining and processing information about other properties, or about statistical variations of properties, or about the processes and results of events other than the current action. Examples includes comparing the current motion with several previous ones, or making a plan for the next motion based on some risk assessment calculated from the results of prior motions.

Figure 8.1 is a diagram that combines these concepts and provides examples for six of the possible eight combinations. The missing two imply a contradiction in terms, indicating that the categories, while useful, are imperfectly defined.

While much useful research has been done in these areas, much remains to be done. The following sections of the chapter will discuss each in turn, concluding with some unsolved problems. To give the flavor of the state of the art, we can cite some examples here:

8.1.1 Manipulation

Remote manipulators are manually operated devices used for doing mechanical work in an inaccessible or dangerous location [Sheridan and Ferrell, 1967]. They have been under more or less continuous development since 1948 and now comprise dual arm configuration, stereo visual displays, and direct force feedback to aid the operator. Despite all the years of work,

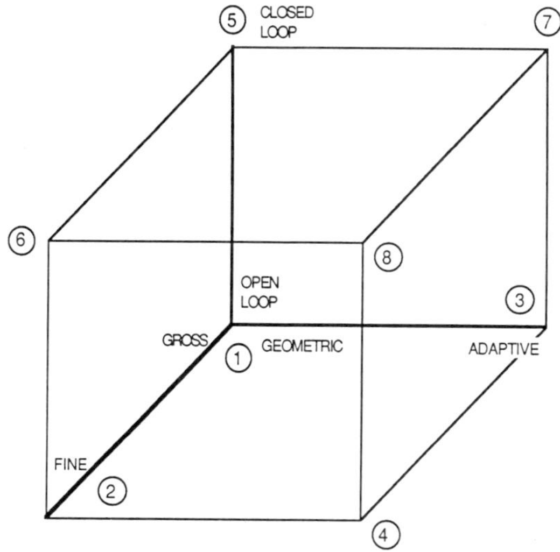

1. CARRYING PARTS
2. PUSHING PARTS AGAINST A FIXED STOP OR PALLETIZING
3. ADJUSTING A TRAJECTORY PLAN BASED ON NEW LOCATIONS OF OBSTACLES
4. DOES NOT EXIST.
5. ?
6. COMPLIANT ASSEMBLY OR FORCE FEEDBACK ASSEBMLY
7. ADJUSTING A TRAJECTORY BASED ON NEW PAYLOAD OR DRIFT OF ROBOT CHARACTERISTICS
8. PRECISION ASSEMBLY OR DEBURRING

Figure 8.1
Some kinds of possible robot motions.

these devices are still much slower than the operator using his bare hands, with the time disadvantage being typically a factor to two to ten. In the context of figure 8.1, typical remote manipulation tasks fall into categories 1, 2, 3, and 6. The best industrial robots, programmed in advance and provided with well-fixtured parts and suitable tools, can do such jobs with time ratios between 0.5 and 2. This suggests that part of the problem and the solution must be found in the definition and conditions of the task, not just in the manipulator or robot itself [Whitney, 1986].

8.1.2 Robot Payload

Some time ago, we defined a robot performance criterion called the payload ratio [Whitney, Book, and Lynch, 1974]. This is calculated as the ratio of a robot's payload to the weight of its moving parts. This ratio could approach 0.5 for a person in good condition, whereas for a robot 0.05 to

0.10 is typical and has been for many years. Some of the better payload ratios are exhibited by robots that do not have to support all of their weight on driven axes. Cartesian and SCARA designs have this property. Arms designed like the human arm have more versatility than the SCARA type and intrude less into their workspace than some kinds of Cartesian robots. This suggests that relying on a human model can have advantages and drawbacks.

The remainder of this chapter will try to continue in the spirit of this introduction, providing definitions and examples, plus indications of unsolved problems and possible solutions.

8.2 Gross Motion

Before there were robots, there were remote manipulators and artificial limbs. The earliest powered versions of each date from the 1950s. Goertz [Goertz, 1952] developed remote manipulators for the Atomic Energy Commission to aid radioactive laboratory work. These arms were activated by people, who received force feedback from the slave end through the master end they maneuvered. Although artificial arms have existed since antiquity, some of the earliest attempts to give them outside power and control were made at MIT [Rothchild and Mann] in the early 1960s. Good illustrated summaries of the state of the art of these devices in the early 1970s may be found in [Johnsen and Corliss].

Work in these two technologies began to converge in the late 1960s. At that time, several knowledge gaps were identified. On the analytical side, work began on describing manipulators kinematically and dynamically, so that equations of motion could be written [Kahn and Roth, 1971; Pieper, 1968]. On the control side, efforts began to provide computer control to manipulators, to overcome anticipated transmission delays associated with their use in space [Ferrell and Sheridan, 1967], and to artificial limbs, to extend them from one powered axis to several [Moe and Schwarz; Cavrilorić and Marić; Whitney, 1969].

All of these early efforts focused on gross motion control. This section describes the main issues.

8.2.1 Coordination and Purposeful Motions

Early work in gross motion control occurred in an environment in which kinematic models of arms were rare and difficult to derive. Computer aids were used to derive the equations, resulting in mountains of formulas due to the inability of the derivation programs to simplify the equations they ground out. Computers were big, slow, short of memory, and hungry for power. Attaching them to amputees or sending them into space seemed unlikely.

Inverse kinematic models, in particular, were hard to obtain or non-existent. Since inverse models are the basis for kinematic control to drive an arm to a desired endpoint, the lack of models was a severe drawback (see section 8.2.2).

However, it was not a total blockage in the environment of the artificial arm. Such a device is driven by someone who wears it and can see its end point at all times. His natural way of guiding it through gross motions might well be incremental and relative, if he had the proper control interface. That is, if he could see where the hand was and where he wanted it to go, he could imagine a path along which he could drive the hand to get it there.

A basic problem was to provide a framework and interface language by which an amputee could express his path, as well as modify it as it evolved. This is no problem for an arm with one powered gross motion axis. Typically that axis is the elbow. The challenge is to provide multiaxis control.

But this raises an additional problem: an amputee can control a single-axis powered arm "naturally" by means of electromyeographic (emg) electrodes attached to the skin of his stump. These electrodes pick up signals from the biceps and triceps muscles and he can operate the elbow in much the same way he did his real arm. Such natural control sites are usually missing for the other axes, especially for a full amputee who lacks an effective shoulder joint.

Thus a major challenge was to provide an efficient way to control several joints at once. But simultaneity is not enough. One must achieve coordination, that is, a simultaneous motion of several joints that does something useful in what have come to be called outside (either world or hand) coordinates. Figure 8.2 illustrates some useful motions.

Achievement of simultaneous coordinated control promised an improvement over existing velocity control methods for underwater and large terrestrial manipulators as well. Typical of the state of the art in the 1960s was switch box control, in which each switch turned one arm joint on or off. Controlling an arm in this way is unnatural and confusing to a person,

Figure 8.2

as well as being inefficient due to the lack of simultaneity. The commander of the research diving vessel Alvin told me that its arm had two control speeds: "Slow, and real slow."

Returning to the idea that an amputee can visualize an incremental motion along a path in front of himself, one can imagine providing a hand with "steering directions" that express useful motions. In world coordinates, suitable names might be FORWARD, LEFT, UP, CLOCKWISE, and so on. In hand coordinates, evocative names are REACH, LIFT, SWEEP, TWIST, TURN, AND TILT [Whitney, 1969]. Figure 8.3 shows these names attached to a hand.

Two problems remained: providing the control sites and formulating the equations that convert commands in world coordinates into motions in joint coordinates. The former still has not been satisfactorily solved for amputees. Faced with the lack of direct control sites, researchers on this issue turned toward various methods of interpreting signals from secondary emg sites on the amputee's chest, shoulder, and back [Jacobsen]. But control of multiaxis artificial arms still seems to lie in the future. Coordinated control of manipulators has proved to be much easier, and utilizes a variety of three- and six-axis hand controllers [Nevins, et al., 1972; Hirzinger]. The example of the Space Shuttle Remote Manipulator System (RMS) is described below.

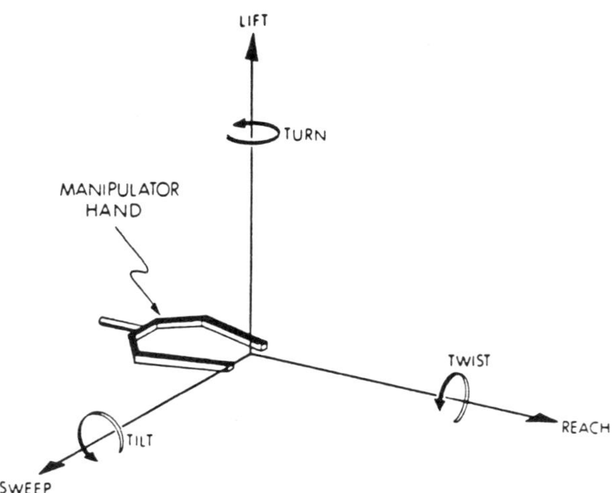

Figure 8.3
Hand with hand-oriented coordinate system.

8.2.2 Resolved Motion Rate Control

My own work at this time was on the mathematics of coordinated motion [Whitney, 1969; Whitney, 1972]. It was inspired by attempts to solve the inverse kinematics problem numerically. The method employed was the Newton-Raphson technique. If the hand is at a world coordinate location $\mathbf{x}_i$ (corresponding to joint coordinate location $\boldsymbol{\theta}_i$) and we desire it to be at $\mathbf{x}_f$ (corresponding to $\boldsymbol{\theta}_f$), then we solve for the unknown $\boldsymbol{\theta}_f$ by writing

$$\mathbf{x}_f \approx \mathbf{x}_i + \partial\mathbf{x}/\partial\boldsymbol{\theta}(\boldsymbol{\theta}_f - \boldsymbol{\theta}_i), \tag{1}$$

or

$$\mathbf{x}_f \approx \mathbf{x}_i + \mathbf{J}(\boldsymbol{\theta}_f - \boldsymbol{\theta}_i), \tag{2}$$

or

$$\boldsymbol{\theta}_f = \boldsymbol{\theta}_i + \mathbf{J}^{-1}(\mathbf{x}_f - \mathbf{x}_i). \tag{3}$$

This is the classic formulation of the Newton-Raphson method. The matrix of partial derivatives, **J**, is called the Jacobian matrix of the arm. It relates small motions in the world coordinate frame to small motions in the joint coordinate frame.

To see how this formulation can be converted into an on-line control scheme, one need only collect the $\boldsymbol{\theta}$'s on the left, divide both sides of (3) by a time interval ΔT, and take the limit as $\Delta T \rightarrow 0$, obtaining the basic equation of Resolved Motion Rate Control:

$$\dot{\boldsymbol{\theta}} = \mathbf{J}^{-1}\dot{\mathbf{x}}. \tag{4}$$

RMRC is by now well-known and has been advanced considerably. Early on it was recognized that the inverse of matrix **J** might not exist everywhere in the manipulator's work space, and much study has been devoted to identifying **J**'s singularities [Waldron, Wang, and Bolin, 1985].

Second, it was known early on that arms with more or less joints than world coordinate directions existed, and especially that control of redundant arms (having extra joints) was interesting. The most important advance was that of Liegeois [Liegeois, 1977] in which the extra degrees of freedom in the arm are utilized to determine its shape or configuration independently of the path of the hand.

Third, calculation of **J**, while simpler than a complete kinematic inverse, still is lengthy. Early on, we implemented an interpolation scheme that utilized precalculated inverses distributed strategically in world coordinate space. The interpolation used was similar to that used by Coons for his CAD work [Coons]. However, no great computation time savings were realized.

Finally, it was known that RMRC could be implemented in many coordinate frames, not just that of the hand or the shoulder. For example [Whitney, 1972], one could use the coordinates of a camera that is looking at the hand. Natural command directions might be along the zoom, pan, and tilt directions of the cameras. Having the camera track the hand presents no technical difficulties. Commands in each of the alternate frames can be calculated by multiplying **J** by the appropriate rotation matrix.

8.2.3 Current Applications of RMRC

The best known application of RMRC is the Space Shuttle Manipulator System (RMS). The RMS is 50 feet long and has 6 powered axes. Master-slave control was never considered practical for this arm because the typical practice of making master and slave the same size was impossible. Making the master 10 or more times smaller, for size and weight reasons, would cause problems due to multiplying all of the master's motions (including a person's hand tremors) by such a large scale factor. Thus RMRC was chosen, and there remained the choice of hand controlled design.

While six-axis hand controllers had been built and used at Draper Laboratory in the early 1970s, NASA chose to control the RMS with two 3-degrees-of-freedom controllers similar to those used by aircraft pilots. The translation commands are on one controller while the rotations are on the other. At various times in the Shuttle program, the software has contained provision for preprogramming the arm to go to certain points. Also, an innovative control coordinate frame was implemented, namely, that of a payload orbiting near the Shuttle. Such a frame was thought useful for satellite retrieval.

However, the RMS has seen limited use as a retriever of satellites, probably for two reasons. First, when the arm is extended, it has two singularities, one in the wrist and one in the shoulder. This keeps the arm from being useful in a large spherical sector above the Shuttle. Safety considerations keep the arm from working with payloads near the vehicle. Thus the useful working region is a narrow annular region over the payload bay [Nevins, Whitney, and Wang, 1978]. In addition, satellite retrieval requires good mechanical interfaces on target satellites. Few satellites have such interfaces, and several that do are not well documented. This problem falls in the class of preparing the robot's work environment, treated elsewhere [Whitney, 1986].

8.2.4 Open Issues

While there are many open research issues, here and elsewhere in the chapter I shall refer only to those to which I have given at least a little thought, making no claim to completeness. With that caveat, there follow discussions of stylized motions and standard motions, hybrid kinematic controls, redundant arms with multiple criteria, and learning of motions by infants.

Stylized and Standard Motions Gross motion programming at present occurs mostly at the level of *motion*, that is, a specification of a sequence of points in one coordinate system or another. If these points are in joint coordinates, then we refer to "teach by showing"; if the points are in world or tool coordinates, then we can refer to "off-line programming." But there are few provisions in current robot programming languages for named functional motions, with the exception of APPROACH or DEPROACH, which first occurred in AL [Finkel; Taylor, 1974]. Some years ago, we at

Draper recognized the need for macros [J. L. Nevins et al., 1977] to reduce the likelihood of errors by the programmer. A typical macro contained the instructions for a complete tool change maneuver, including all the operations of air solenoids that locked and unlocked tool sockets, etc. Yet today there is still no accepted list of moves, such as "pick up part and transfer," that can be expected in any robot control language. By comparison, one an always expect "push stack" or "load accumulator" or something analogous in most computer assembly languages, or a "while" structure in a higher level language.

Hybrid Kinematic Controls The hybrids discussed here include combinations of explicitly commanded and implicitly computed motions, as well as combinations of rate and position control.

Robots often have six degrees of freedom, but the programmer may be interested in only three or some other number. If three, then they may not be all translational, but rather may be a mix, such as X, Y, and Θ_z. Only recently has attention turned to providing options in motion control so that a programmer can specify just some of the motion coordinates while an algorithm fills in the rest. By analogy, when we ask another person to go somewhere, we are implicitly referring to his center of mass, allowing him to determine the other 96 (or whatever) degrees of freedom of his body himself.

A second version of this problem concerns redundant machines, where even a complete specification in the traditionl six degrees of freedom still leaves some aspects of arm control unspecified. Liegeois [Liegeois, 1977] and others have suggested using the redundancy to avoid obstacles or joint limits, but one could imagine obeying additional constraints, such as avoiding singularities or worn regions in the arm.

For example, it is possible to extend the method in [Whitney, 1969], where redundancy was removed by an instantaneous minimum kinetic energy criterion, to cover a problem analyzed by Hanafusa et al. [Hanafusa, Yoshikawa, and Nakamura, 1981]. Here the issue is to accommodate not one but two motion tasks, one specified explicitly and the other a goal to be met as well as possible. Let $\mathbf{r}_1$ be a velocity of the hand that we want obeyed explicitly and let $\mathbf{r}_2$ be a velocity of the hand or other part of the arm that is set as a goal, defined above. Assume that $\mathbf{r}_1$ is related to joint angle rates by

$$\mathbf{r}_1 = \mathbf{J}_1 \dot{\boldsymbol{\theta}}. \tag{5}$$

Let us extend the technique in [Whitney, 1969] for absorbing redundancy, which calls for setting up a synthetic optimization function

$$C = .5\boldsymbol{\theta}^T\mathbf{M}\boldsymbol{\theta} + a\boldsymbol{\theta}^T\mathbf{r}_2 + \boldsymbol{\lambda}^T[\mathbf{r}_1 - \mathbf{J}_1\dot{\boldsymbol{\theta}}^T]. \tag{6}$$

Equation (6) contains three terms. The first is proportional to the arm's kinetic energy. The second expresses our wish to meet the goal, and the third is the constraint that the arm follow $\mathbf{r}_1$ adjoined to the criterion with a Lagrange multiplier $\boldsymbol{\lambda}$. When we carry out the usual differentiations, we obtain

$$\dot{\boldsymbol{\theta}} = \mathbf{M}^{-1}[\mathbf{J}_1^{\#}\mathbf{r}_1 + a(\mathbf{J}_1^{\#}\mathbf{J}_1\mathbf{M}^{-1} - \mathbf{I})\mathbf{r}_2], \tag{7}$$

where $\mathbf{J}^{\#}$ is the pseudo-inverse of $\mathbf{J}$ and $\mathbf{I}$ is the identity matrix. If $a = 0$ and $\mathbf{M} = \mathbf{I}$, this reduces to (4).

A third version of the hybrid motion control problem is more relevant to manual control of manipulators and was evident from the earliest days of research on Resolved Motion Rate Control at Draper. If the arm is controlled from a hand-oriented control system, the hand and the controller mechanism will in general be oriented differently from each other. The Reach direction, for example, will continue to point ahead of the operator, but he may have oriented the manipulator's hand to point to the left, and that is the way the hand's Reach direction points. This disjunction of command and response directions robs the method of some of its natural feeling. A remedy that was tried briefly at Draper consists of a hybrid control in which the translations are commanded in rate mode as before but the orientations are commanded in position mode [J. L. Nevins et al., 1973]. This was mechanized in a six-axis hand controller in which the translations were on a three-axis rate controller, which was in turn carried in a three-axis gimbal mechanism. The positions of the gimbal axes were used to control the positions of the hard orientation axes, by means of an integrating rate controller.

Learning of Motions by Infants I did all the kinematics research cited above during the time when my two boys were babies, and I learned a lot by watching them learn to use their hands and arms. A lot is known about how babies acquire the ability to talk, recognize faces, and so on, but less is known about acquisition of manipulative skills. I noticed, for example, that babies can move their arms from the shoulder at birth, but control nothing else between shoulder and fingers at that time except the palm grasp reflex. Within a few days or weeks, they bring their elbows under control and succeed in reliably getting their thumbs in their mouths. Yet it takes them about three years (based on my own observations, admittedly not scientific) to learn to coordinate wrist actions with arm motions so that they can hold a glass level until it reaches their lips. Much milk is spilled before then. (See figures 8.4, 8.5 and 8.6.)

All of the above discussion concerns gross motions. We now turn to fine motions, which differ fundamentally in size and time scales.

8.3 Fine Motion

Robot motion can often be divided up into alternating *gross motions*, in which large changes in position or orientation are made preparatory to actual work, and *fine motions*, which comprise the work itself. Fine motions are fundamentally different from gross motions and often must be carried out by different mechanisms designed specifically for them. The major differences are

time scale,
size scale,
force scale, and
resolution of motion.

8.3.1 Time and Size Scale Issues

It is no surprise, given the name, that fine motions are expected to be small, but it is also true, at least in industrial applications, that fine motions must also be fast. The combination of these has certain implications for designers of fine motion devices.

To illustrate the problem, let us do a simple dimensional analysis. We shall compare an arm (A) and a hand (H), each consisting of a rigid massless hinged link of length L, a mass M on the free end, and a torque source at the hinged end. Let τ be the required torque and T be the time needed for a move. Then the torque can be expressed in proportion to M, L, and T as

$$\tau \propto ML^2/T^2 \tag{8}$$

and

$$(\tau_H/\tau_A) = (M_H/M_A)(L_H/L_A)^2(T_A/T_H)^2. \tag{9}$$

Figure 8.4
Child learning to coordinate several degrees of freedom.

Figure 8.5
Stiff wrist drinking algorithm. This procedure succeeds as soon as parents learn to fill the glass no more than half-full.

Figure 8.6
Failure of stiff wrist drinking algorithm.

Clearly, if T for a hand is to be much less than T for an arm, then M and especially L for the hand must be much less than M and L for the arm, or else very large torques will be needed. The mass of the hand must also be added to that carried by the arm, of course.

Nature has solved this problem in people by providing us with a small hand, that is, a hand having low mass, low length, and low range of motion. Its bandwidth, or frequency of fastest oscillatory motion, is about three times higher than that of the arm, but it depends on the arm to carry it from place to place.

Again, we can illustrate the problem with a simple analysis. In figure 8.7 is shown a workpiece about to enter a chamfered hole. If the workpiece is moving toward the hole with velocity V and strikes the chamfer a distance E to the left of the leftmost unobstructed path, then it must move to the right as it continues toward the hole. That is, it must accommodate a lateral jog in its path of size E during the time it takes to travel a comparable distance (say E again if the chamfer angle is 45°) toward the hole. Let us consider this jog as a time function (Figure 8.8) and assume that a control system must command such a move. What bandwidth is required to follow such a command? The Fourier series expansion of this function is

$$f(t) = \sum \alpha_n e^{jn\omega t}, \tag{10}$$

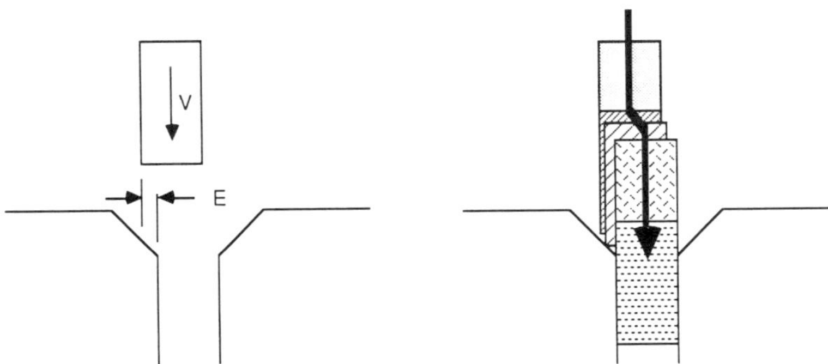

Figure 8.7
Peg entering hole with error E to the left.

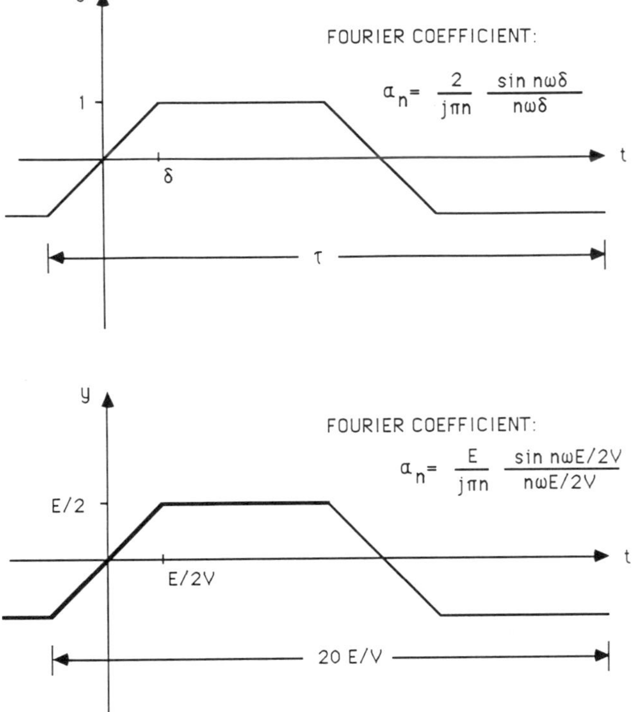

Figure 8.8
Time function equivalent to the jog in the trajectory of the peg in figure 8.7.

where

$$\alpha_n = \begin{cases} (E/j\pi n)(\sin(n\omega E/v))/(n\omega E/v), & n \text{ odd} \\ 0, & n \text{ even} \end{cases} \tag{11}$$

and

$$\omega\tau = 2\pi. \tag{12}$$

Here we must make an assumption to scale the problem, in order to define the total time τ. We shall assume that the hole is $10E$ deep; thus the peg reaches the bottom of the hole in time $10E/V$. Let this time be $\tau/2$. On this basis,

$$\omega = \pi V/10E \tag{13}$$

and the base frequency is

$$f = V/20E. \tag{14}$$

If we assume that the fifth Fourier component is a sufficient approximation to this bandwidth, we find for an example of $V = 2''$/sec (5 cm/sec) and $E = 0.05''$ (1.2 mm) that the fifth harmonic is 10 Hz. For a more economic $V = 10''$/sec (25 cm/sec), the fifth harmonic is 50 Hz, well beyond the bandwidth of gross motions of current robots.

8.3.2 Force and Motion Resolution Issues

Fine motion devices can make small moves but are limited in how small a force increment they can deliver. The issues can again be summarized with simple relationships. We repeat (4) in the form

$$\dot{\mathbf{x}} = \mathbf{J}\dot{\mathbf{e}} \tag{15}$$

and note the dual relation [Nevins and Whitney, 1973]

$$\boldsymbol{\tau} = \mathbf{J}^T\mathbf{F}, \tag{16}$$

where $\boldsymbol{\tau}$ is the joint torque vector and $\mathbf{F}$ is the vector of end point forces. In simple terms, large arms have large values in matrix $\mathbf{J}$. Equation (15) thus predicts that even small changes in joint angle will cause large changes in end point position in a large arm. Conversely, (16) says that even a small change in torque will cause a large change in end point force in a small arm. This means that small fine motion devices may have trouble making small end point force changes if there is a little friction in the joints. By contrast, it should be relatively easy to make a small change in the end point force of the Space Shuttle Manipulator, which is 50 feet (15.25 m) long, but difficult to make a small motion. The joint motors work through high gear ratios to overcome these problems, but friction and nonlinear complicance in these gears cause other difficulties.

8.3.3 Open Issues

Some important unsolved problems in fine motion can be classified into design and control issues. Design, as mentioned above, includes actuators and materials. Section 8.1 noted that most arms have a very low value of payload to weight ratio. The same is true of powered fine motion devices.

Robot hand designers face a difficult design problem if they try to duplicate or approximate the agility and degrees of freedom of the human hand. A major reason is the lack of actuators that are small, light, efficient, and strong. A second problem is the weight of hand mechanisms. The best current designs employ novel materials like kevlar and carbon-reinforced plastic, plus novel actuators like shape memory alloy wires or air pistons. These devices are discussed elsewhere in this book.

I shall defer until a later section a review of existing nonanthropomorphic fine motion devices, except to say that all are small and have limited motion range. The fastest are passive; that is, they contain no actuators or sensors. Thus they are light and capable of high bandwidth. They can easily achieve the motions called for in the above analysis, putting pegs in holes at 10″/sec, although the actuating forces, which arise directly from part-part contact, can be large for short instants if impacts occur.

Control issues are of two sorts: managing the transition between gross motion and fine motion, and creating and executing fine motion strategies. Since the latter is dependent on task knowledge, I shall again defer the discussion to a later section where tasks are treated.

The transition from gross to fine motion can happen very suddenly following contact. Both the gross motion and the fine motion devices must alter their behavior. Several possibilities exist:

a. The gross motion device halts and all motion is done by the fine motion device.
b. The gross motion device grounds itself onto the work surface, and then the fine motion device moves.

c. The two devices move in tandem, or the gross motion device moves when the fine motion device reaches the end of its useful range and needs to move farther.
d. The gross motion device begins to slow down when the fine motion device makes initial contact, so that no large impact forces are transmitted to the gross motion device, or the fine motion device is initially compliant and stiffens during impact, acting as a shock absorber.

Actively powered fine motion devices can participate in any of the above strategies, but passive fine motion devices operate only under method (c). Such devices get their motive power and actuating forces from the gross motion device and the surfaces they contact, respectively, so they cannot move actively as the other strategies require.

8.3.4 Summary

The previous two sections outlined gross and fine motion issues somewhat out of context. That is, the issues were considered from the point of view of the robot alone. If robots are to be relevant to real tasks, however, the issues must be restated in terms that include those tasks. The remaining sections approach robot manipulation and assembly from that combined point of view.

8.4 Open Loop and Force-Touch Mediated Motion

8.4.1 Gross Motion + Find Motion = Assembly

Previous sections discussed gross motions and fine motions. An important example of the combination of these in a task environment is mechanical assembly. The typical operating cycle is

gross motion to a part,
fine motion to fetch it,
gross motion to bring the part to the assembly point,
fine motion to insert or place it, and
repeat.

If we include within fine motion at least a part of the time needed to slow down just before it and speed up after it, then fine motions can take up a sizable portion of the above cycle, perhaps as much as 30%. It may be that robots moving short distances per move may never reach their top speed due to limits on their maximum acceleration, so that the average speed over a whole cycle can be quite small compared to quoted "maximum" speeds. Moreover, it may be counterproductive to try to push a robot to higher gross motion speeds because the controller may be unable to stop the arm smoothly. If it overshoots, time will be lost waiting for it to settle down within an acceptable distance from the target. This settling time will be proportional to the top speed reached during gross motion. The result is that *small* stop-to-stop time may be achieved by *reducing* gross motion speed. An analysis of control options for the Space Shuttle Manipulator in [Nevins, Whitney, and Wang] comes to this conclusion.

Combining these ideas, we conclude that a designer of a robot task must be quite careful when estimating total task time. This total time is usually tightly constrained in real assembly applications. Thus the methods chosen for assembly fine motions must be fast. The next subsection introduces some of the available methods.

8.4.2 Errors and Error-Removal Strategies in Assembly

Assembly is dominated by geometry. The shapes of parts, the clearances between their mating surfaces, the paths of robots, and the positions and orientations of parts determine most of the effects that are observed. If all of these determinants could be predicted accurately or controlled tightly enough, then assembly would occur with perfect success 100% of the time. The required accuracy is roughly the part-to-part clearance, which is often on the order of 0.1% of the size of the part. Typical clearances for parts handled by typical robots range from 0.0004″ (10 μm) to 0.02″ (0.5 mm). Such clearances are usually less than 0.01% of the size of typical robots. This order of fractional error is usual for machine tools but is well below that achievable by economical servo-controlled robots. Alternate approaches include fixed stop robots (fast, inexpensive, few taught points, difficult to reprogram) and small robots (possibly too small to do several tasks at one station). Thus we can see right away that it is not easy to achieve the accuracies required for assembly.

Note that it is not enough to have an accurate robot. As figure 8.9 shows, there are many sources of error, and they add in complex ways to produce the final error Σ between the parts. (The figure depicts only translational errors, but it should be understood that angular errors can occur as well.)

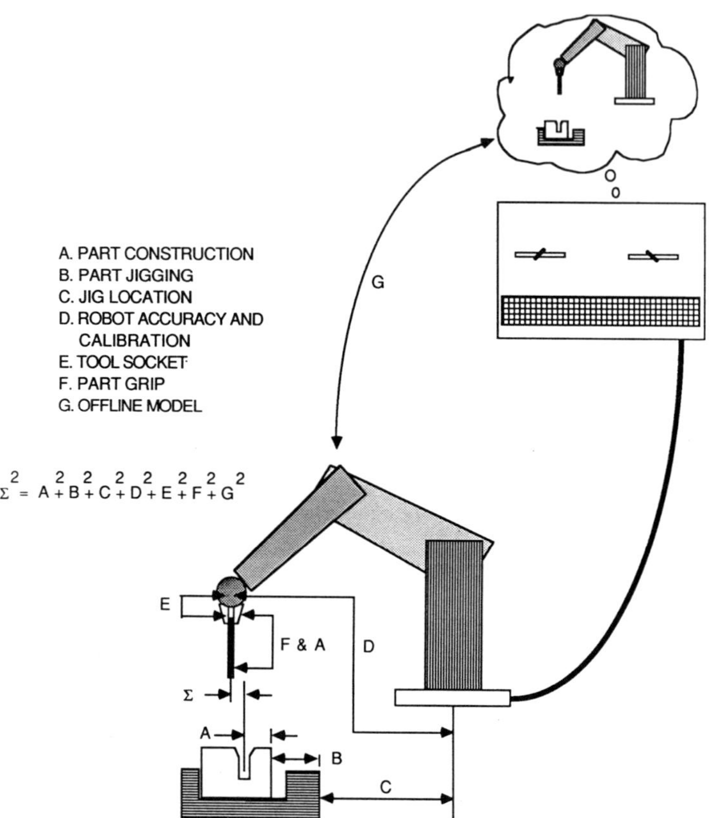

Figure 8.9
Some of the many sources of error that must be controlled.

Thus we can state the assembly problem heuristically as one of trying to keep the total part-to-part error Σ within some limit when several independent sources contribute to it. The generic name for the limit on Σ is the "error budget." The concept is well-known in the field of navigation. Since Σ and its components are random variables, they are often dealt with statistically be the methods of Statistical Quality Control. In SQC terms, Σ is a measure of "process capability" and it is customary to deem a process acceptable if Σ or its standard deviation is 6 times smaller than the budget. This level of performance can be interpreted as a percent of assembly attempts in which Σ will exceed the limit. If the errors are assumed to be Gaussian random variables with zero mean, then Σ will exceed 6 times its standard deviation very rarely.

Again, speaking heuristically, there are two ways of dealing with errors. One is to eliminate them from the start, while the other is to eliminate them during assembly. The former is usually expensive, especially since such control must be exerted on the parts as well as the robot and its tools. One must consider every surface on a part that draws dimensional reference from a gripper or jig, plus corresponding surfaces on the tools and jigs. Making all of these things to better tolerances than the functional part mating surfaces themselves is very costly, and more so if it must be repeated for many parts.

The second alternative comprises discovery and correction on a case basis. Implementing it may or may not involve the usual type of discovery, namely, via sensors. Instead, it may combine many approaches, depending on their cost per unit of accuracy, their feasibility in each case, the absolute size of Σ, and, very important, the amount of time available in which to do each assembly cycle. Factors such as the time value of money and the cost of assembly failure will dictate whether error elimination or discovery/correction will be chosen. Most practical solutions involve a combination of the two methods.

Figure 8.10 depicts some of the available strategies. At the top of the figure, unassembled parts are assumed to have a lot of error with respect to one another. The task of assembly is to remove that error. The figure shows that one can use different methods to reduce the uncertainty, and that it can be reduced in stages, with each stage being accomplished by different means, having different costs and execution speeds, and being capable of removing different amounts of error.

For example, bowl feeders vibrate parts into the correct orientation and present them at the end of a feeder track. Thus bowls can convert errors the size of the bowl itself down to a small fraction of an inch. But bowls cannot perform assembly, so even this great error reduction is not enough. A similar error reduction can be achieved by buying parts directly from the manufacturer on carriers. Electronic parts are often supplied in this fashion.

As another example, one can design high-quality grippers suited to gripping a particular part with high accuracy. However, this is costly and also implies that time may be needed to change tools in order to grip different parts. (Or else, multisurface grip fingers can be designed so that one tool can grip many parts.) Instead of making specialized grippers, one could make more general but less accurate ones. This would save some money and time but requires the use of additional error-removal methods.

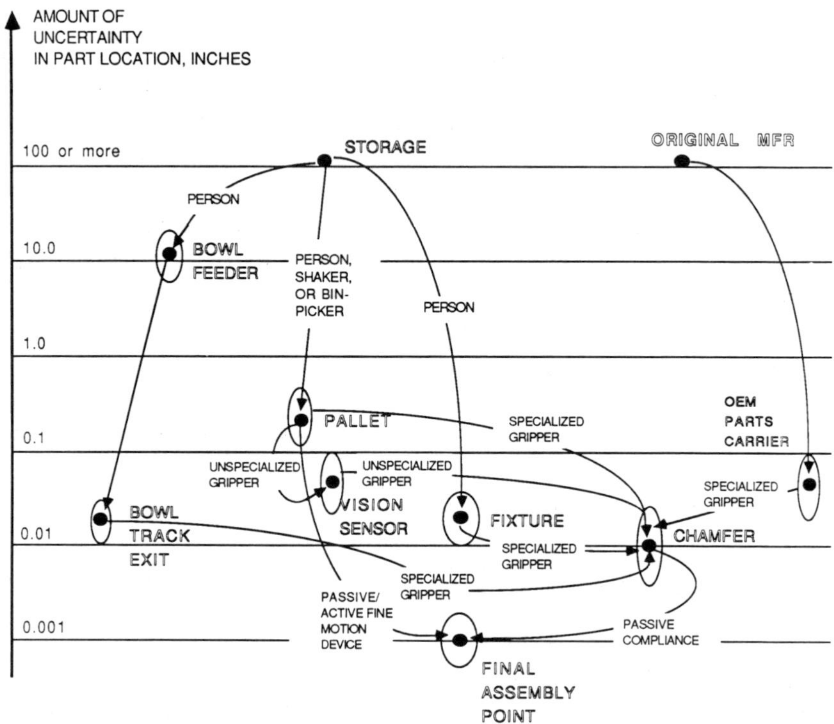

Figure 8.10
Different error removal strategies that result in parts being assembled successfully.

For example [Gordon], one can use a vision sensor to find the part in the gripper.

Even these means still may not get the parts near enough in terms of their clearances. In section 8.5, I discuss what the real combined lateral and angular error limits are. Here it suffices to say that the method normally used to bridge the last error gap is the chamfer. Chamfers may be applied to one or both mating parts. As implied in section 8.3, chamfer sizes often scale with part size, so difficult close clearance assemblies on small parts mean small chamfers, too. On parts with 50-microinch (1.2-μm) clearances, a typical chamfer is 0.004″ (0.1 mm) wide!

Thus error removal strategies can be split into two stages—getting parts close enough to mate their chamfers, and getting parts together from that point on. The first stage depends on a priori error control, employing part and gripper design, robot calibration, and so on. Two generic approaches to the second stage exist, called *force-touch mediated* and *compliant assembly*. These each have their advantages and disadvantages, especially in terms of execution time.

8.4.3 Force-Touch Mediated Motion

Force-touch mediated motion (FTMM) often goes by the names force feedback or force control. I prefer the first name because it emphasizes the relationship between force and motion rather than implying that force alone is at issue. The essence of force-touch mediated motion is that a motion task is to be performed with the help of force information. That is, we intend to use force sensing to steer or modify some motion.

To accomplish this we need several things:

1. A model of the task we want done, so that we can tell what forces will arise in various situations.
2. A method for inverting the above model so that sensed forces can be interpreted to reveal the current geometric situation. This step may involve additional sensing means.
3. A strategy for deciding what motions to call for in response to various sensed forces. This comprises the method for deciding how to convert the detected situation into the desired one.
4. A control means for exerting the desired motion without causing instability, for example by promoting extraneous forces.

This paradigm for force-touch mediated motion has been successfully applied to assembly [Simunovic; Whitney 1977; Whitney 1982] and grinding [Whitney, 1985].

A general review of robot force control appears in [Whitney, 1987]. In that paper, I describe two types of force feedback, called *logic branching* and *continuous*. Logic branching consists of a program of "if" statements and alternative motion commands. This program follows various paths depending on the occurrence of various force contacts. Often only the contact itself is noted, not its magnitude, duration, direction, or evolution over time. A well-known example of logic branching in assembly is the Hi-Ti Hand developed by Hitachi in the mid 1970s [Gotoh, Takeyasu, and Inoyama].

Our own work in FTMM has been primarily of the continuous type. Assembly usually involves continuous or extended contact between objects, during which an entire time history of vector forces and torques is generated. The magnitude, direction, and evolution of this force trajectory contains valuable information about the progress of the task that has been recruited in our force strategies.

Two concepts are central to our FTMM work. One is the idea of the *coordinate frame of reference*, and the other is *stability*.

Our work in Resolved Motion Rate Control enabled us to look upon the required motions of assembly as small corrections commanded in hand or tool coordinates. Work by others at Draper [Nevins and Whitney, 1973; Drake and Watson, 1975] had recognized the need for sensors close to the hand to sense forces and torques directly in hand coordinates, and had created such sensors.

It was a natural step to express our first FTMM strategies directly in tool coordinates. The technique adopted was to think of the desired response motion as a vector (either of incremental motion $\Delta\mathbf{x}$ or of velocity $\mathbf{V}$) that is related to the sensed force vector $\mathbf{F}$. A good way to express the relationship is via matrix multiplication:

$$\Delta\mathbf{x} = \mathbf{K}\mathbf{F}, \tag{17}$$

or

$$\mathbf{V} = \mathbf{K}\mathbf{F}. \tag{18}$$

The first expression creates spring like behavior, whereas the second creates dashpot behavior, essentially the time integral of the former. The matrix $\mathbf{K}$ has to be constructed so as to generate the desired motion commands in response to the sensed forces and torques.

Figure 8.11 is a generic diagram of FTMM. It assumes that the task is driven from the outside by a nominal position or velocity trajectory, which will be modified by force information. Both the nominal trajectory and the sensed forces are in world or tool coordinates, but the arm must be driven in joint coordinates. Thus some coordinate transformations are needed. Note that the figure contains a model of how the sensed forces arise, namely via deflection of an environmental whose stiffness is $\mathbf{K}_E$. This stiffness actually comprises everything that might deflect during contact between the arm and the world, including the arm itself, tools, parts, and jigs.

Using this particular structure for FTMM we can address the two issues of *strategy* and *stability*. Matrix $\mathbf{K}$ contains the strategy. A nonzero entry in $\mathbf{K}_{i,j}$ indicates that we want motion along component i when force/torque is sensed along component j. A zero entry means that we want to control

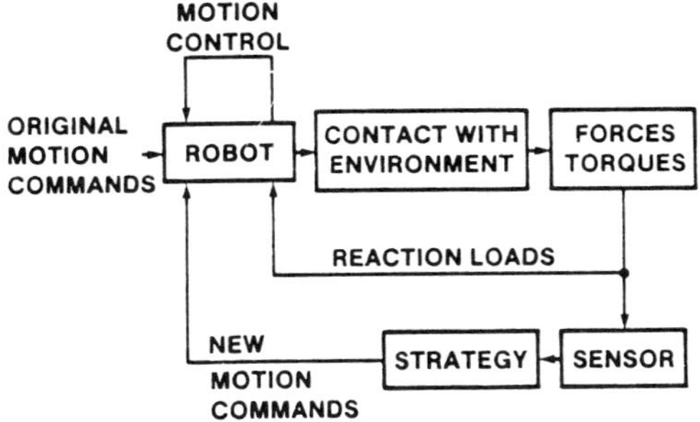

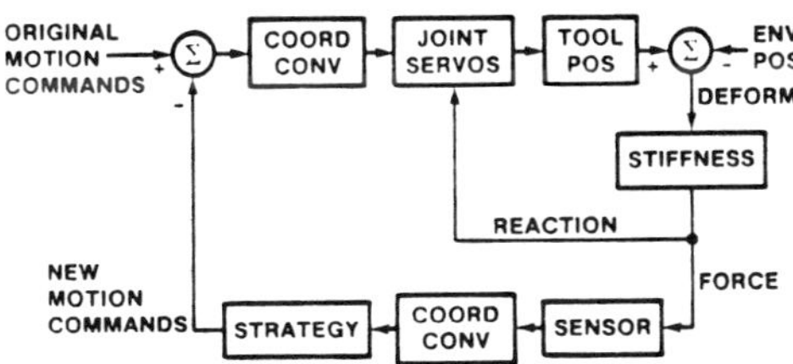

Figure 8.11
Architecture of FTMM.

that direction via the nominal position or velocity, not by force. Thus this technique is a primitive version of hybrid force-motion control [Mason, 1981; Raibert-Craig, 1981]. Several examples of **K** for box packing and edge following can be found in [Whitney, 1977; Whitney, 1986].

Of importance to us here is the **K** for putting pegs in holes. Recall that we can put the center of tool coordinates where we want, and a good place conceptually is at the tip of the peg. An analysis of the assembly process [Simunivic, 1975; Whitney, 1982] shows that lateral errors of the tip during chamfer crossing and entry into the hole cause lateral forces that point in the direction the tip should translate, while angular errors during entry cause torques that, like the forces, turn in the direction the part should reorient, using the tip as a rotation center. While lateral errors are being

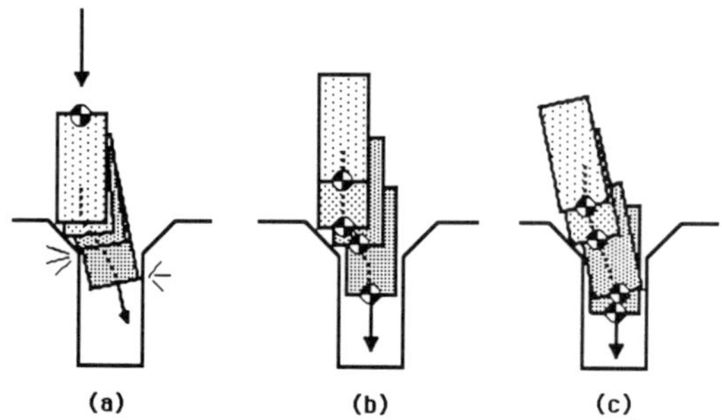

Figure 8.12
(a) Peg hits hole wall because of lateral error and wrong location of compliance center. (b) Peg with lateral error slides down chamfer parallel to itself because compliance center is at tip. (c) Lateral and angular errors are removed one at a time. Neither is made worse while the other is being removed.

corrected, the part should translate only and should not rotate; correspondingly, there should be no spurious translation while angular errors are being removed (see figure 8.12). The absence of such forces and torques indicates that there is no lateral or angular error. Since corrective motions should also occur with respect to the tip of the peg, the way to obtain them via **K** is to place entries on **K**'s diagonal so that X force will call for X motion, and so on. The off-diagonal entries in **K** can be zero. When **K** is formed in this way, lateral and angular errors will be removed independently as desired, without spurious motions.

The magnitude of the diagonal entries affects the stability of the entire system. Both sampled-data and continuous analyses of such systems [Whitney, 1977; Whitney, 1986; Roberts, Paul, and Hillberry, 1985; Eppinger and Seering, 1986] show that feedback gain **K**, environmental stiffness $\mathbf{K}_E$, other stiffnesses in the arm, sampling interval, and arm inertia all play a role in stability. Exact conclusions depend on the amount of detail in the model, but generally reducing **K** and/or $\mathbf{K}_E$ is beneficial. Note, however, that $\mathbf{K}_E$ may not be under the designer's control, and reducing **K** makes the system unresponsive. Stability of FTMM systems remains an active problem area.

Understanding that **K** for peg-holing should be diagonal when expressed in tool coordinates attached to the tip of the peg was an important conceptual step. It led directly to the concept of passive compliant assembly, which I discuss next.

8.4.4 Passive Compliant Assembly

One can make the jump intuitively from diagonal **K** to a passive spring-mounted robot wrist. Or one can invoke a principle from mechanics that says that passive linear systems can be expressed in terms of symmetric, positive semidefinite matrices. The **K** matrices we developed for assembly fit this definition. A passive system contains no sources of energy. Masses, springs, and dashpots, as well as linked combinations of them, are examples of passive systems. The Hi-Ti hand, which contains sensors, logics, and motors, is an active system. Thus in principle, there should be a passive realization of our diagonal **K**, which will have positive or zero entries on the diagonal if the resulting FTMM system is to be stable. If such a passive **K**-system could be created, it would enable us to do assembly via the energy inherent in the nominal open loop arm motion, as long as we could get the parts' chamfers to meet.

An obvious but unrealizable way to get such a **K** would be to attach linear and rotary springs to the tip of the peg, and attach their other ends to the gripper. A way had to be found to achieve the same effect with springs that attached to the peg somewhere else so that they would not be in the way during insertion. The Draper Laboratory solution to this problem is called the Remote Center Compliance or RCC [Watson, 1976; Drake, 1977]. A detailed explanation of the RCC appears in [Whitney, 1988], and design equations for several types of RCC's appear in [Drake, 1977; Whitney and Rourke, 1986]. An intuitive explanation of how the RCC works is shown in figure 8.13. A more rigorous explanation of the RCC in relation to part mating theory can be found in [Simunovic, 1975; Whitney, 1982]. Section 8.5 contains an introductory discussion.

The important point to note here is that the RCC is a true fine motion device in the spirit of section 8.3. It is capable of small motions and can do them rapidly because it and the items it carries are of low mass. It is activated by the contact forces that arise due to assembly errors, and it

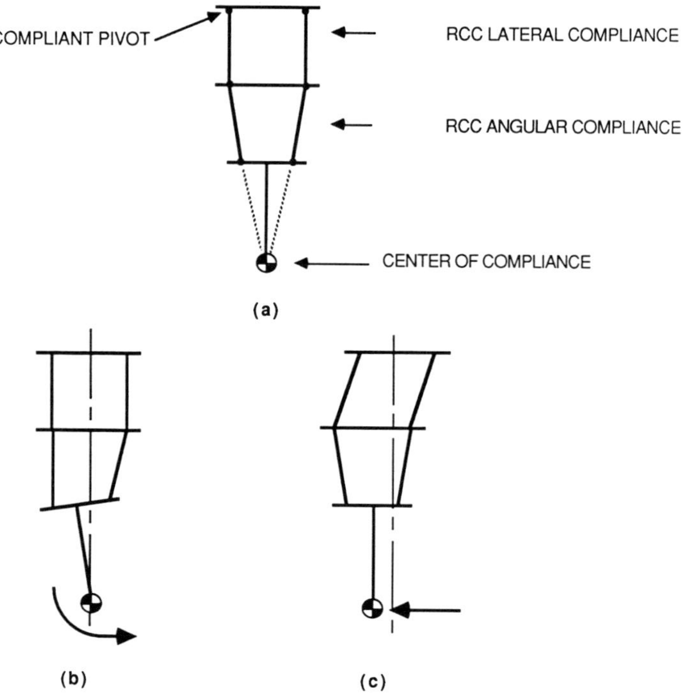

Figure 8.13
(a) The RCC consists of two parts: a lateral or translational part and an angular part. (b) The angular part responds to torques at the compliance center by rotating without translating. (c) The lateral part responds to lateral forces by translating without rotating.

automatically executes the desired motions to correct those errors. It directly mechanizes in passive hardware the diagonal **K** strategy described above.

The typical RCC is rather stiff in the direction of assembly motion. If there are substantial initial errors, a collision or wedging can occur that the RCC cannot mitigate. Also, when assembly is complete, large axial forces can occur. A way to prevent damage in such circumstances is to add additional axial compliance. This is usually accompanied by a preload so that normal insertion forces can be sustained, as well as by a sensor that detects excess axial force. The resulting system is actually a hybrid of continuous and logic-branching, as well as between active and passive.

In addition, sensing can be added to monitor the lateral and angular corrections executed by the RCC, and these sensor outputs can be used to

United States Patent [19] [11] **4,324,032**

Gustavson et al. [45] **Apr. 13, 1982**

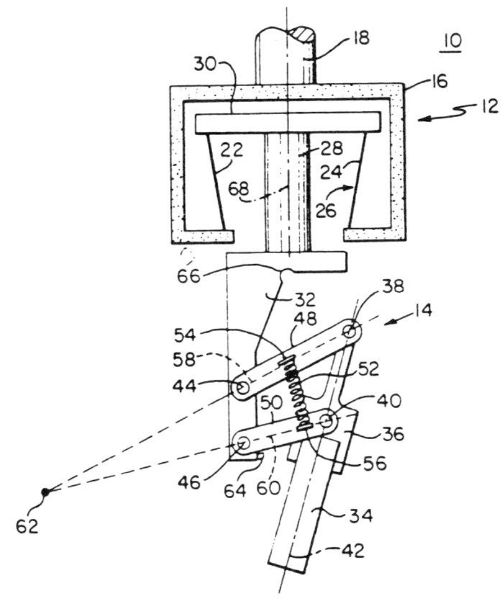

Figure 8.14
A passive chamferless insertion device. Linkage 14 permits the peg to tilt counterclockwise upon hitting the hole mouth. When the peg is vertical, it locks at point 66, and the inserter becomes an ordinary RCC. The peg is in the hole mouth at this point, and the rest of the insertion proceeds as if there had been a chamfer.

reteach or steer the arm to correct errors. RCCs with such sensors are called Instrumented RCCs (IRCCs) and are described in [De Fazio, Seltzer, and Whitney, 1984]. This reference contains additional references to applications of the IRCC. The combination of IRCC and instrumented *z*-axis compliance is by now a well-developed one in our laboratory and has been used on numerous assembly robot projects for experiment and for delivery to industry.

A more specialized passive device has been invented for performing chamferless insertions. Since an insertion involving errors larger than the chamfer is in principle a chamferless insertion (if it succeeds!), the device explained here has application in large error situations regardless of the existence of chamfers. The device is depicted in figure 8.14 and consists of an RCC plus an additional linkage. This extra linkage can be thought of as providing a rotation center to one side of the peg. The peg's axis is initially tilted with respect to the hole's axis and the tip of the peg is driven into the mouth of the hole. Axial contact force on the peg rotates it up to a position almost parallel to the hole's axis. The top of the peg is then in a fixed position and can be steered into the hole by the RCC.

RCCs, IRCCs, and chamferless inserters are covered by numerous patents and are in use in industry.

8.4.5 Open Issues

The twin concepts of FTMM and passive compliant assembly are still actively under development. While the next two sections deal with some of the extensions, other more speculative ones are mentioned here briefly.

I alluded to the fact that practical developments of FTMM and passive compliant assembly are hybrids of both, but there is little theory about the design and behavior of such system. They are put together quite ad hoc right now. Only very simple events are planned for, and the force-mediated motions are generally slow and deliberate. In addition, there is little theory for design of matrix **K** except the engineer's intuition. Design and stabilization are generally two separate steps.

Another hybrid for which there is no theory combines gross and fine motion. At present, one merely detects a collision and switches control modes, or else one relies on the passive compliance to perform while gross motion charges ahead. Similar techniques are used to land an airplane and bring a ship to stop at a pier. People let their fingers go limp in anticipation of a collision they cannot predict, which is analogous to keeping $\mathbf{K}_E$ small, or else they suffer some pain. Since toes are too short to form into a curved limp shape without impairing walking, stubbed toes occur.

The models of FTMM in this section concentrate on compliance or stiffness as the main converter of force into motion. In general, force-motion relations are called impedances, and one can imagine an environment that is primarily massive rather than springy, for example. Invoking heuristically the concept of impedance matching, it might be worth investigating alternative ways of expressing FTMM. The work of Hogan on impedance control is an example of this approach [Hogan, 1985]. In his work, (17) is replaced by a differential equation containing springs, dashpots, and masses.

An example environment for such a system is space, where mass is the dominant property of objects like satellites that a robot arm must grasp.

The way an arm with FTMM control reacts to inertial forces will be important.

Finally, there is the issue of stability. The problem remains much as it was when we first studied FTMM. It is difficult to obtain high-bandwidth, low-contact force, high sensitivity to contact forces, stiff objects, and stable behavior, especially using only gross motion devices. A long-standing problem is impact: How does the arm behave during the transition between gross motion in free space and fine motion in contact with an environment?

8.4.6 Summary

This section has dealt with fine motion in the context of a task, namely, assembly, developing the idea that assembly constitutes elimination of error between pairs of parts. Numerous strategies exist for removing the error in stages. Regardless of the situation, all of these strategies are available, and the task of the designer or researcher is to devise ways to decide what combination is suitable in each situation.

Error removal strategies fall into two classes, a priori and detect/correct. A priori methods include proper design of parts, jigs, and grippers. Detect/correct techniques can be either active or passive. If active, they involve sensors and actuators. If passive, they consist of structures, linkages, and springs arranged to execute an error-correction stragtegy by mustering the energy in the arm's gross motion. Both active and passive methods utilize predictions of what contact forces will arise in different error conditions, and mechanize techniques for converting those forces into response motions that will reduce the error.

The next two sections deal in more detail with motion tasks, dividing them into those that are geometrically determined and those that are stochastic or adaptive. The difference lies in the properties of the task or in the environment in which the task exists.

8.5 Geometrically Determined Motion

As we proceed through the sections of this chapter, we are gradually shifting our focus from the motion of the arm to the tasks on which the motion acts. This is reasonable, since motion is only a means to an end, and the task defines the end.

To execute a task by some automated means, we need several things:

1. *A process model to the task*, that is, a mathematical or engineering statement about how the task and its measurable states will evolve in response to changes in the inputs. Only some of those inputs are under our control, of course, and our control may be imperfect. Our ability to measure the task's response may also be imperfect.
2. *A strategy for directing the evolution of the task to the state we desire.* This consists of a plan for manipulating the variables under our control and monitoring the states we can measure.
3. *An environment in which the process is able to operate.* This environment includes the equipment that will implement the process as well as other conditions that may affect the likelihood that the process attempt will succeed.

In another article [Whitney, 1986] I give an example of a robot that washes dishes. Instead of using brushes and soap containers to wash dishes one at a time as people do, it has a watertight chamber in which it sprays all the dishes at once with a powerful jet of scalding hot water. This is a process that people could not use. The "robot" is called an automatic dishwasher, sold in any department store. In terms of the above paradigm, we have

1. *Process model:* Hot water will melt the dirt on dishes, and high enough shear forces will push the dirt off.
2. *Strategy:* Place dishes near each other and spray scalding hot water at them parallel to their surfaces.
3. *Environment:* A watertight chamber with racks to hold the dishes upright while the spray comes from above and below.

This section and the next deal with tasks that exist in different environments or for which we have models and strategies of different types. In this section we deal with the easier type, namely, tasks that are dominated by the shapes of parts, under the assumption that those shapes either do not change or change in predictable ways.

8.5.1 Geometry, Compliance, and Friction

Assembly of parts was referred to in the previous section as being primarily a problem of geometry. In fact, geometry, compliance (or stiffness of things), and friction are together the determinants of behavior in assembly.

Assembly can be classified into that of *rigid* parts and *compliant* parts. The above determinants still apply, but compliant parts deform in expected and acceptable ways during assembly, whereas rigid parts do not or should not. The theory of quasi-static (i.e., negligible accelerations) rigid part assembly is reviewed in [Whitney, 1982] for the case of round pegs and holes modeled in two dimensions as flat tabs and slots. This reference contains a bibliography of related work on this problem. Gustavson [Gustavson, 1985] presents aspects of the three-dimensional problem.

The essence of the tab-slot model can be seen in figure 8.15. The part is assumed to be grasped compliantly by a gross motion device that moves on a nominal trajectory that is generally in the insertion direction but that does not deviate during assembly. If the part changes position or orientation as a result of contact with the other part, then the springs in the compliance will deflect, exerting forces and torques on the part. These forces and torques must balance the forces exerted by contact and friction between the mating parts. The direction in which these reaction forces point depends on the orientation of the parts and the shapes of their mating surfaces.

The place where the part is gripped compliantly is marked with a target symbol ⊕. This point is called the center of compliance. It indicates the place where the compliance behaves as pure linear and rotational springs. Not every compliant structure has such a place, but such structures can be built. The RCC is such a structure. In the context of figures 8.15, the RCC behaves in such a way as to make the length $L_g = 0$. That is, the compliance center is at or near the tip of the peg. The significance of this is that insertion and contact forces are minimized, and the likelihood of assembly success is greatest.

The theory behind the above assertions is explained in detail in [Whitney, 1982], and the behavior of the RCC and the concept of the center of compliance are explained in [Whitney, 1988]. Here we limit the discussion to the basic results. In particular, we assume that the compliant interface between the part and the gross motion device has a compliance center.

Rigid part mating theory divides assembly into two stages, as discussed in section 8.4. These stages comprise

1. getting the peg into the chamfer opening and from there part way into the hole, and
2. getting the peg the rest of the way into the hole.

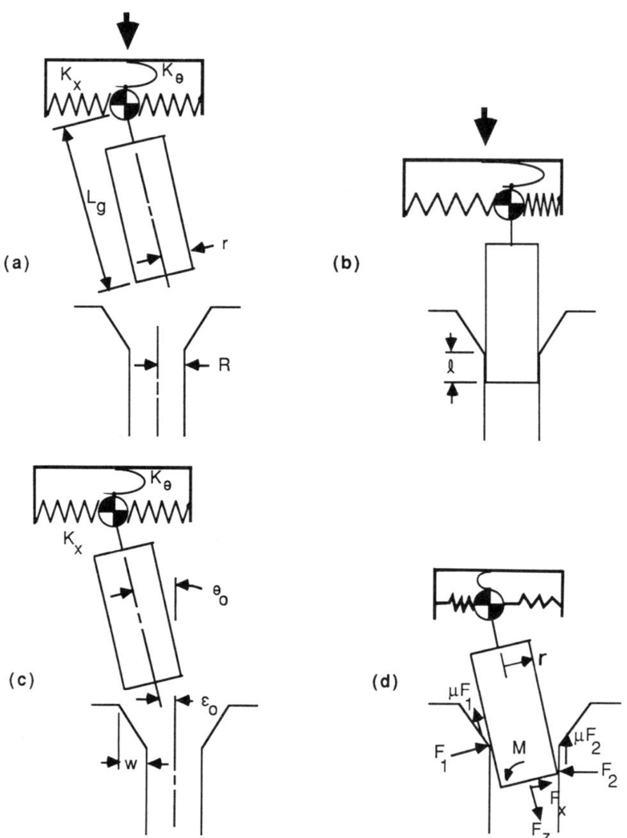

Figure 8.15
Nomenclature for rigid part mating theory: (a) prior to insertion; (b) during insertion; (c) initial errors; (d) applied forces F_x, F_z, and M, and reaction forces F_1, F_2, and friction.

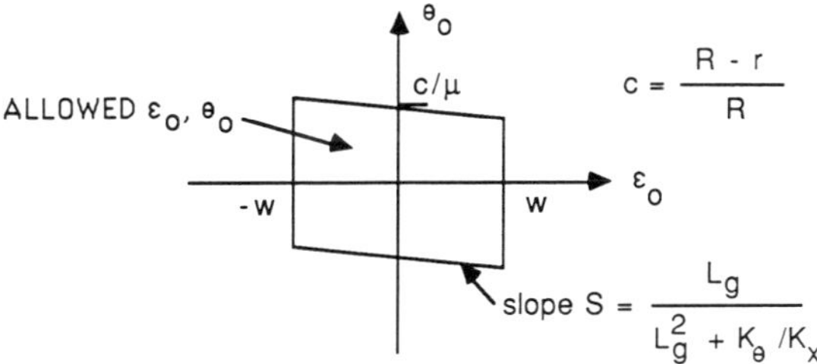

Figure 8.16
Conditions on initial errors that permit chamfer entry and avoid wedging.

For each stage there are well-defined conditions for success. For stage 1, the requirements are that the lateral error of the tip of peg must be small enough to permit the peg to arrive within the width of the chamfer opening, and the angular error must be small enough to prevent *wedging* from occurring. Wedging is an event in which the contact forces between the peg and hole fall within their respective friction cones, so that the peg cannot advance into the hole. The reaction forces exactly balance the insertion forces. (Wedging cannot occur if the parts are mathematically rigid, so we must allow for some compliance at the contact points. But the deformations are quite small and we still may speak of the parts as rigid.) When all the forces, both reaction forces and those provided by the compliance, are taken into account, the conditions for successfully crossing the chamfer, entering the hole, and avoiding wedging can be expressed by the diagram in figure 8.16. When the peg is far enough into the hole so that the dimensionless insertion depth exceeds the coefficient of friction,

$$l/d > \mu, \tag{19}$$

wedging is no longer possible, and the peg enters the second stage of assembly.

In the second stage, "jamming" can occur. In this event, the peg may fail to advance into the hole because the insertion force points too far off the axis of the hole. The insertion force as referred to here includes both the linear force and the torque, which we assume are provided by the compliance. The diagram in figure 8.17 expresses the conditions on these forces and torques so that jamming will be avoided. As explained in [Whitney, 1982; Simunovic, 1975], the RCC automatically satisfies these conditions.

Reviewing the conditions for avoiding wedging and jamming, respec-

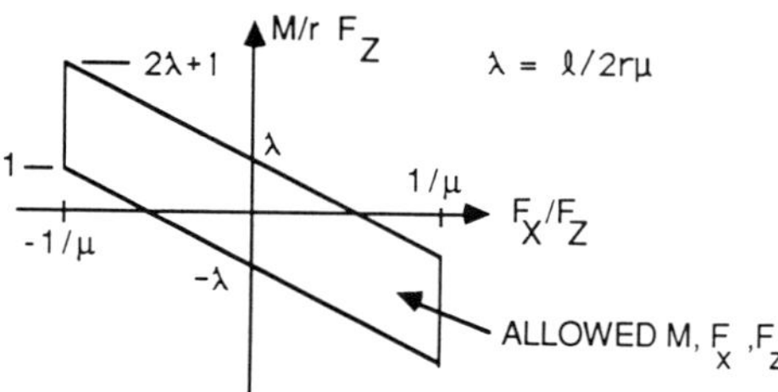

Figure 8.17
Conditions on applied forces and moments to avoid jamming.

tively, we can see that a priori error control is essential for avoiding wedging, whereas the RCC, operating as a discover/correct device, avoids jamming. Also, unless a device like that shown in figure 8.14 or some other method is used, at least one of the parts must have a chamfer.

Together these conditions comprise a *strategy* and an *environment* in which rigid part assembly can occur with good reliability. That is, proper design of the parts is an essential element of the environment.

8.5.2 Part Shape as a Determinant in Assembly

We have spoken so far about geometry, compliance, and friction, but geometry has entered in simple ways only, usually in terms of the location of compliance centers or the clearances between mating parts. Several years ago we studied the mating of compliant parts such as electrical connectors and learned the importance of the *shapes* of mating surfaces. Later we learned that these shapes have the same relevance to rigid part mating.

In compliant part mating, one or both of the mating parts deforms. We usually model the deforming portion as rigid, assuming that it pivots about some point. Then we model this point as another compliance center, allowing us to use the existing rigid part mating theory to analyze the problem. Figure 8.18 shows this approach applied to mating of an electronic dual-in-line package (DIP) into a socket with sloping walls. These walls can be considered chamfers and the compliant legs of the DIP as rigid parts suspended from compliance centers. Typically such DIPs have a dozen or more legs, and the insertion forces are quite large. A relevant question is then, Is there a better shape for the chamfers other than straight? It should be clear that this question applies to rigid part mating as well. But it is especially important in some kinds of compliant part mating where

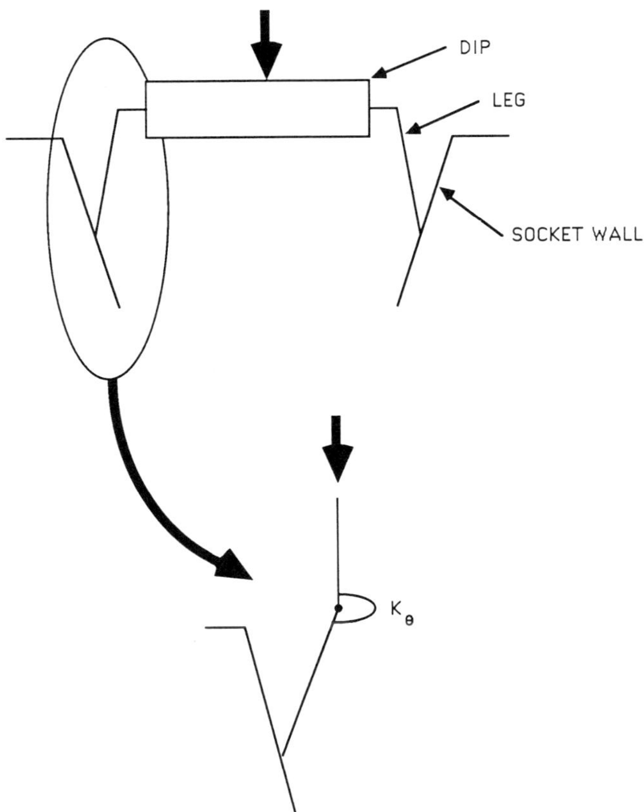

Figure 8.18
Modeling compliant assemby of a DIP with flexible legs using a rigid leg hinged at the top and a coil spring at the hinge.

the stiffnesses can be large and chamfer-crossing is the only event of significance in assembly.

In [Whitney, Gustavson, and Hennessey, 1983] this problem is treated in some detail, with the general result that chamfers should be convex up in order to reduce the insertion force. The exact shape suggested by the theory depends on how the problem is posed, as well as on the geometry and friction coefficient. No reasonable formulation of the problem yields concave chamfers.

Two basic results emerged from this study. First, the slope of the base line of the chamfer is the most important determinant of insertion force, given the friction and stiffness. The baseline is the line joining the entry lip and the hole mouth, as shown in figure 8.19. The baseline slope must exceed the chamfer angle $\tan^{-1}\mu$ plus the angle θ at which the part is tipped. If this

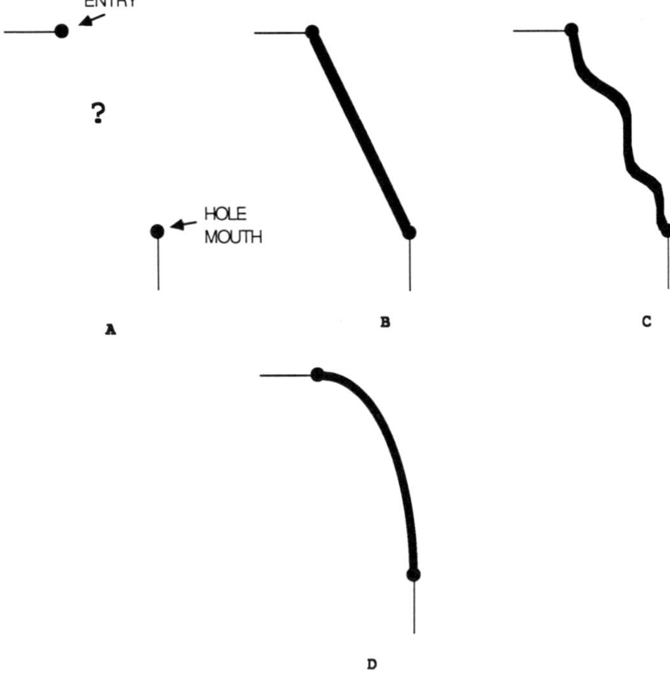

Figure 8.19
Options for chamfer design. (A) What shape should the chamfer be? (B) A straight chamfer whose slope is the baseline slope. (C) An arbitrarily shaped nonstraight chamfer. (D) A typical optimal mini-max chamfer. First law: There will be an insertion force. Second law: Compared to a straight chamfer between the same two points, the optimal peak insertion force will be at least half the peak insertion force from the straight chamfer.

condition is not met, the part will wedge somewhere on the chamfer, regardless of whether the chamfer is a straight one lying on the baseline or a curved one joining the lip and hole mouth in some arbitrary way. Steeper baselines yield lower insertion force, but they invade the hole, taking space that may not be available. Thus there is an upper limit on the baseline slope.

Given a baseline slope, what *shape* should the chamfer be? One interesting case emerges from seeking the chamfer whose peak insertion force is minimized. The mathematics for this case is very easy, and one learns that a chamfer yielding *constant* force also yields mini-max force. If the coefficient of friction is zero, the result is that the chamfer should be a section of a parabola and the constant force is exactly half the peak force of the straight chamfer lying on the baseline. If the coefficient of friction is greater than zero, the shape is similar, though not a parabola, and the insertion force is more than half the straight chamfer's peak force. Test chamfers were constructed according to this theory, and the results appear in figure 8.20.

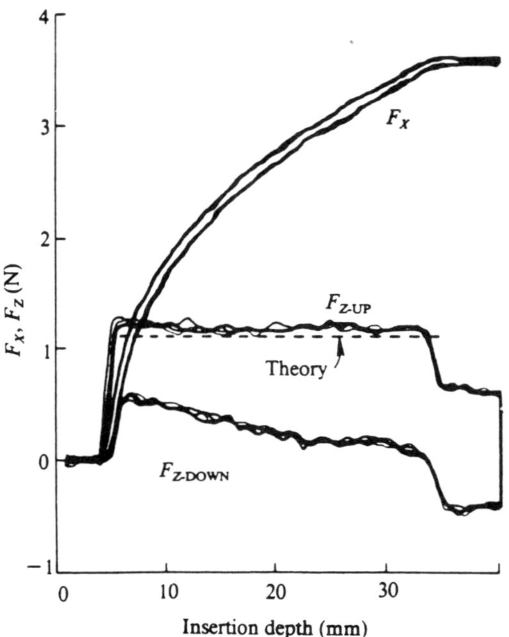

Figure 8.20
Experimental results: logarithmic chamfer, with $\mu = 0.1$, $S = 3$, and $R = 2.4$ The straight line is the theory, with $\mu = 0.1$, $S = 3$, and $R = 2.4$.

8.5.3 State of Part Mating Theory

From this and other figures depicting part mating experiments in [Whitney, 1982] one may see how closely theory and experiment can be made to agree. The geometry, compliance, and friction must be described accurately, of course, but if that is done, one can get as good agreement between theory and experiment as one wishes. This means that geometry-dominated assembly is a well-understood, well-conditioned process.

In addition to the literature cited above on part mating models, I can also cite [Whitney and Rourke, 1986], which contains design equations for commonly available RCCs plus [Drake, 1977], which has the equations for early all-metal RCCs. These references together provide information on how to predict the behavior of combinations of part size and clearance, initial lateral and angular errors, friction, and RCC stiffnesses and compliance center locations. Thus the problem in any situations is one of

1. determining how much error is likely to be encountered,
2. deciding how to reduce that error enough so that the entry conditions (figure 8.17) are met—perhaps by widening chamfers, tightening part or jig tolerances, adding a vision sensor, etc.—
3. deciding how much error will still remain,
4. deciding how much contact force the parts can be subjected to, and
5. setting the stiffnesses of the RCC so that actual contact forces developed at maximum remaining error do not exceed these limits.

The essence of this strategy is depicted in figure 8.21.

The above procedure may seem dry in comparison to the usual image of "robotics," but industrial assembly occurs in a controlled environment, and sound engineering dictates that every advantage should be sought to use this environment to enhance the efficiency and reliability of the process. We call this the act of "structuring" the problem so that orderly, rational, predictable behavior will occur. We have all the engineering models and tools necessary to accomplish this goal. The example of the dishwasher at the beginning of this section was intended to contrast a "robotic" approach with a rationalized one. The reader should draw his own conclusions.

What about environments like space? Must we assume that the environment is uncontrollable? By no means! Part of the environment, such as vacuum and lack of gravity, is indeed uncontrollable. But many factors remain under our control in such areas as satellite retrieval, docking, and

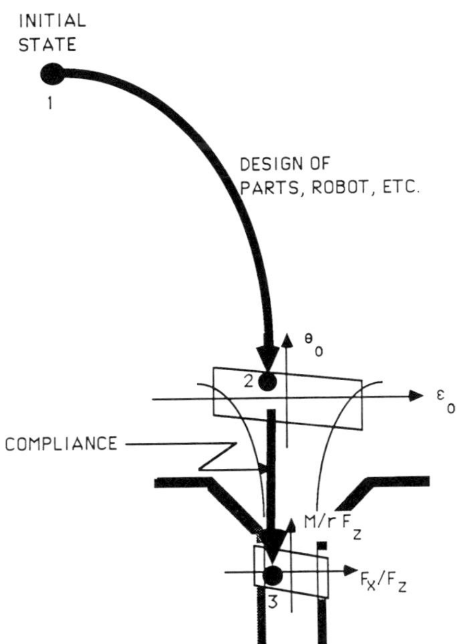

Figure 8.21
A consistent assemby strategy—from state 1, achieve state 2 (ready to enter) by means of a priori error control. From state 2, achieve state 3 (fully assembled) by means of compliance or, if necessary, force control.

space station construction. In particular, the design of parts, grippers, jigs, fasteners, and control and sensing strategies can still be guided by existing theories that have served so well in the industrial environment. Indeed, in space the need for reliability is paramount, because mistakes can be fatal. Also, fuel and time are in short supply, so errors, rework, and wasted motion must be avoided. Structuring space assembly tasks using the above approach will help make them reliable and fast.

8.5.4 Summary

This section described the theoretical base for geometrically determined motion in the context of assembly. The base is quite strong and has high predictive value. It provides a consistent strategy of error analysis, reduction, and elimination that allows an engineer to select, by technical and economic means, the necessary combination of methods, including a priori and detect/correct, to accomplish particular assembly tasks.

There is a need for systematic design procedures, as contrasted with analysis procedures, for *synthesizing* economic solutions to this problem. In additon, the theory needs to be extended to include impacts between parts, especially during chamfer crossing, and more detailed modeling of contacts is necessary for the case of extremely close clearances (1 μm) where surface finish is a factor and simple friction models break down.

The last section of this chapter deals with a class of problems where we do not have such a solid theoretical base—where the issue is to obtain any strategy at all that has a chance of accomplishing the task.

8.6 Stochastically Mediated or Adaptive motion

8.6.1 Motivation

The previous section dealt with rather clean problems where the theory is well developed and one can control and measure the important variables. Tasks are well behaved, outcomes are fairly certain, error sources are known, models for the tasks exist, and real tasks conform well to the models. In this section we deal with motion problems that do not conform to these benign conditions. They are distinguished from those of the previous section in some or all of the following ways:

1. They evolve quantitatively or qualitatively. For example, a grinding wheel may become dull or even break.
2. They vary spatially, or even randomly. For example, the hardness of material being ground may be different in different regions of the workpiece.
3. We do not have complete models of them. Existing models are incomplete in the sense that task evolution is influenced by variables we have not identified or by variables whose influence we do not understand even if we know of their existence. We may be unable to measure important variables, or variables we attempt to manipulate do not have the expected effect.
4. Our efforts to manipulate or measure the task are clouded by the very acts of measuring or manipulating, due to unmodeled effects or noise in our tools. For example, a robot used to measure a surface may itself waver while scanning, thus adding spurious variations to the surface measurement.

Tasks like these are more typical of the "real world" and less like those that robots do now in industry. The tendency in the field of robotics, when confronting uncertain taks, has been to try to determine how people deal with them, and seek to embody robots with the same capabilities. The reader may well expect that I favor a different approach. Rather than study people in search of a solution, I prefer to study the tasks as far as possible first. This approach served us very well in the area of assembly. To extend it to more complex tasks requires a more complex approach and more sophisticated processing, but not necessarily—and not right away—human mimicry.

Another way of describing the challenge of such problems is in terms of "autonomous systems." The assembly robot, provided with a suitably constructed environment, can be allowed to operate on its own. We can be confident that it will not meet circumstances beyond its capabilities. Such circumstances probably represent a violation of the assumptions that underlie high-quality assembly anyway. They should occur too rarely to make it worth our while providing a built-in response. In effect, the assembly robot does not need much autonomy.

By contrast, a grinding robot faces an environment we cannot completely structure. It must be able to deal with a wider range of circumstances than the assembly robot does, or else it will need help so often that it will be impractical. These varying circumstances are not always outside the range of normal operation, so we are justified in attempting to build a robot that can accommodate them.

The autonomous robot is usually thought of in terms of locomotion, and most of the literature is about this kind of robot. However, I hope the foregoing argument makes it clear that autonomy can be studied in the context of stationary robots, too.

The paradigm we have followed in our research on such robots is, like the rest of our research, *model*-based rather than *rule*-based, even though the models are necessarily less precise. The advantages of having a model are many:

1. Models can be quite terse, whereas rule sets can be large. Imagine the rules that would be necessary to describe optimal chamfers: "If the friction isn't too great and if the spring isn't too stiff, then make the chamfer arched a little bit...."
2. Models can be tested for consistency, whereas rule sets, if not controlled, can grow and can contain internal contradictions. This is especially true if one yields to the temptation to "add another rule."

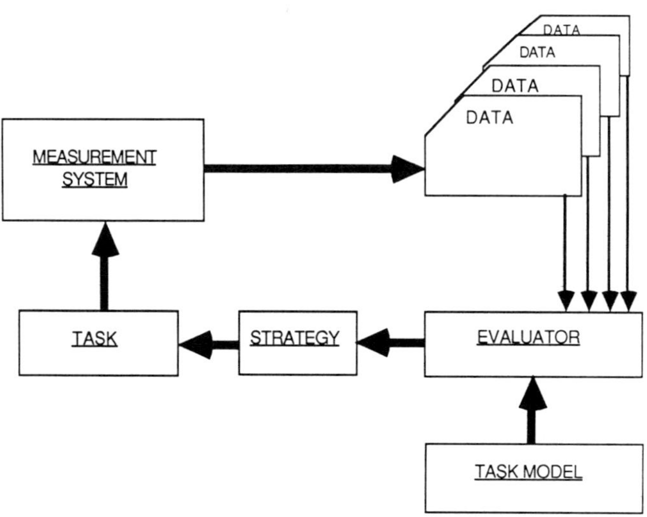

Figure 8.22
Paradigm for stochastically mediated or adaptive motion.

3. A rule set is a model of the problem solver, rather than a model of the problem itself. In many cases, the problem is inaccessible and only the problem solver is available to model. However, the ability of people to deal with problems does not mean that those problems cannot themselves be modeled. It may mean only that no one has tried to make a model yet. Or, in their present state, the models lack crucial components or variables, so that attempts to use them fail. Finally, a true model may take too long or cost too much to construct, and so a rule set is used instead.

The model-based paradigm is illustrated in figure 8.22. The figure shows a system comprising a task and a controller to deal with it. The controller has, as before, a task model, a strategy, and a measurement system. In addition, it has an evaluator, which contains at minimum a memory by which it can compare current with past events. How we equip this evaluator is a key to the system's behavior. In our work the models and evaluators contain probabilistic, often empirical, components.

The probabilistic element is appropriate because robots often repeat the same task, or almost the same task. This repetition allows some comparison to previous occurrences, but recognizes that the comparison will be imperfect. If the robot can function when events do not repeat exactly, then it has some autonomy. Probabilistic modeling also allows us to cover situations that we cannot model exactly. We know approximately what will happen, and we may know how the variability itself will evolve. In the grinding example described below, we surmise that the harder we grind, the more uncertain will be the resulting cut depth.

This structure does not comprise a general theory, so the best way to discuss it is in terms of two examples from our current work. These are *edge tracing* and *grinding.*

8.6.2 Edge Tracing

I was motivated to study edge tracing because of a conviction that there is more information in force signals than we were currently using in assembly tasks. Simunovic [Simunovic, 1979] showed that the evolution of position signals over time could be interpreted to show the state of an assembly task. Different task states have different geometric constraints, leading to predictable correlations between different coordinates of a part's motion. For example, a part pivoting on its end at the mouth of a hole would display a correlation between Z position and X rotation. A filter constructed to detect such correlation could determine whether the task was in this state or not. An array of filters, each trying to correlate to a different presumed state, could determine which state was currently occupied. Since assembly tasks possess only a few possible states, such an approach is possible in principle. The main barrier is that a low signal-to-noise ratio can sometimes make different states indistinguishable. In such a situation, a deliberate test motion may be used, one whose outcome will tell unambiguously which state is occupied.

The array of filters idea is actually a form of pattern recognition. The pattern detectors operate on time series, that is, on signals that extend over time. In Simunovic's example, existence of two or more signals along different coordinate axes simultaneously is also an essential element. Note, too, that the filters represent memory in two forms. First, they remember the recent past of the signals they are processing. Second, they represent the system's memory of knowledge of what behavior is possible. If the system lacks a filter for a possible behavior, then it will become confused when that behavior occurs. It is beyond the scope of this chapter to deal with the issue of learning the new behavior, but that issue is certainly important.

The analog of a time series of position signals is a time series of force signals. Originally, "force feedback" consisted of detecting occasional contacts without regard to magnitude or direction. Our work on FTMM used continuously sensed forces in many directions. Still, each measurement made at a particular time was used at that time and then was forgotten. Our recent work on edge tracing has utilized the history of force readings.

The essence of the idea is that the surface of an object is correlated spatially; the position of a point on the surface cannot vary purely at random from the locations of its neighbors. In fact, the pattern of correlation may comprise a signature that describes the surface, making it possible to recognize the surface, inspect it, find flaws, or characterize it in other ways. If a robot scans the surface using a touch sensor (or possibly a grinding tool), then a time series of force signals will be obtained. What can be learned from such a signal?

The technique we chose is called Linear Predictive Coding (LPC), a technique widely used in speech synthesis. It is also used for speech recognition and is closely related to typical processing methods like Kalman Filtering. In this technique, a signal is represented approximately by a differential equation whose impulse response matches the signal in some root mean square sense over a portion of time or space. A long signal will require segmentation and a different differential equation for each segment.

In our case, we used finite difference equations, in which samples of the signal are represented. The form of the representaton is

$$x_k = \sum a_i x_{k-i}, \qquad i = 1, \ldots, N. \tag{20}$$

That is, the signal at point k is assumed to be related to that at N previous points. The coefficients a_i are obtained by cross correlating the signal with itself. This gives the method its root mean square behavior and enables it to represent noisy signals or signals that contain inherently some uncertain behavior. If a signal has no correlation, all the a_i are zero. Periodic signals can be represented accurately with two coefficients. The literature and our own experience indicate that few such representations have more than 7 coefficients.

If a surface can be represented via (20), then the robot system can tell if it is scanning that surface or another by comparing the coefficients it calculates with those it expects. At least, that is the intent. Unfortunately, the robot would have to be able execute a perfect reference trajectory in order to provide a signal that represented only the surface. In fact, the robot executes a random, correlated path of its own, so we have the task of detecting two signals and separating them in order to see the surface. As explained in [Edsall and Whitney, 1984], this was done by operating two detection systems in parallel, one looking for the robot's signature, the other looking for the surface's. The idea and its behavior are shown schematically in figure 8.23.

In this figure we see that the system detects only the difference between the robot's position and that of the surface. In principle, it is impossible to tell from this difference whether the surface has moved up or the robot has moved down. That is, the signals cannot be separated by looking at individual measurements. However, they can be separated if a series is looked at. This is the same as recognizing that each set of coefficients $\{a_i\}$ has a corresponding Fourier transform representation. In the frequency domain, the signals are different and thus can be separated. Were they not different in the frequency domain—and there is no fundamental reason why they should be—the technique would not work. The other requirement is that we have previously measured what the signal would be if the surface were ideally smooth (i.e., $a_1 = 1$, all other a's $= 0$), so that we have a pure sample of the robot's signal.

One may well ask, Why not just store the robot's signal and subtract it from the input to obtain the surface. There are several reasons. First, the coefficients are a very compact means of storage. Five or six coefficients can store about 50 msec of speech, which may contain harmonics up to 2 KHz, or 100 times the bandwidth of the segmentation. Second, using the raw robot signal requires perfect registration and repetition of the robot's path and its behavior, none of which is likely. Third, our goal was to capture the statistical essence of the robot and the surface so that the method would be robust against minor changes in circumstances. This was our way of giving it some autonomy.

8.6.3 Intelligent Robot Grinding

The grinding tasks of interest to us here are those typically done by people, such as smoothing off welds, for which robots are often considered suitable. Robot grinding is a much more complex problem than merely tracing a surface because grinding is inherently stochastic: we do not understand the process very well and we cannot predict perfectly what will happen, even if we could control the robot and the grinder perfectly. Furthermore, the criteria for task success and completion are much less clear than in assembly.

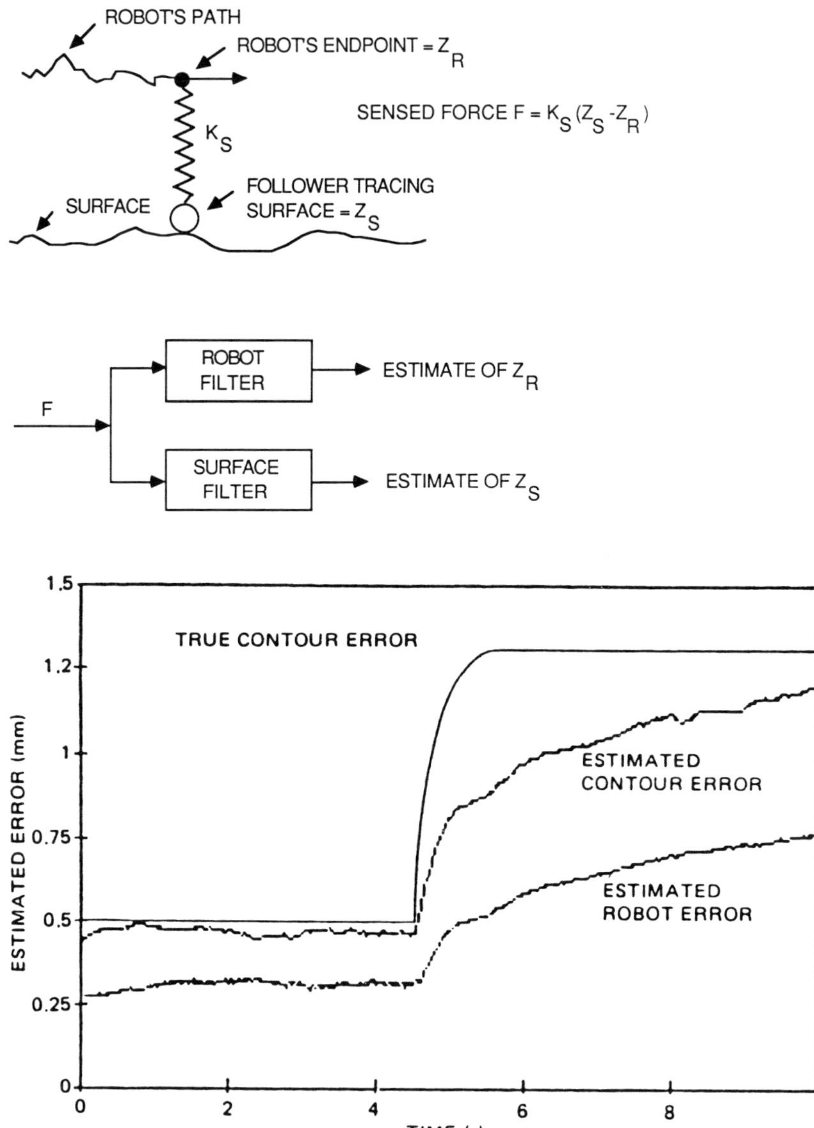

Figure 8.23
(Top) Schematic of robot tracing a surface. (Middle) Simplified diagram of filter. (Bottom) Example results, detecting a step height increase.

By and large, assembly tasks have a unique finished state, which is easy to distinguish from any unfinished states that really matter. By contrast, grinding is multidimensional, in the sense that numerous factors contribute to the final state, such as surface shape, subsurface microstructure condition, surface finish, and blend of the ground region into a surrounding region. As mentioned above, grinding wheels and ground material change with time, perhaps suddenly, though not necessarily catastrophically. Finally, grinding is a one-way process: Except at considerable cost and with different tools, one cannot "ungrind" if too much material has been ground, whereas one can sometimes disassemble to correct an assembly error. Yet economical grinding must finish in a reasonable time and meet reasonable tolerances on the final state.

For these reasons a robot grinder must have considerable autonomy in the sense defined above—the ability to deal with a variety of possibly changing circumstances. Although our research in this area is immature, we have made progress by following the paradigm shown in figure 8.22. An architecture has been adopted and the strategy elements have been identified. Figure 8.24 shows the architecture. It contains three elements in a hierarchy, comprising the ability to do long-term planning under uncertainty (the Process Planner), short-term planning under negligible uncertainty (the Pass Planner), and short-term plan execution (the Disk Controller). Short-term evaluation is done by comparing the lower level's performance with the short-term plan, and corresponding for long-term evaluation.

The job of the Process Planner is to plan an entire grinding job, say the removal of a weld bead. Figure 8.25 defines the terms and sketches a robot grinder. Depending on the size and length of the bead and the time allowed, the planner must decide how many passes to use. We assume that each pass costs something and takes some time, and that its actual outcome is uncertain. The uncertainty in actual amount removed is assumed to be a fixed fraction of the amount requested to be removed. There is a small penalty for leaving too much metal and a larger one for taking off too much. The decision variables available to the planner are the power P delivered to the metal (proportional to rpm times downward force), the feed rate V_f, the number of passes, and the amount to be removed during each pass. The volume rate of metal removal is assumed to be proportional to P. The power delivered must be chosen carefully because too much applied in one location will damage the metal. The maximum power that avoids

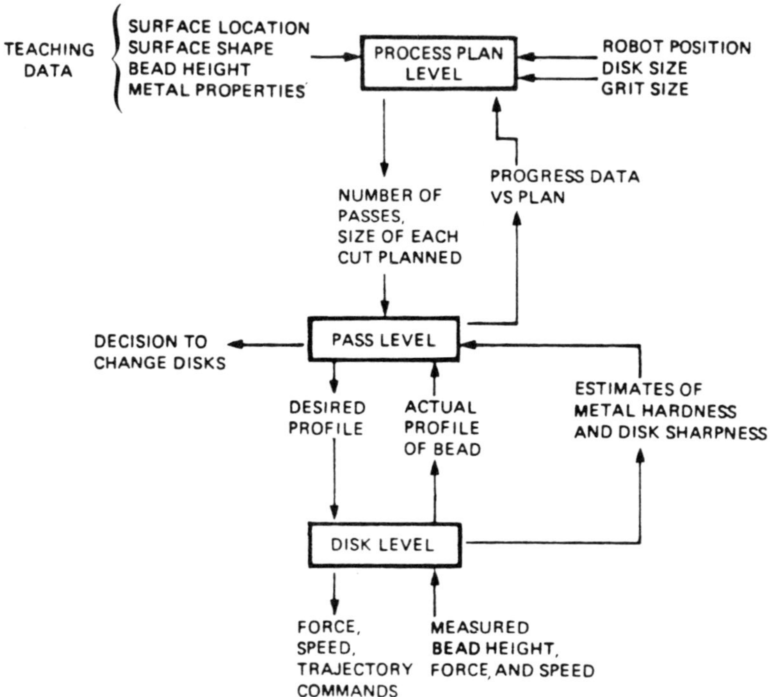

Figure 8.24
Three-level grinding controller.

metal damage is proportional to the half-power of V_f. Finally, the surface finish or smoothness is approximately proportional to V_f. The relation between these variables is diagrammed in figure 8.26.

The planner's problem is thus to apply enough power to finish the job in a limited amount of time while not risking overgrinding near the end, not overheating the work, and not leaving a rough finish. These goals conflict, and the planner must find an optimal strategy, dividing up the remaining time and material into segments and determining how much material should be removed during each time segment. A plan may have a few deep cuts or many shallow ones.

Stochastic Dynamic Programming has been chosen for this task [Whitney and Brown, 1987]. The algorithm finds the policy that minimizes the expected cost to go given the current state (time and material remaining). Figure 8.27 is an example of the output, showing a few of the optimal

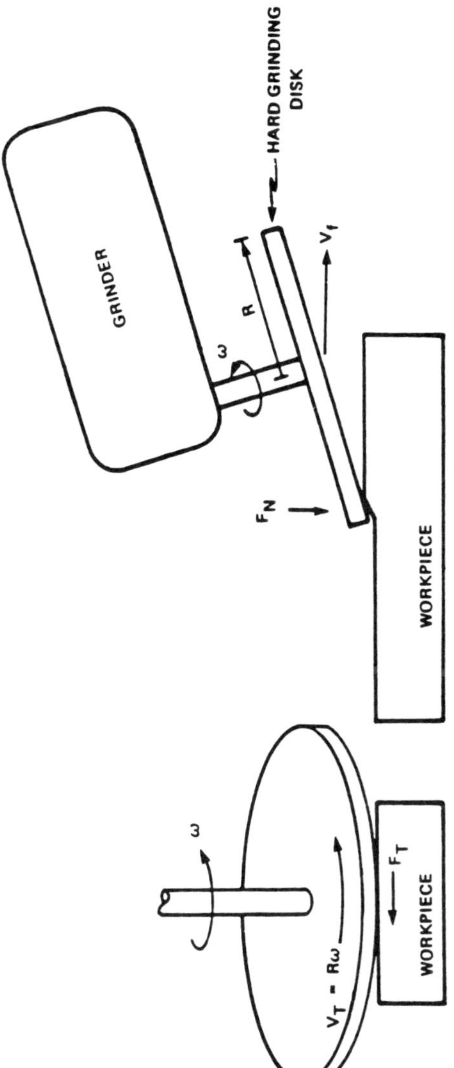

F_T = TANGENTIAL FORCE ON WORKPIECE

F_N = NORMAL FORCE ON WORKPIECE

$F_T V_T$ = POWER DELIVERED TO WORKPIECE (ASSUMING $V_T >> V_f$)

$\dot{Q}$ = METAL REMOVAL RATE

Figure 8.25
Mechanics of grinding.

POWER
METAL REMOVAL MODEL
METAL BURNS
?
HEAT TRANSFER MODEL
1
2
MRR
K-1
V max
Vf

Ra
SURFACE FINISH MODEL
SMOOTH
?
K
Vf-1
1 = METAL REMOVAL REGION
2 = SURFACE FINISH REGION
? = UNCERTAINTY

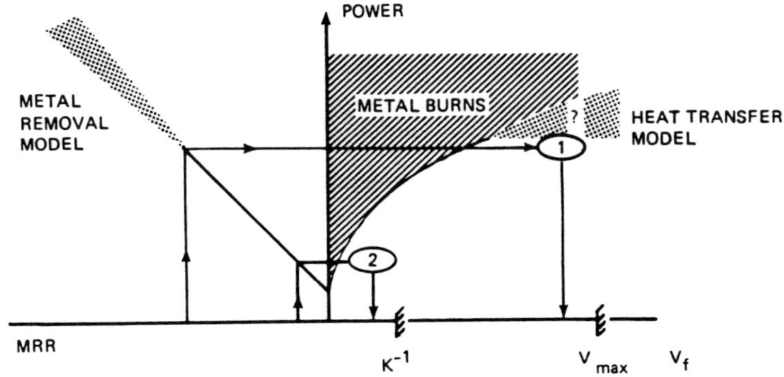

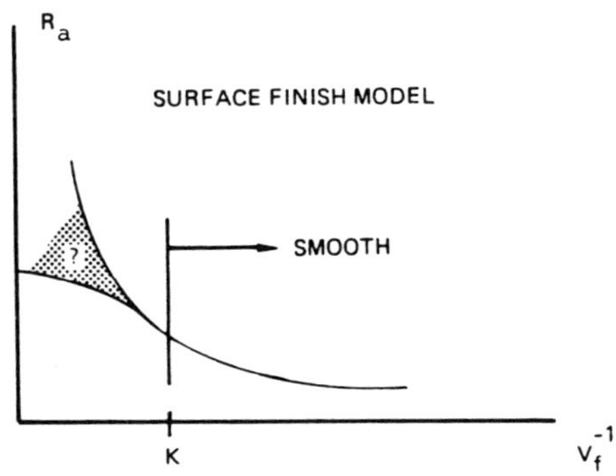

Figure 8.26

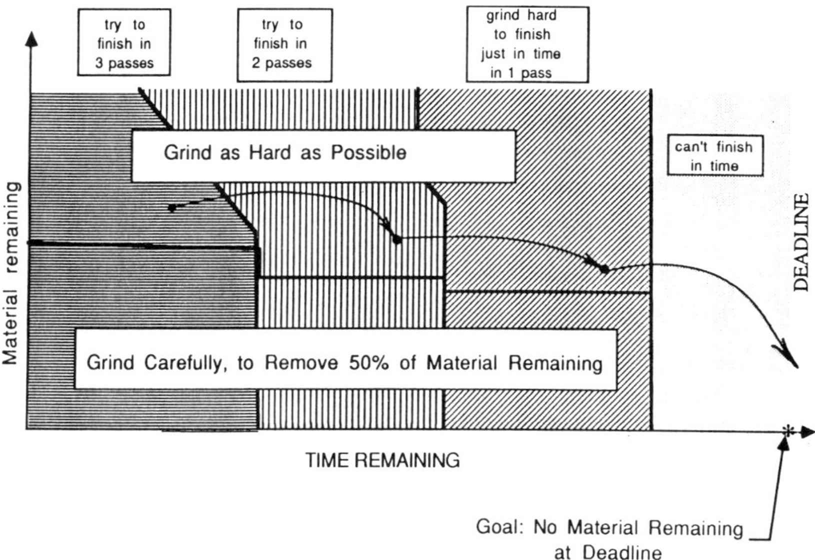

Figure 8.27
Optimal grinding plan for varying time and material remaining.

policies. For example, if there is a lot of material remaining and not much time, the planner may choose not to grind at all, preferring that to the risk of overgrinding. On the other hand, if there is a lot of material and a lot of time, the planner will try several passes, reducing the metal removal rate near the end where overgrinding risk is the highest. If time is cheap and resetting the robot between passes is rapid, the planner will take many small passes, minimizing risk and maximizing information from interpass measurements.

The job of the Pass Planner is to convert the actual state of the work after pass_i into a detailed positon or force profile for the robot to execute during pass_{i+1} as commanded by the Process Planner. The Pass Planner needs a dynamic model of metal removal and must know or estimate the metal hardness and the sharpness of the grinder. The dynamic metal removal model that we are currently evaluating is derived in [Whitney, 1985] and is based on assuming that the steady state metal removal rate model (metal removal rate proportional to input power) holds during every Δt. When this requirement is combined with conservation of metal, one obtains a nonlinear first-order differential equation for the height of the grinding wheel's tip. Model verification appears in [Kurfess].

This equation has several interesting properties. First, like a linear first-order differential equation, its "step response" looks like a first-order exponential lag. A step response is obtained by running the grinder into a sample of metal with a step increase in its height. Second, unlike a linear differential equation, if we run the grinder into a sample whose height varies sinusoidally, the result is again a sinusoidal profile, but the new sinusoid has been shifted ahead on the workpiece, opposite the direction of the grinder's feed velocity. This amounts to a phase *lead*, whereas a linear system would be expected to create a phase *lag*.

Other nonlinearities occur in real grinding that this model ignores. For example, if we grind with a paper disk and a rubber backing, when we push down harder, more of the grinding disk engages the metal, and the newly applied grits are often sharper than the ones that bore the load alone just before. Thus the metal removal rate rises more rapidly than linearly with increasing delivered power.

Recently, Tate [Tate, 1986] applied force control to a PUMA and did some flexible disk grinding. Among his experiments was one in which he programmed the robot to exert a sinusoidally varying force trajectory while tracking a straight sample. The resulting force profile was quite a good sinusoid, but the resulting profile of the metal was rather far from sinusoidal, further emphasizing the nonlinearity of the process. That is, achieving force control is only one part of achieving controlled robot grinding.

8.6.4 Open Issues

The above discussion should have convinced the reader that we have a long way to go in establishing a grinding robot that has some intelligence and autonomy. I hope, too, that the reader appreciates the importance of a process model of grinding to the ultimate goal.

Among the generic open issues common to stochastically mediated or adaptive motion are these:

1. Obtaining any strategy at all. While grinding is difficult to model, we have hope of doing so. There are other tasks where a model may be quite difficult to obtain. This leaves the strategy designer with no guidelines based on the task itself. So far, the only remedy is to ask people we hope "know what to do." The result is optimistically called an Expert System.

2. Describing the undescribable. We have something of a description of the grinding task in terms of shape (a fair curve joining the unground surrounding regions), temperature, and surface smoothness, plus some criteria for guiding the task planning processes. Other tasks have less precise terminal or intrajob criteria. Perhaps the problem is that no one has tried hard enough to establish criteria. Imagine telling a robot "Rub until your thumb feels hot," or "Be careful not to apply too much glue," or "Look for defects." Each of these is an actual instruction that we found being used in industry to explain a job to a person.
3. Interfacing with sudden breaks in routine. The problem here is not boredom, such as would cause a person to ignore the break. Instead it is one of setting alarm levels economically and handling dynamic transients. The conflict is between being too conservative and suffering many false alarms, or being careless and possibly suffering a crash. This is a classic problem in Operations Research and can be solved when costs and criteria are clear, and when overreaction transients are not at issue. A grinding robot not only must be careful of broken disks, but it must also not cause breaks by grinding too hard and must not cause worse damage by pushing harder in the first few milliseconds after a break occurs.
4. Is there such a thing as learning? We can program the robot to estimate key parameters such as wheel sharpness and metal hardness. The robot has a model of what should happen when it applies a certain amount of power. It can look back later and measure how much metal actually was removed, and adjust its parameters. It can also gather statistics and update its stochastic decision algorithm. Finally, it could tell us how well its models correlate to the measurements, essentially commenting on how well we have educated it. Does this constitute learning? Will it ever tell us anything we did not already know, or did not prepare it to tell us?

8.7 Conclusions

This chapter has followed a historical track that illustrates a point, namely, that consideration of what robots *are* has evolved over time into consideration of what robots *do*. The robot is not a pure generic device able to do any task without preparation or modification, but instead is tailored to its task for reasons of efficiency and technical esthetics. This fact forces us to think of robot and task together.

A symmetric evolution has also occurred with respect to the tasks. That is, where we once thought any task could be done by a robot, we now realize that speed, reliability, energy, or other considerations make some tasks ineligible, at least until they are modified or reconceived. There is a growing realization that only by integrating task design and robot design will large strides in robot applications be achieved. Such considerations also apply to other kinds of machines, such as airplanes and milling machines. Neither can do the other's job. In this sense, robots may no longer be viewed as special but must be recognized as merely another kind of machine, and dealt with as such.

References

[Coons] S. A. Coons, "Surfaces for Computer-Aided Design of Space Forms," MIT Project MAC Report TR-41, June, 1967.

[De Fazio, Seltzer, and Whitney] T. L. De Fazio, D. S. Seltzer, and D. E. Whitney, "The Instrumented Remote Center Compliance," *The Industrial Robot*, Vol. 11, No. 4, December, 1984, pp. 238–242.

[Drake, 1977] S. H. Drake, "Using Compliance in Lieu of Sensory Feedback for Automatic Assembly," *IFAC Symposium of Information and Control Problems in Manufacturing Technology*, Tokyo, 1977.

[Drake and Watson, 1975] S. H. Drake and P. C. Watson, "Pedestal and Wrist Force Sensors for Industrial Assembly," *5th International Symposium on Industrial Robots*, Chicago.

[Edsall and Whitney] A. C. Edsall and D. E. Whitney, "Modelling Robot Contour Processes," Proceedings *American Control Conf.*, San Diego, 1984.

[Eppinger and Seering, 1986] S. D. Eppinger and W. P. Seering, "On Dynamic Models of Robot Force Control," *Proceedings, 1986 IEEE Conf. on Robotics and Automation*, San Franciso, April, pp. 29–34.

[Ferrell and Sheridan, 1967] W. R. Ferrell and T. B. Sheridan, "Supervisory Control of Remote Manipulators," *IEEE Spectrum*, Vol. 4, pp. 81–88, October.

[Finkel, Taylor, Bolles, Paul, and Feldman, 1974] R. Finkel, R. Taylor, R. Bolles, R. Paul, and J. Feldman, "AL, A Programming System for Automation," Stanford University Artificial Intelligence Laboratory Memo No. AIM-177, November.

[Gavrilovic and Maric, 1969] M. M. Gavrilovic and M. R. Maric, "An Approach to the Organization of the Artificial Arm Control," *Proc. 3rd Ann. Int'l. Symp. on External Control of Human Extremities*, Dubrovnic.

[Goertz, 1952] R. C. Goertz, "Fundamentals of General Purpose Remote Manipulators," *Nucleonics*, Vol. 10, Nov., pp. 36–42.

[Gordon, 1987] S. J. Gordon "Automated Assembly Using Feature Localization," MIT Mech. Eng. Ph.D. Thesis, February.

[Gotoh, Takeyasu, and Inoyama] T. Gotoch, K. Takeyasu, and T. Inoyama, "Control Algorithm for Precision Insert Operation Robots," *IEEE Trans., Systems, Man and Cybernetics*, Vol. SMC-10, No. 1, January, 1980, pp. 19–25.

[Gustavson] R. E. Gustavson, "A Theory for the Three-Dimensional Mating of Chamfered Cylindrical Parts," *ASME J. Mechanisms, Transmissions, and Automation in Design*, Vol. 107, pp. 112–122, 1985.

[Hanafusa, Yoshikawa, and Nakamura, 1981] H. Hanafusa, T. Yoshikawa, and Y. Nakamura, "Analysis and Control of Articulated Robot Arms with Redundancy," *Preprints, 8th Triennial IFAC World Congress*, Kyoto, pp. 78–83.

[Hirzinger, 1983] G. Hirzinger, "Direct Digital Robot Control Using a Force-Torque Sensor," *IFAC Symp. on Real Time Digital Control Applications*.

[Hogan] N. Hogan, "Impedance Control: An Approach to Manipulation," *ASME J. Dyn. Sys., Meas and Control*, March, 1985, pp. 1–24.

[Jacobsen] S. C. Jacobsen, "Control Systems for Artificial Arms," MIT Mech. Eng. Ph.D. Thesis, January, 1973.

[Johnsen and Corliss] E. G. Johnsen and W. R. Corliss, *Teleoperators and Human Augmentation*, NASA report SP-5047, December, 1967.

[Kahn and Roth, 1971] M. Kahn and B. Roth, "The Near Minimum Time Control of Open Loop Articulated Kinematic Chains," *ASME J. Dun. Sys., Meas. and Control*, Vol. 93, pp. 164–172.

[Kurfess 1987] T. Kurfess, "Verification of a Dynamic Grinding Model," MIT SM Thesis, Mech. Eng. Department, May, 1987.

[Liegeois, 1977] A. Liegeois, "Automatic Supervisory Control of the Configuration and Behavior of Multibody Mechanisms," *IEEE Trans. Systems, Man, and Cybernetics*, Vol. SMC-7, pp. 868–871.

[Mason] M. T. Mason, "Compliance and Force Control for Computer Controlled Manipulators," *IEEE Tr. Systems, Man and Cybernetics*, Vol. SMC-11, No. 6, 1981, pp. 418–432.

[Moe and Schwartz] M. I. Moe and J. T. Schwartz, "A Coordinated Proportional Motion Controller for an Upper-Extremity Orthotic device," *Proc. 3rd Int. Symp. on External Control of Human Extremities*, Dubrovnik, 1969.

[Nevins and Whitney, 1973] J. L. Nevins and D. E. Whitney, "The Force Vector Assembler Concept," *Proceedings, First CISM IFToMM Symposium on Theory and Practice of Robots and Manipulators*, Udine, September, pp. 273–288 (pub. by Springer, 1974).

[Nevins et al. 1973] J. L. Nevins et al., "Annual Report No. 1 on Development of Multi-Mode Remote Manipulator Systems," C. S. Draper Laboratory Report C-3790, January.

[Nevins et al., 1973] J. L. Nevins et al., "Annual Report No. 2 on Development of Multi-mode Remote Manipulator Systems," C. S. Draper Laboratory Report C-3901, January.

[Nevins et al., 1977] J. L. Nevins et al., "Exploratory Research on Industrial Modular Assembly," C. S. Draper Laboratory, Report R-1111, August.

[Nevins, Whitney, and Wang, 1978] J. L. Nevins, D. E. Whitney, and S. J. Wang, *Satellite Retrieval Study*, C. S. Draper Laboratory Report R-1186, October.

[Pieper, 1968] D. L. Pieper, "The Kinematics of Manipulators under Computer Control," Stanford University Comp Sci. Ph.D. Thesis.

[Raibert and Craig, 1981] M. H. Raibert and J. J. Craig, "Hybrid Position/Force Control of Manipulators," *IEEE J. Dyn. Sys. Means. and Control*, Vol. 102, June, pp. 126–133.

[Roberts, Paul, and Hillberry, 1985] R K. Roberts, R. P. Paul, and B. M. Hillberry, "Effect of Wrist Force Sensor Stiffness on the Control of Robot Manipulators," *IEEE Conf. on Robotics and Automation*, St. Louis, pp. 269–274.

[Rothchild and Mann] R. A. Rothchild and R. W. Mann, "An EMG-Controlled Force Sensing Proportional Rate Elbow Prosthesis, *Proc. 1966 Symp. on Biomedical Eng.*, Milwaukee, 1966.

[Simunovic, 1975] S. Simunovic, Force Information in Assembly Processes," *5th Int'l. Symp. on Industrial Robots*, Chicago.

[Simunovic 1979] S. Simunovic, "An Information Approach to Part Mating," MIT Mech. Eng. Dept. Ph.D. Thesis, Apr.

[Tate, 1986] A. R. Tate, "Closed Loop Force Control for a Robotic Grinding System," MIT Mech. Eng., SM Thesis, June.

[Waldron, Wang, and Bolin, 1985] K. J. Waldron, S. L. Wang, and S. J. Bolin, "A Study of the Jacobian Matrix of Serial Manipulators," *ASME J. Mechanisms, Transmissions, and Automation in Design*, Vol. 107, No. 2, pp. 230–238.

[Watson 1976] P. C. Watson, "A Multidimensional System Analysis of the Assembly Process as Performed by a Manipulator," *1st North American Robot Conf.*, Chicago.

[Whitney, 1969] D. E. Whitney, "Resolved Motion Rate Control of Manipulators and Human Prostheses," *IEEE Trans. Man-Machine System*, Vol. MMS-10, No. 2, June, pp. 47–53.

[Whitney, 1972] D. E. Whitney, "The Mathematics of Coordinated Control of Prosthetic Arms and Remote Manipulators," *ASME J. Dyn. Sys., Meas. and Control*, Vol. 93, Series G, No. 4, December, pp. 303–309.

[Whitney, Book, and Lynch, 1974] D. E. Whitney, W. J. Book, and P. M. Lynch, "Design and Control Considerations for Industrial and Space Manipulators," *Proceedings, Joint Automatic Control Conference*, Austin, TX.

[Whitney, 1977] D. E. Whitney, "Force Feedback Control of Manipulator Fine Motions," *ASME J. Dyn. Sys., Meas. and Control*, Vol. 99, No. 2, June, pp. 91–97.

[Whitney, 1982] D. E. Whitney, "Quasi-Static Assembly of Compliantly Supported Rigid Parts," *ASME J. Dyn. Sys., Meas. and Control*, Vol. 104, No. 1, March, pp. 65–77.

[Whitney, Gustavson, and Hennessey] D. E. Whitney, R. E. Gustavson, M. A. Hennessey, "Designing Chamfers," *Int'l. J. of Robotics Research*, Vol. 2, No. 4, December, 1983, pp. 3–18.

[Whitney, 1985] D. E. Whitney, "Elements of an Intelligent Robot Grinding System," *3rd Int'l. Symp. of Robotics Research*, Gouvieux, October.

[Whitney, 1986] D. E. Whitney, "Real Robots Don't Need Jigs," Proceedings, *IEEE Conference on Robotics and Manipulation*, San Francisco.

[Whitney and Rourke, 1986] D. E. Whitney and J. M. Rourke, "Mechanical Behavior and Design Equations for Elastomer Shear Pad Remote Center Compliances," *ASME J. Dyn. Sys., Meas. and Control*, Vol. 108, No. 3, Sept, pp. 223–232.

[Whitney 1987] D. E. Whitney, "Historical Perspective and State of the Art in Robot Force Control," *Intl. J. Robotics Research*, Vol. 6, No. 1, Spring, pp. 3–14.

[Whitney and Brown, 1987] D. E. Whitney and M. L. Brown, "Metal Removal Models and Process Planning for Robot Grinding," *Proceedings, 17th Int'l. Symp. on Industrial Robots*, Chicago, April.

[Whitney, 1988] D. E. Whitney, "The Remote Center Compliance," in *The Encyclopedia of Robotics*, ed. S. Y. Nof, New York: John Wiley and Sons.

9 Control

Suguru Arimoto

9.1 Introduction

The motion of a robot manipulator is naturally subject to Newton's laws of motion even if its structure and dynamics are complicated. Therefore, certain aspects of the robot motion must be described in terms of fundamental and natural properties of classical Newtonian mechanics, such as Hamilton's principle, Hamilton's canonical equation, and Lagrange stability of motion. This implies some possibility of utilizing effectively those favorable properties of Nature in the design of motion control systems for robot manipulators. This point of view is therefore essential not only in problems of motion control but also in basic problems of mechanical manipulation where there arise mechanical interactions between the manipulator and the object being manipulated.

On the other side, however, it is commonly claimed in much of the literature on robot control that the dynamics of a robot manipulator is highly nonlinear and has strong couplings between joints. This is true if the manipulator is of revolute type and its dynamics is described in a detailed differential equation of motion. To avoid these obstacles, many control methods have been proposed, among the important ideas of which have been (1) RMRC (Resolved Motion Rate Control) [1], (2) inverse kinematics [2, 3], (3) the computed torque method [4, 5], (4) the nonlinear decoupling method [6], (5) applications of nonlinear control theory [7], and (6) applications of linear control theory based on linearized dynamics. All these methods, except (1), rely considerably on a large amount of computation in real time and hence need a novel design of microprocessor-based control systems based on parallel decentralized techniques.

However, are high nonlinearity and existence of strong couplings between joints in robot dynamics real obstacles? Actually, most present industrial robots do work well, even if they are of revolute type and their control systems are relatively simpler. In fact, most of actual robot control systems used in industry have only independent local feedback loops of PD-type; that is, each joint servo has feedback of its own position (Proportional) and velocity (Differential) signals and does not refer to any signal at other joints. In addition, a pulse-incremental control scheme is still used at each joint.

In this chapter, we first explain in section 9.2 the present status of control methodology for actual industrial robot manipulators. Then, after giving a short exposition of robot dynamics, we show in section 9.4 that a simpler classical control system consisting of independent local feedback loops can work well provided that it is used as a joint controller in conjunction with a PTP (Point to Point) control scheme and a Teaching/Playback control scheme. This is demonstrated not only experimentally but also theoretically by pointing out natural and fundamental properties of Newtonian mechanics of robot manipulators [8, 9]. Second, we review in section 9.5 a class of more complicated control methods, such as PD feedback control with variable gain, nonlinear decoupling control, and nonlinear theoretical control. It is pointed out that another class of advanced control technologies, such as "inverse kinematics" and "inverse dynamics (or computed torque method)," which are considered to be a feedforward compensation technique, may play an important role in the case of trajectory-tracking and path-tracking control, corresponding to joint-coordinates and task-oriented representations, respectively.

From another viewpoint, control of mechanical manipulation, the importance of "force control," "hybrid control," and "impedance control" is emphasized [10–14] when mechanical interactions occur to some extent between the robot and the object being manipulated. In connection with this, we also discuss relatively new problems of cooperation control of multirobots, such as teleoperation of master and slave robot systems and manipulation by using multifingers. Finally, in section 9.7, we discuss a class of more advanced control technologies for robot motion such as adaptive control [15–17] and learning control [18–20], which are introduced to improve the quality of dynamic performance in manipulator systems. In conclusion, further applications of these methods to more complicated tasks imposed on a robotic system consisting of mechanical arms, hands, and/or fingers are suggested and some possibilities of development of a kind of knowledge-based control based upon the learning control scheme are pointed out.

9.2 Present Status of Control Methodology for Industrial Robots

Operation of most of the industrial robots in use at present is still subject to the control methodology called "teaching and playback." For predefined tasks, a robot is taught the desired trajectory either by using a teaching joystick or by programming. During the teaching mode of operation, the posture data of the robot are picked up pointwise along the desired trajectory by direct measurement of joint variables (joint angle or displacement).

The control law of movement from a point (one posture) to another point (another posture) is predetermined or can be selected among a finite set of preprogrammed laws in the CPU computer of the robot control system, and thereby generation of the whole joint trajectory is programmed in it. This is called the PTP (Point to Point) control scheme. When continuous path generation is required, data points (postures) are chosen densely or point-to-point interpolation is employed. During the task-repeat operation, the measured joint trajectories are compared with the desired joint trajectories and each joint error is picked up by its corresponding servo controller for compensation.

In each joint actuator servo, conventional linear servo control using position and velocity feedback loops is used, a typical example of which is shown in figure 9.1A. The servo unit in each joint consists of typically a servo module card called a "driver" that generates a signal of control action, a servo amplifier, and a servo actuator such as a dc servo motor. Joint sensors such as resolvers, encoders, potentiometers, and tachometers are attached directly to the servo motor. Since the deviation counter in the driver circuit plays a role of integral operation, the block diagram described by figure 9.1A can be rewritten into the alternative expression described by figure 9.1B by changing the viewpoint from velocity reference to position reference. The block diagram of figure 9.1B is nothing but a PID (Proportion-Integral-Derivative) feedback control system, which is widely

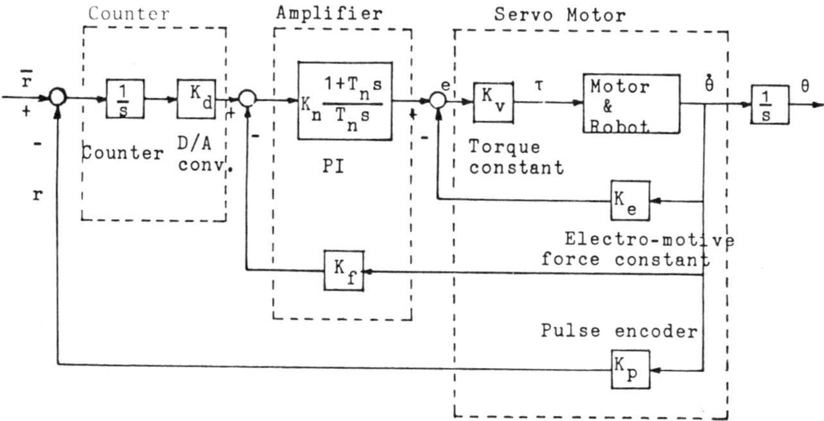

Figure 9.1A
Block diagram of a typical servo unit.

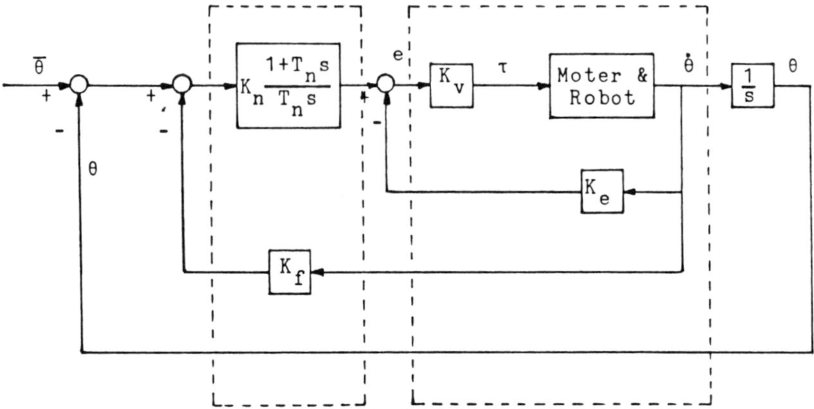

Figure 9.1B
Equivalent block diagram seen from the viewpoint of positional reference.

used in control of servomechanisms. In practice, the amplifier gain K_n is set high and simultaneously the time constant T_n is set relatively large. This means that the effect of the I-part in the servo loop is reduced relative to the PD-part, and hence the overall control system plays the role of a standard high-gain PD (position and velocity as physical variables) feedback controller.

9.3 Dynamics of a Mechanical Manipulator System

We consider a manipulator system with n actuators and n rigid links, which are connected serially by either rotational or prismatic joints. It is well-known that the dynamics of the rigid link system is expressed by the form

$$\frac{d}{dt}\left(\frac{\partial L}{\partial \dot{q}_i}\right) - \frac{\partial L}{\partial q_i} = f_i(q_i, \theta_i) - b_{1i} q_i, \tag{3.1}$$

where L is the Lagrangian, q_i and $\dot{q}_i$ are the position and velocity of the ith component of joint coordinates $q = (q_i, \ldots, q_n)^{\mathrm{T}}$, θ_i is the rotational angle of the ith actuator, f_i is the joint torque transmitted from the ith actuator through transmission mechanisms, and b_{1i} is the coefficient of frictional damping. The Lagrangian L is defined as

$$L = T - P, \tag{3.2}$$

where T is the kinetic energy expressed as

$$T = \tfrac{1}{2}\dot{q}^{\mathrm{T}} H(q)\dot{q} \tag{3.3}$$

and $P = P(q)$ is the potential energy. In (3.3), $H(q)$ is a symmetric and positive definite matrix called the inertia matrix. Substitution of (3.3) into (3.1) yields

$$H(q)\ddot{q} + (\dot{H}(q) + B_1)\dot{q} - \frac{\partial T}{\partial q} + g(q) = f, \tag{3.4}$$

where

$$\frac{\partial T}{\partial q} = \left(\frac{\partial T}{\partial q_1}, \ldots, \frac{\partial T}{\partial q_n}\right)^{\mathrm{T}}, \qquad f = (f_1, \ldots, f_n)^{\mathrm{T}},$$

$$g(q) = (g_1, \ldots, g_n)^{\mathrm{T}} = \left(\frac{\partial P}{\partial q_1}, \ldots, \frac{\partial P}{\partial q_n}\right)^{\mathrm{T}},$$

and

$$B_1 = \begin{bmatrix} b_{11} & & 0 \\ & \ddots & \\ 0 & & b_{in} \end{bmatrix}.$$

Next we suppose that the dynamics of the drive system is expressed as

$$J\ddot{\theta} + B_0\ddot{\theta} = -f + \tau, \tag{3.5}$$

where $\tau = (\tau_1, \ldots, \tau_n)^{\mathrm{T}}$, τ_i is the control torque generated by the ith actuator, $\theta = (\theta_1, \ldots, \theta_n)^{\mathrm{T}}$, and

$$J = \begin{bmatrix} J_1 & & 0 \\ & \ddots & \\ 0 & & J_n \end{bmatrix}, \qquad B_0 = \begin{bmatrix} B_{01} & & 0 \\ & \ddots & \\ 0 & & b_{0n} \end{bmatrix}.$$

Here, J_i and b_{0i} denote the inertia and the damping coefficient of the ith driving system, respectively.

At the first stage, we deal with the case that the robot dynamics is directly connected to the dynamics of the drive system through reduction gears (see figure 9.2). In other words, it is assumed that $q = \theta$ and thereby, according to (3.4) and (3.5), the system is subject to

$$(J + H(q))\ddot{q} + (B + \dot{H}(q))\dot{q} - \frac{\partial T}{\partial q} + g(q) = \tau, \tag{3.6}$$

where

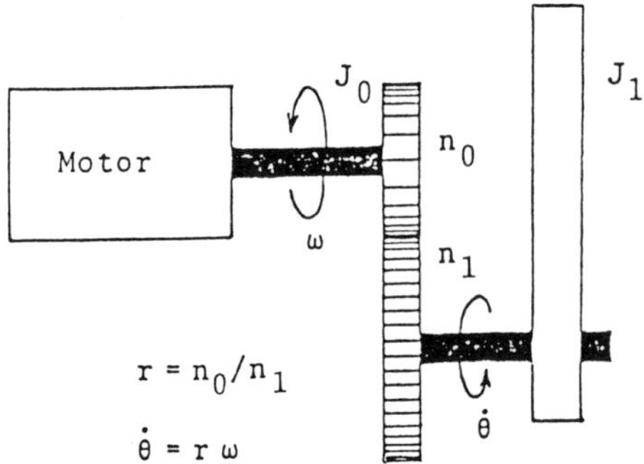

Figure 9.2
Reduction gear attached to motor shaft.

$$B = B_1 + B_0.$$

As is always pointed out in the literature, this differential equation of robot motion is nonlinear and has dynamic couplings between joints. However, it should be pointed out that each diagonal component of the inertia matrix is considerably smaller than the corresponding motor shaft inertia referred to the angular velocity of the axis of load inertia if high-gear ratios are employed in drive units (for example, harmonic drives). In fact, the ith diagonal component J_i of matrix J can be expressed as

$$J_i = r_i^{-2} J_{0i} = (n_{1i}/n_{0i})^2 J_{0i},$$

which is considerably larger than J_{1i} or $h_{ii}(q)$, the (i, i)th entry of the inertia matrix $H(q)$. This fact also causes the resulting cross-coupling torque effects as reflected back to the motor shafts to be relatively small.

At the second stage we shall take flexibility of the drive system into consideration and assume that the stiffening spring characteristics of the drive system can be approximated by a linear relation of the form

$$f_i(q_i, \theta_i) = K_i(\theta_i - q_i). \tag{3.7}$$

Then the total dynamics of the manipulator system is expressed as

$$\begin{cases} H(q)\ddot{q} + (\dot{H}(q) + B_1)\dot{q} - \dfrac{\partial T}{\partial q} + g(q) = K(\theta - q), \\ J\ddot{\theta} + B_0\dot{\theta} = -K(\theta - q) + \tau, \end{cases} \tag{3.8}$$

where K is a diagonal matrix with K_i as the ith diagonal element.

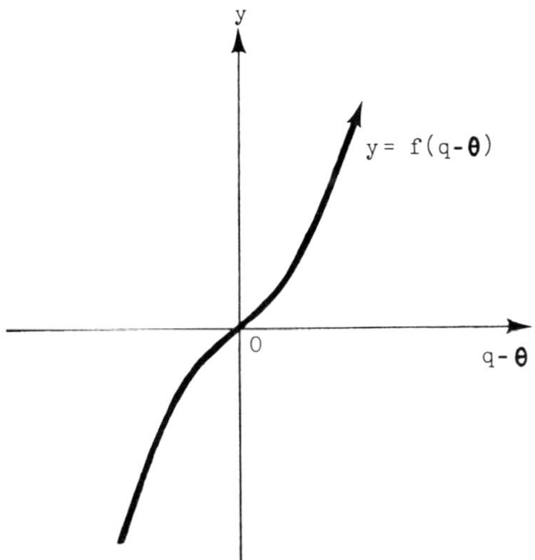

Figure 9.3
A typical profile of stiffening spring characteristics $y = f(q - \theta)$.

Finally we shall discuss a more general situation where the restoring force f_i is nonlinear with respect to $\theta_i - q_i$ but satisfies certain physically reasonable conditions, namely,

$$f_i(q_i, \theta_i) = f_i(q_i - \theta_i), \qquad (q_i - \theta_i)f_i(q_i - \theta_i) > 0 \qquad \text{for} \quad q_i \neq \theta_i.$$

A profile of characteristic function f_i is illustrated in figure 9.3. In this case, the total dynamics of the manipulator system is expressed as

$$\begin{cases} H(q)\ddot{q} + (\dot{H}(q) + B_1)\dot{q} + \dfrac{\partial T}{\partial q} + g(q) = f(\theta - q), \\ J\ddot{\theta} + B_0\dot{\theta} = -f(\theta - q) + \tau, \end{cases} \tag{3.9}$$

where $f = (f_1, f_2, \ldots f_n)^{\mathrm{T}}$.

9.4 PD Feedback with Gravity Compensation

In most present control systems of industrial robots an ordinary PD (Proportional plus Differential) feedback law is implemented; that is, an independent servomechanism with position and velocity feedback loops is constructed in each individual axis. This is expressed as

$$\tau_i = -k_{1i}\dot{q}_i - k_{0i}(q_i - \bar{q}_i), \tag{4.1}$$

where $\bar{q}_i$ denotes the reference input to the ith actuator. Substituting a previous equation yields

$$(J + H(q))\ddot{q} + (B + K_1 + \dot{H}(q))\dot{q} - \frac{\partial T}{\partial q} + g(q) + K_0 q = K_0 \bar{q}, \tag{4.2}$$

where

$$K_1 = \begin{bmatrix} k_{11} & & 0 \\ & \ddots & \\ 0 & & k_{1n} \end{bmatrix}, \quad K_0 = \begin{bmatrix} k_{01} & & 0 \\ & \ddots & \\ 0 & & k_{0n} \end{bmatrix}.$$

It is said that present industrial robots are overdesigned [21] in the following sense: (1) high reduction-gear ratios and (2) high-gain feedback loops are adopted. In addition, in the design of the arm the effect of gravity terms is minimized by balancing the mass distribution of its links and motor locations. In view of these points, the magnitudes of the nonlinear cross-coupling terms become considerably small in relation to the dominant dynamics of each individual axis. In other words, the robot dynamics of (4.2) can be well approximated by the following decoupled linear differential equation:

$$J\ddot{q} + (B + K_1)\dot{q} + K_0 q = K_0 \bar{q}. \tag{4.3}$$

Here, it should be noted that all coefficient matrices are diagonal. In other words, a servo controller at each joint actuator can be designed independently only if it refers to its own joint variables (position and velocity). This is one of the reasons why such a classical PD feedback servo is still adopted as an actuator controller at each joint in industrial robots in use at present. However, we must bear in mind that the use of driving mechanisms with a high reduction-gear ratio, such as harmonic drives, induces a torsional oscillation due to the flexibility of driving mechanisms, and that the use of high-gain feedback discloses many of the parasitic terms in the motion equation of the robot, such as frictional forces, backlashes, and effects of nonlinear characteristics of servo amplifiers.

Nevertheless, the current design of robot arms tends to lessen the problems pointed out above. One example of this tendency is the use of direct drive motors [22], and another is the adoption of timing belts or chains as drive mechanisms, instead of harmonic drives, since motor locations are

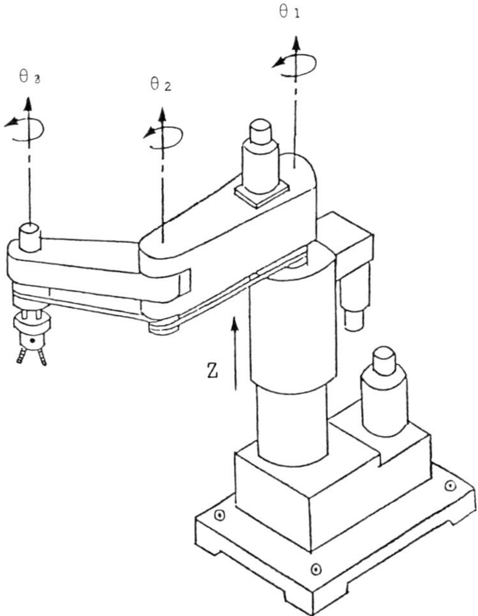

Figure 9.4
SCARA robot.

concentrated around the basement link. In those cases the magnitudes of motor shaft inertias J_i are matched to corresponding components of the inertia matrix $H(q)$, and hence nonlinear terms in (4.2) cannot be ignored in the analysis of robot motion and control. However, as will be discussed later, fortunately we know that such a linear independent PD feedback law is also effective even in those cases, provided that the effect of gravity is carefully avoided in the design of the arm or can be compensated for in real-time control. The motion of a SCARA-type robot restricted in a horizontal plane (see figure 9.4) is such an example. Or we can reasonably assume that the gravity term $g(q)$ can be compensated by computing it in real time because it is of simpler form than the other nonlinear terms in (3.6). Hence, we consider a linear independent PD feedback law with real-time compensation for the gravity term, which implies (see figure 9.5)

$$\tau = -K_1\dot{q} - K_0(q - \overline{q}) + g(q). \tag{4.4}$$

Substitution of this into (3.6) yields the following closed-loop form of the

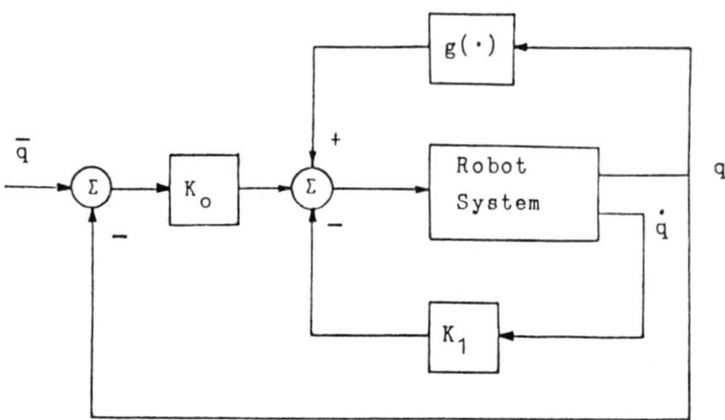

Figure 9.5
Linear independent PD feedback with gravity compensation.

motion equation:

$$(J + H(q))\ddot{q} + (B + K_1 + \dot{H}(q))\dot{q} - \frac{\partial T}{\partial q} + K_0 q = K_0 \bar{q}. \tag{4.5}$$

Then the effectiveness of (4.4) as a practical PTP (Point to Point) control scheme is assured theoretically by proving the asymptotic stability of the equilibrium state $(q, \dot{q}) = (\bar{q}, 0)$ for the system of (4.5), where $\bar{q}$ is a given desired position of the arm in configuration space.

The proof was first given by Takegaki and Arimoto [8] on the basis of Hamilton's canonical form. To gain an insight into essential parts of the argument, we point out the importance of the following relation:

$$\dot{q}^{\mathrm{T}} \frac{\partial T}{\partial q} = \dot{q}^{\mathrm{T}} \frac{\partial}{\partial q}\left(\frac{1}{2}\dot{q}^{\mathrm{T}} H \dot{q}\right) = \frac{1}{2}\sum_{i=1}^{n} \frac{\partial}{\partial q_i}(\dot{q}^{\mathrm{T}} H \dot{q})\dot{q}_i = \frac{1}{2}\dot{q}^{\mathrm{T}} \dot{H} \dot{q}. \tag{4.6}$$

Bearing this in mind, we introduce a Lyapunov function of the form

$$V(q, \dot{q}) = \frac{1}{2} q^{\mathrm{T}} (J + H(q))\dot{q} + \frac{1}{2}(q - \bar{q})^{\mathrm{T}} K_0 (q - \bar{q}). \tag{4.7}$$

Differentiation of this with respect to time t yields

$$\begin{aligned} \dot{V} &= \dot{q}^{\mathrm{T}}\left\{(J + H(q))\ddot{q} + \frac{1}{2}\dot{H}\dot{q} + K_0(q - \bar{q})\right\} \\ &= -\dot{q}^{\mathrm{T}}(B + K_1)\dot{q} + \dot{q}^{\mathrm{T}}\left(\frac{\partial T}{\partial q} - \frac{1}{2}\dot{H}\dot{q}\right). \end{aligned} \tag{4.8}$$

The second term on the right-hand side vanishes due to the relation of (4.6), and hence

$$\dot{V} = -\dot{q}^{\mathrm{T}}(B + K_1)\dot{q} \leq 0. \tag{4.9}$$

Although the derivative of V is only semidefinite, the asymptotic stability of the equilibrium point $(\bar{q}, 0)$ follows from LaSalle's theorem [23], because the maximal invariant set of (4.5) in state space satisfying $V = 0$ consists of a single point $(q, 0)$.

Next we discuss the speed of convergence to the equilibrium point. It is well-known that in the case of a time-invariant system described by (4.3) the equilibrium point $(\bar{q}, 0)$ becomes exponentially asymptotically stable. Fortunately, it is possible to prove [9] that this property holds in the case of the nonlinear dynamics of (4.5) after the robot motion approaches a neighborhood of the equilibrium state. In other words, this means that even if the robot manipulator is governed by the nonlinear dynamics with strong couplings between joints, it can behave by using a linear independent PD feedback like a second-order linear dynamical model governed by (4.3).

When the payload weight changes or is unknown or there exists a noteworthy evaluation error of the gravity term, then a PID-type feedback control may be suggested to clear the steady state error. This gives rise to another stability problem, which is discussed in our previous paper [24] with the aid of a more complicated Lypunov function. We further point out that such a local joint-based PD or PID controller serves as a basic control tool in case of force control or hybrid position/force control of robotic systems.

Finally, consider the situation in which the flexibility of drive mechanisms cannot be ignored, and hence the total dynamics of the manipulator system is considered to be governed by (3.9). In this case, the stability problem of positioning must be described in terms of the equilibrium state $(\bar{q}(=\bar{\theta}), 0, \bar{\theta}, 0)$ in $4n$-dimensional state space $(q, \dot{q}, \theta, \dot{\theta})$. Since in this case each PD feedback at its corresponding joint servo must be described in terms of the variable θ_i inside its joint actuator, that is,

$$\tau = -K_1\dot{\theta} - K_0(\theta - \bar{\theta}), \tag{4.10}$$

it is impossible to compensate for the gravity term directly in (3.9). However, if the effect of gravitation can be ignored, as is likely in the case of the motion of a SCARA robot restricted in a horizontal plane, then it is possible to prove the asymptotic stability of equilibrium state $(\bar{q}(=\bar{\theta}), 0, \bar{\theta}, 0)$. In fact, define a Lyapunov function as

$$W(q, \dot{q}, \theta, \dot{\theta}) = \frac{1}{2}\left[\dot{q}^{\mathrm{T}} H(q)\dot{q} + \dot{\theta}^{\mathrm{T}} J\dot{\theta} + \frac{1}{2}(\theta - \bar{\theta})^{\mathrm{T}} K_0(\theta - \bar{\theta})\right] + \sum_{i=1}^{n} F_i(\theta_i - q_i), \tag{4.11}$$

where each potential function F_i is defined by

$$F_i(u) = \int_0^u f_i(p)\, dp.$$

Since $F_i(u) > 0$ for all $u(\neq 0)$ owing to its characteristics as shown in figure 9.3, W is positive definite with respect to the $4n$-dimensional state variable $(q, \dot{q}, \theta, \dot{\theta})$. Further, the minimal value of W is attained at $(q, \dot{q}, \theta, \dot{\theta}) = (\bar{\theta}, 0, \bar{\theta}, 0)$, which is the equilibrium point of the system of (3.9) where $g(q) = 0$ and τ is replaced by (4.10). On the other hand, it is easy to check that

$$\dot{W} = -\dot{q}^{\mathrm{T}} B_1 \dot{q} - \dot{\theta}^{\mathrm{T}}(B_0 + K_1)\dot{\theta} \leqq 0. \tag{4.12}$$

Thus, it is possible to conclude again the asymptotic stability of positioning with the aid of LaSalle's theorem [25].

From a practical point of view, we must note the importance of damping factors in (3.9). In general, damping coefficients $b_{1i}(i = 1, \ldots, n)$ are considerably smaller relative to the corresponding diagonal components of B_0 and K_0. Therefore, the derivative of the Lyapunov function W may be approximately represented by

$$\dot{W} \approx -\dot{\theta}^{\mathrm{T}}(B_0 + K_1)\dot{\theta}.$$

This implies, in other words, that some missing terms in (3.9), like parasitic forces due to nonrigidity of link elements and backlashes, may cause W to be nonnegative even if $\dot{q} \neq 0$. Therefore, in order to increase the margin of stability, it is necessary to increase each damping factor b_{1i} as much as possible.

9.5 Nonlinear Control and Feedforward Compensation

Next we discuss the problem of trajectory tracking control for robot manipulators. In this case it is natural to expect that the same PD feedback control locally closed at each joint as given in (4.1) still works well provided that the magnitude of velocity $\dot{q}$ of the given desired trajectory is relatively small. In fact, it is possible to prove with the aid of a similar Lyapunov function method that the actual trajectory follows in a neighborhood of the desired one by using a high-velocity feedback gain [26]. However, it is claimed in much of the literature that in case of fast movement for trajectory or path tracking the performance of such a classical servo control becomes unsatisfactory. As a result of this argument, a variety of "advanced" control techniques has been proposed. The majority of these may be classified into (1) PD feedback with nonlinear gains, (2) applications of nonlinear control theory, and (3) the so-called computed torque method.

In category (1) Samson [27] has proposed a set of stable and robust controls for nonlinear mechanical systems. His idea can be traced back to an original paper presented by Leitmann [28]. A recent contribution of Shoureshi et al. [29] applies the same idea to the path tracking control for robot manipulators. Both papers show the usefulness of such a nonlinear feedback scheme by simulations [27] or experiments [29]. This is not surprising from a theoretical point of view since such a PD feedback scheme with nonlinear and variable gains can be regarded as a natural extension of classical linear PD or PID feedback control.

In much of the literature belonging to category (2) there are three main streams: (i) feedback linearization, (ii) nonlinear decoupling, and (iii) the robust controller. The original idea of feedback linearization for a class of nonlinear dynamical systems is found in the paper of Hunt et al. [30]. It has been shown [7, 31] that the nonlinear robot dynamics described by (3.6) can be globally linearized and decoupled by nonlinear feedback, and, as a result, it is transformed into a set of double integration equations that can be controlled by adding an "outer loop" control. An extension of this procedure to robot dynamics with elastic joints [described by (3.8)] was recently investigated by Cesareo and Marino [32] and Spong [33]. Nonlinear decoupling control of category (ii), proposed by Freund [6], is similar to the method of feedback linearization in the sense that it uses a nonlinear transformation in order to decouple completely the nonlinear model of robot dynamics. Since both nonlinear transformations use complete knowledge of the inertia matrix, an elaborate computation of the input driving signal is required. A robust controller of category (iii) has been proposed recently by Mills and Goldenberg [34] for trajectory tracking in the presence of large unmodeled disturbances, such as unknown payload mass/inertia or viscous joint friction. It consists of two separate components: (1) a stabilizing compensator, which is an ordinary linear PD or PID feedback, and (2) a robust compensator, which causes asymptotic regulation to occur for a class of disturbance signals. Nevertheless, all these nonlinear control methods have the same drawback—the amount of computation required to formulate the driving torque is enormous and almost equivalent to that needed for the "computed torque method." The remaining problem is to know what improvement in the control performance can be achieved by these advanced control methods at the expense of an enormous amount of required computation.

If a model of robot dynamics as described by (3.6) can be computed in real time (within every period of servo cycle time) along the desired trajectory, then each required input torque can be generated at the joint according to command data sent from the CPU control computer. This is called a "computed torque" or "inverse dynamics" method. Two different on-line computation algorithms with almost the same computational complexity were proposed by Hollerbach [5] and Luh et al. [4]. However, in this method there must arise a noteworthy discrepancy between the computed torque and the real required one due to the presence of large parameter perturbations, unknown joint frictions, and unknown payload mass and inertia properties [35]. There may also exist some unknown effects due to the flexibility of joints and the nonrigidity of link elements. In spite of these, the computed torque method will be quite important in obtaining the first approximation to the required input torque that can grossly compensate for main contributions of nonlinear terms of manipulator dynamics. Then, the remaining perturbed system can be controlled by an ordinary PD feedback.

For tasks where the end positions are of primary concern, the on-line computation of the inverse kinematics is needed in addition to the on-line computation of the inverse dynamics. Unfortunately, computation of the transformation of position, velocity, and acceleration from the end effector coordinates to the joint coordinates cannot be sped up much by only using multiprocessors, since it is hard intrinsically to parallelize the algorithm. In contrast to this, the RMRC scheme realizes a simple inverse transform of velocity as described by [1],

$$\frac{dq}{dt} = J^{-1}(q)\frac{dx}{dt}, \tag{5.1}$$

and therefore can be easily implemented as far as the Jacobian matrix is nonsingular during the robot motion. Here, $x = (x_1, \ldots, x_n)^{\mathrm{T}}$ denotes the end effector coordinates. Note that in this case a PD feedback control law can be devised of the form

$$\tau = -J^{\mathrm{T}}(q)L_0(x - \bar{x}) - K_1\dot{q}, \tag{5.2}$$

where $\bar{x}(t)$ means a desired path given in terms of the end effector coordinates [8, 9]. It should be noted that in (5.2) it is necessary to compute only $J(q)$ and not its inverse, which is a considerably easier problem. The stability problem of this method in PTP control was also studied in our previous papers [8, 9].

9.6 Force Control and Impedance Control

In order to expand the feasible applications of robots, it is necessary to control not only positional variables of joint or end point coordinates but also the force in certain directions. For example, such tasks as opening a door, rotating a handle, grinding, driving a screw, inserting a part into a hole, and handling an object may be carried out by means of cooperation of multirobots or by interaction of a robot with the environment. An essential idea of the hybrid technique that combines force or torque information with positional data to satisfy simultaneous position and force trajectory constraints specified in task-oriented coordinates is given by Raibert and Craig [12]. Also, Mason [13] has pointed out that a natural way to apply force control to complex handling and assembly tasks is to divide the various Cartesian degrees of freedom between force and position control.

The necessity of control of the manipulator impedance was stressed by Hogan [14], who also gave a unified approach to the control of target acquisition, kinematically constrained motion, and dynamic interaction. The principal objective of impedance control is to control simultaneously position and force by setting the mechanical impedance of the end effector with respect to the external force exerted by the environment.

An elaborated experimental work on force and compliant (or impedance) control has been carried out by Maples and Becker [36] using a PUMA 560 manipulator and an Adept One robot. The results achieved with various servo algorithms using position and force sensors are quite interesting. However, in this chapter we shall omit further comments on this subject but instead point out an important survey on the status of the art of robot force control conducted by Whitney [37].

Before ending this section, we emphasize potential applications of the learning control scheme to force and impedance control in situations in which the modeling of dynamics of interactions between the robot and its environment is hard. Such an illustrative application was carried out in our previous paper [43], where an iterative learning control scheme similar to that explained in section 9.7 is applied to relatively simple tasks such as polishing and grinding.

9.7 Adaptive Control and Leaning Control

Since an adaptive scheme was first applied by Dubowsky [15] to the trajectory following control of robots, much literature on this has accumulated. Most of it is grossly classified into (1) classical MRAC [15, 38] (Model Reference Adaptive Control), (2) MRAC based on Lyapunov's method [16], (3) MRAC based on hyperstability [17], and (4) self-tuning [39, 40]. As described in figure 9.6, an MRAC system includes a "reference model" that is usually chosen as a set of second-order underdamped linear dynamical systems. The adaptive scheme is designed by devising a regulation law so

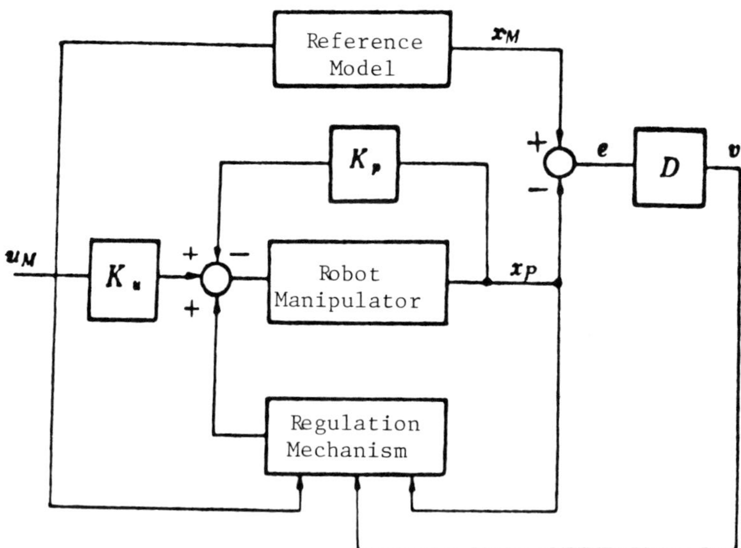

Figure 9.6
Adaptive robot control of MRAC-type.

that the difference between the output of the reference model and the actual response of the robot is minimized in a certain sense. However, in the system of adaptive robot control proposed by Takegaki and Arimoto [16], there is no reference model, but a desired trajectory is specified instead. The stability of regulation law that ensures the convergence of trajectory following was first attacked by Takegaki and Arimoto [16] for a system of linearized dynamics. The convergence for a nonlinear system of robot dynamics has been proved by Balestrino et al. [17] by using Popov's hyperstability. All proposed adaptive algorithms based on the MRAC scheme have the same burden of computational complexity. The adaptive scheme of the self-tuning type for robot control has been proposed by Koivo and Guo [39] and Leininger [40], where the controller consists of two devices: estimation of model parameters and determination of control gains or control input. Usually, a discrete-time linear autoregressive model or a continuous-time linear second-order time-varying system is employed as a model of robot dynamics. This scheme does not need much computation, but the convergence of trajectory following is obscure.

To overcome those drawbacks, a new adaptive robot control algorithm was recently proposed by Slotine and Li [41]; it consists of a PD feedback part and a full dynamics feedforward compensation part. This algorithm is computationally simple but cannot assure the ability to follow exactly any given trajectory.

To reduce the trajectory following error greatly, a new concept of learning control has been proposed by the author and his colleagues [18–20]. This concept differs from that of conventional classical or modern control techniques, in that in the latter case a control law designed for a controlled system is given in advance and fixed during operation of the system. In fact, actual operation data of input and output will never be used in modifying the control law once it is implemented in the control system. In contrast, the learning control concept stands for the repeatability of operating a given objective system and the possibility of improving the control input on the basis of previous actual operation data.

Repeatability of operation is one of main technical features of industrial robot manipulators. Even in cases where the robot arm is subject to the teaching and playback control mode, a human operator must teach the robot how to move in space and fulfill a desired task by handling a joystick or manipulating switches on a teaching box and improve its form of movement and performance through repeated operation. Here arises a problem, whether mechanical robots can or cannot learn autonomously (without the help of human operators) from measurement data of previous operations and improve their performance in future operations. More restrictedly, given a desired output trajectory $y_d(t)$ over a finite time duration $[0, T]$, is there any efficient control scheme that updates the control input based on previous operation data that at the next operation will yield a better performance in a certain sense? To respond to this question, we proposed two types of learning control scheme (see figures 9.7 and 9.8):

$$\text{(i)}\quad u_{k+1}(t) = u_k(t) + \left(\Gamma\frac{d}{dt} + \Phi\right)(y_d(t) - y_k(t)), \tag{7.1}$$

$$\text{(ii)}\quad u_{k+1}(t) = u_k(t) + \left(\Phi + \Psi\int dt\right)(y_d(t) - y_k(t)), \tag{7.2}$$

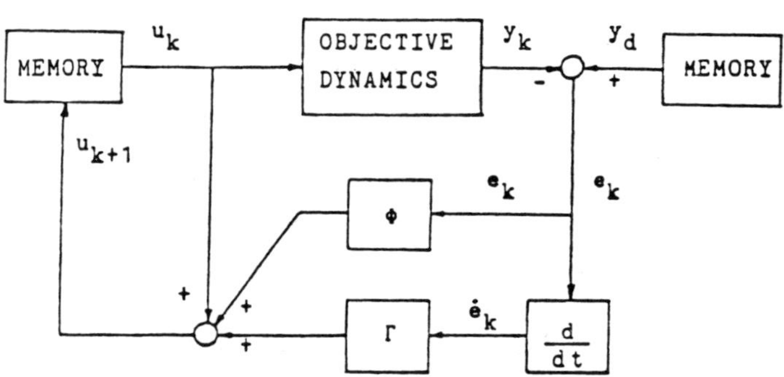

Figure 9.7
PD-type learning control law.

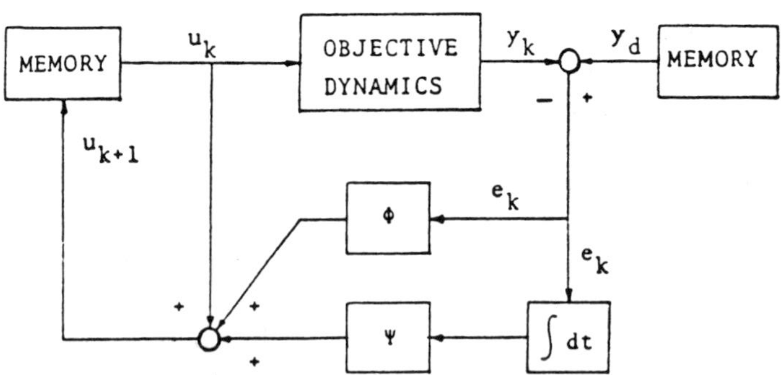

Figure 9.8
PI-type learning control law.

where u_k denotes the control input at the kth trial operation, y_k is the measured output trajectory at the kth trial, and Γ, Φ, and Ψ are constant gain matrices. For a class of robot manipulators y_d or y_k is a velocity vector of joint coordinates. It has been shown [18–20] that the PD-type learning control law shown in figure 9.7 with appropriate gain matrices Γ and Φ is convergent in a sense that the output trajectory approaches the desired one with the repetition of operation, namely, $y_k(t) \to y_d(t)$ uniformly in $t \in [0, T]$ as $k \to \infty$. A similar convergence of the PI-type leaning scheme shown in figure 9.8 has also been assured [42–46], but the argument of its proof is based on the linearized dynamics of the robot arm around the desired motion trajectory and the ignorance of higher terms. In addition, it is implicitly assumed in both cases that the arm must take the same initial position at every operation trial. This condition can be nearly satisfied by most industrial robots in use at present because of their fine characteristics of repeatability precision. However, to certify that such a learning control scheme is technically sound and robust for practical use, it is necessary to prove its convergence under more relaxed assumptions and show its stability regardless of the existence of some bounded errors in the initial setting and disturbances during operation. This problem has been discussed in our recent paper [47].

The principles that underlie the proposed learning control can be summarized in the following set of postulates:

9.7.1 Paradigm of Learning Control

1. Every operation ends in a fixed time interval $T > 0$.
2. A desired output $y_d(t)$ is given a priori over that time $t \in [0, T]$.
3. Repeatability of the initial setting is satisfied; i.e., the initial state $x_k(0)$ of the objective system can be set the same at the beginning of each operation

$$x_k(0) = x^0 \qquad \text{for} \quad k = 1, 2, \ldots. \tag{7.3}$$

4. Invariance of the system dynamics is assured throughout this repeated training.
5. Each output trajectory $y_k(t)$ can be measured, i.e., the error signal

$$e_k(t) = y_d(t) - y_k(t) \tag{7.4}$$

can be utilized in construction of the next control input.
6. The next control input $u_{k+1}(t)$ must be composed of a simple and fixed recursive law,

$$u_{k+1}(t) = F(u_k(t), e_k(t)). \tag{7.5}$$

It should be noted that (7.1) or (7.2) may be one of the simplest recursive forms.

For the practical use of this approach, the simpler the recursive law for (7.5), the better it is [like (7.1) and (7.2)]. Furthermore, in order to ascertain the practical effectiveness of this concept, we had better prove the convergence of such a learning control scheme under an appropriate choice for gain matrices Γ and Φ in (7.1) or Φ and Ψ in (7.2) without referring to a precise description of system dynamics. In line with these points, we proposed a learning control law given by (7.1) for a class of robot manipulators and proved the convergence in our recent papers [42–47].

As was pointed out previously, the dynamics of most industrial robots in use at present can be well approximated by its linear part as in (4.3), provided a high reduction gear ratio and a high feedback gain are employed at each joint. Then, if it is easy to know the coefficients in (4.3), it is possible to design a learning law of the form

$$u_{k+1}(t) = u_k(t) + \left\{ J\frac{d}{dt} + (B + k_1) + k_0 \int dt \right\} \{y_{\rm d}(t) - y_k(t)\}, \tag{7.6}$$

where $y_{\rm d}(t)$ and $y_k(t)$ stand for velocity vectors $\dot{q}_{\rm d}(t)$ and $\dot{q}_k(t)$, respectively. When this PID learning law is applied to the closed-loop system shown in figure 9.9, the joint coordinates at the kth trial must be subject to the following equation as is shown by (4.5):

$$(J + H(q_k))\ddot{q}_k + (B + K_1 + \dot{H}(q_k))\dot{q}_k - \frac{\partial T}{\partial q_k} + K_0 q_k = K_0 q_{\rm d} + u_k. \tag{7.7}$$

Since the linear part of this equation is dominant (in comparison with nonlinear terms), it is reasonable to rewrite (7.7) as

$$J\ddot{q}_k + (B + K_1)\dot{q}_k + K_0 q_k + \varepsilon\{\bar{H}(q_k)\ddot{q}_k + \bar{f}(q_k, \dot{q}_k)\} = K_0 q_{\rm d} + u_k, \tag{7.8}$$

where ε is a small number. Then, putting $k \leftarrow k + 1$ in (7.8) and substituting (7.6) in the resulting equation yield

$$J\dot{e}_{k+1} + (B + K_1)e_{k+1} + K_0 \int_0^t e_{k+1}\,dt$$
$$= \varepsilon\{H(q_{k+1})\ddot{q}_{k+1} - H(q_k)\ddot{q}_k + f(q_{k+1}, \dot{q}_{k+1}) - f(q_k, \dot{q}_k)\}. \tag{7.9}$$

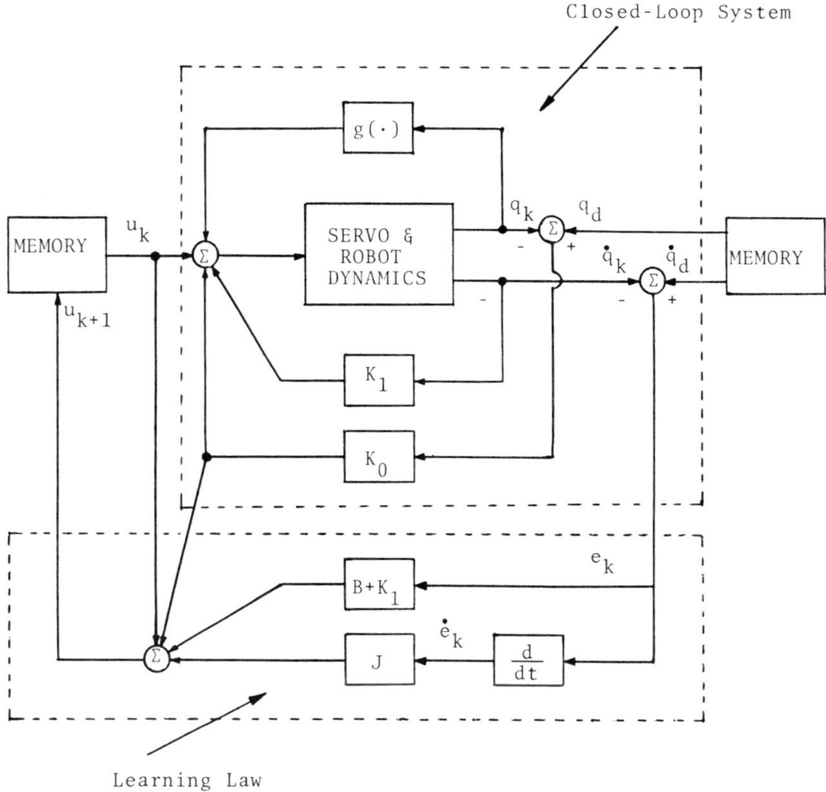

Figure 9.9
Learning control law when the linear part of robot dynamics is known.

Since all of the coefficient matrices on the left-hand side of this equation are diagonal with positive diagonal elements and the right-hand side is of order ε, the solution $e_{k+1}(t)$ must be of order ε, too. Furthermore, it is easy to check that $e_{k+1} = O(\varepsilon) \cdot e_k$, because the term within curly brackets in (7.9) can be rewritten as the sum of terms e_k and e_{k+1}, their derivatives $\dot{e}_k$ and $\dot{e}_{k+1}$, and their integrals $\int_0^t e_k \, d\tau$ and $\int_0^t e_{k+1} \, d\tau$. Thus, it is expected that the learning control scheme of (7.6) converges rapidly and hence the desired trajectory is realized after several trials. A similar result can be obtained by the same argument when a great part of the nonlinearities is compensated by the aid of the "computed torque" method [48].

Before concluding, we show some experimental results obtained by using a commercial robot (Mitsubishi Movemaster II). The overall experimental system is given in figure 9.10, where a felt pen is attached at the end of the third link of Movemaster II. Figure 9.11 shows the process of learning by the robot when a desired trajecting (to draw a circle on a flat plate) is given in terms of joint coordinates. Only measurement signals from joint encoders are used in the experiment of figure 9.11. Figure 9.12 shows the process of learning by the robot when a desired trajectory is described in terms of Cartesian coordinates. In this case two PSD signals, as shown in figure 9.10, are used to determine the 3-dimensional position of the tip of the manipulator. From figures 9.11 and 9.12, it is seen that the learning scheme based on task-oriented coordinates is more effective than that based on joint-coordinates. Figure 9.13 shows the robot's process of learning to draw a letter "R," in which a desired pattern "R" is given by tracking the movement of an LED attached at the tip of the manipulator driven by a human operator. By means of this method, the robot can learn how a human being writes letters or can carry out specified tasks.

9.8 Conclusion

It is shown in this chapter that the learning scheme based on repeatability of an operating robot is considerably effective in path or trajectory tracking. This scheme is capable of constructing inverse dynamics without knowing the model of robot dynamics.

One of the interesting problems to be attacked is how to use the knowledge base of input patterns acquired by learning. In other words, given a new desired trajectory, is it possible to find an efficient way to construct the input torque that realizes the desired trajectory by using a set of previous input patterns without need of repetitive operation of the robot? One of the promising ideas in this matter will be presented by the authors in a future paper [49].

Another interesting application of learning control is in the broad area of practical robot tasks encountered in factories, such as picking, grasping, handling, and assembling, where there may exist mechanical interaction between the robot and the environment. In such a case, it is in general hard to measure instantaneous contact force and interactive force or to know physical parameters such as stiffness and compliance. An application of learning control to such more complicated tasks will be presented in our future paper [50].

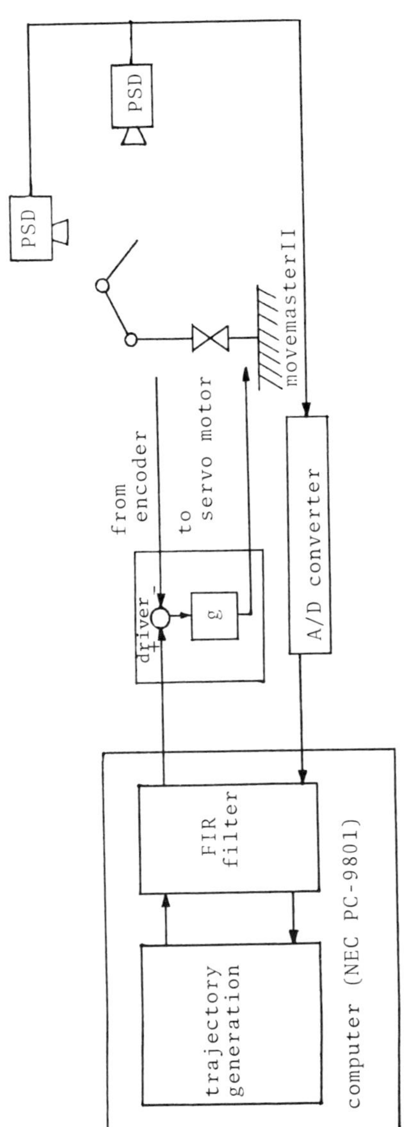

Figure 9.10
Experimental system.

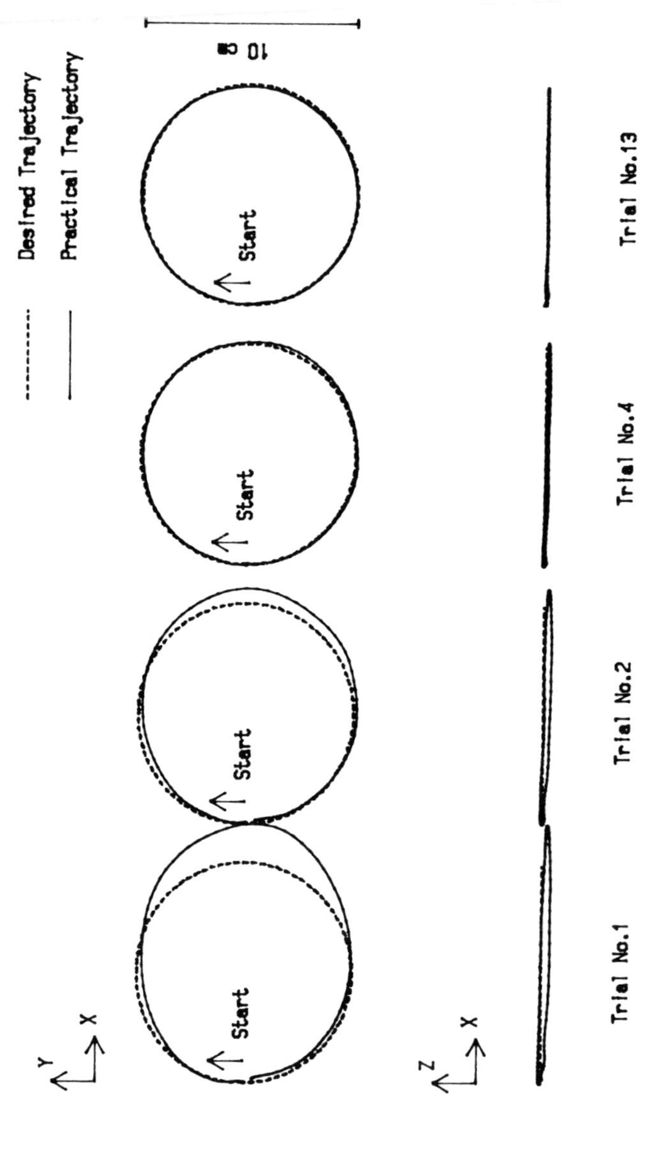

Figure 9.11
Experimental results (task-coordinates).

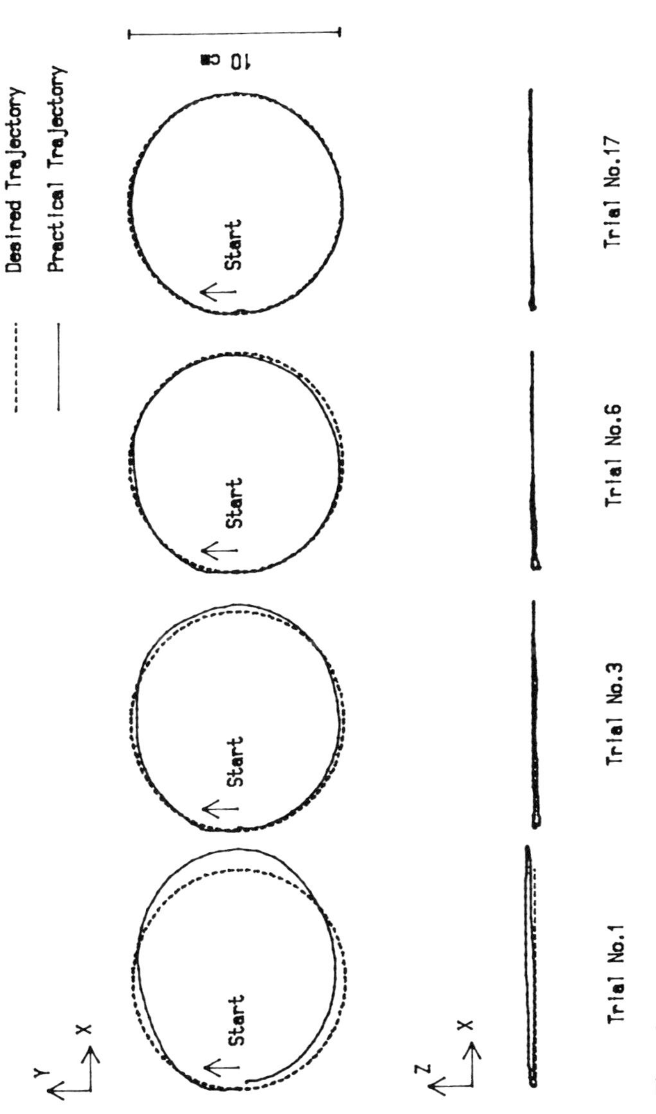

Figure 9.12
Experimental results (joint-coordinates).

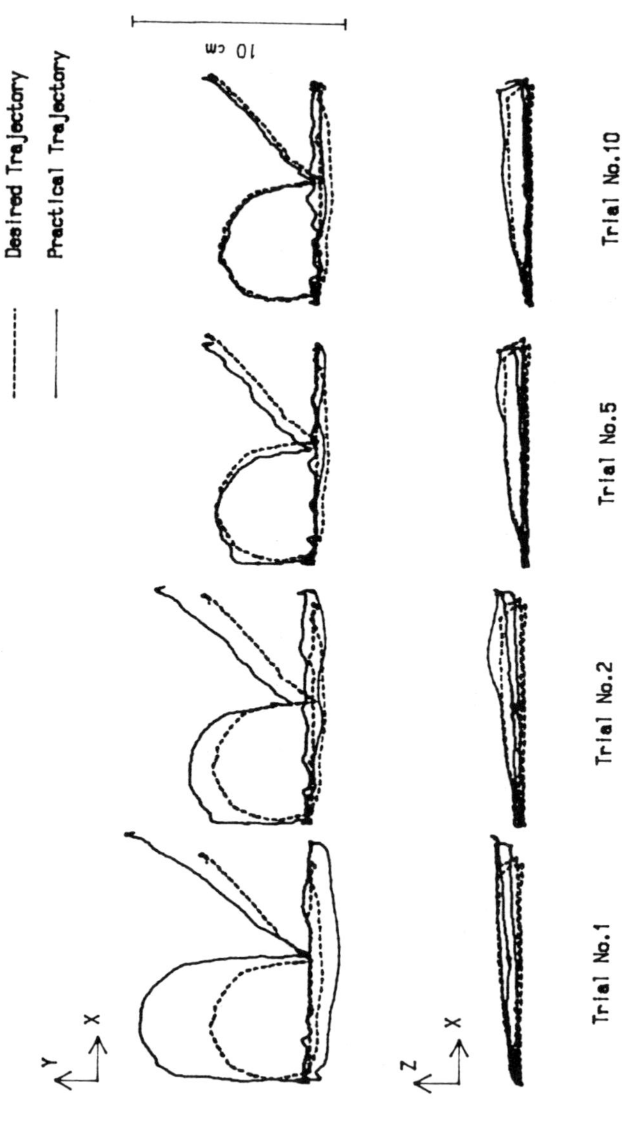

Figure 9.13
Experimental results.

References

[1] Whitney, E. D., "Resolved Motion Rate Control of Manipulators and Human Protheses," *IEEE Trans. Man-Machine Systems*, Vol. 1 MMS-10, pp. 47–53 (1969).

[2] Paul, R. P., *Robot Manipulators*, The MIT Press, Cambridge, Massachusetts (1981).

[3] Vukobratovic, P., *Scientific Fundamentals of Robotics 1 and 2*, Springer-Verlag, Berlin (1982).

[4] Luh, J. Y. S., Walker, M. W., and Paul, P. C., "Online Computation Scheme for Mechanical Manipulators," *ASME Trans. J. Dynam. Syst. Measure. Control*, Vol. 102, pp. 69–76 (1980).

[5] Hollerbach, J. M., "A Recursive Lagrangian Formulation of Manipulator Dynamics and a Comparative Study of Dynamic Fomulation Complexity," *IEEE Trans. Syst. Man. Cybernet.*, Vol. SMC-10, pp. 730–736 (1980).

[6] Freund, E., "Fast Nonlinear Control with Arbitrary Pole-Placement for Industrial Robots and Manipulators," *Int. J. Robotics Research*, Vol. 1, No. 1, pp. 65–78 (1982).

[7] Tarn, T. J., et al., "Nonlinear Feedback in Robot Arm Control," *Proc. IEEE 23rd Conference on Decision and Control*, Las Vegas, pp. 736–751 (1984).

[8] Takegaki, M., and Arimoto, S., "A New Feedback Method for Dynamic Control of Manipulators," *Trans. ASME, J. of DSMC*, Vol. 103, pp. 119–125 (1981).

[9] Arimoto, S., and Miyazaki, F., "Asymptotic Stability of Feedback Control Laws for Robot Manipulators," *Proc. of IFAC Symp. on Robot Control*, Barcelona, Spain (1985).

[10] Whitney, D. E., "Force Feedback Control of Manipulator Fine Motions," *Trans. ASME, J. DSMC*, Vol. 99, pp. 91–97 (1977).

[11] Hanafusa, H., and Asada, H., "Stable Prehension by a Robot Hand with Elastic Fingers," *Proc. 7th ISIR*, pp. 361–368 (1977).

[12] Raibert, M. H., and Craig, J. J., "Hybrid Position/Force Control of Manipulator," *Trans. ASME, J. of DSMC*, Vol. 103, pp. 126–133 (1981).

[13] Mason, M. T., "Compliance and Force Control for Computer Controlled Manipulators," *IEEE Trans. on Systems, Man, and Cybernetics*, SMC-11, No. 6, pp. 418–432 (1981).

[14] Hogan, N., "Mechanical Impedance Control in Assistine Devices and Manipulators," *Proc. of JACC*, Vol. 1, San Francisco (1980).

[15] Dubowsky, S., and Desforges, D. T., "The Application of Model Reference Adaptive Control to Robotic Manipulators," *Trans. ASME, J. of DSMC*, Vol. 101, pp. 193–200 (1979).

[16] Takegaki, M., and Arimoto, S., "An Adaptive Trajectory Control of Manipulators," *Int. Jour. Contr.*, Vol. 34, No. 7, pp. 219–230 (1981).

[17] Balestrino, A., De Maria, G., and Sciavicco, "An Adaptive Model Following Control for Robot Manipulators," *Trans. ASME, J. of DSMC*, Vol. 105, pp. 143–151 (1983).

[18] Arimoto, S., Kawamura, S., and Miyazaki, F., "Bettering Operation of Robots by Learning," *J. of Robotics Systems*, Vol. 1, No. 2, pp. 123–140 (1984).

[19] Idem, "Can Mechanical Robots Learn by Themselves?" In H. Hanafusa and H. Inoue (eds.), *Robotics Research: The Second International Symposium*, pp. 127–134, MIT Press (1985).

[20] Idem, "Bettering Operation of Dynamic Systems by Learning: A New Control Theory for Servomechanism or Mechatronics Systems," *Proc. of 23rd IEEE CDC*, Las Vegas (1984).

[21] Sweet, L. M., and Good, M. C., "Redefinition of the Robot Motion-Control Problem," *IEEE Control Systems Magazine*, Vol. 5, No. 3, pp. 18–25 (1985).

[22] Asada, H., Kanade, T., and Takeyama, I., "Control of a Direct-Drive Arm," *Trans. ASME, J. of DSMC*, Vol. 105, pp. 136–142 (1983).

[23] LaSalle, J. P., "Some Extensions of Lyapunov's Second Method," *IRE Trans. on Circuit Theory*, Vol. CT-7, pp. 520–527 (1960).

[24] Arimoto, S., and Miyazaki, F., "Stability and Robustness of PID Feedback control for Robot Manipulators of Sensory Capability," in M. Brady and R. P. Paul (eds.), *Robotics Research: First International Symposium*, pp. 783–799, MIT Press (1984).

[25] Arimoto, S., and Miyazaki, F., "Stability and Robustness of PD Feedback Control with Gravity Compensation for Robot Manipulator," in F. W. Paul and K. Youcef-Toumi (eds.), *Robotics: Theory and Applications*, DSC-Vol. 3 (presented at the winter Annual Meeting of ASME), pp. 67–72 (1986).

[26] Kawamura, S., Miyazaki, F., and Arimoto, S., "Stability of Trajectory Tracking PD Feedback Control for Robot Manipulator," to be presented to *SICE '87 in Hiroshma*, Japan (1987).

[27] Samson D., "Robust Nonlinear Control of Robotic Manipulators: Implementation Aspects and Simulations," *Robotics Research: The Next Five Years and Beyond*, MS 84-481, Society of Manufacturing Engineers, Deaborn, MI, pp. 1–17 (1984).

[28] Leitmann, G., "Guaranteed Asymptotic Stability for Some Linear Systems with Bounded Uncertainties," *Trans. ASME, J. of DSMC*, Vol. 101, pp. 212–216 (1979).

[29] Shoureshi, R., Roesler, M. D., and Corless, M. J., "Control of Industrial Manipulators with Bounded Uncertainties," to appear in *IEEE Trans. on Robotics and Automation* (1987).

[30] Hunt, L. R., Su, R., and Meyer, G., "Design for Multi-input Nonlinear Systems," *Differential Geometric Control Theory Conf.*, Birkhauser, Boston, pp. 268–298 (1983).

[31] Spong, M. W., and Vidyasagar, M., "Robust nonlinear Control of Robot Manipulator," *Proc. 24th IEEE CDC*, pp. 1767–1772, Fort Lauderdale, (1985).

[32] Casareo, G., and Marino, R., "On the Controllability Properties of Elastic Robots," *Sixth Int. Conf. on Analysis and Optimization of Systems*, INRIA, Nice (1984).

[33] Spong, M. W., "Modeling and Control of Elastic Joint Robots," in F. W. Paul and K. Youcef-Toumi (eds.), *Robotics: Theory and Applications*, DSC-Vol. 3 (presented at the winter Annual Meeting of ASME), pp. 57–65 (1986).

[34] Mills, J. K., and Goldenberg, A. A., "Robust Control of a Robotic Manipulator with Joint Variable Feedback," *ibid.*, pp. 43–49 (1986).

[35] Gilbert, E. G., and Ha, I. J., "An Approach to Nonlinear Feedback Control with Applications to Robotics," *Proc. 22nd IEEE CDC*, San Antonio (1983).

[36] Maples, J. A., and Becker, J. J., "Experiments in Force Control of Robotic Manipulators," *Proc. of the 1986 IEEE Int. Conf. on Robotics and Automation*, pp. 695–702, San Francisco (1986).

[37] Whitney, D. E., "Historical Perspective and State of the Art in Robot Force Control," *Proc. of the 1985 IEEE Int. Conf. on Robotics and Automation*, pp. 262–268 (1985).

[38] Horowitz, R., and Tomizuka, M., "An Adaptive Control Scheme for Mechanical Manipulators-Compensation of Nonlinearity and Decoupling Control," ASME paper 80 WA/DSC-6 (1980).

[39] Koivo, A. J., and Guo, T. H., "Adaptive Linear Controller for Robotic Manipulators," *IEEE Trans on Automatic Control*, Vol. AC-28, pp. 162–171 (1983).

[40] Leininger, G., "Adaptive Control of Manipulators Using Self-Tuning Methods," in M. Brady and R. P. Paul (eds.), *Robotics Research: First International Symposium*, MIT Press (1984).

[41] Slotine, J. J. E., and Li, W., "On the Adaptive Control of Robot Manipulators," *ASME WA/DSC-Vol. 3, Robotics: Theory and Applications*, pp. 51–56 (1986).

[42] Kawamura, S., Miyazaki, F., and Arimoto, S., "Iterative Learning Control for Robotic Systems," *Proc. of IECON '84*, Tokyo (1984).

[43] Idem, "Hybrid Position/Force Control of Manipulators Based on Learning Method," *Proc. of '85 Inter. Conc. on Advanced Robotics*, Tokyo (1985).

[44] Idem, "Applications of Learning Methods for Dynamic Controls of Robot Manipulators," *Proc. of 24th IEEE CDC*, Fort Lauderdale (1985).

[45] Arimoto, S., Kawamura, S., Miyazaki, F., and Tamaki, S., "Learning Control Theory for Dynamical Systems," *ibid.* (1985).

[46] Arimoto, S., "Mathematical Theory of Learning with Application to Robot Control," in K. S. Narendra (ed.), *Adaptive and Learning Systems*, pp. 379–388, Plenum Publishing Co. (1986).

[47] Arimoto, S., Kawamura, S., and Miyazaki, F., "Convergence, Stability, and Robustness of Learning Control Schemes for Robot Manipulators," in M. Jamshidi, L. Y. S. Luh, and M. Shahinpoor (eds.), *Recent Trends in Robotics: Modeling, Control, and Education*, pp. 307–317 (1986).

[48] Atkeson, C. G., and McIntyre, J., "Robot Trajectory Learning through Practice," *Proc. of 1986 IEEE Conf. on Robotics and Automation*, pp. 1737–1742 (1986).

[49] Kawamura, S., Miyazaki, F., and Arimoto, S., "Intelligent Control of Robot Motion Based on Learning," *Proc. of 1987 IEEE Inter. Symp. on Intelligent Control*, Philadelphia (1987).

[50] Arimoto, S., Miyazaki, F., and Kawamura, S., "Cooperative Motion Control of Multiple Robot Arms or Fingers," *Proc. of 1987 IEEE Conf. on Robotics and Automation* (1987).

10 Kinematics and Dynamics for Control

John M. Hollerbach

10.1 Introduction

This chapter reviews progress in the kinematics and dynamics for control. It discusses efficiency and efficacy: efficiency of kinematic and dynamic transformations from a standpoint of real-time control, and efficacy in terms of improvement in control. The past few years have seen remarkable progress in these areas, primarily from a theoretical standpoint. Lagging are actual experimental demonstrations incorporating advanced kinematics and dynamics, but some preliminary results have emerged. Benchmark papers in these areas are identified, and representative works are reviewed.

To provide a framework for the role of kinematics and dynamics in control, a hybrid position/force feedforward controller is outlined in figure 10.1 for trajectory control. This controller is only one way in which kinematics and dynamics can be incorporated, but its transparency is suitable for discussion purposes. The discussion in this chapter focuses on three parts of this figure: the inverse kinematics transformation, the inverse dynamics transformation, and the role of dynamics in feedback control.

As a general discussion of this figure, the first part is a trajectory planner, which specifies the motion of the robot in terms of the desired position and orientation $\mathbf{x}_d$ of the hand. For this controller, the trajectory planner also specifies the desired hand velocity $\dot{\mathbf{x}}_d$ and acceleration $\ddot{\mathbf{x}}_d$. By trajectory control, the most general definition is intended, where position and force are simultaneously controlled. Thus the trajectory plan may also specify the desired force $\mathbf{f}_d$, when the manipulator is in contact with the environment. The desired force may depend on the position variables, such as in stiffness control or in impedance control. One reason for planning in task coordinates is the ease of partitioning variables into position controlled $\mathbf{x}_d$ versus force controlled $\mathbf{f}_d$ variables.

In order to execute this plan, task coordinates are converted to joint coordinates by an inverse kinematics transformation. To indicate how this is done, it is easiest first to write the direct kinematics relationship between the position variables and the corresponding joint variables:

$$\mathbf{x}_d = \mathbf{f}(\boldsymbol{\theta}_d), \tag{1}$$

$$\dot{\mathbf{x}}_d = \mathbf{J}\dot{\boldsymbol{\theta}}_d, \tag{2}$$

$$\ddot{\mathbf{x}}_d = \mathbf{J}\ddot{\boldsymbol{\theta}}_d + \dot{\mathbf{J}}\dot{\boldsymbol{\theta}}_d. \tag{3}$$

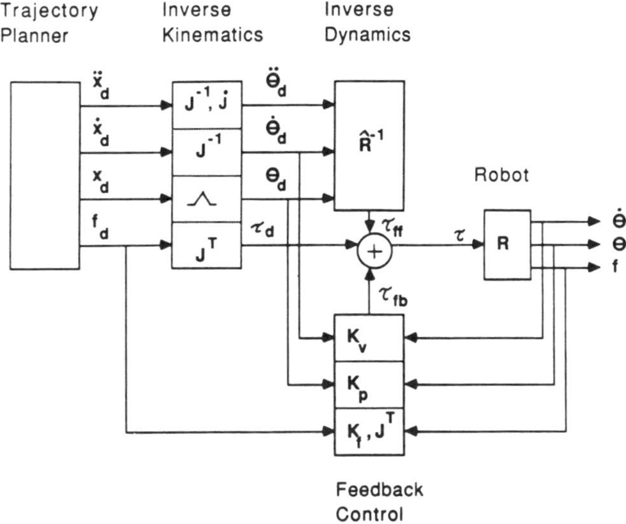

Figure 10.1
A hybrid position/force feedforward controller.

The function **f** is a nonlinear transformation from the joint angles $\boldsymbol{\theta}_d$ to the end point positions $\mathbf{x}_d$, and depends on the kinematic parameters of the robot. The fidelity of these kinematic parameters determines how accurately the robot may be positioned, and kinematic calibration is one of the basic model-building procedures to be discussed. The direct kinematic velocities and accelerations express linear relationships, with the aid of the Jacobian matrix $J_{ij} = \partial f_i / \partial \theta_j$.

The inverse kinematics transformation is required to convert the end point trajectory plan to joint variables, by inverting the direct kinematic relationships (1)–(3):

$$\boldsymbol{\theta}_d = \boldsymbol{\Lambda}(\mathbf{x}_d), \tag{4}$$

$$\dot{\boldsymbol{\theta}}_d = \mathbf{J}^{-1}\dot{\mathbf{x}}_d, \tag{5}$$

$$\ddot{\boldsymbol{\theta}}_d = \mathbf{J}^{-1}(\ddot{\mathbf{x}}_d - \dot{\mathbf{J}}\dot{\boldsymbol{\theta}}_d). \tag{6}$$

The nonlinear function $\boldsymbol{\Lambda} = \mathbf{f}^{-1}$ is a one-to-many mapping, and is problematical. For a six-degrees-of-freedom manipulator, unless the manipulator has the proper kinematic structure, $\boldsymbol{\Lambda}$ cannot be expressed analytically. If there are redundancies, then some method must be chosen to resolve the redundancies. We now know that there are fundamental difficulties with instantaneous redundancy resolution methods. Robust methods for handling singularities, where $\mathbf{J}^{-1}$ does not exist, are required. The expressions (5) and (6) would not actually be used to evaluate $\dot{\boldsymbol{\theta}}_d$ and $\ddot{\boldsymbol{\theta}}_d$ for efficiency reasons, but rather customized computations would be formed that take advantange of regularities in the robot's kinematic structure.

The desired endpoint force $\mathbf{f}_d$ is directly transformed to joint torques τ_d by the fundamental mechanical relationship:

$$\tau_d = \mathbf{J}^T \mathbf{f}_d. \tag{7}$$

This relation is one of statics rather than kinematics, but is grouped with kinematics because the transformation involves just the transpose Jacobian matrix $\mathbf{J}^T$.

After the inverse kinematics and statics transformation, the next step in making the robot follow a desired trajectory is supplying appropriate commands to the actuators. The simplest approach is feedback control, which generates these commands by measuring the difference between where the arm is and where it is supposed to be at any instant in time and using some function (usually a linear function) of this error as the drive signal to the actuators. In figure 10.1, feedback control is indicated by a box that contains feedback gains $\mathbf{K}_p$ for position error, $\mathbf{K}_v$ for velocity error, and $\mathbf{K}_f$ for force error, to produce an output:

$$\tau_{fb} = \mathbf{K}_p(\boldsymbol{\theta}_d - \boldsymbol{\theta}) + \mathbf{K}_v(\dot{\boldsymbol{\theta}}_d - \dot{\boldsymbol{\theta}}) + \mathbf{J}^T \mathbf{K}_f(\mathbf{f}_d - \mathbf{f}). \tag{8}$$

Note that the force error is computed in task coordinates, and after application of the gain $\mathbf{K}_f$ must be converted to joint torques by the transpose Jacobian $\mathbf{J}^T$.

Feedback control is useful and necessary to compensate for unpredicted disturbances. In particular, when linear feedback control is used alone, the rigid body dynamics of the manipulator are considered as disturbances. These dynamics may cause substantial trajectory errors for faster motions, unless gains in the feedback control are made correspondingly higher. Yet there are practical limits to how high gains can be set, given actuator saturation and stability problems.

To reduce the errors that need to be corrected by feedback, one approach is feedforward control, which uses a dynamic model $\hat{R}$ of the robot to predict actuator commands corresponding to a desired motion. This model $\hat{R}$ hopefully represents the actual robot dynamics R fairly accurately, so that the unmodeled dynamics will not cause significant perturbations. Besides the kinematic parameters, the inertial parameters of the links go into the model $\hat{R}$, and their accurate estimation is useful in reducing trajectory errors.

The direct dynamics R is the transformation from the robot input (joint torques) to the robot output (joint motion); in control terms, R represents the plant dynamics. In figure 10.1 the calculation of driving torques from a model of the robot and a desired trajectory is the inverse dynamics R^{-1}, obtained by inverting the robot plant R. Since the plant dynamics are not known exactly, only the estimated inverse $\hat{R}^{-1}$ may be used to predict the feedforward torques τ_{ff}. Thus an important issue is how well the robot dynamics can be identified.

In practice, there are always unexpected disturbances or modeling errors that make feedforward control imperfect, and a feedback controller is also included to compensate for unpredicted disturbances. Thus the total output τ to the robot is given by

$$\tau = \tau_{fb} + \tau_{ff} + \tau_d \tag{9}$$

Note that the desired torque τ_d based on the planned end point force $\mathbf{f}_d$ is also included in this calculation, and can also be thought of as a feedforward command.

10.2 Kinematics

This section reviews the inverse kinematics transformation, beginning with nonredundant, six-degrees-of-freedom manipulators. Efficient algorithms for inverse kinematic positions, velocities, and accelerations are presented, as are methods for handling singularities. Redundant manipulators are discussed next, focusing particularly on the feasibility of local inverse functions. Last, the timely topic of kinematic calibration, namely, how an accurate kinematic model of the robot can be estimated, is discussed.

10.2.1 Inverse Kinematics of Nonredundant Manipulators

Much of the work on inverse kinematics has focused on six-degrees-of-freedom manipulators, which are commonly available commercially. Examples include the Unimation PUMA robot, the Cincinnati-Millicron T3 robot, and the ASEA robot. Since position and orientation of an object in space are specified by six numbers, the six-degrees-of-freedom manipulators have been referred to as general purpose. In the following sections, we review progress in the inverse kinematics transformations for position and for velocity and acceleration. We then consider the serious problem of singularities.

10.2.1.1 Inverse Kinematics for Position Ever since the pioneering work of Pieper (1968), we have known that unless a six-degrees-of-freedom manipulator is structured in one of several ways, the calculation of inverse kinematics for position cannot be expressed analytically. Two particular classes of manipulators that can be solved analytically are

- manipulators containing the equivalent of a spherical joint anywhere in the kinematic chain and
- manipulators containing the equivalent of a planar pair anywhere in the kinematic chain.

When a manipulator has one of these two structures, the robot can be computationally broken apart into pieces that are analytically tractable.

The most common example of a manipulator containing a spherical joint is the spherical-wrist rotary manipulator (henceforth referred to as the 6R arm), depicted in figure 10.2A without joint offsets in the shoulder and elbow joints. The PUMA and ASEA robots have structures similar to this. A common example of a manipulator containing a planar pair is the rotary

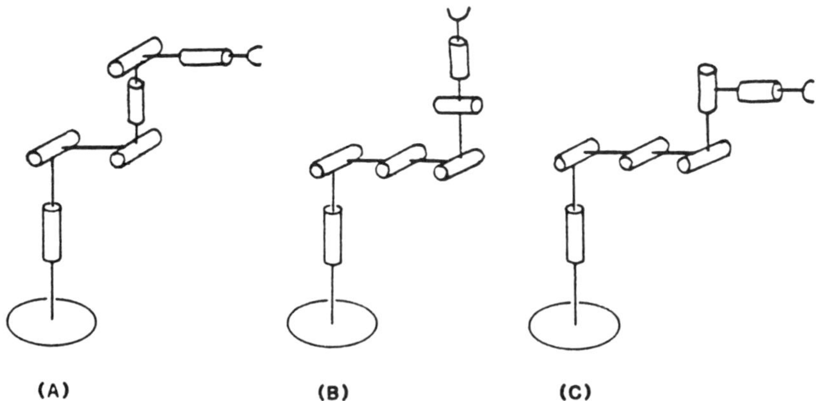

Figure 10.2
(A) Spherical-wrist rotary manipulator without joint offsets. (B), (C) Planar-pair rotary manipulators with differing wrist structures.

manipulator of figure 10.2B, where the neighboring shoulder, elbow, and wrist axes are parallel. Planar pair refers to a three-joint arrangement that produces planar motion, and can be realized by several combinations of rotary and prismatic joints. The Cincinnati T3 robot is geometrically similar to the manipulator in figure 10.2B. A slightly different wrist arrangement on a planar pair arm is depicted in figure 10.2C, similar to the Hitachi A6030 robot.

For the 6R arm, an efficient inverse kinematics computation begins by decomposing the manipulator into two parts at the wrist (Featherstone, 1983). If $\mathbf{p}$ is the given position of a reference point on the hand and $\mathbf{R}$ is the given orientation matrix of the hand, then the position $\mathbf{p}_w$ of the wrist is given simply by (figure 10.3)

$$\mathbf{p}_w = \mathbf{p} - \mathbf{R}\,{}^h\mathbf{p}_h, \tag{10}$$

where ${}^h\mathbf{p}_h$ is a vector from the wrist to the hand reference point expressed in the hand coordinate system, and is presumably known. The first three joint angles can then be solved analytically from the wrist position. The absolute orientation $\mathbf{R}_f$ imparted to the forearm by the first three joint angles is computed, and is removed from the absolute orientation of the hand to leave a relative hand orientation $\mathbf{R}_f^T\mathbf{R}$ that the last three joint angles alone must realize. The beauty of the 6R arm is that any relative hand orientation can be realized, after the forearm orientation has been factored out. An analogous albeit slightly more involved algorithm can be posed for the planar pair arm.

There has also been a fair amount of activity in the inverse kinematics of more general six-degrees-of-freedom manipulators that do not contain one of the simplifying structures above (e.g., Tsai and Morgan, 1985).

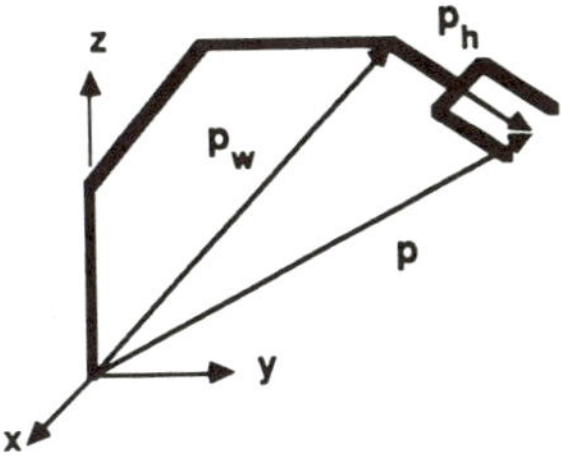

Figure 10.3
Decomposing a spherical-wrist manipulator.

So far, it has not been convincingly argued why anyone would build a manipulator that is not analytically solvable, and why this problem is more than just of theoretical interest. Rationales for nonanalytically solvable manipulators could include the following:

- *Workspace.* Presumably one could encounter a workspace where a special manipulator geometry is required that cannot be realized with a spherical joint or planar pair. So far, such a requirement has not been clearly articulated.
- *Modular design.* Design constraints may lead to combinations of actuators, transmission elements, and structures that do not readily lend themselves to a simplifying geometry. Presumably the design advantages are so significant as to outweigh the computational disadvantages. The Robotics Research arm is such a manipulator, where a modular two-joint structure leads to nonintersecting rotary joint axes. Nevertheless, it is not clear that complete consistency in a modular design should be pursued.
- *Kinematic calibration.* Kinematic calibration could lead to a nonanalytically solvable manipulator. Although the original intent might have been to build a 6R arm, for example, the manufacturing process is imperfect. A subsequent kinematic calibration might then reveal slight joint offsets that were not planned for and that make the wrist nonspherical.

The kinematic calibration argument is serious, and its possible solution is briefly discussed. Presumably the corrected kinematic structure would only differ slightly from the nominal analytically solvable kinematic structure. The nominal model $\mathbf{x} = \mathbf{f}_N(\boldsymbol{\theta})$ can then be used in conjunction with the calibrated model $\mathbf{x} = \mathbf{f}(\boldsymbol{\theta})$ for solution by iteration (Hayati and Roston, 1986; Sugimoto and Okada, 1985). Suppose $\mathbf{x}_d$ is the desired end point position. Then the nominal joint angle solution $\boldsymbol{\theta}_N = \mathbf{f}_N^{-1}(\mathbf{x}_d)$ can be found since the manipulator was designed to be analytically solvable. The actual end point location that would result from the nominal joint angle solution is $\mathbf{x} = \mathbf{f}(\boldsymbol{\theta}_N)$, based on the calibrated model. The endpoint error $\Delta\mathbf{x} = \mathbf{x}_d - \mathbf{x}$ can be corrected to first order by using (5):

$$\Delta\boldsymbol{\theta} = \mathbf{J}^{-1}(\boldsymbol{\theta}_N)\,\Delta\mathbf{x}, \tag{11}$$

$$\boldsymbol{\theta} = \boldsymbol{\theta}_N + \Delta\boldsymbol{\theta}. \tag{12}$$

Equation (11) is valid because the error $\Delta\mathbf{x}$ is presumably small, and the Jacobian matrix $\mathbf{J}$ relates incremental end point motion to incremental joint motion. The updated joint angle estimate (12) can be iterated if the first correction $\Delta\boldsymbol{\theta}$ is not adequate, but presumably this would not be necessary if the nominal and calibrated models are very close. Thus this inverse kinematics position calculation should still be efficient, especially considering that the inverse Jacobian matrix is based on the nominal model and the efficient inverse kinematic velocity algorithms discussed in the next section could be used for transformation (11) instead.

So far, actual experimental results in kinematic calibration have shown that wrist off-sets are not significant (Shamma and Whitney, 1987), leaving still a solvable manipulator from a standpoint of geometric link parameters. Nevertheless, nongeometric factors such as gear eccentricity, backlash, and joint compliance may also prevent the formulation of an analytic inverse. The procedure outlined above could still be useful in the latter case.

10.2.1.2 Inverse Kinematics for Velocity and Acceleration Until just a few years ago, it was thought that inverse kinematics transformations for velocity and acceleration were too computationally expensive for real-time control. For inverse kinematic velocities, a number of shortcuts were proposed to approximate these computations. The reason for this viewpoint is that investigators took the inverse kinematic velocity relation (5) literally and attempted to invert the Jacobian matrix. As seen in table 10.1, the method of inverting the Jacobian matrix for the Stanford manipulator requires almost 300 multiplications to derive the joint velocities. Since the inverse Jacobian matrix is by itself usually not of intrinsic interest, Gaussian elimination can be used instead to cut the number of arithmetic operations in half.

More recently, Paul (1981) proposed a method particularized to the Stanford manipulator that is somewhat more efficient than Gaussian elimination (table 10.1); the numbers given are slightly larger than stated in

Table 10.1
Inverse kinematic velocities for the Stanford manipulator

Method	Multiplications	Additions
6×6 matrix inversion	287	193
Gaussian elimination	141	98
Paul (1981)	94	55
Featherstone (1983)	36	20

(Paul, 1981). Paul's method involves manipulation of 4×4 transformation matrices, and takes implicit advantage of the spherical-wrist configuration of the Stanford manipulator through algebraic manipulations.

Featherstone (1983) in a benchmark paper took more explicit advantage of the spherical-wrist configuration of manipulators to propose a highly efficient method for computation of the inverse kinematic velocities. We presume that the hand linear velocity $\dot{\mathbf{p}}$ and hand angular velocity $\boldsymbol{\omega}$ have been specified by the trajectory planner. The method is composed of four steps, and is entirely analogous to the inverse kinematics position calculation for spherical wrist manipulators.

1. Find the linear velocity of the wrist from the hand linear and angular velocity:

$$\dot{\mathbf{p}}_w = \dot{\mathbf{p}} - \boldsymbol{\omega} \times \mathbf{p}_h. \tag{13}$$

2. Find the first three joint rates from the wrist linear velocity $\dot{\mathbf{p}}_w$. The joint angles of the 6R arm are defined according to the Denavit-Hartenberg link parameters (figure 10.4a; taken from Hollerbach and Sahar, 1983):

$$\dot{\theta}_1 = -\frac{{}^2\dot{p}_{wz}}{{}^1p_{wx}}, \tag{14}$$

$$\dot{\theta}_2 = \frac{{}^2\dot{p}_{wy}\cos\theta_3 - {}^2\dot{p}_{wx}\sin\theta_3}{a_2\cos\theta_3}, \tag{15}$$

$$\dot{\theta}_3 = \frac{{}^2\dot{p}_{wx} - \dot{\theta}_2(s_4\cos\theta_3)}{s_4\cos\theta_3}, \tag{16}$$

where components of the wrist position and velocity have been expressed in link coordinate systems 1 and 2, respectively.

3. Find the angular velocity of the hand $\boldsymbol{\omega}_h$ relative to the forearm:

$$\boldsymbol{\omega}_h = \boldsymbol{\omega} - \mathbf{z}_0\dot{\theta}_1 - \mathbf{z}_1\dot{\theta}_2 - \mathbf{z}_2\dot{\theta}_3. \tag{17}$$

4. Find the last three joint rates from the relative angular velocity of the hand (figure 10.4b):

$$\dot{\theta}_5 = {}^4\omega_{hz}, \tag{18}$$

$$\dot{\theta}_4 = -{}^4\omega_{hx}/\sin\theta_5, \tag{19}$$

$$\dot{\theta}_6 = {}^4\omega_{hy} - \dot{\theta}_6\cos\theta_5, \tag{20}$$

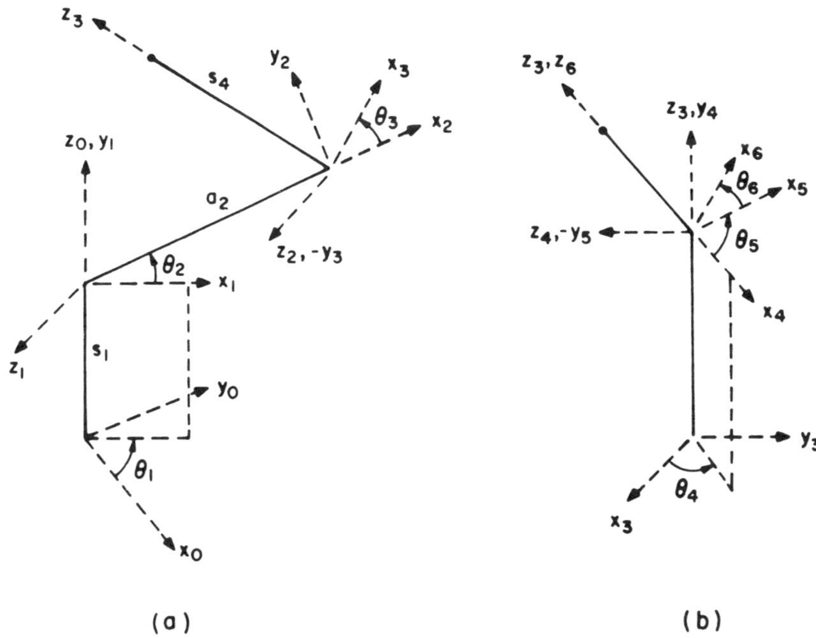

Figure 10.4
Denavit-Hartenberg parameters for the 6R arm. (a) The first three degrees of freedom. (b) The wrist degrees of freedom.

where the relative hand angular velocity has been expressed in link coordinate system 4.

Featherstone's method applied to the Stanford manipulator is almost 3 times more efficient than Paul's method.

For inverse kinematic accelerations, Hollerbach and Sahar (1983) applied Featherstone's method to a six-degrees-of-freedom rotary manipulator without joint offsets (figure 10.2A). Table 10.2 shows a considerable saving over the Jacobian methods. Paul, Renaud, and Stevenson (1984) recast the homogeneous transformation method to arrive at a similar efficiency for both inverse velocities and accelerations. Elgazzar (1985) applied Featherstone's method to the PUMA 560 robot, which differs from the rotary manipulator discussed in (Featherstone, 1983; Hollerbach and Sahar, 1983) because of joint offsets at the shoulder and elbow joints, to arrive at a similar efficiency also.

Table 10.2
Inverse kinematic accelerations for a six-degrees-of-freedom rotary manipulator

Method	Multiplications	Additions
6×6 matrix inversion	394	260
Gaussian elimination	213	162
Hollerbach and Sahar	78	57

Table 10.3
Total computational complexity for inverse kinematics

Kinematic parameter	Multiplications	Additions
Joint angles	64	38
Joint velocities	37	25
Joint accelerations	78	57
Combined total	179	120

The computational complexity of inverse kinematic positions, velocities, and accelerations considered together is presented in table 10.3. With modern microcomputers and engineering workstations, these numbers of arithmetic operations are readily implemented for real-time control. Suh (1983) simulated the algorithm in (Hollerbach and Sahar, 1983) on a VAX 11/750 with a floating point accelerator and achieved a cycle time of 1.9 msec, of which 1 msec was used for transcendental function calls. Given this level of efficiency with standard computer hardware, there is no strong impetus to devise special purpose computational architectures for inverse kinematics, although such architectures have been reported (Orin and Tsai, 1986).

10.2.1.3 Handling Singularities A singularity for the nonredundant six-degrees-of-freedom manipulator is a kinematic alignment of the joints that reduces the degrees of freedom below six. The singularities of the 6R arm (figure 10.5) can be labeled as the shoulder, elbow, and wrist singularities (Hollerbach, 1985).

- For the shoulder singularity, the wrist point is above the shoulder and a motion of the wrist normal to the forearm-upper arm plane is precluded.
- For the elbow singularity, the forearm and upper arm are collinear and a radial motion of the wrist is precluded.

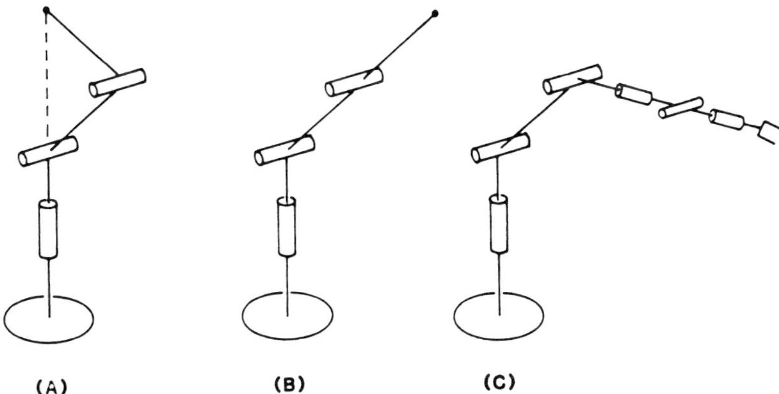

Figure 10.5
Singularities of the 6R arm, labeled shoulder (A), elbow (B), and wrist (C) singularities.

- For the wrist singularity, the hand is collinear with the forearm and an angular motion is lost.

The shoulder and wrist singularities are generally interior to the workspace boundary, and are much more serious than the exterior elbow singularity. After all, any workspace has an external boundary, and external singularities will always exist. Of the two interior singularities, that of the wrist is the more serious because the first three degrees of freedom sweep this singularity throughout the workspace. The shoulder singular points are confined to a line coincident with the first rotation axis.

Singularities are manifested in inverse kinematic velocity and acceleration problems, with the Jacobian matrix in (5)–(6) becoming noninvertible. Desired Cartesian motions at the singular configuration become impossible to realize, manifested mathematically by the joint velocities evidently going to infinity. Near singularities the movements are also ill-conditioned, in the sense that while infinite joint velocities do not result, still very large joint velocities are required. Thus there are forbidden regions represented by cones of ill-conditoned joint angles centered at the singular joint values (Paul and Stevenson, 1983).

Singularity problems for six-degrees-of-freedom manipulators have long been recognized, and two main approaches have evolved to deal with them. One approach is to avoid singularities and associated ill-conditioned regions entirely by planning trajectories that stay away from them. Unfortunately it is not always possible to keep away from singularity points, and even if it were, a considerable burden is placed on the trajectory planner given the difficulty of relating the hand Cartesian space and joint space (Lozano-Pérez, 1982). The second approach is to violate the Cartesian motion specification, in effect achieving some compromise between desired hand motion and possible joint motion. Near singularities, for example, one can switch from Cartesian trajectories to joint-interpolated motion (Taylor, 1979). A disadvantage of joint interpolation is loss of end point control in a complicated way.

More recently, a number of new approaches toward singularities have been suggested. At the shoulder and wrist singularities, the nonredundant manipulator can actually execute a self motion, which is an internal motion of the linkage that does not move the end point, because the six-degrees-of-freedom robot has an excess freedom with regard to the five-degrees-of-freedom end point motion possible at these singularities. Sugimoto and Duffy (1981) point out that a self-motion at the singularity can be used to find an alternative configuration for which the permissible end point velocities happen to coincide with the desired one. The drawback is that the manipulator must essentially stop for this self-motion to occur, before heading off in the new direction. They also point out that a restricted class of end point motions is still possible at a singularity, and as long as the end point trajectory specification falls into this class, then the singularity does not cause a problem.

Nakamura, Hanafusa, and Yoshikawa (1987) suggest making use of the self-motion at a singularity by partitioning variables into high-priority variables $\mathbf{x}_h$, which must be realized, and low-priority variables $\mathbf{x}_l$, which are sacrificed to avoid the singularity but that should be realized insofar as is possible. For example, in spray painting the control of rotation about the spray direction is not as important as the control of other positioning variables, and could be sacrificed at a singularity. They proceed by decomposing the direct kinematic velocity relation (2) into two parts:

$$\dot{\mathbf{x}}_h = \mathbf{J}_h\dot{\boldsymbol{\theta}}, \tag{21}$$

$$\dot{\mathbf{x}}_l = \mathbf{J}_l\dot{\boldsymbol{\theta}}. \tag{22}$$

The joint velocities that realize the high-priority variables can be solved using the generalized inverse $\mathbf{J}_h^{\dagger}$ of the Jacobian matrix:

$$\dot{\boldsymbol{\theta}} = \mathbf{J}_h^{\dagger}\dot{\mathbf{x}}_h + (\mathbf{I} - \mathbf{J}_h^{\dagger}\mathbf{J}_h)\dot{\boldsymbol{\phi}}, \tag{23}$$

where $\dot{\boldsymbol{\phi}}$ is an arbitrary vector of joint velocities, $\mathbf{I}$ is the identity matrix, and $(\mathbf{I} - \mathbf{J}_h^{\dagger}\mathbf{J}_h)\dot{\boldsymbol{\phi}}$ is the null space vector corresponding to the self-motion. The null space vector can be chosen to approximate the low-priority variables in a least squares sense by combining these relations:

$$\dot{\boldsymbol{\phi}} = (\mathbf{J}_l(\mathbf{I} - \mathbf{J}_h^{\dagger}\mathbf{J}_h))^{\dagger}(\dot{\mathbf{x}}_l - \mathbf{J}_l\mathbf{J}_h^{\dagger}\dot{\mathbf{x}}_h). \tag{24}$$

Another recent approach is based on damped least squares, equivalent to a Levenberg-Marquardt algorithm and ridge regression (Wampler, 1986; Nakamura and Hanafusa, 1986). The goal is to find the $\dot{\boldsymbol{\theta}}$ that minimizes $|\dot{\mathbf{x}} - \mathbf{J}\dot{\boldsymbol{\theta}}|^2$ at or near singularities, but that prevents $\dot{\boldsymbol{\theta}}$ from becoming too large. That is to say, one wishes to minimize $|\dot{\mathbf{x}} - \mathbf{J}\dot{\boldsymbol{\theta}}|^2 + \alpha^2|\dot{\boldsymbol{\theta}}|^2$, where α is some weighting constant. The solution is

$$\dot{\boldsymbol{\theta}} = (\mathbf{J}^T\mathbf{J} + \alpha^2\mathbf{I})^{-1}\mathbf{J}^T\dot{\mathbf{x}}. \tag{25}$$

The constant α can be chosen to realize any joint velocity constraints that may exist, and efficient methods for computing (25) have been proposed (Wampler, 1986). When to apply damped least squares can be tracked by the smallest eigenvalue in the singular value decomposition of the Jacobian $\mathbf{J}$. The advantage of this approach is the graceful degradation of end point control near a singularity.

Two other approaches will be mentioned to conclude this section that are designed to handle wrist singularities. At or near this singularity, joint 4 of the 6R arm wishes to assume a very high velocity in order to flip the hand around. Aboaf and Paul (1987) apply an a priori upper velocity limit to this wrist motion, and adjust the end point motion to be realized by the other joint rates.

In another approach, Asada and Cro Granito (1985) note that the wrist singularity unfortunately occurs at a highly desirable configuration, when the wrist is straight or in a polar orientation. They suggest a diagonal tool mount that places the singular region away from the polar region, because when the tool is pointing at the pole the wrist is actually bent. They also note that if the hand-held object is rotationally symmetric, a diagonal tool mount can yield an alternative kinematic solution that is not singular but that flips the object 180°.

10.2.2 Inverse Kinematics of Redundant Manipulators

Limitations in workspace and in flexibility of six-degrees-of-freedom manipulators have given impetus to the development of redundant manipulators.

The major flaw of six-degrees-of-freedom manipulators is the presence of singularities in the interior of the workspace, which must be eliminated from general use because certain motion directions are precluded. Singularities populate a substantial portion of the workspace due to proximal joints that sweep the singular points through large volumes and to ill-conditioned regions around singular points that magnify the influence of singularities. It is a kinematic nightmare to plan trajectories that do not pass through or near singularities, given the complex transformation between end effector locations and joint angles. Because a substantial fraction of the workspace is lost to singularity regions, Paul and Stevenson (1983) characterize a kinematic six-degrees-of-freedom manipulator as functionally a five-degrees-of-freedom manipulator.

For this reason, Hollerbach (1985) has argued that a six-degrees-of-freedom geometry can no longer be considered a general-purpose manipulator, but rather a seven-degrees-of-freedom geometry. By adding an extra degree of freedom, all interior workspace points can be made functional in the sense that a nonsingular kinematic configuration can be found corresponding to the workspace point. Singular configurations will still arise, but they will be avoidable through exercise of a self-motion to arrive at a new configuration. The trajectory planner must still be on guard to the presence of interior singularities, but upon arriving at one it could backtrack to evolve a different configuration at the singular point. Thus a seven-degrees-of-freedom geometry is the first time that all interior workspace points are usable, providing impetus for nomination as the true general-purpose geometry.

An extra degree of freedom can also aid obstacle avoidance, through selection of a configuration by exercise of the self-motion. The major plane of motion of the 6R arm is vertical, corresponding to flexion at the shoulder and elbow. If an obstruction also is primarily vertical, then the robot will not be able to reach around it. A self-motion of a redundant mechanism will typically allow the major plane of motion to be rotated away from vertical.

Seven degrees of freedom are advocated instead of eight because the situation with singularities cannot be improved upon. More degrees of freedom bring more singularities, and the effect of making all interior workspace points useful has already been achieved with seven. Though workspace can be improved by making the arm more snakelike, extra degrees of freedom might profitably be applied instead toward making the robot mobile or increasing the dexterity of the end effector. In considering a variety of possible seven-degrees-of-freedom geometries, generated by adding a joint to the 6R arm, Hollerbach (1985) argued that an arm with a spherical shoulder joint, kinematically equivalent to the human arm, offers the best design from a standpoint of elimination of internal singularities, simplicity, and workspace.

10.2.2.1 Inverse Velocity Methods The vast majority of research into the control of redundant arms has involved the instantaneous resolution of the redundancy at the velocity level through use of the pseudoinverse $\mathbf{J}^{\dagger}$ of the Jacobian matrix $\mathbf{J}$:

$$\dot{\boldsymbol{\theta}} = \mathbf{J}^{\dagger}\dot{\mathbf{x}} + (\mathbf{I} - \mathbf{J}^{\dagger}\mathbf{J})\dot{\boldsymbol{\phi}}. \tag{26}$$

where $\dot{\boldsymbol{\phi}}$ is an arbitrary vector of joint velocities and $(\mathbf{I} - \mathbf{J}^{\dagger}\mathbf{J})\dot{\boldsymbol{\phi}}$ is its projection into the null space of $\mathbf{J}$, corresponding to a self-motion of the linkage that does not move the end effector. The pseudoinverse $\mathbf{J}^{\dagger} = (\mathbf{J}^T\mathbf{J})^{-1}\mathbf{J}^T$ is also known as the Moore-Penrose generalized inverse. The attractiveness of this approach is twofold. First, the pseudoinverse is one of the types of generalized inverse that has a least squares property, in the present case minimizing $|\dot{\boldsymbol{\theta}}|^2$. Presumably any joint is prevented from moving too fast, leading to a more controllable motion. It is also presumed that squared velocities are approximately related to kinetic energy, which would then also be approximately minimized (Whitney, 1972).

Second, the redundancy available beyond that required for the tip motion is succinctly characterized by the null space of the Jacobian, which may be freely utilized to assist in the realization of some chosen objective. Liegeois (1977) developed a general formulation that tends to minimize a position-dependent scalar performance criterion $p = g(\boldsymbol{\theta})$. The null space vector $\dot{\boldsymbol{\phi}} = k\,\partial g/\partial\boldsymbol{\theta}$, where k is an arbitrary constant, projects the gradient of p onto the joint motion in such a way as to reduce p through subsequent motion:

$$\dot{\boldsymbol{\theta}} = \mathbf{J}^{\dagger}\dot{\mathbf{x}} + (\mathbf{I} - \mathbf{J}^{\dagger}\mathbf{J})k\frac{\partial g}{\partial\boldsymbol{\theta}}. \tag{27}$$

Liegeois demonstrated a way of avoiding joint limits, by minimizing the scalar function

$$p = \sum_{i=1}^{n}\left(\frac{\theta_i - \theta_i^{\mathrm{mid}}}{\theta_i^{\mathrm{mid}} - \theta_i^{\mathrm{max}}}\right)^2, \tag{28}$$

where $\theta_i^{\max}$ and $\theta_i^{\min}$ are the upper and lower joint angle limits for joint i and $\theta_i^{\mathrm{mid}} = (\theta_i^{\min} + \theta_i^{\max})/2$.

The null space vector has also been used in singularity avoidance. Yoshikawa (1985) proposed minimizing a dexterity measure w, called by him the manipulatability measure, given by

$$w = \sqrt{\det(\mathbf{J}\mathbf{J}^T)}. \tag{29}$$

Since at a singularity $w = 0$, then the scalar function $p = -w(\boldsymbol{\theta})$ instantaneously maximizes w and tends to keep the arm away from singularities. Klein and Blaho (1987) compared various dexterity measures including w and decided instead on the minimum singular value.

Maciejewski and Klein (1985) used the null space vector to aid obstacle avoidance. Suppose $\mathbf{x}_o$ is some point on the arm closest to an obstacle, $\mathbf{J}_o$ is the Jacobian such that $\dot{\mathbf{x}}_o = \mathbf{J}_o\dot{\boldsymbol{\theta}}$, and $\dot{\mathbf{x}}_o$ is a movement of this point away from the obstacle that would be desirable. In a procedure like that in (23), $\dot{\boldsymbol{\phi}}$ is found as

$$\dot{\boldsymbol{\phi}} = [\mathbf{J}_o(\mathbf{I} - \mathbf{J}^\dagger\mathbf{J})]^\dagger(\dot{\mathbf{x}}_o - \mathbf{J}_o\mathbf{J}^\dagger\dot{\mathbf{x}}). \tag{30}$$

10.2.2.2 Problems with Generalized Inverses The previous redundancy resolution schemes are purely kinematic, despite the presumption that since the sum of squares of joint velocities is minimized by the pseudoinverse, the kinetic energy is approximately minimized. The kinetic energy is

$$T = \tfrac{1}{2}\dot{\boldsymbol{\theta}}^T\mathbf{H}\dot{\boldsymbol{\theta}}, \tag{31}$$

where $\mathbf{H}$ is the inertia matrix. It can be shown (Whitney, 1972) that the actual generalized inverse that instantaneously minimizes T is the inertia-weighted pseudoinverse $\mathbf{J}_H^\dagger$:

$$\mathbf{J}_H^\dagger = \mathbf{H}^{-1}\mathbf{J}^T(\mathbf{J}\mathbf{H}^{-1}\mathbf{J}^T)^{-1}. \tag{32}$$

Hence the relation to kinetic energy of an unweighted pseudoinverse is not clear, and cannot be used as a justification for its use.

When the null space vector $\dot{\boldsymbol{\phi}}$ in (26) is chosen, the least squares property of the pseudoinverse is lost. Thus there is no particular reason to use this form of generalized inverse, and more computationally convenient forms can be used instead (Lovass-Nagy and Shilling, 1985).

A serious problem with the pseudoinverse is that its application is nonconservative (Klein and Huang, 1983). If the manipulator end point executes a closed path repetitively from a given starting point, then a different joint configuration will result each time the starting point is encountered. This drift in configuration is highly undesirable, since, for example, collisions with obstacles could result. Another serious problem with the pseudo-inverse is that it does not avoid singularities (Baillieul, 1985), even though the sum of squares of joint velocities is instantaneously minimized.

To attempt to solve these problems of singularities and nonconservative motions, Baillieul (1985) proposed the introduction of an additional constraint function $g(\boldsymbol{\theta}) = 0$ that is to be satisfied along with the kinematic function $\mathbf{x} = \mathbf{f}(\boldsymbol{\theta})$. The two functions are solved together differentially:

$$\mathbf{J}_{\text{ex}}\dot{\boldsymbol{\theta}} = \begin{bmatrix} \dot{\mathbf{x}} \\ 0 \end{bmatrix}, \qquad \mathbf{J}_{\text{ex}} = \begin{bmatrix} \mathbf{J} \\ \dfrac{\partial g}{\partial \boldsymbol{\theta}} \end{bmatrix}, \tag{33}$$

where $\mathbf{J}_{\text{ex}}$ is the extended Jacobian matrix as defined by Baillieul. Assuming that $\mathbf{J}_{\text{ex}}$ is nonsingular, this relation may be inverted directly to find $\dot{\boldsymbol{\theta}}$. It is possible to recast this method in terms of a generalized inverse with the null space vector. In actuality, the extended Jacobian matrix does not generate conservative motions or avoid singularities (Baker and Wampler, 1987). Although the constraint function can be defined to avoid kinematic singularities, it introduces algorithmic singularities instead.

More generally, Baker and Wampler (1987) showed in a benchmark paper that local inverse functions will always suffer from problems of singularities and nonconservative motion. A local inverse function $\boldsymbol{\Lambda}(\mathbf{x}) = \boldsymbol{\theta}$ is a continuous mapping over the whole workspace $\mathbf{x}$ to the joint angles $\boldsymbol{\theta}$, and includes as special cases the generalized inverse methods and the extended Jacobian method. The fundamental problem with defining an inverse function is the difference in topology between a sphere, representing the pointing directions of the end point, and a torus, representing the workspace of a number of rotary joints (Gottlieb, 1985). No continuous mapping from a sphere to torus is possible.

This very general result is bad news for robotics, because it essentially rules out good methods for instantaneously resolving redundancies. Any local inverse function will generate singularities, and conservative motion is impossible. Nevertheless, Baker and Wampler do suggest some possible solutions to this dilemma, which, however, require further work.

- *Restricted workspace.* If the local inverse function is defined on only part of the workspace, then in principle it could be fashioned to avoid singularities and make repetitive motions conservative (Wampler, 1987). Intuitively, if a hole is punched in a sphere, it can be mapped into a torus. Since joint limits ordinarily restrict workspace anyhow, it may not represent a severe problem to eliminate the associated workspace region. A local inverse function can also be realized mechanically by a closed-loop linkage.

- *Discontinuous function.* Different local inverse functions could be defined on different parts of the workspace, and a discrete switching, which makes the combined functions discontinuous, could be used between them.
- *Global redundancy resolution.* If the redundancy is resolved by optimizing some function over the whole path, then the motion can be made conservative and singularities can be avoided. Examples of such methods include the work of Uchiyama, Shimizu, and Hakomori (1985), who parameterized joint angles by polynomials and optimized manipulability w integrated across the whole trajectory by gradient adjustment of polynomial coefficients; Nakamura and Hanafusa (1987), who used Pontryagin's Maximum Principle to minimize a cost function combining manipulability w plus $|\dot{\boldsymbol{\theta}}|^2$; and Suh and Hollerbach (1987), who minimized the sum of squared torques.

10.2.3 Kinematic Calibration

Recently, there has been an explosion in the literature on the topic of kinematic calibration, reflecting its current importance and development in robotics. Kinematic calibration concerns the determination of an accurate relationship between joint angles and end point positions. Kinematic calibration encompasses primarily geometrical effects, such as variations in link lengths, joint axis orientation, and base location, but also nongeometrical effects, such as gear eccentricity, backlash, and joint compliance. To provide bounds on this topic, kinematic calibration is restricted to a static analysis of end point position, leaving such time-varying or dynamic effects as servo and trajectory errors, link flexibility during motion, and end point vibration to other topics. Nevertheless, kinematic calibration does include such static effects as joint compliance due to gravity loading.

As often stated in the literature, one major need for kinematic calibration is to correct the kinematic model of the robot after its manufacture. Variations in the kinematic model arise from imprecisions in the manufacturing process, but improving the precision in the manufacturing process is costly. Thus every robot needs to be calibrated after it is built, and the corrected kinematic parameters incorporated into the software for accurate posi-

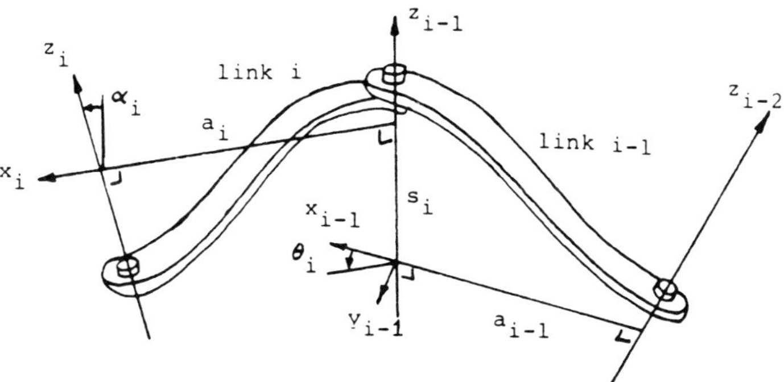

Figure 10.6
The Denavit-Hartenberg link parameters.

tioning. For the PUMA 560 robot, for example, accuracies before such calibration are only on the order of 1 cm, whereas after kinematic calibration some of the papers that follow report accuracies of 0.2–0.3 mm. This is a drastic improvement of almost two orders of magnitude.

Kinematic calibration ordinarily begins by forming a parametric model of the manipulator kinematics in the case of geometrical factors, and of the joint backlash, compliance, and gear eccentricity in the case of non-geometrical factors. Dealing just for the moment with the geometrical factors, a typical kinematic calibration procedure is outlined below. Most papers have assumed the Denavit-Hartenberg (1955) link parameters (figure 10.6), where

- $\Delta\boldsymbol{\theta}$ is the vector of joint angle offsets,
- $\boldsymbol{\alpha}$ is the vector of skew angles between neighboring joint axes,
- $\mathbf{a}$ is the vector of the link lengths, and
- $\mathbf{s}$ is the vector of joint offsets.

In (1), the end point position was written only as a function of the joint angles, namely, $\mathbf{x} = \mathbf{f}(\boldsymbol{\theta})$. For kinematic calibration, this relation becomes

$$\mathbf{x} = \mathbf{f}(\boldsymbol{\theta}, \boldsymbol{\alpha}, \mathbf{a}, \mathbf{s}). \tag{34}$$

The calibration typically proceeds by iteration of the linearized equations (e.g., Wu, 1984). Taking the first difference of (34),

$$\Delta\mathbf{x} = \frac{\partial \mathbf{f}}{\partial \boldsymbol{\theta}}\Delta\boldsymbol{\theta} + \frac{\partial \mathbf{f}}{\partial \boldsymbol{\alpha}}\Delta\boldsymbol{\alpha} + \frac{\partial \mathbf{f}}{\partial \mathbf{a}}\Delta\mathbf{a} + \frac{\partial \mathbf{f}}{\partial \mathbf{s}}\Delta\mathbf{s}. \tag{35}$$

The term $\partial\mathbf{f}/\partial\boldsymbol{\theta}$ is just the ordinary Jacobian matrix $\mathbf{J}$, defined in (2). The other partial derivatives of $\mathbf{f}$ with respect to the remaining three Denavit-Hartenberg parameters are also different Jacobian matrices.

The difference $\Delta\mathbf{x}$ may be interpreted as the error in position of the end point, obtained by subtracting the computed from the measured position. The differences $\Delta\boldsymbol{\theta}$, $\Delta\boldsymbol{\alpha}$, $\Delta\mathbf{a}$, and $\Delta\mathbf{s}$ may be viewed as the corrections to the kinematic parameters. These corrections may be solved for by combining many readings throughout the workspace and inverting (35):

$$[\Delta\boldsymbol{\theta} \quad \Delta\boldsymbol{\alpha} \quad \Delta\mathbf{a} \quad \Delta\mathbf{s}] = \begin{bmatrix} \dfrac{\partial\mathbf{f}}{\partial\boldsymbol{\theta}} & \dfrac{\partial\mathbf{f}}{\partial\boldsymbol{\alpha}} & \dfrac{\partial\mathbf{f}}{\partial\mathbf{a}} & \dfrac{\partial\mathbf{f}}{\partial\mathbf{s}} \end{bmatrix}^{*} \Delta\mathbf{x}, \tag{36}$$

where the * indicates the generalized inverse of the enclosed matrix, giving a least squares solution for the parameter corrections. Since kinematic calibration is a nonlinear estimation problem, one must iterate this procedure by reevaluating the Jacobians at the updated parameter values.

Several investigators have pointed out a fatal problem with the Denavit-Hartenberg parameters when neighboring joint axes are nearly parallel (e.g., Hayati, 1983; Mooring, 1983). The common normal is extremely sensitive to slight variations in axis orientation, and its location will vary wildly. Hayati (1983; Hayati and Mirmirani, 1985) proposed a modification of the Denavit-Hartenberg parameters in this case, by restricting the distal origin to lie in the previous origin's XY plane and replacing the joint offset s with an additional angular alignment parameter β (figure 10.7).

Another problem with the procedure presented above is that singularities in the parameters may be encountered that stop the iteration (Hollerbach and Bennett, 1987). Singularities are particularly likely to occur in case

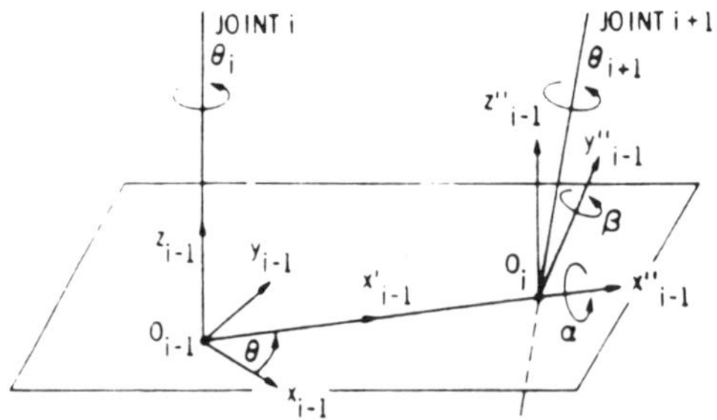

Figure 10.7
Hayati modification to Denavit-Hartenberg parameters in case of nearly parallel joint axes.

the initial guesses for the geometric parameters are far from the correct solutions. This situation also is likely because some parameters are very hard to guess at, such as joint angle offsets. One way to solve this problem is to use the Levenberg-Marquardt algorithm to generate good initial values, but then to use iteration of the linearized equations for any updating.

Nongeometric parameters are particularly important in geared robots. In a benchmark paper, Whitney, Lozinski, and Rourke (1986) showed for the PUMA 560 robot that the nongeometric parameters are as important as the geometric parameters in determining end point accuracy.

The most difficult aspect of kinematic calibration is obtaining accurate measurements of the end point position. Past work in kinematic calibration has involved calibration fixtures with precision points (Hayati and Roston, 1986), special apparatuses (Stone, Sanderson, and Neuman, 1986), or external position measurements with theodolites (Whitney, Lozinski, and Rourke, 1986). In particular, the latter work achieved extremely good results, but one drawback is the extensive human involvement in making the theodolite measurements; one calibration experiment was said to take 5 days. Special calibration fixtures also require extensive human involvement. A challenge for the future will be to develop apparatus that are fast, accurate, and convenient to use. Three-dimensional motion tracking systems appear to represent one good avenue (Hollerbach and Bennett, 1987), as well as laser range finders (Lau, Hocken, and Haynes, 1985).

Finally, the parametric models result in accuracies on the order of a few tenths of a millimeter. If higher accuracies on the order of tens of microns are required, one may have to rethink kinematic calibration or the problem being addressed. One possibility is to elaborate the parametric models, taking into account, for example, thermal effects and link flexibility. Since it becomes quite difficult and complex to cast a parametric model for all conceivable effects at this high resolution, another possibility is to employ nonparametric calibration (Shamma and Whitney, 1987). Real-time end point tracking can also be considered.

10.3 Dynamics

This section reviews progress in the formulation and computation of the inverse dynamics for control purposes. It begins by critically analyzing recent claims about dynamics formalisms said to be more efficient than the recursive Newton-Euler equations. Then improvements to the recursive Newton-Euler formalism are examined, due to customization to a particular manipulator structure and to parallelism. The modeling and role of flexible link dynamics is briefly considered. Last, the important issue is considered of how the inertial parameters of a manipulator's links, and any load it may pick up, can be accurately determined.

10.3.1 Efficiency of Inverse Dynamics Computation

Since 1980, we have known that the inverse dynamics of a rigid-link manipulator is a tractable computation of linear complexity (Hollerbach, 1980; Luh, Walker, and Paul, 1980a). Since 1982, we have come to understand that the dynamics formalism used to derive the manipulator dynamics is not important for computational efficiency, but rather whether a recursive or closed form is used and whether a matrix or vector representation of angular velocity is used (Silver, 1982).

10.3.1.1 Kane's Equations More recently, there has been a proliferation of papers that apply Kane's equations to manipulator dynamics, ostensibly because they lead to more efficient equations (e.g., Huston and Kelly, 1982; Kane and Levinson, 1983). In actuality, Kane's equations make no difference whatsoever to dynamics efficiency, since they are equivalent to the Newton-Euler dynamics. The three main formalisms that have been applied to manipulator dynamics are

- Lagrange equations:

$$\tau_i = \frac{d}{dt}\left(\frac{\partial L}{\partial \dot{\theta}_i}\right) - \frac{\partial L}{\partial \theta_i}, \qquad i = 1, \ldots, n, \tag{37}$$

where $L = K - P$, K is the manipulator kinetic energy, and P is the potential energy.

- Newton-Euler equations:

$$\mathbf{f}_i = m_i \ddot{\mathbf{r}}_i, \tag{38}$$

$$\mathbf{n}_i = \mathbf{I}_i \dot{\boldsymbol{\omega}}_i + \boldsymbol{\omega}_i \times \mathbf{I}_i \boldsymbol{\omega}_i. \tag{39}$$

In Newton's equation, $\mathbf{f}_i$ is the net force on link i, m_i is its mass, and $\ddot{\mathbf{r}}_i$ is the acceleration of the center of gravity. In Euler's equation, $\mathbf{n}_i$ is the net torque, $\mathbf{I}_i$ is the inertia, and $\boldsymbol{\omega}_i$ is the angular velocity.

- Kane's equations:

$$\tau_i = \sum_{j=1}^{n} \left(\mathbf{f}_j \cdot \frac{\partial \dot{\mathbf{r}}_j}{\partial \dot{\theta}_i} + \mathbf{n}_j \cdot \frac{\partial \omega_j}{\partial \dot{\theta}_i} \right). \tag{40}$$

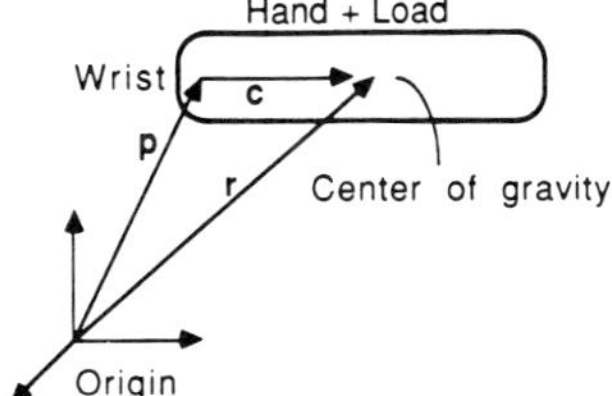

Figure 10.8
Locating the joints and center of gravity of a link.

To see the equivalence of Kane's equations to the Newton-Euler formulation, the static relationships in the Newton-Euler dynamics are written:

$$\mathbf{f}_{i-1,i} = \mathbf{f}_i + \mathbf{f}_{i,i+1}, \tag{41}$$

$$\mathbf{n}_{i-1,i} = \mathbf{n}_i + \mathbf{n}_{i,i+1} + \mathbf{c}_i \times \mathbf{f}_i + \mathbf{p}_i^* \times \mathbf{f}_{i,i+1}, \tag{42}$$

$$\tau_i = \mathbf{z}_{i-1} \cdot \mathbf{n}_{i-1,i}, \tag{43}$$

where $\mathbf{f}_{i-1,i}$ is the force and $\mathbf{n}_{i-1,i}$ is the torque exerted on link i by link $i-1$, $\mathbf{p}_i^*$ connects joint i to joint $i+1$ of link i (figure 10.8), and $\mathbf{c}_i$ is the vector from joint i to the link i center of mass. A closed form equation can be written from above:

$$\tau_i = \mathbf{z}_{i-1} \cdot \sum_{j=1}^{n} (\mathbf{c}_i \times \mathbf{f}_i + \mathbf{n}_i), \tag{44}$$

where the initial condition is of the end point not touching.

The equivalence of Kane's equation is now transparent. In Kane's equation, the partial linear velocity $\partial \dot{\mathbf{r}}_j / \partial \dot{\theta}_i$ is nothing more than the moment arm by which the force of acceleration from Newton's equation $\mathbf{f}_j$ is projected onto the axis of rotation:

$$\frac{\partial \dot{\mathbf{r}}_j}{\partial \dot{\theta}_i} = \mathbf{z}_{i-1} \times \mathbf{c}_i, \tag{45}$$

where use has been made of the relation $a \cdot (b \times c) = (a \times b) \cdot c$. The partial angular velocity $\partial \omega_j / \partial \dot{\theta}_i$ is nothing more than the axis of rotation $\mathbf{z}_{i-1}$ of joint i. In fact, Silver (1982) used Kane's equation as an intermediate form to demonstrate the equivalence of the Lagrangian and Newton-Euler formulations. Thus the claim in (Kane and Levinson, 1983) that their formulation leads to a more efficient computation than the Newton-Euler equations is wrong.

Despite Silver's benchmark paper, a flood of papers on dynamics efficiency has continued unabated. None of them make any difference, and their authors would more profitably spend their time worrying about how dynamics can make control better than on nonexistent incremental improvements in dynamics efficiency.

10.3.1.2 Customized Dynamics The recursive Newton-Euler equations, which is the most efficient dynamics formalism, apply to any general manipulator structure. Yet in the previous section on the inverse kinematics of nonredundant manipulators, it was seen that unless manipulators have a particular kind of simple substructure, inverse kinematic positions could not be derived analytically. Thus the manipulators that inverse dynamics is to be applied to are likely to be kinematically simple. Aside from a spherical wrist or planar pair, another kinematic simplicity is that a manipulator is almost always designed so that neighboring joint axes are either parallel or perpendicular.

These forms of kinematic simplicity make the dynamic equations more efficient (Hollerbach and Sahar, 1983; Kanade, Khosla, and Tanaka, 1984; Horak, 1985) when the dynamics are customized to a particular manipulator. The dynamics efficiency results from simpler coordinate transformations and the intersection of neighboring joints. For the 6R arm with a spherical wrist, table 10.4 shows that a computational savings of 40% exists in the recursive Newton-Euler dynamics in that over a general six-degrees-of-freedom rotary manipulator of arbitrary structure (Hollerbach and Sahar, 1983). A minor 10% computational savings results if the inverse kinematics computation for position, velocity, and acceleration is conducted at the same time, since linear and angular velocities and accelerations are common to both computations.

Finally, it might happen that the inertial parameters are simple as well. The principal axes of inertia of a link are aligned with the link's joint axis,

Table 10.4
Dynamics computation complexity for a six-degrees-of-freedom manipulator under various conditions

Condition	Multiplications	Additions
General rotary manipulator	688	558
Spherical-wrist manipulator	408	324
Precomputed inverse kinematics	368	296
Simplified inertial parameters	194	138

and the center of gravity lies on a line between the two joints. Table 10.4 shows an additional reduction to just 194 multiplications, a low number on par with the inverse kinematics computation (table 10.3).

This work customizes dynamics in the context of a recursive dynamics computation, whereas past work has customized dynamics in closed form:

$$\tau = \mathbf{H}\ddot{\theta} + \dot{\theta}\cdot\mathbf{C}\cdot\dot{\theta} + \mathbf{g}, \tag{46}$$

where $\mathbf{H}$ is the inertia matrix, $\mathbf{C}$ is an $n \times n \times n$ matrix of position-dependent Coriolis and centrifugal coefficients, and $\mathbf{g}$ is the gravity vector. This equation is then ordinarily linearized by eliminating the velocity product terms and by specializing the inertia matrix $\mathbf{H}$ to a particular manipulator, omitting also the least significant terms (Bejczy, 1979; Izaguirre and Paul, 1986). The rationale for linearization is to apply linear control theory. Omitting the velocity product terms has been questioned in (Hollerbach, 1984), where fundamental time scaling properties of dynamics were shown to leave the relative contribution from inertial and velocity product terms the same for all speeds.

Some authors refer to the recursive form dynamics as a numerical form and to the closed form dynamics as a symbolic form. The recursive form is called numerical because it can ostensibly only be used to generate numbers in an on-line computation, while a closed form shows the interactions from all the links at a particular joint, and hence can be used for analysis. A closed form can be used, for example, to identify the linearly independent link inertial parameters, discussed later. Burdick (1986) presented an algorithm for simplifying the exact closed form equations for a particular manipulator. It is sometimes wrongly stated that particularizing a symbolic or closed form to a manipulator can lead to a more efficient computation than a customized recursive form. In making this mistake, authors typically allow a simplification in the closed form due to manipulator structure, but forget to allow this simplification in the recursive form.

10.3.1.3 Parallel Dynamics A reformulation of manipulator dynamics that is computationally significant was presented in a benchmark paper by Lathrop (1985). He offered several parallel formulations that drastically reduce the computation time. Elements of the recursive Newton-Euler equations were given in (38)–(39) and in (42)–(43). The following direct kinematics relations complete this formulation:

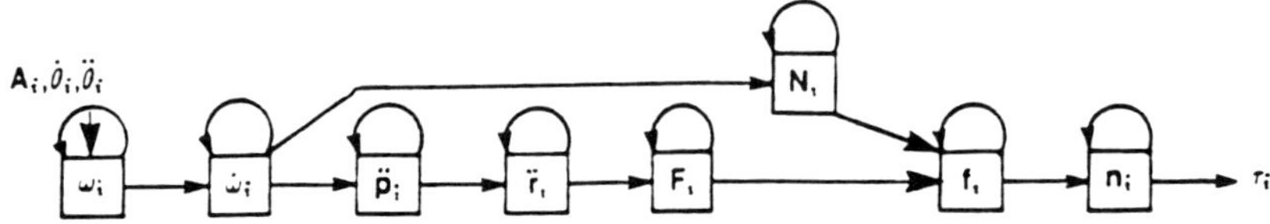

Figure 10.9
Serial evaluation of recursive Newton-Euler equations.

$$\boldsymbol{\omega}_i = \boldsymbol{\omega}_{i-1} + \dot{\theta}_i \mathbf{z}_{i-1}, \tag{47}$$

$$\dot{\boldsymbol{\omega}}_i = \dot{\boldsymbol{\omega}}_{i-1} + \ddot{\theta}_i \mathbf{z}_{i-1} + \dot{\theta}_i \boldsymbol{\omega}_i \times \mathbf{z}_{i-1}, \tag{48}$$

$$\ddot{\mathbf{p}}_i = \ddot{\mathbf{p}}_{i-1} + \dot{\boldsymbol{\omega}}_i \times \mathbf{p}_i^* + \boldsymbol{\omega}_i \times (\boldsymbol{\omega}_i \times \mathbf{p}_i^*), \tag{49}$$

$$\ddot{\mathbf{r}}_i = \ddot{\mathbf{p}}_i + \dot{\boldsymbol{\omega}}_i \times \mathbf{r}_i^* + \boldsymbol{\omega}_i \times (\boldsymbol{\omega}_i \times \mathbf{r}_i^*), \tag{50}$$

where the position vectors have been previously defined in figure 10.8.

Ordinarily, the recursive Newton-Euler equations are computed serially (figure 10.9): first all the angular velocities $\boldsymbol{\omega}_i$ are computed, then the angular accelerations $\dot{\boldsymbol{\omega}}_i$, etc. In his first parallel architecture (figure 10.10), Lathrop pointed out dependences in the computation that allowed various quantities to be computed in parallel. Thus after $\boldsymbol{\omega}_1$ has been computed, both $\boldsymbol{\omega}_2$ and $\dot{\boldsymbol{\omega}}_1$ can be computed, etc. Lathrop showed that each step could be executed in a time proportional to one matrix-vector multiplication plus one vector addition. For an n-joint manipulator, the total computational time is proportional to $2n + 3$ multiplications and $6n + 7$ additions, a drastic improvement over serial evaluation (table 10.5).

Lathrop then proposed a more efficient systolic pipeline architecture that uses buffering for different time instances. Remarkably, this architecture allows the dynamics to be computed in a time proportional to 1 multiplication and 3 additions, after the pipeline has generated the first torque values. The initial startup of the pipeline still takes the same amount of time as the linear parallel architecture, which limits its use in some real-time controllers such as computed torque since the servo errors will require a new state each time.

To remedy this situation, Lathrop proposed a logarithmic parallel formulation that truly breaks the linear dynamics complexity (table 10.5). The logarithmic parallel architecture works by computing all elements in a recursion in pairs: angular velocity is computed as $\boldsymbol{\omega}_1 + \boldsymbol{\omega}_2$, $\boldsymbol{\omega}_3 + \boldsymbol{\omega}_4$, etc. Lathrop also proposed a robot chip design to implement these architectures.

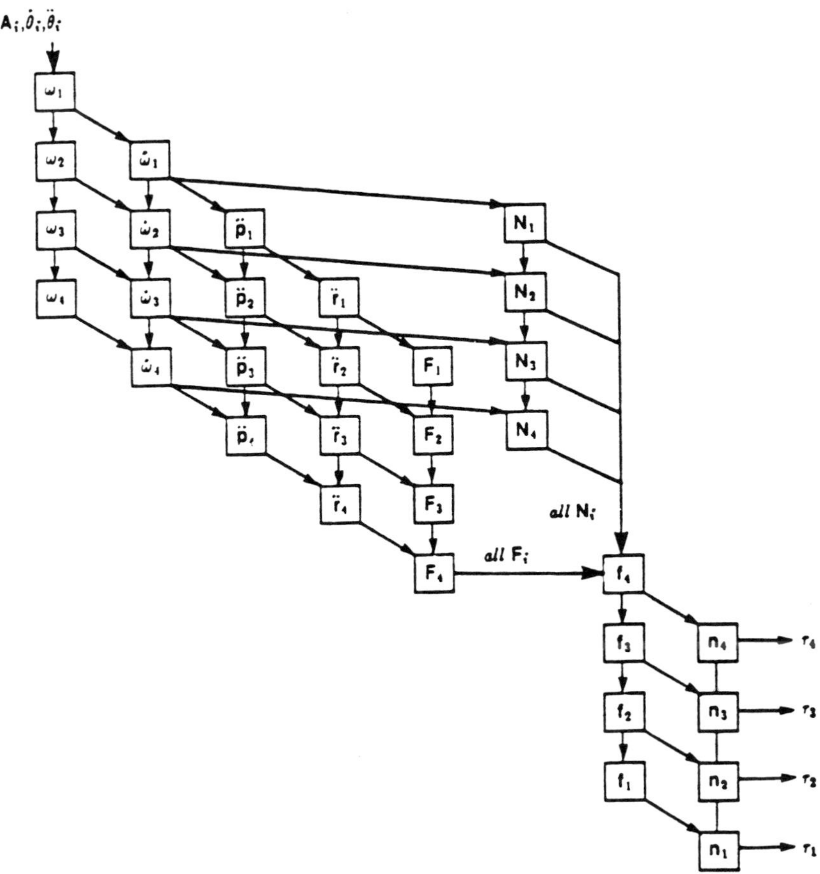

Figure 10.10
Linear parallel Newton-Euler formuation.

Table 10.5
Dynamics complexity of parallel Newton-Euler forms

Method	Multiplications	Additions
Recursive Newton-Euler	$150n - 48$	$131n - 48$
Linear parallel Newton-Euler	$2n + 3$	$6n + 7$
Logarithmic parallel Newton-Euler	$2[\log_2(n + 1)] + 5$	$6[\log_2(n + 1)] + 10$
Systolic pipeline Newton-Euler	1	3

10.3.1.4 Flexible Link Dynamics A topic of great activity in the past few years has been flexible link manipulators (e.g., Book, 1984; Rauh and Schielen, 1986). All manipulators, to be sure, have links that bend, but there is a question of how much and whether manipulators should be purposely designed to have flexible links. The most common rationale is that manipulators need to be made lighter, stronger, and faster, and that decreasing the link weight addresses this goal but leads to link flexibility.

It is not clear how seriously this argument should be taken. Manipulator links can be designed to be quite light and rigid using composite material technology. To make manipulators stronger and faster, better actuation and transmission systems are alternative avenues. Powerful direct drive motors and improved hydraulic and pneumatic actuators are emerging in new robot designs, while parallel drive linkages reduce the peripheral arm weight (Asada and Youcef-Toumi, 1987).

To date, actual demonstrations of modeling and control of flexible arms have been primarily restricted to single beams (Cannon and Schmitz, 1984; Daniel et al., 1987). A more likely source of flexibility is joint flexibility due to transmission elements, such as tendons (Hollars and Cannon, 1985) or gears (Good, Sweet, and Strobel, 1985). Joint flexibility, which can be isolated to single joint variables, is very much easier to model and compensate than link flexibility, which leads to horrible coupled dynamic equations.

10.3.2 Dynamics Calibration

Given that the inverse dynamics can be computed efficiently, they will only be useful for control if a dynamic model of the robot can be determined accurately. A dynamic model is comprised of the link inertial parameters and the inertial parameters of any load picked up. Modeling the dynamics of the actuator system is a separate topic not addressed here.

There are conceptually many different ways in which these components can be determined. Link inertial parameters can be determined from CAD modeling from blueprints, or by taking apart the robot and weighing, counterbalancing, and swinging the pieces. These procedures are error prone and require significant human involvement. An alternative procedure that has been pursued vigorously by a number of investigators is for the robot to estimate automatically these parameters as a result of motion. As mentioned earlier, this is also the goal that kinematic calibration should ascribe to.

Table 10.6
Using sensing to decouple system identification problems

Subsystem	Input	Output
Kinematic parameters	Joint angles	External position sensing
Link inertial parameters	Joint torques	Joint rates
Load inertial parameters	Joint rates	Wrist force/torque sensing

An insight that simplifies automatic calibration is to use sensing to decouple different parameter estimation problems. Different estimation procedures can then be applied to the different subsystems (table 10.6). For example, although kinematic and inertial parameters could in principle be identified with the same procedure, in practice it is easier to identify these parameters separately. Kinematic calibration can be accomplished by combining vision and joint angle sensing, while inertial parameter estimation combines joint torque sensing and joint angle sensing.

Widely different rates and types of system change call for different system identification algorithms to track the changes. In the case of an arm with variable loads, we expect the arm dynamics to be constant or only slowly varying, while load dynamics change rapidly as loads are picked up or put down. Arm identification can be based on a fixed model structure, while a complete load identification system must handle many different model structures. A wrist force/torque sensor can be used to estimate load parameters independently of the arm dynamics.

10.3.2.1 Load Estimation Since a load is essentially a part of the last link, the knowledge of the inertial parameters of manipulator loads is important for accurate control of manipulators. One alternative is to use object models and precise knowledge of grasping to predict how the load affects the inertial parameters of the last link. A more general alternative would not rely on object models or exact grasping, but would allow a robot to use sensing and an appropriate system identification procedure to identify the load itself.

A general procedure for rigid-body load estimation based on the Newton-Euler equations was presented in (Atkeson, An, and Hollerbach, 1986). Beginning with Newton's equation (38), since the position of the load's center of gravity is not generally known, we decompose $\mathbf{r}$ into the position of the wrist $\mathbf{p}$ plus the position of the load's center of gravity relative to the wrist $\mathbf{c}$ (figure 10.8). Then

$$\ddot{\mathbf{r}} = \ddot{\mathbf{p}} + \dot{\boldsymbol{\omega}} \times \mathbf{c} + \boldsymbol{\omega} \times (\boldsymbol{\omega} \times \mathbf{c}). \tag{51}$$

Substituting into (38),

$$\mathbf{f} = m\ddot{\mathbf{p}} + \dot{\boldsymbol{\omega}} \times m\mathbf{c} + \boldsymbol{\omega} \times (\boldsymbol{\omega} \times m\mathbf{c}). \tag{52}$$

In Euler's equation (39), the moment of inertia is expressed about the load's center of gravity. Since a wrist force/torque sensor is used in the estimation procedure, it is more convenient to express the inertia of the load with respect to the wrist point:

$$\mathbf{n}_w = (\mathbf{g} - \ddot{\mathbf{p}}_w) \times m\mathbf{c} + \mathbf{I}_w\dot{\boldsymbol{\omega}} + \boldsymbol{\omega} \times \mathbf{I}_w\boldsymbol{\omega}, \tag{53}$$

where $\mathbf{I}_w = \mathbf{I} + m(\mathbf{c}^T\mathbf{c}\mathbf{1} - \mathbf{c}\mathbf{c}^T)$ is the inertia of the load about the wrist, derived from the parallel axis theorem (Symon, 1971), and $\mathbf{1}$ is the identity matrix. The Newton-Euler equations may then be rewritten symbolically as

$$(\mathbf{f}, \mathbf{n}_w) = \mathbf{A}(\ddot{\mathbf{p}}, \boldsymbol{\omega}, \dot{\boldsymbol{\omega}})(m, m\mathbf{c}, \mathbf{I}). \tag{54}$$

Note that the inertial parameters of mass m, mass moment $m\mathbf{c}$, and inertia $\mathbf{I}$ appear linearly in the Newton-Euler equations, even though the dynamic equations as a whole are nonlinear. The nonlinearity appears only in the kinematic terms, combined in the matrix $\mathbf{A}(\ddot{\mathbf{p}}, \boldsymbol{\omega}, \dot{\boldsymbol{\omega}})$, which is known as a result of measurement during the motion.

By measuring force and torque at the wrist, linear least squares estimation can then be used by combining a number of readings and inverting (54). Setting $\boldsymbol{\phi} = (m, m\mathbf{c}, \mathbf{I})$ as the vector of unknown inertial parameters and $\mathbf{w} = (\mathbf{f}, \mathbf{n}_w)$ as the wrench,

$$\boldsymbol{\phi} = [\mathbf{A}]^*[\mathbf{w}], \tag{55}$$

where the brackets indicate that many readings have been combined to form a large set of observations, and $*$ once again denotes the generalized inverse.

This identification scheme was implemented on a PUMA 600 robot and on the MIT Serial Direct Drive Arm (DDArm), using commercial wrist force/torque sensors (Atkeson, An, and Hollerbach, 1986). For the PUMA 600 robot, it was found that the load mass and center of mass could be accurately identified (merely reflecting the calibration of the force sensors and robot kinematics), but that inertia could not because of

- the limited acceleration capability of the PUMA, leading to a low signal-to-noise ratio of the effect of inertia on wrist torque;

• the limited position resolution of the joint encoders, and the noise resulting from doubly differentiating this position data to derive acceleration; and

• the parallel axis component $m(\mathbf{c}^T\mathbf{c}\mathbf{1} - \mathbf{c}\mathbf{c}^T)$, which is typically an order of magnitude larger than the intrinsic load inertia $\mathbf{I}$ about the load's center of mass. Since torques are being sensed at the wrist, the wrist-referenced inertia $\mathbf{I}_w$ is dominated by the parallel axis component.

These difficulties could be ameliorated somewhat by larger loads and special test motions, but basic limitations of the PUMA robot limited the success of this procedure. On the other hand, if the goal in determining load inertial parameters is improved control rather than description of the load, then these difficulties are less serious. If the load inertia does not affect wrist torque much, then it will not require compensation by the controller. In effect, if it cannot be accurately identified, then it may not be important.

It is also possible to recast the estimation procedure by integrating the Newton-Euler equations, so that only velocity but not acceleration need be derived. Since the force/torque readings are then integrated, large errors can result from small low frequency errors such as bias. In fact, our experience with the integrated equations showed that the estimation results were not as good as with the original equations (Atkeson, An, and Hollerbach, 1986). Better performance may be achieved in principle by applying some optimal filter whose shape is an integrator at high frequencies but a differentiator at lower frequencies (Atkeson, 1986).

Inertia estimation was much more successful with the DDArm, because of (1) its high acceleration capability, and (2) more accurate acceleration estimates because of velocity sensing at the joints with high-quality tachometers. Special test motions were not necessary and did not lead to better estimates.

Load estimation has been pursued vigorously recently by other investigators, although in general experimental results have not been presented. Olsen and Bekey (1986) and Mukerjee and Ballard (1985) presented approaches similar to that above. Other investigators specify the use of joint torques rather than wrist force/torque sensing (Coiffet, 1983; Khalil, Gautier, and Kleinfinger, 1986), but as mentioned above this couples the dynamics of the arm into the dynamics of the load and makes the estimation less accurate. In addition, Coiffet (1983) requires special test motions, whereas the method above identifies the load parameters simultaneously as a result of a single motion.

10.3.3 Link Estimation

The inertial parameters, i.e., the mass, the location of center of mass, and the moments of inertia of each rigid body link of a robot, are usually not known even to the manufacturers of the robots. Robots are usually designed to satisfy kinematic specifications, but the inertial parameters are incidental attributes. Since commercial robots are invariably controlled by simple independent-joint PID control, there is no impetus by the manufacturer to determine the inertial parameters accurately since a dynamic model is not used.

As mentioned above, link inertial parameters can be determined experimentally by disassembling the robot, and then weighing the pieces for mass, counterbalancing for center of mass, and swinging for moments of inertia (Armstrong, Khatib, and Burdick, 1986). Besides requiring intensive human involvement, this procedure introduces considerable measurement difficulties. Counterbalanced points have to be referred somehow to the joint axes, while some components of inertia are difficult to determine by pendular motion.

Another approach is CAD modeling of the parts, where computerized geometric information can be combined with specific gravities of materials to estimate the inertial parameters. Again, this approach requires intensive human involvement, and is also subject to modeling errors.

An approach that emphasizes the robot calibrating itself was presented in (Atkeson, An, and Hollerbach, 1986). Link inertial parameter estimation is similar to load estimation, in that each link can be viewed as a load to the proximal joints. The inertial parameters again appear linearly in the dynamic equations, even though the multilink rigid body dynamics are nonlinear. A major difference from load estimation is that there is not a six-axis force/torque sensor at each joint, and only the component of torque about the joint axis can be measured.

Suppose for the moment that there is a six-axis force/torque sensor at each joint. Then the wrench w_{ii} seen at the proximal joint i of link i, due to motion of link i alone, is simply taken from (54), using the more compact notation:

$$\mathbf{w}_{ii} = \boldsymbol{A}_i \boldsymbol{\phi}_i, \tag{56}$$

where $\boldsymbol{\phi}_i$ is the vector of link i's inertial parameters, and $\boldsymbol{A}_i$ describes the motion of link i. A more distal link j will also exert a wrench $\mathbf{w}_{ij}$ at joint i

due to acceleration of link j, transmitted to joint i by the intervening linkages:

$$\mathbf{w}_{ij} = \mathbf{U}_{ij}\mathbf{w}_{jj}, \tag{57}$$

where $\mathbf{U}_{ij}$ is a wrench transmission matrix and is purely a function of the linkage geometry. The total wrench $\mathbf{w}_i$ seen at joint i is obtained by summing all the distal wrenches:

$$\mathbf{w}_i = \sum_{j=i}^{n} \mathbf{U}_{ij}\mathbf{w}_{jj} = \sum_{j=i}^{n} \mathbf{U}_{ij}A_j\boldsymbol{\phi}_j. \tag{58}$$

Unfortunately, no robot has a six-axis force/torque sensor at each joint, and only the component of torque about the joint axis can be measured. Thus this equation must be reduced by projecting the torque component along the joint i axis $\mathbf{z}_{i-1}$:

$$\tau_i = \sum_{j=i}^{n} K_{ij}\boldsymbol{\phi}_i, \tag{59}$$

where the joint torque $\tau_i = (\mathbf{0}, \mathbf{z}_{i-1})\cdot\mathbf{w}_i$ and $\mathbf{K}_{ij} = (\mathbf{0}, \mathbf{z}_{i-1})\cdot\mathbf{U}_{ij}A_i$. In terms of all the joints, this relation may be written as

$$\boldsymbol{\tau} = \mathbf{K}\boldsymbol{\phi}, \tag{60}$$

where $\boldsymbol{\tau} = (\tau_1, \ldots, \tau_n)$, $\boldsymbol{\phi} = (\boldsymbol{\phi}_1, \ldots, \boldsymbol{\phi}_n)$, and the elements of $\mathbf{K}$ are $\mathbf{K}_{ij}$.

To solve this equation, one again takes a number of readings throughout a trajectory, and augments (60):

$$[\boldsymbol{\tau}] = [\mathbf{K}]\boldsymbol{\phi}, \tag{61}$$

where the square brackets again denote the augmented readings. Unfortunately, it is not possible to invert (61) as before to obtain the link inertial parameters. Aside from the limited sensing at each joint, the proximal links do not undergo general motion because of restricted degrees of freedom. The manifestation of limited sensing and movement in estimation is twofold.

1. Not all parameters can be identified, since they do not influence the joint torques. For example, only the link 1 inertia about the first joint is reflected in the joint 1 torque for an arm on a stationary base; all other link 1 inertial parameters cannot be identified.
2. Other inertial parameters can only be determined in linear combinations.

Several ways of solving this rank-deficient least squares problem were presented in (Atkeson, An, and Hollerbach, 1986), including ridge regression [or damped least squares as in (25)], singular value decomposition, and explicit expansion of the closed-form dynamic equations. All methods were implemented and gave identical results. The ridge regression solution is

$$\boldsymbol{\phi} = (\mathbf{K}^T\mathbf{K} + d\mathbf{1})^{-1}\mathbf{K}^T\boldsymbol{\tau}, \tag{62}$$

where the constant d must be chosen small enough not to perturb the solution.

The results of applying this algorithm to the DDArm showed that good estimates of the link inertial parameters were obtained, as determined from the match of predicted torques to measured torques. A set of CAD-modeled inertial parameters was also available for this arm, but predicted the measured torques less accurately. This lends viability to the dynamic estimation procedure above, which evidently produced a better set of parameters than CAD modeling, and which certainly is less labor intensive.

Like load estimation, link estimation has been pursued recently by a number of other investigators; again, few of these studies involved actual experimentation. Mayeda, Osuka, and Kangawa (1984) required special test motions to estimate the coefficients of a closed-form Lagrangian dynamics formulation, which lumps the 10 inertial parameters of each link into these numerous, redundant coefficients. Mukerjee and Ballard (1985) directly applied their load identification method to link identification, by requiring full force/torque sensing at each joint. Olsen and Bekey (1986) presented a two-step link identification procedure using joint torque sensing based on the Newton-Euler equations. In their procedure for estimating the center of mass for each link, the angular accelerations were approximated to be 0's, restricting the movements to be slow for this phase. Khalil, Gautier, and Kleinfinger (1986) used a Lagrange formulation in presenting an identification model for link inertial parameters. Khosla (1986) developed a link estimation algorithm similar to ours, and implemented it on the CMU DDArm II.

Slotine and Li (1987) and Hsu et al. (1987) developed manipulator control algorithms that include on-line adaptation for rigid body link and load inertial parameters, without using the measurements of joint accelerations. Armstrong (1987) developed a method of choosing optimal, persistently exciting trajectories for such identification experiments. In a principled examination of the work of (Atkeson, An, and Hollerbach, 1986), he devised trajectories that were more persistently exciting than those used in this reference. On the other hand, Armstrong found that trajectory space is sparsely populated with good trajectory choices, and Atkeson, An, and Hollerbach (1986) serendipitously picked a relatively good one.

10.4 Dynamics in Position and Force Control

Is a rigid-body dynamic model of a robot manipulator important in control? This question has been a source of controversy over the past few years. The basic source of the controversy is the structure of the manipulator: geared commercial robots are said not to have significant nonlinear dynamic interactions. More recently, with the emergence of direct drive arms, experimental results have become available that show the computation of the dynamics is important for control for this class of manipulators. This section reviews the evidence for the importance of dynamic interactions in control for geared robots and for direct drive robots.

While the argument about dynamic models has taken place primarily in the context of position control, there is also an issue about the role of dynamics in force control. A variety of different Cartesian-based force control strategies has been proposed, some with and some without a dynamic model. This section ends by considering the evidence for any advantage of including dynamic models in force control.

10.4.1 Position Control of Geared Robots

In this section we consider why the structure of geared robots may make the computation of the rigid body dynamics less important for control. Many commercial manipulators have gears, which amplify the motor torque by a factor α equal to the gear ratio, allowing the robot designer to use smaller motors. Gears create the following problems:

• *Friction and backlash.* These nonlinear effects are due to preloading, tooth wear, misalignment, and gear eccentricity. They are extremely difficult to model, although parametric models have been attempted in the context of kinematic calibration (e.g., Whitney, Lozinski, and Rourke, 1986). Rather than modeling, another approach is to minimize backlash and friction by mechanical tuning techniques (Dagalakis and Myers, 1985).

Friction torques in many geared robots can be so large as to dominate link dynamics.

• *Joint flexibility.* Particularly for robots with harmonic drives, such as the ASEA robot, the gear elements act like springs and will deflect varying amounts depending on the load and link configuration. Joint flexibility will cause loss of accuracy at the end point, particularly complicating kinematic calibration. It also adds undesirable transmission dynamics, causing difficulties in designing a wide bandwidth controller (Good, Sweet, and Strobel, 1985).

• *Speed limitations.* All electric motors have a maximum speed at which they can operate, due to back emf and characteristics of the power amplifiers. Commercial robots often operate near this limit, but the resulting joint speed is not very fast due to the gear reduction. Moreover, the amplifiers impose a slew rate limitation, so that joint acceleration is limited. The end result is that geared robots are relatively slow, and their dynamics are dominated by gravity and friction.

• *Dominance of rotor inertias.* A gear ratio of α multiplies an electric motor's rotor inertia by α^2. Many commercial robots are designed with gear ratios that cause rotor inertia to match or dominate link inertias. For example, a typical gear ratio of 100 : 1 reduces the inertial effects of the links by 10^{-4}. The end result is that the dynamics of commercial robots are well approximated by single-joint dynamics, and the nonlinear rigid-body dynamic interactions can be ignored (Goor, 1985; Good, Sweet, and Strobel, 1985).

When these points are taken together, geared robots do not conform to the rigid-body dynamic models hypothesized in most theoretical robot controllers. Dynamic interactions between moving links are insignificant, because (1) rotor inertia dominates link inertia, (2) friction torques dominate inertial torques, and (3) gravity torques dominate inertial torques. Hence robot controllers for commercial robots are designed as parallel single-input, single-output system.

A more severe problem with commercial robots is the inability to control joint torques, yet virtually all advanced control strategies are predicated on this capability. There are two reasons why joint torque control is difficult to implement on commercial robots.

1. The nonlinear joint dynamics due to gear friction and backlash make the measurement and specification of joint torque very difficult. Motor current cannot then be used to infer joint torque, and the alternative of

mounting joint torque sensors at the output side of a gear train is problematical and seldom done (Luh, Fisher, and Paul, 1983).
2. Commercial motor servos are typically position controllers, to which one can only send setpoints. To get at the motor currents, it is often necessary to reverse engineer the controllers. Robot manufacturers are often unwilling to provide specifications to assist in this endeavor for proprietary reasons and also because of safety issues.

There has been some excellent work on improving the control of existing commercial robots. Goor (1985) designed an independent-joint adaptive controller for the PUMA robot, and improved its ability to move rapidly and accurately. He found that a third-order differential equation best described the motor characteristics. As a result, third-order polynomial trajectories are at least required, because the typical second-order polynomial trajectories supplied to the PUMA represent a step input to the servo system with concomitant oscillation and overshoot. He also found that the servo system was too slow, and that the infrequent position updates contributed to oscillation. Goor found that electronic gains in the motor servos needed to be frequently tuned because of daily variations. Chen (1987) improved Goor's controller by discretization and implementation of a lag-lead compensator. Good, Sweet, and Strobel (1985) demonstrated that the ASEA robot could be well modeled by single-joint dynamics, and that it was necessary to model and compensate for the joint flexibility induced by the harmonic drives.

The work of Goor (1985) and Good, Sweet, and Strobel (1985) indicates that in robotics it is not an issue of whether to use a model, but of choosing a suitable model. Their models of commercial robots as single-input, single-output systems are suitable, while the nonlinear models in most published robot control papers may not be. All too often in the past, authors have stated that a particular control design is intended for commercial robots, without considering that their models do not apply to these robots. Sophisticated nonlinear controllers have even been implemented on these robots, with questionable benefits. Thus an air of realism has crept into the manipulator control field. Today, few talk any longer about rigid-body dynamics in the control of PUMA robots.

10.4.2 Position Control of Direct Drive Arms

Direct drive arms are being developed to overcome some of the performance limitations of highly geared robots. Direct drive refers to the elimination of gearing and the direct coupling of links to motor axes. Since there is no gearing to amplify the motor torques, the motors have to be large and powerful to exert large torques. Motor technology has recently produced such motors, making direct drive arms possible. By eliminating gearing, backlash effects are eliminated and joint friction is minimized. Hence joint torques can be controlled and measured much more accurately, because the torque a motor produces is virtually exactly that produced at the joint (Asada, Youcef-Toumi, and Lim, 1984). The joints are also directly back-drivable against the motors, without dissipative losses due to friction.

These characteristics give direct drive arms the potential for serving as good experimental devices for testing advanced arm control strategies. Rigid-body dynamics are a good model for these arms, because of the reduction of friction and backlash. Rotor inertias of motors do not overwhelm link dynamics, and the full nonlinear dynamic interactions between moving links are manifested. Hence these manipulators are suitable for testing control algorithms based on link dynamics, and indeed it is necessary to use such sophisticated control algorithms since the full coupled dynamics of the links are reflected directly to the actuators.

There is a vast literature on the position control of robots, reflecting the variety of approaches developed in control theory. Linear control, adaptive control, sliding mode control, and nonlinearity cancellation are some of the broad categories of control theory applied to robotics. It is well beyond the scope of this chapter to review this literature, especially considering that very little of it has been experimentally tested to verify that the proposed controllers actually work well or make a difference relative to simpler alternatives.

Instead, we shall concentrate on three basic control alternatives with regard to incorporating a dynamic model, which have been experimentally tested on direct drive arms.

- Independent-joint proportional-derivative (PD) control:

$$\tau = \mathbf{K}_p(\boldsymbol{\theta}_d - \boldsymbol{\theta}) - \mathbf{K}_v(\dot{\boldsymbol{\theta}}_d - \dot{\boldsymbol{\theta}}). \tag{63}$$

- Feedforward controller (Liégeois, Fournier, and Aldon, 1980):

$$\tau_{ff} = \hat{\mathbf{H}}(\boldsymbol{\theta}_d)\ddot{\boldsymbol{\theta}}_d + \dot{\boldsymbol{\theta}}_d \cdot \hat{\mathbf{C}}(\boldsymbol{\theta}_d) \cdot \dot{\boldsymbol{\theta}}_d + \hat{\mathbf{g}}(\boldsymbol{\theta}_d), \tag{64}$$

$$\tau = \tau_{ff} + \mathbf{K}_v(\dot{\boldsymbol{\theta}}_d - \dot{\boldsymbol{\theta}}) + \mathbf{K}_p(\boldsymbol{\theta}_d - \boldsymbol{\theta}), \tag{65}$$

where the hat ($\hat{}$) refers to the modeled values.

• Computed torque control (Paul, 1972):

$$\ddot{\boldsymbol{\theta}}^* = \ddot{\boldsymbol{\theta}}_d + \mathbf{K}_v(\dot{\boldsymbol{\theta}}_d - \dot{\boldsymbol{\theta}}) + \mathbf{K}_p(\boldsymbol{\theta}_d - \boldsymbol{\theta}), \tag{66}$$

$$\tau = \hat{\mathbf{H}}(\boldsymbol{\theta})\ddot{\boldsymbol{\theta}}^* + \dot{\boldsymbol{\theta}} \cdot \hat{\mathbf{C}}(\boldsymbol{\theta}) \cdot \dot{\boldsymbol{\theta}} + \hat{\mathbf{g}}(\boldsymbol{\theta}). \tag{67}$$

Independent-joint PD control is by far the most popular feedback controller for robots, and is distinguished from the other two controllers in that the latter use a dynamic model of the robot. The feedforward controller and computed torque control differ in how the dynamic model is used in conjunction with a feedback loop.

• In the feedforward controller, the dynamics computation is based on the planned trajectory, and a PD controller acts independently and in parallel. Thus the dynamics computation can be done off-line.
• In computed torque control, the feedback law is processed through the dynamics, which must be computed in real time.

That the dynamics computation in the feedforward controller is done on the basis of the planned trajectory, and hence can be done off-line, is an advantage over computed torque control. This may have been an important issue in the past, but it is much less of one today due to increases in computational power of real-time control systems and in the efficiency of dynamics computation. A disadvantage of the feedforward controller is that the PD portion of the controller acts independently of the dynamics and produces perturbations at neighboring joints. That is to say, a corrective torque at one joint pertubs the other joints, whereas ideally the corrective torques would decouple joint interactions. Decoupling is done by computed torque control, which should therefore be more accurate than the feedforward controller.

Experimental results for these controllers were obtained on the MIT Serial Link Direct Drive Arm (An et al., 1987), under a sampling frequency of 133 Hz (figure 10.11). The computed torque and the feedforward controllers both showed a significant reduction in tracking errors for joints 1 and 2 compared with the PD controller, with no clear distinction between feedforward and computed torque. The trajectory errors for joint 3 were the same for all three controllers, indicating that the model for the third link may not be very good.

Experimental results on the CMU Direct Drive Arm II showed that computed torque control was more accurate than PD control (Khosla and

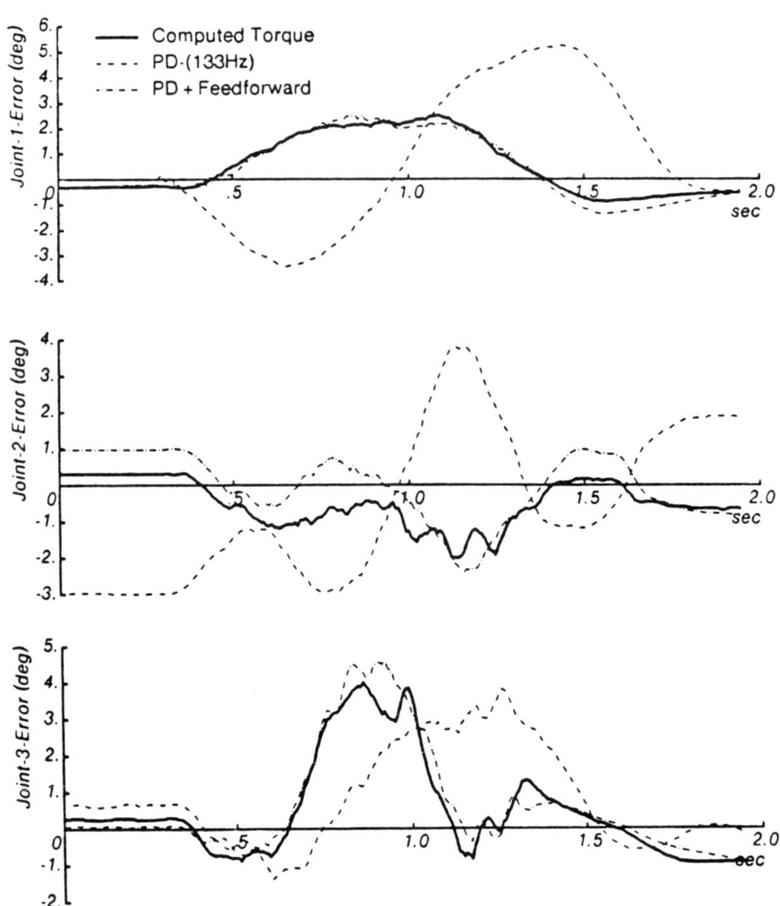

Figure 10.11
Trajectory errors of the three controllers for a 1.3-sec motion.

Kanade, 1986). Khosla (1987) showed that sampling frequency affected the accuracy of either controller, since the maximum gain depends directly on the sampling frequency. Khosla (1986) also showed that the feedforward controller and computed torque control were about equally accurate.

Thus these experiments on two different direct drive arms yielded similar conclusions. Using a dynamic model improved position control. The expectation that computed torque control would be more accurate than the feedforward controller was not met, as perhaps surprisingly both were about equally accurate.

These conclusions probably do not represent the final word on this topic, and more experimentation will be required with newer and better direct drive robots. In the experiments of An et al. (1987), the accuracy of specifying joint torque was limited due to inherent problems with the motors. This was especially apparent for joint 3, where the dynamic models made no difference in control accuracy. Also, higher sampling rates than 133 Hz are desireable and may affect the results.

10.4.3 Cartesian Force Control

A variety of force control schemes has been devised based on Cartesian coordinates of the end point or of an external reference frame. These include hybrid control (Raibert and Craig, 1981), stiffness control (Salisbury, 1980), resolved acceleration force control (Luh, Walker and Paul, 1980b; Shin and Lee, 1985), operational space method (Khatib, 1987), and impedance control (Hogan, 1985a–c; Kazerooni, Sheridan, and Houpt, 1986a–c). These force control schemes fall into two classes: those that do not use a dynamic model (hybrid control, stiffness control), and those that do (resolved acceleration force control, operational space method, impedance control).

Hybrid control. The Cartesian positions and the velocities are computed from the joint positions and velocities, respectively, by direct or forward kinematics (Raibert and Craig, 1981). Neglecting the integral terms,

$$\tau = \mathbf{K}_{pj}\mathbf{J}^{-1}\mathbf{S}(\mathbf{x}_d - \mathbf{x}) + \mathbf{K}_{vj}\mathbf{J}^{-1}\mathbf{S}(\dot{\mathbf{x}}_d - \dot{\mathbf{x}}) + \mathbf{K}_f\mathbf{J}^T(\mathbf{I} - \mathbf{S})(\mathbf{f}_d - \mathbf{f}), \tag{68}$$

where $\mathbf{K}_{pj}$ and $\mathbf{K}_{vj}$ are the position and velocity gain matrices in joint coordinates. $\mathbf{I}$ is the identity matrix, and $\mathbf{S}$ is the diagonal selection matrix. The (i, i) entry of $\mathbf{S}$ is 1 if the ith axis is to be position controlled, and 0 if it is to be force controlled.

Stiffness control. Proposed by Salisbury (1980), stiffness control was originally cast to compute kinematic errors in joint coordinates $\boldsymbol{\theta}$:

$$\tau = \mathbf{J}^T\mathbf{K}_p\mathbf{J}(\boldsymbol{\theta}_d - \boldsymbol{\theta}) + \mathbf{K}_{vj}(\dot{\boldsymbol{\theta}}_d - \dot{\boldsymbol{\theta}}), \tag{69}$$

where $\mathbf{K}_p$ is the stiffness matrix in Cartesian coordinates. By slightly modifying the stiffness control algorithm, errors can be computed in Cartesian coordinates as in hybrid control:

$$\tau = \mathbf{J}^T(\mathbf{K}_p(\mathbf{x}_d - \mathbf{x}) + \mathbf{K}_v(\dot{\mathbf{x}}_d - \dot{\mathbf{x}})). \tag{70}$$

Since the stiffness, not the pure force, is to be controlled, the above controller equations are shown without any force feedback term. A force term, however, can be added if the stiffness matrix alone does not provide enough force resolution or if pure force control is desired in some direction.

Resolved acceleration. Although Luh, Walker, and Paul (1980b) originally formulated resolved acceleration for position control only, it can be simply reformulated to control force and position simultaneously (Shin and Lee, 1985). The modified resolved acceleration controller is

$$\mathbf{x}_m^* = \ddot{\mathbf{x}}_d + \mathbf{K}_v(\dot{\mathbf{x}}_d - \dot{\mathbf{x}}) + \mathbf{K}_p(\mathbf{x}_d - \mathbf{x}), \tag{71}$$

$$\tau = \mathbf{H}\mathbf{J}^{-1}[\mathbf{S}\mathbf{x}_m^* - \dot{\mathbf{J}}\dot{\boldsymbol{\theta}}] + \dot{\boldsymbol{\theta}}\cdot\mathbf{C}\cdot\dot{\boldsymbol{\theta}} + \mathbf{g} + \mathbf{J}^T(\mathbf{I} - \mathbf{S})\mathbf{f}^* \tag{72}$$

$\mathbf{f}^*$ is the command vector for active force control, which is the only modification from the original formulation.

The modified resolved acceleration control heads a class of essentially similar methods, including the operational space method (Khatib, 1987) and impedance control (Hogan, 1985a–c). The relation among these methods was not clearly articulated in the past, and has perhaps led to some unnecessary confusion. The operational space method is in fact exactly identical to resolved acceleration control; the casting of dynamics in the operational space method into Cartesian coordinates obscures this relation. The equivalence of the operational space method was pointed out in (An and Hollerbach, 1987b; DeSchutter, 1986), and is demonstrated in the appendix. For impedance control, the only difference from resolved acceleration control is that there is an additional force feedback term to alter the apparent inertia of the robot.

Hybrid control, stiffness control, and resolved acceleration control differ in how the kinematic transformation is accomplished and in whether the dynamics of the manipulator are incorporated. In hybrid control and stiffness control, the dynamic model of the robot is not included, whereas in resolved acceleration control it is. The inverse Jacobian appears in hybrid control and in resolved acceleration control, but only the Jacobian transpose appears in stiffness control.

10.4.3.1 Kinematic Instability in Force Control A new and surprising result is that hybrid control (Raibert and Craig, 1981) is fundamentally unstable for rotary manipulators. An and Hollerbach (1987b) termed this instability a kinematic instability, since it depends on the kinematic structure of the manipulator and on its location in the workspace. Experiments and simulation show that in certain nonsingular configurations, the feedback around certain rotary joints effectively becomes positive and causes the manipulator to drift to a singular configuration, where the inverse Jacobian is undefined. Altering the nominal feedback gains does not change the inherent unstable property, although the region of instability may become smaller.

At other nonsingular configurations, hybrid control is stable. The regions of stability can be predicted fairly well from the kinematic and dynamic structure and on the choice of feedback gains. The instability evidently arises from the interaction of the inverse Jacobian matrix with the selection matrix and the inertia matrix.

With regard to kinematic structure, An and Hollerbach (1987b) have shown that hybrid control is actually stable for a polar manipulator. This confirms the results of Raibert and Craig (1981), who demonstrated that hybrid control was stable on two joints of the Stanford manipulator, equivalent to a polar manipulator. Large friction in the Stanford manipulator may also have contributed to stability. Thus there is a curious situation where hybrid control is stable for polar manipulators, but unstable for rotary manipulators.

Stiffness control and resolved acceleration control do not show this kinematic instability for either revolute or polar manipulators. Stiffness control is stable because it uses the Jacobian transpose for coordinate transformations, rather than the inverse Jacobian.

Resolved acceleration control is stable because the inertia matrix is included and cancels the destabilizing effects of the inverse Jacobian. If the dynamic modeling is exact, the motion of the manipulator would be completely decoupled at the end effector in Cartesian coordinates. Then the response under the pure position control mode would be that of a unit mass along each Cartesian degree of freedom. Even if the dynamic modeling is in error by 50%, simulations show that resolved acceleration is still stable.

Nevertheless, there is a limit to the tolerance to model inaccuracy. Note that if the inverse dynamics transformation is modeled as the identity matrix, then resolved acceleration control essentially reduces to hybrid control. Since hybrid control is unstable, this says that the identity matrix is not a good model for the manipulator dynamics. Thus a reasonably accurate dynamic model of the manipulator is as important in force control as it is in position control.

10.4.3.2 Cartesian Force Control with a Direct Drive Arm In implementing Cartesian force control on robots, one is faced with the same problems of mechanical structure as were considered earlier in position control. In commercial robots where joint torque cannot be controlled, one must implement force control with position controllers, by sensing end point force and sending position adjustments to the servos. Position resolution of commercial manipulators is usually not adequate to permit fine force gradations, and the response speed of the position servos is typically too slow. Relying on end point force sensing, moreover, brings potential stability problems during hard contact (An and Hollerbach, 1987a).

High gear ratios in commercial robots also restrict force control capability. The high rotor inertia presents a large impedance to the environment that is difficult to modulate, and the manipulator will appear stiff and unresponsive. Friction, and in particular static friction or stiction, also reduces the capability of the manipulator to sense and respond to external forces.

Direct drive arms are well suited toward force control. Since the actuators can be treated as torque sources, they are more suitable for controlling forces and torques at the tip of the manipulator. Conversely, contact forces are directly reflected to the motors without the masking effects of friction. The motors also present a low impedance to the environment, because their inertia has not been multiplied by the squared gear ratio. Hence the robots will be less stiff and more responsive to environmental contact.

Using the MIT Serial Link Direct Drive Arm, An and Hollerbach (1987b) experimentally compared stiffness control to resolved acceleration control. Stiffness control and resolved acceleration control are both stable, but differ in whether a dynamic model is incorporated. The results showed that resolved acceleration control was considerably more accurate than stiffness control, again demonstrating the importance of a dynamic model in force control. Since in Cartesian force control there are positions being generated as well as forces, these results are consistent with those on position control mentioned earlier.

For resolved acceleration control, additional experimental tests in force step tracking, force sinusoid tracking, and constant force application against a sinusoidally undulating contact point showed good results. An example is the force step test (figure 10.12). The x axis is force controlled with the square wave inputs of 10 N and 15 N, and the y axis is position controlled at a stationary $y = -0.669$ m. The step response is stable and fast, with a response time (delay plus rise time) of approximately 20 msec. The two axes are minimally coupled, as verified by the small y position error.

10.5 Conclusion

There has been substantial progress in the past few years in our understanding and implementation of kinematics and dynamics for control. For the most part, computational considerations are no longer a major issue in the implementation of kinematics or dynamics algorithms, due both to improvements in microprocessors and to efficient formulations. The advent of direct drive technology has been a big boost to experimental robotics, since meaningful control studies incorporating dynamic models are being carried out.

Looking to the future, the driving force in the study of kinematics and dynamics for control must be the design of better manipulators. Direct drive technology will continue to improve, and other alternatives to geared robots should be considered. The goal should be to design a robot that facilitates the implementation of advanced control concepts, and in particular the control of force.

Redundant manipulators will be built to overcome inherent limitations of current six-degrees-of-freedom manipulators. In controlling these robots, investigators will seek good ways to overcome the inherent limitations in using local inverse functions. Finally, arms will be attached to multifingered hands, and issues of coordination between them will be addressed.

Robot manipulators will be attached to moving bases, such as mobile robots, and new issues of control under these circumstances will arise. Real-time end point tracking will be pursued more vigorously. In general, hand-eye coordination using sophisticated vision systems will become possible due to improvements in vision systems.

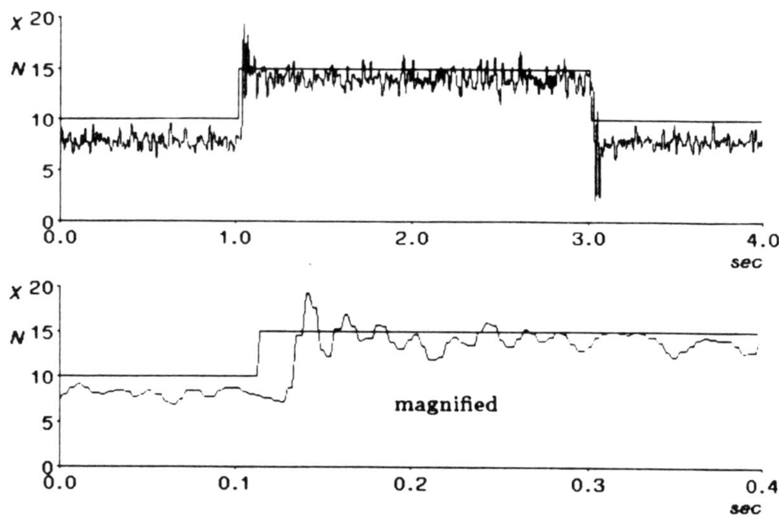

FORCE and TORQUE SENSING: X-AXIS STEP FORCE

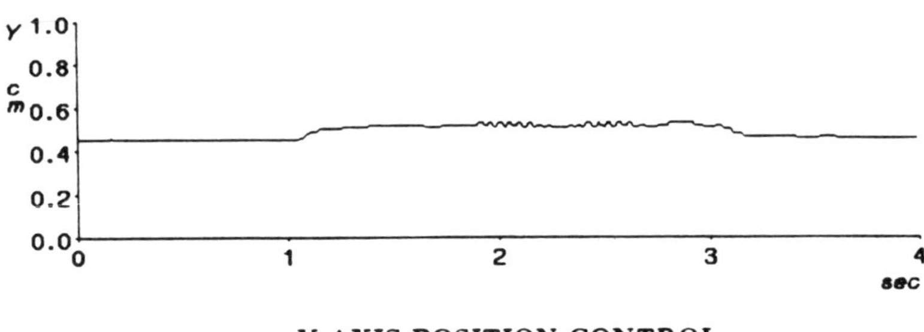

Y-AXIS POSITION CONTROL

Figure 10.12
Resolved acceleration: x axis = force steps, y axis = constant position.

Appendix

In this appendix, the equivalence of the operational space method and the modified resolved acceleration control is demonstrated. In the reports by Khatib (Khatib, 1987; Khatib and Burdick, 1986), the dynamics of a manipulator in operational space or end effector Cartesian space are described by

$$\mathbf{\Lambda}(\mathbf{x})\ddot{\mathbf{x}} + \boldsymbol{\mu}(\mathbf{x}, \dot{\mathbf{x}}) + \mathbf{p}(x) = \mathbf{f}. \qquad [1]$$

The equation numbers in square brackets refer to the equation numbers in (Khatib and Burdick, 1986). In joint coordinate system, the same dynamics are described by

$$\mathbf{H}(\mathbf{q})\ddot{\mathbf{q}} + \dot{\mathbf{q}}\cdot\mathbf{C}(\mathbf{q})\cdot\dot{\mathbf{q}} + \mathbf{g}(\mathbf{q}) = \boldsymbol{\tau}. \qquad (73)$$

$\mathbf{H}(\mathbf{q})$ and $\mathbf{\Lambda}(\mathbf{x})$ are related by

$$\mathbf{H}(\mathbf{q}) = \mathbf{J}^T(\mathbf{q})\mathbf{\Lambda}(\mathbf{x})\mathbf{J}(\mathbf{q}) \qquad [3]$$

or

$$\mathbf{J}^{-T}(\mathbf{q})\mathbf{H}(\mathbf{q})\mathbf{J}^{-1}(\mathbf{q}) = \mathbf{\Lambda}(\mathbf{x}) = \mathbf{\Lambda}(\mathbf{q}). \qquad (74)$$

Also, the operational space force vector $\mathbf{f}$ and the joint torque vector $\boldsymbol{\tau}$ are related by

$$\boldsymbol{\tau} = \mathbf{J}^T(\mathbf{q})\mathbf{f}. \qquad [4]$$

For a decoupled end effector motion commanded by $\mathbf{f}_m^*$,

$$\begin{aligned}\boldsymbol{\tau} &= \mathbf{J}^T(\mathbf{q})\mathbf{\Lambda}(\mathbf{q})\mathbf{f}_m^* + \mathbf{J}^T\boldsymbol{\mu}(\mathbf{x}, \dot{\mathbf{x}}) + \mathbf{J}^T\mathbf{p}(\mathbf{x}) \\ &= \mathbf{J}^T(\mathbf{q})\mathbf{\Lambda}(\mathbf{q})\mathbf{f}_m^* + \dot{\mathbf{q}}\cdot\tilde{\mathbf{C}}(\mathbf{q})\cdot\dot{\mathbf{q}} + \mathbf{g}(\mathbf{q}), \qquad [7]\end{aligned}$$

where

$$\dot{\mathbf{q}}\cdot\tilde{\mathbf{C}}(\mathbf{q})\cdot\dot{\mathbf{q}} = \dot{\mathbf{q}}\cdot\mathbf{C}(\mathbf{q})\cdot\dot{\mathbf{q}} - \mathbf{J}^T\mathbf{\Lambda}(\mathbf{q})\mathbf{h}(\mathbf{q}, \dot{\mathbf{q}}), \qquad [8]$$

$$\mathbf{h}(\mathbf{q}, \dot{\mathbf{q}}) = \dot{\mathbf{J}}(\mathbf{q})\dot{\mathbf{q}}. \qquad [9]$$

Typically, for position or trajectory control, a linear second order behavior is commanded so that

$$\mathbf{f}_m^* = \ddot{\mathbf{x}}_d + \mathbf{K}_v(\dot{\mathbf{x}}_d - \dot{\mathbf{x}}) + \mathbf{K}_p(\mathbf{x}_d - \mathbf{x}). \qquad (75)$$

Then, for a hybrid position/force control, the joint command vector with the operational space method is

$$\boldsymbol{\tau} = \mathbf{J}^T(\mathbf{q})[\boldsymbol{\Lambda}(\mathbf{q})\mathbf{S}\mathbf{f}_m^* + (\mathbf{I} - \mathbf{S})\mathbf{f}_a^*] + \dot{\mathbf{q}} \cdot \tilde{\mathbf{C}}(\mathbf{q}) \cdot \dot{\mathbf{q}} + \mathbf{g}(\mathbf{q}), \qquad [18]$$

where $\mathbf{f}_a^*$ is the command vector for force control.

Substituting (74) and [8] into [18],

$$\begin{aligned} \boldsymbol{\tau} = {} & \mathbf{J}^T(\mathbf{q})\mathbf{J}^{-T}(\mathbf{q})\mathbf{H}(\mathbf{q})\mathbf{J}^{-1}(\mathbf{q})\mathbf{S}\mathbf{f}_m^* + \dot{\mathbf{q}} \cdot \mathbf{C}(\mathbf{q}) \cdot \dot{\mathbf{q}} + \mathbf{g}(\mathbf{q}) \\ & - \mathbf{J}^T\mathbf{J}^{-T}(\mathbf{q})\mathbf{H}(\mathbf{q})\mathbf{J}^{-1}(\mathbf{q})\mathbf{h}(\mathbf{q}, \dot{\mathbf{q}}) + \mathbf{J}^T(\mathbf{q})(\mathbf{I} - \mathbf{S})\mathbf{f}_a^*. \end{aligned} \qquad (76)$$

Simplifying the above equation,

$$\boldsymbol{\tau} = \mathbf{H}(\mathbf{q})\mathbf{J}^{-1}(\mathbf{q})[\mathbf{S}\mathbf{f}_m^* - \mathbf{h}(\mathbf{q}, \dot{\mathbf{q}})] + \dot{\mathbf{q}} \cdot \mathbf{C}(\mathbf{q}) \cdot \dot{\mathbf{q}} + \mathbf{g}(\mathbf{q}) + \mathbf{J}^T(\mathbf{q})(\mathbf{I} - \mathbf{S})\mathbf{f}_a^*. \qquad (77)$$

This is identical to equation (72) of the modified resolved acceleration controller.

Acknowledgments

This report describes research done at the Artificial Intelligence Laboratory of the Massachusetts Institute of Technology. Support for the laboratory's artificial intelligence research is provided in part by the Systems Development Foundation and the Defense Advanced Research Projects Agency under Office of Naval Research contracts N00014-80-C-050, N00014-82-K-0334 and N00014-85-K-0124. Partial support for John Hollerbach is provided by an NSF Presidential Young Investigator Award.

References

Aboaf, E. W., and Paul, R. P. 1987 (March 31–April 3). Living with the singularity of robot wrists. *Proc. IEEE Int. Conf. Robotics and Automation.* Raleigh, NC: pp. 1713–1717.

An, C. H., Atkeson, C. G. Griffiths, J. D., and Hollerbach, J. M. 1987 (Mar. 31–Apr. 2). Experimental evaluation of feedforward and computed torque control. *Proc. IEEE Int. Conf. Robotics and Automation.* Raleigh, N.C.: pp. 165–168.

An, C. H., and Hollerbach, J. M. 1987a. Dynamic stability issues in force control of manipulators. *Proc. IEEE Int. Conf. Robotics & Auto.* Raleigh: pp. 890–896.

An, C. H., and Hollerbach, J. M. 1987b (March 30–April 3). Kinematic stability issues in force control of manipulators. *Proc. IEEE Int. Conf. Robotics and Automation.* Raleigh: pp. 897–903.

Armstrong, B. 1987 (March 31–April 3). On finding "exciting" trajectories for identification experiments involving systems with non-linear dynamics. *Proc. IEEE Int. Conf. Robotics and Automation.* Raleigh, NC: pp. 1131–1139.

Armstrong, B., Khatib, O., and Burdick, J. 1986 (April 7–10). The explicit dynamic model and inertial parameters of the PUMA 560 arm. *Proc. IEEE Int. Conf. Robotics and Automation.* San Francisco: pp. 510–518.

Asada, H., and Cro Granito, J. A. 1985 (Mar. 25–28). Kinematic and static characterization of wrist joints and their optimal design. *Proc. IEEE Conf. Robotics and Automation.* St. Louis: pp. 244–250.

Asada, H., and Youcef-Toumi, K. 1987. *Direct Drive Robots: Theory and Practice.* Cambridge, MA: MIT Press.

Asada, H., Youcef-Toumi, K., and Lim, S. K. 1984 (Dec. 12–14). Joint torque measurement of a direct-drive arm. *Proc. 23rd IEEE Conf. Decision and Control.* Las Vegas: pp. 1332–1337.

Atkeson, C. G. 1986 (Sept.). Roles of knowledge in motor learning. Ph.D. thesis, MIT, Dept. of Brain and Cognitive Sciences.

Atkeson, C. G., An, C. H., and Hollerbach, J. M. 1986. Estimation of inertial parameters of manipulator links and loads. *Int. J. Robotics Research.* 5(3):101–119.

Baillieul, J. 1985 (Mar. 25–28). Kinematic programming alternatives for redundant manipulators. *Proc. IEEE Conf. Robotics and Automation.* St. Louis: pp. 722–728.

Baker, D. R., and Wampler, C. W. 1987 (March 31–April 3). Some facts concerning the inverse kinematics of redundant manipulators. *Proc. IEEE Int. Conf. Robotics and Automation.* Raleigh, NC: pp. 604–609.

Bejczy, A. K. 1979 (Nov. 30). Dynamic models and control equations for manipulators. 715–19. Jet Propulsion Laboratory.

Book, W. J. 1984. Recursive Lagrangian dynamics of flexible manipulator arms. *Int. J. Robotics Research.* 3(3):87–101.

Burdick, J. W. 1986 (April 7–10). An algorithm for generation of efficient manipulator dynamic equations. *Proc. IEEE Int. Conf. Robotics and Automation.* San Francisco: pp. 212–218.

Cannon, R. H., Jr., and Schmitz, E. 1984. Initial experiments on the end-point control of a flexible one-link robot. *Int. J. Robotics Research.* 3(3):62–75.

Chen, Y. 1987 (March 31–April 3). Frequency response of discrete-time robot systems—limitations of PD controllers and improvements by lag-lead compensation. *Proc. IEEE Int. Conf. Robotics and Automation.* Raleigh, NC: pp. 464–472.

Coiffet, P. 1983. *Robot Technology: Interaction with the Environment. Vol. 2.* Englewood Cliffs, N.J.: Prentice-Hall.

Dagalakis, N. G., and Myers, D. R. 1985. Adjustment of robot joint gear backlash. *Int. J. Robotics Research.* 4(2):65–79.

Daniel, R. W., Irving, M. A., Fraser, A. R., and Lambert, M. 1987 (August 9–14). The control of compliant manipulator arms. *4th Int. Symp. of Robotics Research.* Santz Cruz, CA: pp. 225–232.

Denavit, J., and Hartenberg, R. S. 1955. A kinematic notation for lower pair mechanisms based on matrices. *J. Applied Mechanics.* 22:215–221.

De Schutter, J. 1986 (February). Compliant robot motion: task formulation and control. Ph.D. thesis, Katholieke Universiteit Leuven, Department Werktuigkunde.

Elgazzar, S. 1985. Efficient kinematic transformations for the PUMA 560 robot. *IEEE J. Robotics and Automation.* 1:142–151.

Featherstone, R. 1983. Position and velocity transformations between robot end-effector coordinates and joint angles. *Int. J. Robotics Research.* 2(2):35–45.

Good, M. C., Sweet, L. M., and Strobel, K. L. 1985. Dynamic models for control system design of integrated robot and drive systems. *ASME J. Dynamic Systems, Meas., Control.* 107:53–59.

Goor, R. M. 1985 (Nov. 17–22). A new approach to minimum time robot control. *ASME Winter Annual Meeting: Robotics and Manufacturing Automation, PED-Vol. 15.* Miami Beach: pp. 1–11.

Gottlieb, D. H. 1986 (April 7–10). Robots and their topology. *Proc. IEEE Int. Conf. Robotics and Automation.* San Francisco: pp. 1689–1691.

Hayati, S. A. 1983 (Dec. 14–16). Robot arm geometric link parameter estimation. *Proc. 22nd IEEE Conf. Decision and Control.* San Antonio: pp. 1477–1483.

Hayati, S. A., and Mirmirani, M. 1985. Improving the absolute positioning accuracy of robot manipulators. *J. Robotic Systems.* 2:397–413.

Hayati, S. A., and Roston, G. P. 1986. Inverse kinematic solution for near-simple robots and its application to robot calibration. *Recent Trends in Robotics: Modeling, Control, and Education.*, ed. M. Jamshidi, J. Y. S. Luh, and M. Shahinpoor. Elsevier Science Publ. Co., pp. 41–50.

Hogan, N. 1985a (March). Impedance control: an approach to manipulation: part I—theory. *ASME J. of Dynamic Systems, Measurement, and Control.* 107:1–7.

Hogan, N. 1985b (March). Impedance control: an approach to manipulation: part II—implementation. *ASME J. of Dynamic Systems, Measurement, and Control.* 107:8–16.

Hogan, N. 1985c (March). Impedance control: an approach to manipulation: part III—applications. *ASME J. of Dynamic Systems, Measurement, and Control.* 107:17–24.

Hollars, M. G., and Cannon, R. H., Jr. 1985 (Nov. 17–22). Initial experiments on the end-point control of a two-link manipulator with flexible tendons. *ASME Winter Annual Meeting.* Miami Beach.

Hollerbach, J. M. 1980. A recursive Lagrangian formulation of manipulator dynamics and a comparative study of dynamics formulation complexity. *IEEE Trans. Systems, Man, Cybern.* SMC-10:730–736.

Hollerbach, J. M. 1984. Dynamic scaling of manipulator trajectories. *ASME J. Dynamic Systems, Meas., Control.* 106:102–106.

Hollerbach, J. M. 1985. Optimum kinematic design for a seven degree of freedom manipulator. *Robotics Research: The Second International Symposium.*, ed. H. Hanafusa and H. Inoue. Cambridge, Mass.: MIT Press, pp. 215–222.

Hollerbach, J. M. 1985 (Nov. 17–22). Evaluation of redundant manipulators derived from the PUMA Geometry. *ASME Winter Annual Meeting: Robotics and Manufacturing Automation, PED-Vol. 15.* Miami Beach: pp. 187–192.

Hollerbach, J. M., and Bennett, D. J. 1987 (August 9–14). Automatic kinematic calibration using a motion tracking system. *4th Int. Symp. of Robotics Research.* Santa Cruz, CA.

Hollerbach, J. M., and Sahar, G. 1983. Wrist-partitioned inverse kinematic accelerations and manipulator dynamics. *Int. J. Robotics Research.* 2(4):61–76.

Horak, D. T. 1984. A simplified modeling and computational scheme for manipulator dynamics. *ASME J. Dynamic Systems, Meas., Control.* 106:350–353.

Hsu, P., Bodson, M., Sastry, S., and Paden, B. 1987 (Mar 31–April 3). Adaptive identification and control for manipulators without using joint accelerations. *Proc. IEEE Int. Conf. Robotics and Automation.* Raleigh: pp. 1201–1215.

Huston, R. L., and Kelly, F. A. 1982. The development of equations of motion of single-arm robots. *IEEE Trans. Systems, Man, Cybern.* SMC-12:259–266.

Izaguirre, A., and Paul, R. P. 1986 (April 7–10). Automatic generation of the dynamic equations of the robot manipulators using a LISP program. *Proc. IEEE Int. Conf. Robotics and Automation.* San Francisco: pp. 220–226.

Kanade, T., Khosla, P. K., and Tanaka, N. 1984 (Dec. 12–14). Real-time control of CMU Direct-Drive Arm II using customized inverse dynamics. *Proc. 23rd Conf. Decision and Control.* Las Vegas: pp. 1345–1352.

Kane, T. R., and Levinson, D. A. 1983. The use of Kane's dynamical equations in robotics. *Int. J. Robotics Research.* 2(3): 3–21.

Kazerooni, H., Houpt, P. K., and Sheridan, T. B. 1986a (April 7–10). The fundamental concepts of robust compliant motion for robot manipulators. *Proc. IEEE Int. Conf. Robotics and Automation.* San Francisco: pp. 418–427.

Kazerooni, H., Sheridan, T. B., Houpt, P. K. 1986b (June). Robust compliant motion for manipulators, Part I: The fundamental concepts of compliant motion. *IEEE Journal of Robotics and Automation.* RA-2: 83–92.

Kazerooni, H., Houpt, P. K., and Sheridan, T. B. 1986c (June). Robust compliant motion for manipulators, Part II: Design method. *IEEE Journal of Robotics and Automation.* RA-2: 93–105.

Khalil, W., Gautier, M., and Kleinfinger, J. F. 1986. Automatic generation of identification models of robots. *International Journal of Robotics and Automation.* 1(1): 2–6.

Khatib, O. 1987. A unified approach for motion and force control of robot manipulators: the operational space formulation. *IEEE J. Robotics and Automation.* RA-3: 43–53.

Khatib, O., and Burdick, J. 1986 (April 7–10). Motion and force control of robot manipulators. *Proc. IEEE Int. Conf. Robotics and Automation.* San Francisco: pp. 1381–1386.

Khosla, P. K. 1986 (August). Real-time control and identification of direct-drive manipulators. Ph.D. thesis, Carnegie-Mellon Univ., Department of Electrical and Computer Engineering.

Khosla, P. K. 1987 (March 31–April 3). Choosing sampling rates for robot control. *Proc. IEEE Int. Conf. Robotics and Automation.* Raleigh, NC: pp. 69–174.

Khosla, P. K., and Kanade, T. 1986 (April 7–10). Real-time implementation and evaluation of model-based controls on CMU DD Arm II. *Proc. IEEE Conf. on Robotics and Automation.* San Francisco: pp. 1546–1555.

Klein, C. A., and Blaho, B. E. 1987. Dexterity measures for the design and control of kinematically redundant manipulators. *Int. J. Robotics Research.* 6(2): 72–83.

Klein, C. A., and Huang, C. H. 1983. Review of pseudoinverse control for use with kinematically redundant manipulators. *IEEE Trans. Systems, Man, Cybern.* SMC-13: 245–250.

Lathrop, R. L. 1985. Parallelism in manipulator dynamics. *Int. J. Robotics Research.* 4(2): 80–102.

Lau, K., Hocken, R., and Haynes, L. 1985. Robot performance measurements using automatic laser tracking techniques. *Robotics & Computer-Integrated Manufacturing.* 2: 227–236.

Liegeois, A. 1977. Automatic supervisory control of the configuration and behavior of multi-body mechanisms. *IEEE Trans. Systems, Man, Cybern.* SMC-7: 868–871.

Liégeois, A., Fournier, A. and Aldon, M. 1980 (August 13–15). Model reference control of high-velocity industrial robots. *Proc. Joint Automatic Control Conf.* San Francisco, CA.

Lovass-Nagy, V., and Schilling, R. J. 1985. Control of kinematically redundant robots using 1-inverse. *IEEE Trans. Systems, Man, Cybern.*: submitted.

Lozano-Pérez, T. 1982. Task planning. *Robot Motion: Planning and Control.*, ed. Brady, J. M., Hollerbach, J. M., Johnson, T. L., Lozano-Perez, T., and Mason, M. T., eds. Cambridge, Mass.: MIT Press, pp. 473–498.

Luh, J. Y. S., Fisher, W. D., and Paul, R. 1983 (Feb.). Joint torque control by a direct feedback for industrial robots. *IEEE Trans. Automatic Control.* AC-28:153–160.

Luh, J. Y. S., Walker, M., and Paul, R. P. 1980a. On-line computational scheme for mechanical manipulators. *J. Dynamic Systems, Meas., Control.* 102:69–76.

Luh, J. Y. S., Walker, M., and Paul, R. 1980b. Resolved-acceleration control of mechanical manipulators. *IEEE Trans. Auto. Contr.* AC-25:468–474.

Maciejewski, A. A., and Klein, C. A. 1985. Obstacle avoidance for kinematically redundant manipulators in dynamically varying environments. *Int. J. Robotics Research.* 4(3):109–117.

Mayeda, H., Osuka, K., and Kangawa, A. 1984 (July 2–6). A new identification method for serial manipulator arms. *Preprints IFAC 9th World Congress.* Budapest: pp. 74–79.

Mooring, B. W. 1983. The effect of joint axis misalignment on robot positioning accuracy. *Proc. Computers in Engineering Conf. and Exhibit.* pp. 151–155.

Mukerjee, A., and Ballard, D. H. 1985 (Mar. 25–28). Self-calibration in robot manipulators. *Proc. IEEE Conf. Robotics and Automation.* St. Louis: pp. 1050–1057.

Nakamura, Y., and Hanafusa, H. 1986. Inverse kinematic solutions with singularity robustness for robot manipulator control. *ASME J. Dynamic Systems, Meas., Control.* 109:163–171.

Nakamura, Y., and Hanafusa, H. 1987. Optimal redundancy control of robot manipulators. *Int. J. Robotics Research.* 6(1):32–42.

Nakamura, Y., Hanafusa, H., and Yoshikawa, T. 1987. Task-priority based redundancy control of robot manipulators. *Int. J. Robotics Research.* 6(2):3–15.

Olsen, H. B., and Bekey, G. A. 1986 (April 7–10). Identification of robot dynamics. *Proc. IEEE Int. Conf. Robotics and Automation.* San Francisco: pp. 1004–1010.

Orin, D. E., and Tsai, Y. S. 1986 (April 7–10). A real-time computer architecture for inverse kinematics. *Proc. IEEE Int. Conf. Robotics and Automation.* San Francisco: pp. 843–850.

Paul, R. Sept. 1972. Modeling, trajectory calculation and servoing of a computer controlled arm. AIM-177. Stanford University Artificial Intelligence Laboratory.

Paul, R. P. 1981. *Robot Manipulators: Mathematics, Programming, and Control.* Cambridge, Mass.: MIT Press.

Paul, R. P., Renaud, M., and Stevenson, C. N. 1984. A systematic approach for obtaining the kinematics of recursive manipulators based on homogeneous transformations. *Robotics Research: The First International Symposium.*, ed. M. Brady and R. Paul. Cambridge, Mass.: MIT Press, pp. 707–726.

Paul, R. P., and Stevenson, C. N. 1983. Kinematics of robot wrists. *Int. J. Robotics Research.* 2(1):31–38.

Pieper, D. L. 1968. The kinematics of manipulators under computer control. Ph.D. thesis, Stanford University, Computer Science.

Raibert, M. H., and Craig, J. J. 1981 (June). Hybrid position/force control of manipulators. *ASME J. of Dynamic Systems, Measurement, and Control.* 102:126–133.

Rauh, J., and Schiehlen, W. 1986 (Sept. 9–12). A unified approach for the modeling of flexible robot arms. *Preprints 6th CISM-IFToMM Symp. on Theory and Practice of Robots and Manipulators.* Cracow, Poland: pp. 74–81.

Salisbury, J. K. 1980 (Dec.). Active stiffness control of a manipulator in Cartesian coordinates. *Proc. 19th IEEE CDC.* pp. 95–100.

Shamma, J. S., and Whitney, D. E. 1987. A method for inverse robot calibration. *ASME J. Dynamic Systems, Meas., Control.* 109:36–43.

Shin, K. G., and Lee, C.-P. 1985. Compliant control of robotic manipulators with resolved acceleration. *Proc. 24th IEEE CDC*. Ft. Lauderdale: pp. 350–357.

Silver, W. 1982. On the equivalence of Lagrangian and Newton-Euler dynamics for manipulators. *Int. J. Robotics Research*. 1(2):60–70.

Slotine, J.-J. E., and Li, W. 1987 (Mar 31–April 3). Adaptive manipulator control: a case study. *Proc. IEEE Int. Conf. Robotics and Automation*. Raleigh: pp. 1392–1400.

Stone, H. W., Sanderson, A. C., and Neuman, C. P. 1986 (April 7–10). Arm signature identification. *Proc. IEEE Int. Conf. Robotics and Automation*. San Francisco: pp. 41–48.

Sugimoto, K., and Duffy, J. 1981 (Oct. 7–9). Special configurations of industrial robots. *Proc. 11th Int. Symp. Industrial Robots*. Tokyo: pp. 309–316.

Sugimoto, K., and Okada, T. 1985. Compensation of positioning errors caused by geometric deviations in robot system. *Robotics Research: The Second International Symposium.*, ed. H. Hanafusa and H. Inoue. Cambridge, Mass.: MIT Press, pp. 231–236.

Suh, K. C. 1983 (June). An implementation of a robot control system. S. B. thesis, MIT, Elec. Eng. and Comp. Sci.

Suh, K. C., and Hollerbach, J. M. 1987 (March 31–April 3). Local versus global torque optimization of redundant manipulators. *Proc. IEEE Int. Conf. Robotics and Automation*. Raleigh, NC: pp. 619–624.

Symon, K. R. 1971. *Mechanics*. Reading, Mass.: Addison-Wesley.

Taylor, R. H. 1979. Planning and execution of straight-line manipulator trajectories. *IBM Journal of Research and Development*. 23:424–436.

Tsai, L.-W., and Morgan, A. P. 1985. Solving the kinematics of the most general six- and five-degree-of-freedom manipulators by continuation methods. *ASME J. Mechanisms, Transmissions, and Automation in Design*. 107:189–200.

Uchiyama, M., Shimizu, K., and Hakomori, K. 1985. Performance evaluation of manipulators using the Jacobian and its application to trajectory planning. *Robotics Research: The Second International Symposium.*, ed. H. Hanafusa and H. Inoue. Cambridge, Mass.: MIT Press, pp. 447–456.

Wampler, C. W., II 1986. Manipulator inverse kinematic solutions based on vector formulations and damped least-squares methods. *IEEE Trans. Systems, Man, Cybern.* SMC-16:93–101.

Wampler, C. W., II 1987 (March 31–April 3). Inverse kinematic functions for redundant manipulators. *Proc. IEEE Int. Conf. Robotics and Automation*. Raleigh, NC: pp. 610–617.

Whitney, D. E. 1972. The mathematics of coordinated control of prosthetic arms and manipulators. *ASME J. Dynamic Systems, Meas., Control.*: 303–309.

Whitney, D. E., Lozinski, C. A., and Rourke, J. M. 1986. Industrial robot forward calibration method and results. *J. Dynamic Systems, Meas., Control.* 108:1–8.

Wu, C.-H. 1984. A kinematic CAD tool for the design and control of a robot manipulator. *Int. J. Robotics Research*. 3(1):58–67.

Yoshikawa, T. 1985. Manipulability of robotic mechanisms. *Int. J. Robotics Research*. 4(2):3–9.

11 The Whole Iguana

Rodney A. Brooks

11.1 Introduction

There has been a flurry of work all over the world in the last three years on autonomous mobile robots. In this chapter we present the philosophy behind the MIT AI Lab mobile robot project and report on the significant departures this work has made from the bulk of current mobile robot work. This is evident right from our starting premise; we set out to build mobile robots in order to study the general problem of how to build artificially intelligent beings; we do this by building robots that autonomously carry out a number of generic task-achieving behaviors without human control or commands. Together, through the dynamics of the world, these behaviors combine to produce useful high-level operation.

Moravec (1984) eloquently argued that mobility has many times been the main impetus that has led to the evolution of intelligence. Dealing in real time with a dynamic and perhaps hostile environment forces the issue of understanding the world and reacting to it appropriately.

Our goal is to understand intelligence and to build intelligent beings. It seems very easy to cheat accidentally in building a sedentary being or program. It is so easy, even unintentionally, for a human to abstract away the truly hard problems. By making sure our programs run a real mobile robot in a dynamic unstructured domain, we believe there is less chance for oversimplifying the problems. More important, however, we have come to realize that this approach makes explicit to us what the real problems are in intelligence. Many of the issues that have been studied over the last 25 years turn out to be irrelevant for a real artificial being in a real world. The true details of interacting with the world are not the same as abstract thinking has led many workers in Artificial Intelligence to believe. Likewise, many problems tackled traditionally by robotics researchers tend to be largely irrelevant if the world is dynamic and unstructured.

Other current mobile robot projects tend to join the worlds of traditional AI with that of traditional robotics. We believe progress suffers from two diseases in these projects. First, from the AI tradition, that a world model must be built and maintained. Second, from the robotics tradition, that this model must be precise.

While we find many useful ideas in these two traditions of research, we maintain a somewhat different mindset of what the purpose of each algorithm or module is. The result is a radically different control structure, with radically different implementation possibilities.

The key thrust of our research is how to create a complete creature that can exist in a dynamic world, completely autonomously for long periods of time.

This chapter, then, is an answer to a challenge posed by Daniel Dennett (1978) in a commentary, titled *Why Not the Whole Iguana?*, on a paper by Pylyshyn (1978). Dennett challenged researchers to build a simulation of a complete creature. He suggested that simulating a person was too hard a problem but perhaps a starfish or a turtle was in reach. But that would require understanding everything about a turtle and its environment. Therefore he suggested *making up* a creature and its environment; perhaps a Martian iguana. His mistake was to suggest simulating the environment. It requires a great deal of work to simulate a rich enough environment to make the problems for the creature realistic; more work, I claim, than in simulating the creature itself. Thus in this chapter we use a real environment, and make up just the creatures—complete creatures, *whole iguanas.*

11.2 Approach

Over the past two years we have developed a new approach to exploring mobile robot control systems. In outline it is

- We decompose the problem based on parallel task-achieving behaviors, rather than on the traditional axis of information-processing modules. See figure 11.1.
- We build each individual task-achieving behavior from a few asynchronous simple finite state machines. Each has inputs and outputs; they are wired together with a fixed topology over low-bandwidth communications lines.
- We build one task-achieving behavior at a time. As we build each one we test and debug it extensively. The idea is that we then freeze that layer or behavior, running on its own hardware, and never have to develop it further. To add capabilities to the robot we simply add new layers of control.
- We are moving toward all onboard computation on our robots including vision. In particular, to achieve real-time vision we are concentrating on low resolution imaging systems.

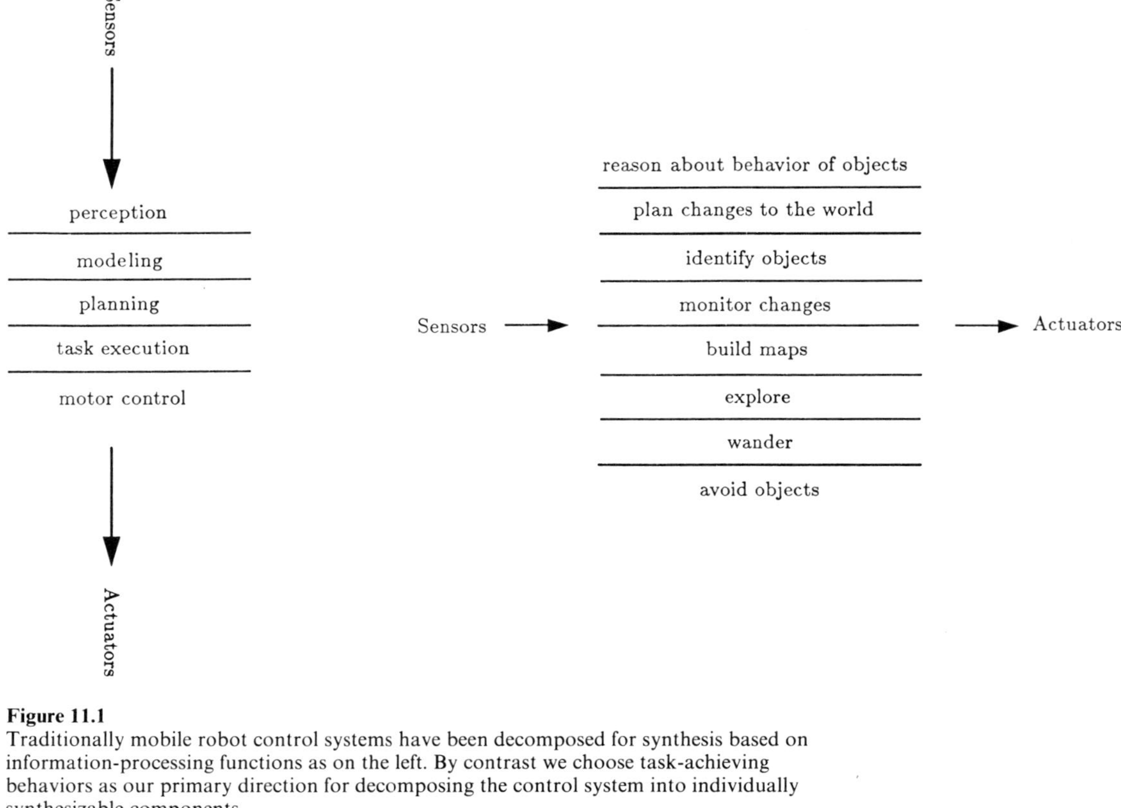

Figure 11.1
Traditionally mobile robot control systems have been decomposed for synthesis based on information-processing functions as on the left. By contrast we choose task-achieving behaviors as our primary direction for decomposing the control system into individually synthesizable components.

11.2.1 The Consequences

Our synthesis method has lead to a number of consequences very different from those found in other mobile robot projects. For instance:

• The very first test of our first mobile robot was in a dynamic environment where it had to avoid collisions with people moving about.
• There is no locus of central control, no central data structures, and no global plan that the robot is following at any time. In particular, there is no way to tell the robot to do something. Instead it senses the world and does what is in its nature to do!

11.2.2 The Approach

The key idea of levels of task-achieving behaviors is that we can build layers of a control system corresponding to each level of competence and simply add a new layer to an existing set to move to the next higher level of overall competence.

We start by building a complete robot control system that achieves a lowest-level task. It is debugged thoroughly. We never alter that system. We call it the zeroth-level control system. Next we build another control layer, which we call the first-level control system. It is able to examine data from the lower-level system and is also permitted to inject data into the internal interfaces of that level, *suppressing* the normal data flow. This layer, with the aid of the lower level, achieves some new task. The lower layer continues to run unaware of the layer above it that sometimes interferes with its data paths.

The same process is repeated to achieve higher levels of competence. See figure 11.2. We call this architecture a *subsumption architecture.*

In such a scheme we have a working control system for the robot very early in the piece—as soon as we have built the first layer. Additional layers can be added later, and the initial working system need never be changed.

But what about building each individual layer? Do we not need to decompose a single layer in the traditional manner? This is true to some extent, but the key difference is that we do not need to account for all desired perceptions, processing, and generated behaviors in a single decomposition. We are free to use different decompositions for different sensor-set task-set pairs.

We have chosen to build layers with a set of small processors that send messages to each other. A prototypical such machine is shown schematically in figure 11.3.

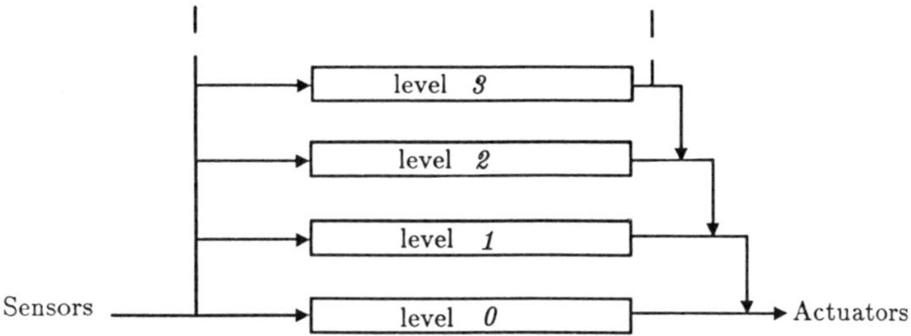

Figure 11.2
Control is layered with higher-level layers subsuming the roles of lower-level layers when they wish to take control. The system can be partitioned at any level, and the layers below form a complete operational control system.

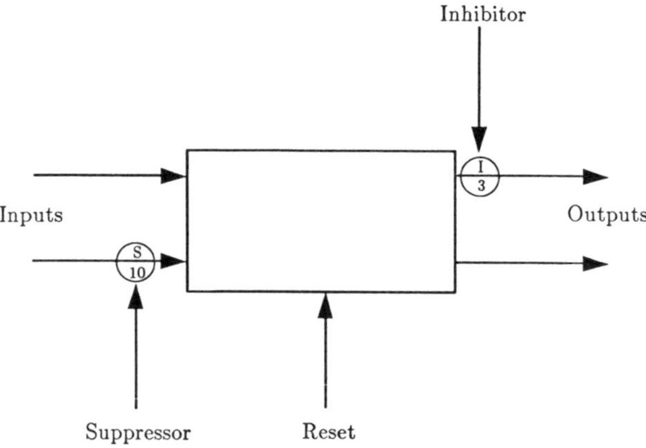

Figure 11.3
A module has input and output lines. Input signals can be suppressed and replaced with the suppressing signal. Output signals can be inhibited. A module can also be reset to state NIL.

Each processor is a finite state machine with the ability to hold some data structures. Processors send messages over connecting "wires." There is no handshaking or acknowledgment of messages. The processors run completely asynchronously, monitoring their input wires, and sending messages on their output wires. It is possible for messages to get lost—it actually happens quite often. There is no other form of communication between processors; in particular, there is no shared global memory.

All processors (which we refer to as modules) are created equal in the sense that within a layer there is no central control. Each module merely does its thing as best it can.

Inputs to modules can be suppressed and outputs can be inhibited by wires terminating from other modules. This is the mechanism by which higher-level layers subsume the role of lower levels. Figure 11.4 shows three layers of control that we have run on our mobile robots.

The lowest-level layer of control makes sure that the robot does not come into contact with other objects. If something approaches the robot, it will move away. If in the course of moving itself it is about to collide with an object, it will halt. Together these two tactics are sufficient for the robot to flee from moving obstacles, perhaps requiring many motions, without colliding with stationary obstacles. The combination of the tactics allows the robot to operate with very coarsely calibrated sonars and a wide range of repulsive force functions. Theoretically, the robot is not invincible, of course, and a sufficiently fast moving object, or a very cluttered environment might result in a collision. Over the course of a number of hours of autonomous operation, our physical robot (see below) has not collided with either a moving or fixed obstacle. The moving obstacles have, however, been careful to move slowly.

The next layer of control, when combined with the lowest, imbues the robot with the ability to wander around without hitting obstacles. This control level relies to a large degree on the zeroth level's aversion to hitting obstacles. In addition it uses a simple heuristic to plan ahead a little in order to avoid potential collisions that would need to be handled by the zeroth level.

The last level is meant to add an exploratory mode of behavior to the robot, using visual observations to select interesting places to visit. A vision module finds corridors of free space. Additional modules provide a means of position servoing the robot along the corridor despite the presence of local obstacles on its path (as detected with the sonar sensing system).

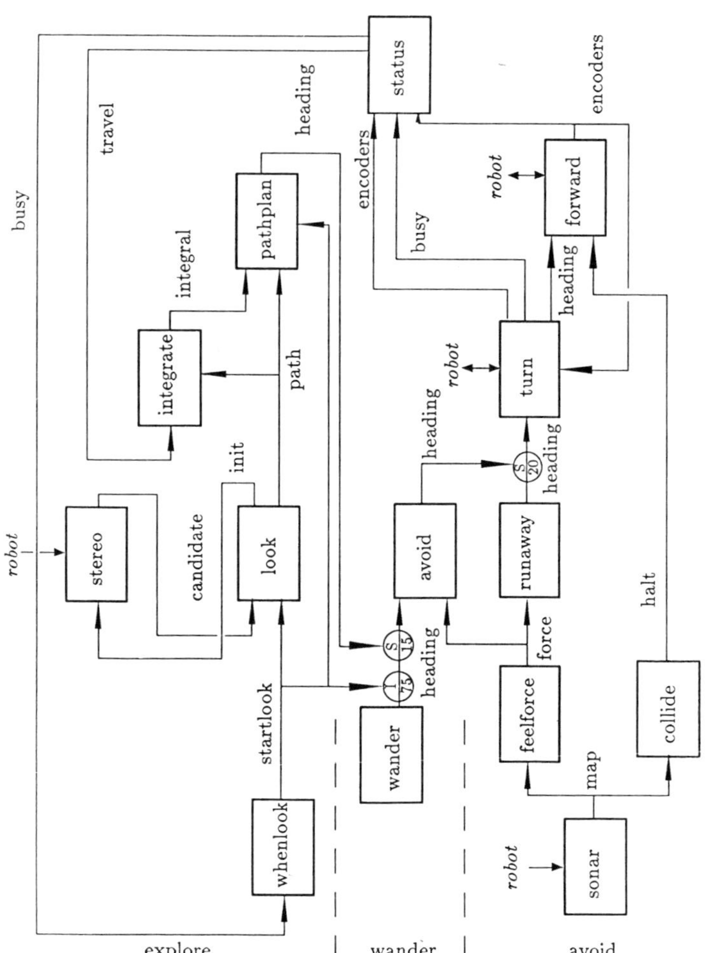

Figure 11.4
We wire finite state machines together into layers of control. Each layer is built on top of existing layers. Lower-level layers never rely on the existence of higher-level layers.

Figure 11.5
The four existing MIT AI laboratory Mobots. Rightmost is the first built Allen, which relies on an offboard Lisp machine for computation support. The leftmost one is Herbert, shown with a 24-node CMOS parallel processor surrounding its girth. In the middle are Tom and Jerry, based on a commercial toy chasis, each with a single PAL (programmable array of logic) as its controller.

The lower-level two layers still play an active role during normal operation of the second layer. (In practice, we have so far only reused the sonar data for the corridor finder, rather than use stereo vision.)

11.3 Some Prototypes

We have built four mobile robots, shown in figure 11.5, named Allen, Herbert, Tom, and Jerry. A fifth robot, named Seymour, is currently under construction. Allen, the first robot, has mostly offboard computers and has been used extensively in experiments for over a year. Herbert, is totally self-contained with an on-board parallel processor. It is just now starting to move around under subsumption control and will be our main experimental workhorse for at least the next year. Tom and Jerry were a diversion to demonstrate the idea of compiling down the control architecture to very simple on-board computation hardware. Tom and Jerry each have only a single 256 gate PAL (programmable array logic) as their brains, and have been operational for a few months. Seymour is an all passive-sensor robot, again with onboard computation.

11.3.1 Allen the Robot

Our first mobile robot consists of a commercial three-wheeled base that can turn in place, a central cardcage for communications and on-board processors, and a sensor platform mounted above that. The cardcage and sensor platforms are coupled to the wheels so that they always point "forward."

The bulk of the processing is done off-board on a Lisp machine that supports the subsumption architecture by simulating individual finite state machines and wires connecting them. Onboard are two CMOS microprocessors: one to servo the drive motors and one to run the sensors and handle off-board communications via a 1,200-baud serial cable.

The sensor platform has a ring of 12 Polaroid sonars and two Sony CCD cameras. All of our real-time experiments to date have only used the sonars. They are arranged in a symmetric circular ring. We ping opposite sensors in parallel, and thus get a complete set of 12 readings in under half a second. The cameras have been used in some static experiments; they are mounted with a fixed parallel geometry on a tilt head (pan can be achieved by spinning the robot base in place). The cameras are fed through a switching box and then to a single TV transmitter. Images are captured off-board by a demodulator and frame grabber on a Lisp machine.

We have implemented precisely the three layers of task-achieving behaviors that are illustrated in figure 11.4; see Brooks (1986) for details. The robot has wandered around a laboratory and a machine room for many hours under control of these layers.

The very first experiments we did were with just the lowest-level layer operatonal. The very first experiments included dynamic obstacles (people), which caused the robot variously to flee and halt. After an initial shakedown period the robot has operated with demos every few days and only hit an obstacle once (a sponsor, as it happened, and we later tracked it down to a loose wire on one of the forward-looking sonars).

The higher levels were added and debugged later. The second layer lets the robot wander around with a drunk's walk. The third layer looks for distant points and then integrates the wheel shaft encoders in order to get to the target in spite of perturbations caused when the lowest level avoids previously unseen obstacles. On occasion, problems with calibration of the shaft encoders have made the upper layer somewhat unreliable. But in all cases, the more primitive lowest-level layer, which does not rely on any but the coarsest of qualitative calibrations, is able to function flawlessly and in parallel. Thus the robot is at least kept out of danger, not colliding with any obstacles. We believe that these accidental experiments have demonstrated the robustness of our decomposition scheme.

Brooks and Connell (1986) report on an alternative set of higher-level layers built on top of exactly the same first level. The second layer let the robot follow walls, skimming past doorways. The third layer explicitly looked for doorways and directed the robot through them.

Recently we have began experiments (Brooks, 1987) with a fourth and a fifth layer built on top of our original three. The fourth layer takes over when it notices that Allen is in a corridor and follows along it. The fifth looks for safe cul-de-sacs at the ends of corridors and sets the robot in a position with its cameras pointing back along the open path.

11.3.2 Herbert the Robot

Our second robot (Brooks, Connell and Flynn, 1986; Brooks, Connell and Ning, 1987) is still under construction, although it does now move under its own computer control, and runs two layers of subsumption architecture. The lowest layer is again *avoid objects*. Above that is a *wall follow* layer.

To demonstrate graphically that there really is no hidden central control or source of synchronization in the subsumption architecture, we decided to build a distributed parallel processor with absolutely no central resource other than power. In particular this means no backplane, no bus among all processors, no switching network for messages, no global clock, and no shared memory.

Additionally, the range of tasks we can demonstrate with Allen is limited since its only form of actuation is to move. A manipulator arm on-board our second robot seemed appropriate.

The parallel processor has 24 independent processor boards. We plan on using them in two modes. At first we shall use one processor to simulate each finite state machine. Later, as we add control layers, we shall partition the wiring diagram and simulate up to six finite state machines with each processor.

Figure 11.6
Each processor board has a single-chip microprocessor plus support chips for extra memory and an optional suppression node. The boards are coupled with absolutely no central resource other than power. There is no backplane, no bus between all processors, no switching network for messages, no global clock, and no shared memory.

The processors are arranged in three layers of 8 arranged in a circle around the body of the robot. Figure 11.6 shows one of the processor boards. It has a CMOS Hitachi 6301 (a version of the well known 6800 family) single-chip processor. The processor chip has 128 bytes of RAM onboard and accepts a piggyback 8K EPROM for programming. Additionally each processor has 2K bytes of off-chip memory. We arrange three serial inputs and three serial outputs for each processor by actively polling some of the parallel port lines available on the chip. The communications protocol over these asynchronous lines makes use of two conductors, one for control and one for data. In our initial protocol we use 24-bit messages. The serial lines are terminated on each board in minature telephone jacks. Wiring diagrams like those in figure 11.4 are implemented by patching the boards together with physical cables using this distributed patch panel.

Each processor board has a little extra room. Some use it to hold a parallel port so that processors can communicate with input/output devices. Others have a hardware suppression node, with two inputs and one output. It is completely independent of the processor with which it shares a board. The suppression node has its own inputs and outputs and can be wired into the network at any position. Still other boards include a serpentine memory for processing last light stripe images.

The manipulator is a lightweight two-degree-of-freedom chain-driven device. There is a parallel jaw gripper which always points directly downward for grasping objects. The gripper can be moved in a plane that is vertical and pointing in the direction the robot moves. There is a 40-inch high and 18-inch long rectangle in which the gripper can be moved. This means the robot can grasp objects at both table top and ground levels, and transfer objects weighing up to 2 pounds between those two levels.

The initial sensors on the robot are a set of up to 48 infrared proximity sensors for local obstacle and intruder detection. A laser light striper is under construction, to be mounted next to the manipulator.

11.3.3 Tom and Jerry the Robots

Our third and fourth robots, Tom and Jerry (Connell, 1987), are identical and are completely operational. Their absolute level of performance is not particularly advanced. They were built to demonstrate the idea of compiling our subsumption architecture down to a network of gates, without any conventional computing element.

Each physical robot is based on a commercial radio controlled toy. We removed the control electronics and replaced it with a self-contained larger board. We mounted four infrared proximity sensors, three forward looking and one rearward looking. The motors can be driven forward and back and the left and right front wheels have steering brakes. There are thus 4 bits of input to and 4 bits of output from the subsumption network.

The subsumption network itself is implemented in a single 256-gate PAL. There are three layers implemented. Some changes had to be made to accommodate 1-bit data paths. One of the many networks we have implemented is shown in figure 11.7. The lowest level is our familar *avoid objects* layer of control, which includes avoidance and halting behaviors. The next layer of control lets the robot explore large areas. The third level gives some extra heuristics for backing out of tight situations.

11.3.4 Seymour the Robot

We have started work on constructing Seymour, and all passive-sensor robots. Our concentration on this robot is to develop reliable real-time

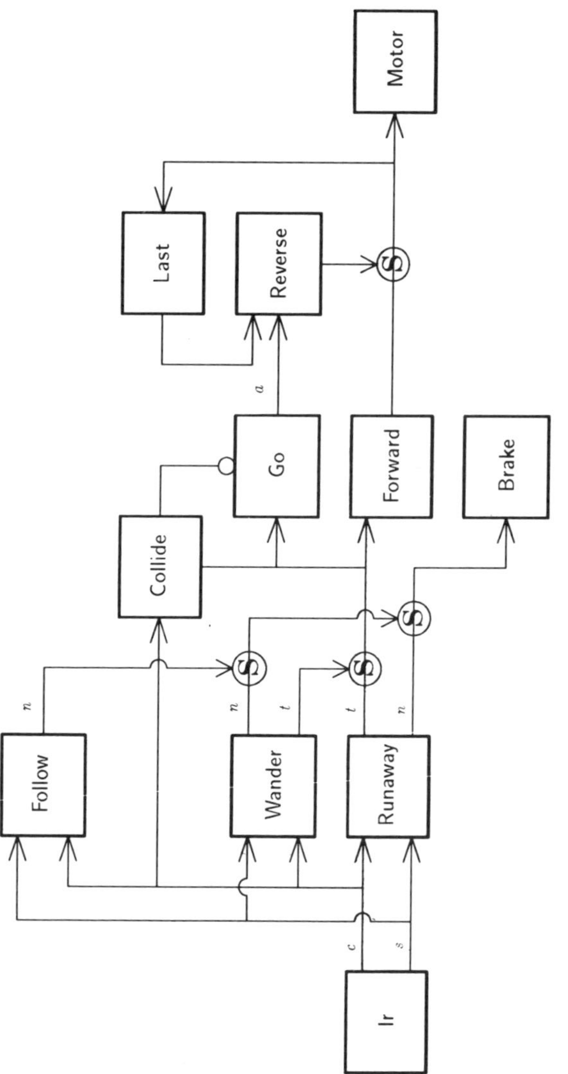

Figure 11.7
The layers of control for Tom and Jerry are all implemented with 1-bit data paths.

vision that is robust over a wide range of circumstances. Brooks, Flynn and Marill (1987) report on some vision algorithms we have developed for this robot.

All vision is to be done on-board. The algorithms are completely task dependent. So far we have concentrated on vision for obstacle avoidance. We feel we can achieve this goal with rather low-resolution cameras. We have constructed some prototypes. Our plan is to mount 8 to 10 cameras on the robot, each a low resolution CCD camera (some 32 × 32 pixels, and some 1 × 1,024). We are constructing a special purpose processor to handle the early processing of the visual data before it gets injected into the subsumption network. There is a Hitachi CMOS 68010 as host processor driving a CMOS Analog Devices DSP2100 slave. The later can do linear convolutions at 8 million pixels per second. The complete processor system draws only 5 watts.

11.4 The AI Problem

This and the next two sections present the reasons we have chosen to build our mobile robots in such a nontraditional way.

Autonomous mobile robots must be intelligent to some degree. Naturally, therefore, they must be *artificially intelligent*. Much of the work in mobile robots during the sixties and seventies was in fact done in Artificial Intelligence Laboratories (e.g., Nilsson, 1984; Moravec, 1983; Giralt et al., 1984). The first of these (Nilsson's robot Shakey) was a source of much inspiration for work in Artificial Intelligence, while the latter two were the first to apply successfully many AI techniques to real world situations.

The Shakey project is often quoted as a smashing success of applied AI. In contrast I feel its methodology was misguided, and caused a misdirection of the bulk of AI research during the seventies and eighties. AI is only just now beginning to recover from this disaster.

Throughout the sixties there had been an implicit assumption that *intelligence* could be separated from *perception* and *action*. See, for example, the collections of papers in Feigenbaum (1963) and Minsky (1968). All of these papers describe disembodied reasoning systems. They are given inputs in terms of a priori databases and strings of characters received from a teletype. They provide outputs in the form of characters sent to a teletype.

Shakey followed this tradition and seemed to validate it in that it demonstrated that a robot could actually exist and operate in the world with very separate perception, action, and intelligence subsystems.

But Shakey cheated.

Shakey's world was a very carefully engineered toy. Walls were specially constructed and painted with matte finishes. Besides walls and Shakey there was a set of large polyhedral blocks, each of whose faces was painted in a solid matte color.

One reason for such a simplified environment was the low perceptual rate achievable with computers (even mainframes) of the day. There were simply no known algorithms that could process visual information reliably in any but highly constrained circumstances. Such algorithms (still in their infancy even today) required the advent of both new theories and faster computers so that they could be tested out in reasonable time frames.

Another reason for the simplified environment was that it simplified the reduction of the total problem into relatively independent pieces with clean interfaces. A simple world meant that the interface language could be simple and descriptions of the world could be short enough to fit into the limited amount of memory available in mainframes of the time. In fact, the interface was not completely straightforward, as a little bit of the real world did manage to creep in, and collusion between intelligence and perception was necessary. As Shakey wandered around, its position estimate from its shaft encoders drifted. It was necessary to refer to its world model for a nearby vertex and then match that with its perceptions to triangulate and correct its position estimate.

Shakey's major contribution to AI thought at the time was the idea of having a complete model of the world. It seemed to show that with an *adequate* model it was sufficient to reason within the model and forget about the world itself. The model, after all, encoded everything of relevance that occurred in the world. I have not seen any papers of the time claiming that these were contributions of Shakey. I do not claim that these were recognized as results of the Shakey experience per se. Rather they were operational lessons that were absorbed into the culture of AI research.

There are two serious drawbacks with Shakey's approach to intelligent systems existing in the world:

1. There is an implicit assumption that the black box of perception will be able to deliver descriptions of the world that are such exact matches to

the world that it is possible to reason with the world model and not be missing any critical facts. Shakey achieved this goal only partially in that it did require an a priori description of the world save the exact positions of the large painted blocks. In any case, even for that simple degree of world model updating, Shaky relied on the class of unknowns being very simple and completely understood a priori.
2. There is an implicit closed world assumption. Shakey knew everything about the world and everything that could influence it. The only exceptions were that a single large block could be magically and grossly displaced from its "known" position. When such a discrepancy was detected Shakey could take its time to update its world model. Things never changed enough that its position estimate update routines would be overloaded. In a real, dynamic, world a robot can not know everything about the world. It must rely on more high-frequency sensing to detect discrepancies and update its world model. When many things can change simultaneously, the simple approach taken by Shakey simply will not work. The changes in the world interact with the robot's use of its world model that is necessary to perceive accurately what needs to be updated.

Unfortunately most of the work in robot planning, for assembly especially, has adopted this idea of having a complete and closed world model. Winograd (1972) constructed a natural language interface (SHRDLU) to a program that simulated a perfect world of colored blocks and pyramids (like Shakey's world, only smaller) where a perfectly behaved robot stacked and unstacked them. Fahlman (1975) brought planning to a more realistic stage in his BUILD program, which generated plans for constructing assemblages in the blocks world by having a robot manipulator grasp blocks, stack them, build scaffolding, and produce complex creations. Again the world was simulated so the true problems of a real world did not creep in. Lozano-Pérez (1984) in a planning system (ATLAS) for assembly tried to make the world more realistic by dealing with bounded position and force uncertainties as would arise from parts tolerances and robot control error but still heavily relied on a closed world assumption. Their proposal remains untested in the real world.

These three examples illustrate how the AI side of robotics has relied on planning off-line in a closed world. Unfortunately the real world is not closed. Much of the planning work done in Artificial Intelligence is irrelevant to mobile robots. Those who argue that the planning work available outstrips in capability the perception work that is available are being confused by a mismatch of expectations from the planning side versus the reality of what perception will ever be able to deliver.

11.5 The Role of Sensing and Perception

Sensors provide raw data bits for perceptual processes. But what should perceptual processes deliver as output? What is the purpose of perception?

Most work in computer perception has been based on visual sensing, although a little has been done on sound (but almost all in the realm of speech understanding), touch, and active sensing (e.g., radar). I believe that the majority of this work has suffered from two drawbacks that make many of its fruits irrelevant to mobile robots. First, there has been a concentration on object shape recovery for recognition, and second, there has been a concentration on extremely accurate shape recovery.

Marr (1982) pointed out that the *purpose of vision* depends on the task the perceiving organism is trying to achieve. His example in this section (p. 32) of his philosophical treatise is the housefly. He then talks about *advanced vision*, which he equates with human vision and uses as his examplar the task of object recognition **with the generality that humans can bring to bear on this task**. The rest of the book develops techniques or ideas for techniques for the transformation of representations from image to primal sketch to $2\frac{1}{2}$-D sketch to 3-D model representation in support of this task.

For object recognition with human level generality as determined by our own introspection, it seems that we need to recover the shape of objects. Indeed this has been a central focus of computer vision research. For instance in Brady (1981) a collection of 14 high-quality AI-flavored papers on vision, fully 8 are explicity concerned with shape recovery, and 3 more are explicity concerned with object recognition but bypass the shape recovery phase. Other monographs (e.g., Lowe, 1985) and textbooks (e.g., Ballard and Brown, 1982) explicitly state that object recognition is the major goal of computer vision and furthermore that explicit high-level world knowledge is an important ingredient in that recognition process.

I believe that such concentration on surface shape recovery and object recognition has been spurred by the belief that complete world models are needed by planning processes. But more insidiously there has also been a tacit belief that the descriptions of objects delivered by vision must be extremely accurate. This may well be due to (1) the traditional use of position control in robotics, because of its computational simplicity and low sensory bandwidth (shaft encoders), and (2) the many simplistic noise-sensitive shape-matching attempts at computer vision. Thus we have seen over the last few years intense amounts of effort going into calibrating (e.g., Moravec, 1983; Faugeras, 1986) visual systems to a 3-D coordinate system.

Suppose now that we back off from this implicit model that computer vision should be like what we imagine human vision is, and reconsider Marr's (1982) point that the purpose of vision depends on the task to be achieved. Then we must examine what it is we want our robots to do, what information they will need to do it, and build vision algorithms that deliver precisely the necessary information. Trying to deliver more general information is a prescription for failure—it will be far too easy to slip into the traps we have seen computer vision fall into already (delivering accurate surface descriptions for their own sake, and worrying about second-order effects and accuracies that are irrelevant in the real world).

In particular, a robot that must avoid obstacles does not need to recognize the obstacles (as chairs, trees, etc.). At most it might need to generate some volumetric description of the obstacles, in order to avoid intersecting those volumes. Except in very cluttered situations the volumes can be generous bounding volumes. This certainly makes the vision problem simpler. But, in fact, there is another approach that simplifies it even more. Instead of using vision to concentrate on where obstacles are, perhaps a better strategy for this particular task is to concentrate on where free, navigable space is, and have the vision systems deliver that as their model of the world. The robot's task determines the most appropriate emphasis of the early vision algorithms being used in support of that task.

11.6 The Role of the World

Since globally consistent world models are hard to build and maintain, and are perhaps not useful, consider the possibility of not building them at all. Instead let us consider the possibility of using the world as its own model. It has the advantage of being a complete model of itself, of being totally accurate, and of always being up to date. The implementation impact of this idea is that rather than try to predict exactly what will happen in a model, we notice aspects of the world from our sensors, and act on those aspects. We rely on the world to integrate all our independently controlled actions to give a cohesiveness to the overall behavior.

We give three examples; the first is another independent piece of work, the second is operational on our robots today, and the third is a scenario we are working toward for our robots.

11.6.1 Hopping Machines Rely on the Physics of the World

Raibert (1983) used this approach in essence in the control systems for his hopping machines. In the case of the one-legged hopper he independently controls three aspects of the robot in the world by controlling two actuators (*1. fluid flow in the leg* and *2. torquing of leg-body joint*) on the basis of five sensors: *a. foot-touching-ground sensor, b. leg extension, c. pressure in leg, d. leg to body angle,* and *e. global body attitude.* The controlled aspects of the robot in the world are

• *hopping height:* sensors *b* and *c* are integrated to determine control of actuator *1*;
• *body attitude:* during stance as determined by sensor *a*, sensors *d* and *e* are integrated to control actuator *2*; and
• *forward running speed:* during flight as determined by sensor *a*, the previous readings during stance of sensors *b*, *d*, and *e* are used to control actuator *2*.

The crucial ideas underlying this approach are

• There is no global model maintained of what the robot is doing. Three aspects of its world behavior are controlled independently and the result is that the world integrates these through physics to produce a running machine.
• Sensors are integrated in groups to provide partial models of what the robot is doing. Some sensors are used more than once to provide independent partial descriptions.

11.6.2 Finding Paths among Dynamic Obstacles

Consider the layers of control described in section 11.2. The robot does not attempt to segment the world into dynamic and static obstacles. Rather it takes instantaneous views of the world and reacts to those. If the world changes, the robot naturally changes its behavior—fleeing from an agressor, for instance. It is its interaction with the world that makes it avoid obstacles, not any internal representation of and reasoning about obstacles.

At the same time, and in parallel, the robot is trying to explore its environment. The exploration layer wants the robot to go in a particular direction. It tells the lower levels so, but then observes the behavior of the robot in the world to see what really happened. The higher-level layer would have to model the lower levels if it was to maintain a halfway accurate model without simply resensing what happens in the world.

11.6.3 Grabbing a Soda Can

Figure 11.8 shows a collection of behaviors that will enable Herbert to wander around office areas collecting empty soda cans from desks and depositing them in some home location. We do not attempt to build a detailed model of the world. In fact we do not even have much interbehavior communication on board the robot. Rather we let the world couple the behaviors.

For instance, one behavior looks for candidate soda cans on tabletops. When it sees one it tells another behavior the location, and that behavior,

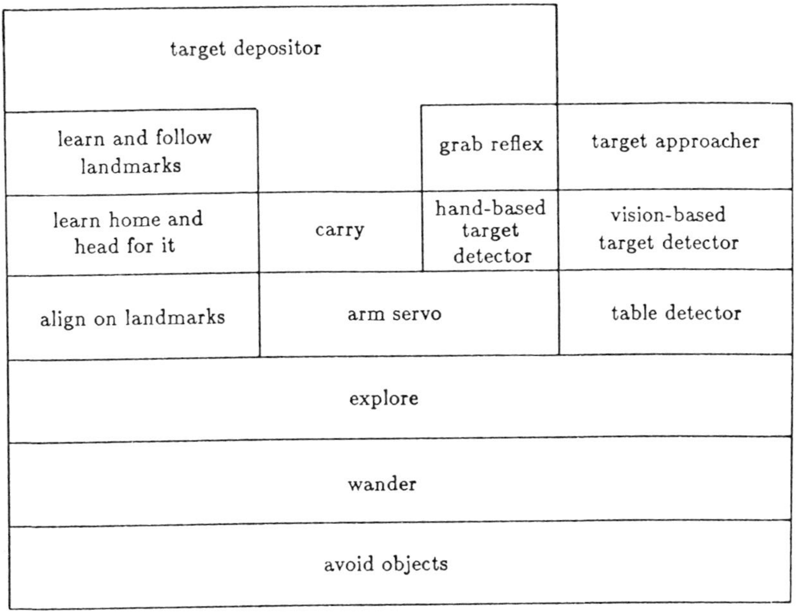

target depositor

learn and follow landmarks | grab reflex | target approacher

learn home and head for it | carry | hand-based target detector | vision-based target detector

align on landmarks | arm servo | table detector

explore

wander

avoid objects

Figure 11.8
We let 14 independent behaviors run in parallel on Herbert. By indexing themselves on sensor values measured in the world, they combine to give the robot a cohesive externally observable behavior—collecting soda cans from offices.

if it is active and has control of the robot, moves the robot so that the gripper is above the target area. Notice that if the robot is busy fleeing or something else important is happening, the robot will not go toward the can. That is no concern of the original soda can detector. It was simply looking for cans and noticing them.

Another behavior is always monitoring sensors on the hand, and whenever it thinks the hand is above a soda-can-like object, it lets the grasp behavior know this fact. If other conditions are met for grasping, it will happen. If the object was incorrectly identified from afar, it will not be grasped. If in the meantime some other high-priority event happens, the soda can will be quietly forgotten. There is no need to program in elaborate abort strategies as each and every behavior is built to be active only when the world is in an appropriate state.

11.7 Dogma

There are a number of aspects of our work that are commonly questioned. In this section we outline these objections and (of course) refute them. Individually each of the questions or objections has merit. We claim, however, that the total picture makes each of our idiosynchratic foibles justifiable.

1. *Why not build regular world models?*
2. *Why not carefully calibrate sensors and actuators before running experiments?*
3. *Why restrict the model to simple networks of such simple machines?*
4. *Why restrict to low-bandwidth communications between processors?*
5. *Why not simulate all the finite state machines on a single large processor?*

The following subsections answer each of these challenges.

1. *No representation* Other projects can explore these topics. We have identified a number of methodological pitfalls in following this practice and have chosen as an object of study to see how far we can push the notion of not having explicit internal models. We want to see what levels of functionality we can reach without them.
2. *No calibration* The desire for calibration comes from the notion that there is an objective reality in the world and *the robot must know this objective reality*. Ignoring the first clause of the conjuction, we take issue

with the second. We suspect that once this idea takes root, researchers search for accurate models of the world for accuracy's sake, not for the sake of the task the robot is to perform. Often they will argue that indeed the accuracy is needed because another algorithm within their robot relies on that accuracy. But we argue that it is that second algorithm that is incorrectly designed.

As an example, consider the problem of visually detecting obstacles in front of the moving robot in order to plan local collision-free paths around them. It is clearly not necessary to extract accurate surface models of the obstacles. One might argue that it is necessary to estimate their depth accurately so that the necessary path can be planned to avoid them. But notice that in carrying out the avoidance maneuver the robot must respond accurately to motion commands. In fact the accurate coordinate system used for recording obstacle position forces accurate electromechanical control of the robot. This internal accuracy merely serves to connect accurate vision and accurate motion, whose accuracy was only demanded by the existence of the internal requirement.

An alternative strategy is to connect vision to motion more directly and let them *calibrate each other*. For instance, if we can determine motion flow (often approximated by optical flow) in a series of forward-looking camera images, and if the robot is traveling with pure translation, then we can estimate the time to collision for a point ahead. It is simply the ratio of the distance of the point in the image from the center of expansion to the velocity of the point away from that center. If we plan to avoid obstacles in time-to-collision space, then we can easily control the robot. Its current velocity determines the scale of the planning space, not through any calibration, but through the physics of imaging in the world. We have removed the internal accurate model by choosing an appropriate coordinate system for a single task that involves both vision and motion. Brooks, Flynn, and Marill (1987) describe experiments using Allen the robot with such a self-calibrating vision algorithm, and in fact use it to calibrate another algorithm, stereo vision, for misalignment of the cameras.

3. *No complex computers* Originally it was not an explicit goal of this project to use simple computation elements. The original conception was that there would a network of 68000s linked by low-bandwidth communication channels. As we worked on the natural decomposition of the system, however, we found repeatedly that very little was demanded of each

parallel process, and that they could all be easily expressed as finite state machines running asynchronously.

After making this leap it was noticed that a new benefit had accrued. Now each of the processors is so simple that it could actually be implemented on a very small piece of silicon. Perhaps it will be possible to fabricate complete mobile robot control systems on a single chip. To explore this, we have started work on a silicon compiler to transform a subsumption network into a single VLSI chip layout.

4. *No high-bandwidth communication* By restricting ourselves to low-bandwidth communication we can ensure that all the data paths in a VLSI implementation of the subsumption architecture can be quite narrow, and thus a large number of them can be implementable. Furthermore, in a more macroscopic scale implementation we eliminate the temptation to link processes with centralized, and ultimately capacity limited (see next subsection), resources, such as a switching communication network.

5. *No nonincremental resources* On robot Allen we simulated the finite state machines on an off-board Lisp machine. Eventually, as we added finite state machines, we overloaded the capacity of that single machine to respond in real time while carrying that simulation load. But well before we reached that saturation point we found we had to spend considerable effort prioritizing the running of the supposedly parallel finite state machines so that nothing critical was starved out.

Once saturation was reached the question of which single processor we should use to simulate all these finite state machines arose. Clearly no matter which machine we chose we would eventually overflow its capacity. Furthermore, we would again have to spend time prioritizing supposedly parallel processes to run on a single processor. Eventually the new processor would become saturated too.

To simplify experiments we therefore decided actually to build an indefinitely extensible parallel processor. The extensibility is ensured by the only global resource being electrical power. Besides being extensible it graphically makes the point of there being no locus of central control in the subsumption architecture.

11.8 Whole Iguanas

Whole iguanas are complex creatures. So are whole houseflies. Even a whole snail is beyond what we can achieve today. It is a completely autonomous system that operates for a long period of time in a dynamic environment, achieving certain built-in goals necessary for its survival.

In this chapter we have argued that for a robot to achieve such levels of behavior as routinely achieved by billions of spieces of animals we need to control the vast complexity of many interacting parts of intelligence (where we extend the definition of intelligence to include the competence demonstrated by a snail, for instance).

The techniques we see as appropriate to controlling this complexity are

- Build systems incrementally.
- Build loosely coupled systems.
- Instead of trying to represent the world explicitly, find mappings from aspects of the world to appropriate actions.

Acknowledgments

This report describes research done at the Artificial Intelligence Laboratory of the Massachusetts Institute of Technology. Support for the research is provided in part by an IBM Faculty Development Award, in part by a grant from the Systems Development Foundation, and in part by the Advanced Research Projects Agency under Office of Naval Research contract N00014-85-K-0124.

References

Ballard and Brown (1982) Dana H. Ballard and Christopher M. Brown, *Computer Vision*, Prentice-Hall, Edgewood Cliffs, NJ, 1982.

Brady (1981) J. Michael Brady (ed.), *Computer Vision*, special issue *Artificial Intelligence* 17, 1981.

Brooks (1986) Rodney A. Brooks, "A Robust Layered Control System for a Mobile Robot," *IEEE Journal of Robotics and Automation*, RA-2, No. 1, 1986, 14–23.

Brooks, Connell, and Flynn (1986) Rodney A. Brooks, Jonathon H. Connell, and Anita M. Flynn, "A Mobile Robot with Onboard Parallel Processor and Large Workspace Arm," *Proceedings AAAI Conference, Philadelphia, PA*, 1986, 1096–1100.

Brooks and Connell (1986) Rodney A. Brooks and Jonathon H. Connell, "Asynchronous Distributed Control System for A Mobile Robot," *Proceedings SPIE, Cambridge, MA*, 1986.

Brooks (1987) Rodney A. Brooks, in preparation.

Brooks, Connell and Ning (1987) Rodney A. Brooks, Jonathon H. Connell, and Peter Ning, in preparation.

Brooks, Flynn, and Marill (1987) Rodney A. Brooks, Anita M. Flynn, and Thomas Marill, "Self Calibration of Motion and Stereo Vision for Mobile Robot Navigation," MIT AI Lab Memo 984, August 1987.

Dennett (1978) Daniel C. Dennett, "Why Not the Whole Iguana?" *The Behavioral and Brain Sciences*, 1978, 103–104.

Fahlman (1975) Scott E. Fahlman, "A Planning System for Robot Construction Tasks," *Artificial Intelligence* 6, 1975.

Faugeras (1986) O. D. Faugeras and G. Toscani, "The Calibration Problem for Stereo," *Proceedings IEEE Computer Vision and Pattern Recognition, Miami, FL*, 1986, 15–20.

Feigenbaum (1963) Edward A. Feigenbaum and Julian Feldman (eds.), *Computers and Thought*, McGraw-Hill, San Francisco, 1963.

Giralt (1984) Georges Giralt, Raja Chatila, and Marc Vaisset, "An Integrated Navigation and Motion Control System for Autonomous Multisensory Mobile Robots," *Proceedings of the First International Symposium on Robotics Research, Bretton Woods, NH*, edited by Michael Brady and Richard Paul, MIT Press, Cambridge, MA, 1984, 191–214.

Lowe (1985) David G. Lowe, "Perceptual Organization and Visual Recognition," Kluwer Academic Publishers, Boston, 1985.

Lozano-Pérez (1984) Tomás Lozano-Pérez and Rodney A. Brooks, "An Approach to Automatic Robot Programming," *Solid Modeling by Computers*, edited by Mary S. Pickett and John W. Boyse, Plenum Press, New York, 1984, 293–327.

Marr (1982) David Marr, *Vision*, W. H. Freeman, San Francisco, 1982.

Minsky (1968) Marvin L. Minsky (ed.), *Semantic Information Processing*, MIT Press, Cambridge, 1968.

Moravec (1983) Hans P. Moravec, "The Stanford Cart and the CMU Rover," *Proceedings of the IEEE*, 71, July 1983, 872–884.

Moravec (1984) Hans P. Moravec, "Locomotion, Vision, and Intelligence," *Proceedings of the First International Symposium on Robotics Research, Bretton Woods, NH*, edited by Michael Brady and Richard Paul, MIT Press, Cambridge, MA, 1984, 215–224.

Nilsson (1984) Nils Nilsson, "Shakey the Robot," SRI AI Center Technical Note 323, April 1984.

Pylyshyn (1978) Zenon W. Pylyshyn, "Computational Models and Empirical Constraints," *The Behavioral and Brain Sciences*, 1978, 93–99.

Raibert (1984) Marc H. Raibert, H. Benjamin Brown, Jr., and Seshashayee S. Murthy, "3-D Balance Using 2-D Algorithms?" *Proceedings of the First International Symposium on Robotics Research, Bretton Woods, NH*, edited by Michael Brady and Richard Paul, MIT Press, Cambridge, MA, 1984, 279–301.

Winograd (1972) Terry Winograd, *Understanding Natural Language*, Academic Press, New York, 1972.

IV DESIGN AND ACTUATION

12 Design and Kinematics for Force and Velocity Control of Manipulators and End-Effectors

Bernard Roth

12.1 Introduction

In robotic devices we often employ a series (or chain) of links where each moving link is attached to the preceding link by a single kinematic joint, which is either a revolute, R, or a prismatic kinematic pair, P. Such systems are characterized by a multiplicity of freedoms, requiring simultaneous, yet independent, control and actuation of the several degrees of freedom. In this chapter we derive the basic equations to design such chains when the design specifications are on either the first-order instantaneous kinematics or the transmitted forces and moments.

Instantaneous kinematic properties are those properties that depend on rates of change at a given position or instant, the so-called differential properties of the motion. The most familiar instantaneous properties are velocity and acceleration; they appear whenever the independent variable is time. Often it is useful and possible to study instantaneous properties that are not time dependent; these are called the geometric instantaneous properties. Tangency and curvature are familiar concepts from this branch of study.

Happily, we get two bonuses when we study the properties which depend on the first derivatives (these are known as the first-order properties). First, it turns out that the time-based properties and the geometric properties are essentially proportional to each other; they differ by scalar multipliers. Second, we can make use of a certain duality between force and velocity: for every result we obtain for the first-derivative properties, there is an analogous result for static force properties if an angular velocity is replaced by a force and a linear velocity is replaced by a moment. In this chapter we shall emphasize this duality whenever we consider static forces.

In a series chain, each moving link has associated with it a motion variable, which measures its position relative to the preceding link. Obviously the motion variable is the rotation in an R joint and the translation in a P joint. The total number of such motion variables in a chain will be taken as the degrees of freedom for the motion of the chain; i.e., we assume all the degrees of freedom are independent.

From the point of view of the instantaneous kinematic properties (including static forces!) it is completely immaterial where the joints are located along the axes; what matters is simply the relative locations of the axes. It requires four parameters to locate a line in three-dimensional space. Thus, we need four parameters to determine a revolute axis. For a prismatic axis we only need two parameters, since any position of a *P* axis parallel to a given direction produces exactly the same motions. The only difference the position makes is in the resisting moments in the bearings of the *P* axis, although this does require consideration before a design is finalized; it has no affect on the type of design problems treated in this chapter.

12.2 One-Degree-of-Freedom Systems

There are two possible one-degree-of-freedom systems, depending on whether the joint is an *R* or a *P*. For an *R* joint the path of every moving point is a circle; all such circles have the same axis, which is the joint axis. For the *P* joint, the paths of all the points are straight lines parallel to the joint axis. Such point paths can always be described analytically and all their differential properties are trivial to obtain. Hence the analysis of such systems is an extremely simple task.

For the kinematic design of such a system it is necessary to locate the joint axis in some fixed reference system. For an *R* jointed system this implies four design parameters. For a prismatic joint there are only two design parameters since only the direction, and not the location of the axis, determines the point paths. Such systems are extremely simple to design, since their basic geometry depends on either circles (for an *R* joint) or straight lines (for a *P* joint). The basic ideas for the position design of such links are described in Roth [1986].

The design of such devices for the transmission of velocities or static forces could proceed as follows. Given a desired velocity $\mathbf{v}_q$ or a desired force $\mathbf{f}_q$, at a point q in the body, it then follows that for an *R* joint, the joint axis must lie in a plane through q perpendicular to the direction of $\mathbf{v}_q$ or $\mathbf{f}_q$. This is obvious from the basic relationship for the velocity of a point rotating about an axis,

$$\mathbf{v}_q = \boldsymbol{\omega} \times \mathbf{r}, \tag{1}$$

and from the equation for the torque required at the joint axis,

$$\boldsymbol{\tau} = \mathbf{f}_q \times (-\mathbf{r}). \tag{2}$$

$\boldsymbol{\omega}$ is the angular velocity of the moving link; $\boldsymbol{\tau}$ is the actuator torque applied to the moving link—this torque is in addition to torques required to

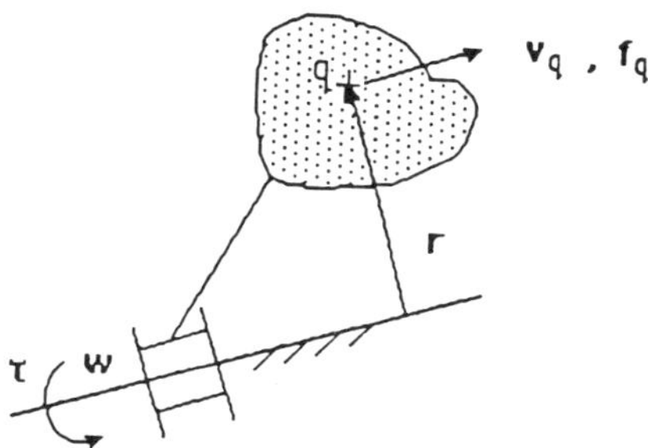

Figure 12.1

balance any gravity loadings that may exist; $\mathbf{r}$ is the position vector of q, measured from any point on the link axis to q (see figure 12.1). The R joint may be chosen any place in the plane defined by (1) or (2); i.e., the joint axis can be any line in the plane through point q normal to $\mathbf{v}_q$ or $\mathbf{f}_q$, provided only that it does not coincide with a line through point q. The magnitude of $\boldsymbol{\omega}$ or $\boldsymbol{\tau}$ depends on the axis choice. If we specify both $\mathbf{v}_q$ and $\mathbf{f}_q$ as a collinear vector pair, we are not adding any additional geometric constraints on the location of the joint axis. However, this implies a contradiction, since in determining (2) we are assuming statics. Although we might justify situations where the inertia effects are low enough to be neglected, in this chapter we shall rule out any specification of both velocity and static forces at the same design position. (There are, of course, other complications in specifying both $\mathbf{v}_q$ and $\mathbf{f}_q$; this would assume we can control the compliance of the environment, which is generally not the case.)

If the axis is a P axis, equation (1) is replaced by the condition

$$\mathbf{v}_q = \dot{\mathbf{d}}, \tag{3}$$

where $\dot{\mathbf{d}}$ is the joint velocity, and (2) is replaced by

$$\mathbf{f} = \mathbf{f}_q. \tag{4}$$

$\mathbf{f}$ is the actuator force applied to the link along the joint; this is the force in addition to any forces required to balance gravity loads. Hence, if we specify $\mathbf{v}_q$ or $\mathbf{f}_q$, (3) or (4) uniquely defines the P axis direction, and since its location is not material, this specifies all the geometry we need to know about the joint (figure 12.2).

For an R joint design it is possible to specify the velocities or forces at one position, q_1, and to also specify a second position q_2, of a point q of

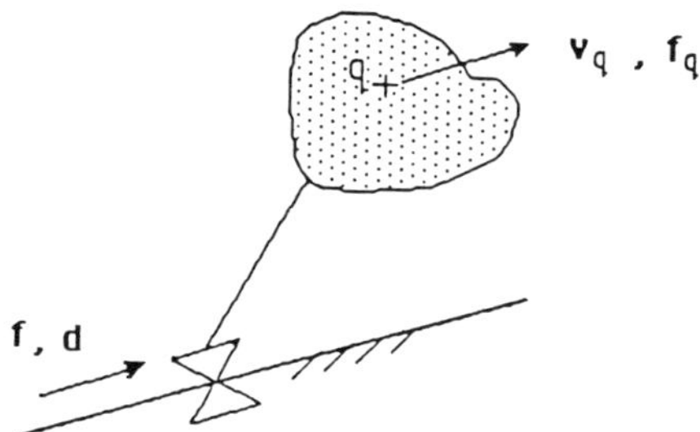

Figure 12.2

the moving link. This implies we would need one equation of the form (1) or one equation of the form (2). Since equations (1) and (2) constrain the joint axis to lie in a defined plane, it is clear that the joint axis must be at the line of intersection of such a plane and the perpendicular bisector plane of the chord q_1, q_2.

If we consider geometric kinematics, then we simply specify the instantaneous direction of the path tangent at point q, and do not associate a velocity magnitude with it. The analogous condition for forces would be the specification of the force direction at q but not its magnitude. In these cases the directional constraints on axis locations remain unaltered from those of the time-based case, but there are no longer constraints on the magnitudes of the velocity or force vectors. This means we have a free parameter in our design equations. Because of this, we can specify conditions at two separate points on the moving link while the link is in one position—say, points q and s. For the path tangent case we would have equation (1) written twice; these would represent two planes, one through q and one through s—their intersection would be the R axis. The magnitudes of $\mathbf{v}_q$, $\mathbf{v}_s$, and $\boldsymbol{\omega}$ are not considered in this case. Similar results hold from (2) in the force direction case.

12.3 Two-Degrees-of-Freedom Systems

There are eight design parameters for the RR type. Instead of equation (1) we have

$$\mathbf{v}_q = \boldsymbol{\omega} \times \mathbf{r} + \dot{\theta}_2 \mathbf{M} \times \mathbf{r}_2, \tag{5}$$

or more symmetrically,

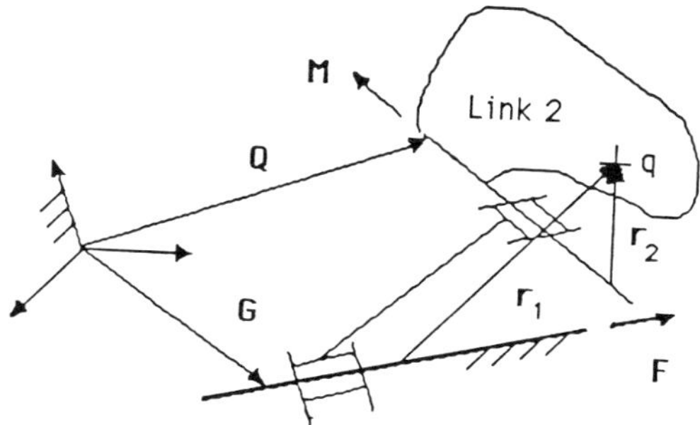

Figure 12.3

$$\mathbf{v}_q = \dot{\theta}_1 \mathbf{F} \times \mathbf{r}_1 + \dot{\theta}_2 \mathbf{M} \times \mathbf{r}_2. \tag{6}$$

Here we have used the well-known principle of carrier velocity to determine the velocity of point q, which is a point embedded in the outermost link, link 2. $\dot{\theta}_1$ and $\dot{\theta}_2$ are the rates of change of the joint motion variables; i.e., the relative angular velocities in joints 1 and 2, and $\mathbf{r}_1$ and $\mathbf{r}_2$ are position vectors of point q with respect to arbitrary reference points on axes 1 and 2, respectively. **F** and **M** are the unit direction vectors for the joint axes 1 and 2, respectively (figure 12.3). Clearly the velocity of q is also equal to the value obtained from the instantaneous screw axis of the motion of link 2 relative to ground:

$$\mathbf{v}_q = U\mathbf{U} \times \mathbf{q} + t\mathbf{U}, \tag{7}$$

where U is the angular velocity magnitude associated with the instantaneous screw axis (ISA) and t is the linear velocity magnitude of link 2 along the ISA. **q** is a position vector of point q relative to an arbitrary reference point on the ISA. **U** is the unit direction vector of the ISA.

In a previous paper [Roth, 1987], it has been shown how to derive design equations for finitely separated position specifications on point q. Those same results can be obtained from equation (6) in the case where the positions are infinitesimally separated. Alternatively, by simply taking the first derivative of the design equation given in [Roth, 1987], we obtain equations to meet up to 11 separate specifications on the velocity at point q.

For the rest of this chapter, instead of dealing with the forces or velocities associated with a single point in the last link, we shall deal with the instantaneous twists and wrenches for the entire link. It is well-known that the ISA and the revolute axes for joints 1 and 2 all have a single common perpendicular, and that the equality of (6) and (7) for every point in the second link requires that

$$U\mathbf{U} = \dot{\theta}_1\mathbf{F} + \dot{\theta}_2\mathbf{M} \tag{8}$$

and

$$t\mathbf{U} = \dot{\theta}_1\mathbf{F} \times (\mathbf{A} - \mathbf{G}) + \dot{\theta}_2\mathbf{M} \times (\mathbf{A} - \mathbf{Q}). \tag{9}$$

Here **A**, **G**, and **Q** are the position vectors from an arbitrary origin to an arbitrary point on, respectively, the ISA, the first (fixed) joint axis, and the second (moving) joint axis when it is in the reference position. **U**, **F**, and **M** are, respectively, the unit direction vectors for these three axes. t and U are the magnitudes of the linear and angular ISA velocities.

Equations (8) and (9) represent six scalar equations. Their use in kinematic design is as follows. We specify the desired velocity of link 2; this means U, **U**, t and **A** are all given. The unknowns are the angular velocities $\dot{\theta}_1$ and $\dot{\theta}_2$, the axis directions **F** and **M**, and the axis locations **G** and **Q**. These represent ten scalar unknowns (since the direction and location vectors each depend on only two independent quantities).

If we were given a second set of specifications for the velocity of link 2, we would write equations (8) and (9) for a second time. We would then have a set of twelve scalar equations, where all the terms on the left-hand sides are known and so are the values of **A** in each set of equations. We have the ten scalar unknowns from the first set and in addition two new unknown angular velocity magnitudes for the values of $\dot{\theta}_1$ and $\dot{\theta}_2$ from the second set. Hence we have a set of twelve equations and twelve unknowns. These can then be used to determine the linkage parameters that will allow a two-link *RR* configuration (figure 12.3) to produce the desired velocities of link 2.

Alternatively, the second design position could be displaced from the first. In that case the vectors **M** and **Q** in the second set of equations would be different from the **M** and **Q** in the first set. However, the two sets of **M** and **Q** are not independent. The second set is obtained from the first by a rotation transformation about axis **F** (through point **G**) by the angle $\Delta\theta_1$. If we express the second **M** and **Q** in terms of the first set we introduce only one additional unknown, $\Delta\theta_1$. If $\Delta\theta_1$ is specified, we again have a twelve by twelve system of equations; otherwise we have one free parameter since we now have thirteen unknowns.

To accomplish a geometric rather than a time-based design we need to eliminate the velocity terms from (8) and (9). Our method here is analogous to the one used ty Tsai [1972, p. 46]. Multiplying (8) by $(\mathbf{F} \times \mathbf{M})\cdot$ yields

$$(\mathbf{F} \times \mathbf{M})\cdot\mathbf{U} = 0. \tag{10}$$

Substituting in (9) for $\dot{\theta}_1$ and $\dot{\theta}_2$ (as obtained by multiplying (8) by, respectively, $\mathbf{M}\times$ and $\mathbf{F}\times$) yields

$$(\mathbf{M} \times \mathbf{F})_k p\mathbf{U} = (\mathbf{M} \times \mathbf{U})_k \mathbf{F} \times (\mathbf{A} - \mathbf{G}) - (\mathbf{F} \times \mathbf{U})_k \mathbf{M} \times (\mathbf{A} - \mathbf{Q}). \tag{11}$$

The subscript k indicates the use of one component of the vector; i.e., either the x, y, or z component can be used. p is the pitch of the ISA; $p = t/U$.

(10) and (11) are equivalent to four scalar equations. If we know the required instantaneous first order geometric motion of link 2, we know the pitch of the ISA and the vectors **U** and **A**. For purposes of design these four equations contain eight unknowns (two each for the vectors **F**, **M**, **G**, and **Q**). Thus we may specify two sets of geometric first-order requirements at a given position by giving two sets of values for p, **U**, and **A**, and using (10) and (11) written twice as a set of eight scalar equations in the eight scalar unknowns contained in **F**, **M**, **G**, and **Q**.

Alternatively, if the two first-order specifications were at different positions of link 2, we would need to express the second set of **M** and **Q** as a linear function of the first set. The argument is as developed above for the time-based motion design, and we again get $\Delta\theta_1$ as either a specified quantity or a free variable.

Now we consider the design when forces rather than velocities are specified. The duals of equations (8) and (9) are, respectively.

$$f\mathbf{U} = f_1\mathbf{F} + f_2\mathbf{M} \tag{12}$$

and

$$n\mathbf{U} = f_1\mathbf{F} \times (\mathbf{A} - \mathbf{G}) + f_2\mathbf{M} \times (\mathbf{A} - \mathbf{Q}), \tag{13}$$

where f, f_1, and f_2 are the force magnitudes along axes with directions **U**, **F**, and **M**, respectively, and **A**, **G**, and **Q** are position vectors to points on these axes. n is the torque magnitude about the axis with direction **U**. Unfortunately (12) and (13) are not valid equations for the RR series linkage (figure 12.3). These equations accurately represent the parallel $\bar{P}\bar{P}$ system shown in figure 12.4. In this chapter, we use the symbol $\bar{P}$ to represent a pure force generator composed of an $SP\bar{S}$ configuration, where the S

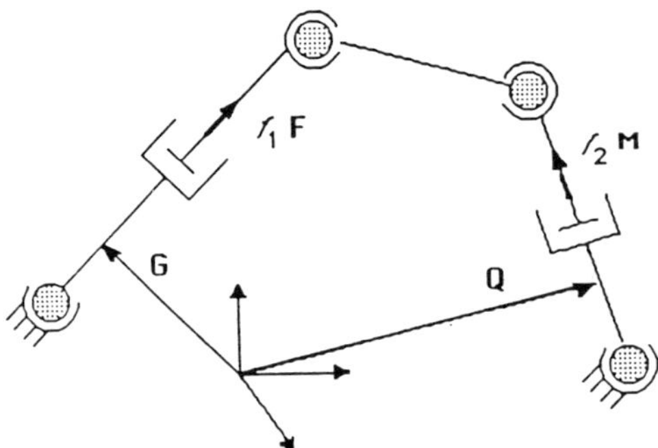

Figure 12.4

denotes a passive spherical joint and the P denotes an actively controlled prismatic joint. $\bar{S}$ denotes a passive joint that connects a crank to ground; it can be spherical, as shown in the figure, or composed of two passive perpendicular revolutes that have their axes perpendicular to the prismatic joint's direction of slide (Waldron and Hunt, 1988). For the system shown in figure 12.4, the force design can be carried out as indicated in the foregoing for the velocity case. Equations (10) and (11) also hold for this case—here $p = n/f$; using these equations allows for designs to constrain the pure geometry of force directions without regard to magnitudes. One problem with this chain is that if only the two prismatic joints are active, it has four degrees of freedom (assuming we discount the coupler rotation about the line connecting the spherical centers, and that we use two sets of orthogonal revolutes instead of spheres as the cranks' grounded joint). In general, anytime we construct a parallel chain that is a force or moment activated "dual" to a series m-degrees-of-freedom (velocity) chain, it will have $6m$ freedoms.

If we pursue the series chain of figure 12.4, the torques at the joints are given by

$$t_2 = \mathbf{M}\cdot(n\mathbf{U} + (\mathbf{A} - \mathbf{Q}) \times f\mathbf{U}) \tag{14}$$

and

$$t_1 = \mathbf{F}\cdot(n\mathbf{U} + (\mathbf{A} - \mathbf{G}) \times f\mathbf{U}). \tag{15}$$

Here t_1 and t_2 are, respectively, the torque magnitudes applied to joint axes 1 and 2. n and f are the torque and force magnitudes applied by link 2 to the environment; these are taken in the form of a wrench acting on an ISA with direction **U** through point **A**.

In order to use (14) and (15) for design we need to specify the wrench applied by link 2 (i.e., n, f, **U**, and **A**) and also the joint torques. Under these circumstances these two scalar equations contain the eight scalar unknowns that define the chain. Hence, if we define four different wrenches and four corresponding sets of joint torques—these torques could, if we wished, all be equal—we would have four sets of two equations. The system of eight equations could then be solved for the eight unknowns.

Instead of specifying all the wrenches at one design position, we can choose different positions. This simply requires that **M** and **Q** in (14) be written as linear functions of their values in the original position. We would proceed as in the velocity case, and the result would be the introduction of a rotation angle $\Delta\theta_1$ for each new design position. If these angles are specified, we can choose at most four positions since we must remain with eight equations in eight unknowns. However, even if they are left as design parameters, if we specify t_1 we can still only write the set (15) four times.

12.4 Systems of Three Degrees of Freedom and More

By analogy to (9) and (10) we have for a series chain with $j + 1$ revolute axes

$$U\mathbf{U} = \dot{\theta}_1 \mathbf{F} + \sum_1^j (\dot{\theta}_{i+1} \mathbf{M}_i) \tag{16}$$

and

$$t\mathbf{U} = \dot{\theta}_1 \mathbf{F} \times (\mathbf{A} - \mathbf{G}) + \sum_1^j (\dot{\theta}_{i+1} \mathbf{M}_i \times (\mathbf{A} - \mathbf{Q}_i)), \tag{17}$$

where the ISA, with axis direction **U** through the point with position vector **A**, is determined by the first order of the motion of link $j + 1$ relative to ground. t is the linear velocity magnitude and U is the angular velocity magnitude of this motion. **F** is the direction of the fixed axis; it passes through the point with position vector **G**. $\mathbf{M}_i$ and $\mathbf{Q}_i$ are, respectively, the direction and position vector for the $i + 1$ axis; the joint rotation rate about this axis is $\dot{\theta}_{i+1}$.

Since the left-hand sides and $\mathbf{A}$ are considered known, (16) and (17) can be regarded as six scalar equations in terms of $5(j+1)$ unknowns. Each time we specify a new instantaneous twist for link $j+1$, we can write (16) and (17) with a new left-hand side and a new $\mathbf{A}$. If these are all at the same linkage position, each new set of equations contains exactly the same axes directions and locations, and only the joint velocities, $\dot{\theta}_i$, change. Thus from k sets of (16) and (17) we have $6k$ scalar equations in terms of $(4+k)(j+1)$ scalar unknowns. If we require that we exactly balance the number of equations and unknowns, then

$$6k = (4+k)(j+1). \tag{18}$$

The RR discussed earlier corresponds to $j=1$ and $k=2$. For an RRR, $j=2$ and therefore $k=4$. For an $RRRR$, $j=3$ and therefore $k=8$. For a five-link chain ($RRRRR$), $j=4$ and therefore $k=20$. For six or more links we always have more unknowns than equations, which is as expected since such systems can in general move their last links to produce any desired instantaneous screw.

So we see that we can design two-, three-, four-, and five-link chains to match, respectively, two, four, eight and twenty specified different instantaneous end-link velocities. If any of the revolute joints are replaced by prismatic joints, these numbers are decreased. The analysis is similar, except that in equation (16) there does not appear an $\mathbf{M}_1$ (or $\mathbf{F}$) term if the axis is prismatic, while in (17) the corresponding term is replaced by simply $\dot{d}_{i+1}\mathbf{M}_i$ (or $\dot{d}_1\mathbf{F}$)—which is to say that the prismatic joint axis location does not appear and we are left with two less unknowns than when the same axis is a revolute. The result is that equation (18) needs to be modified to

$$6k = (4+k)(j+1) - 2P_n, \tag{19}$$

where P_n is the total number of prismatic joints.

If all the velocity design specifications are not for the same position of the linkage chain, the moving axes need to have their vectors $(\mathbf{M}_i, \mathbf{Q}_i)$ transformed to the desired design positions. These vectors can always be expressed as linear functions of their values in a reference position. To do this we simply consider the chain to be in some reference position; all the other positions of $\mathbf{M}_i$, $\mathbf{Q}_i$ can be obtained by rotations about joint axes on the inboard side of the chain. The rotation angles for such linear transformations, $\Delta\theta_i$, can either be specified or left as variable parameters.

12.5 Synthesis for First-Order Geometric Design Specifications

To accomplish geometric, rather than time-based design, we need to eliminate the velocities from (16) and (17).

For a three-degrees-of-freedom RRR system, we can proceed as follows. Multiplying (16) by $(\mathbf{F} \times \mathbf{M}_1)\cdot$, by $(\mathbf{F} \times \mathbf{M}_2)\cdot$, and by $(\mathbf{M}_1 \times \mathbf{M}_2)\cdot$, we obtain, respectively,

$$\begin{aligned}
\dot{\theta}_3 &= U(\mathbf{F} \times \mathbf{M}_1)\cdot \mathbf{U}/[(\mathbf{F} \times \mathbf{M}_1)\cdot \mathbf{M}_2],\\
\dot{\theta}_2 &= U(\mathbf{F} \times \mathbf{M}_2)\cdot \mathbf{U}/[(\mathbf{F} \times \mathbf{M}_2)\cdot \mathbf{M}_1],\\
\dot{\theta}_1 &= U(\mathbf{M}_1 \times \mathbf{M}_2)\cdot \mathbf{U}/[(\mathbf{M}_1 \times \mathbf{M}_2)\cdot \mathbf{F}].
\end{aligned} \tag{20}$$

Substituting these into (17) yields

$$\begin{aligned}
p\mathbf{U} = {} & [(\mathbf{M}_1 \times \mathbf{M}_2)\cdot \mathbf{U}/[(\mathbf{M}_1 \times \mathbf{M}_2)\cdot \mathbf{F}]]\mathbf{F} \times (\mathbf{A} - \mathbf{G})\\
& + [(\mathbf{F} \times \mathbf{M}_2)\cdot \mathbf{U}/[(\mathbf{F} \times \mathbf{M}_2)\cdot \mathbf{M}_1]]\mathbf{M}_1 \times (\mathbf{A} \times \mathbf{Q}_1)\\
& + [(\mathbf{F} \times \mathbf{M}_1)\cdot \mathbf{U}/[(\mathbf{F} \times \mathbf{M}_1)\cdot \mathbf{M}_2]]\mathbf{M}_2 \times (\mathbf{A} - \mathbf{Q}_2).
\end{aligned} \tag{21}$$

Equation (21) is our design equation. It represents three scalar equations, with p, $\mathbf{U}$ and $\mathbf{A}$ known; they contain twelve scalar unknowns. Hence, we can specify up to four arbitrary ISAs (i.e., four different sets of p, $\mathbf{U}$, $\mathbf{A}$). This would result in (21) written four times, and yield a set of twelve scalar equations in the twelve unknowns.

If instead of all three joints being revolutes, one or more is prismatic, for each prismatic joint the corresponding angular velocity term in (16) is missing, and therefore the numerator of the corresponding equation in (20) must be equal to zero. Thus, for example, if axis $\mathbf{M}_1$ is prismatic, we obtain from the $\dot{\theta}_2$ equation in (20)

$$(\mathbf{F} \times \mathbf{M}_2)\cdot \mathbf{U} = 0. \tag{22}$$

Furthermore, remembering that for each prismatic joint (17) would have a linear rather than angular velocity term, we solve for the linear velocities from part of (17) and substitute these values back into the other, as yet unused, equations from (17). For example, if $\mathbf{M}_1$ is prismatic (i.e., the linkage is an *RPR*), the first equation from (17) would yield for $\dot{d}_2$, the translational velocity in the second joint,

$$\dot{d}_2 = [t\mathbf{U} - \dot{\theta}_1 \mathbf{F} \times (\mathbf{A} - \mathbf{G}) - \dot{\theta}_3 \mathbf{M}_2 \times (\mathbf{A} - \mathbf{Q}_2)]_x/\mathbf{M}_{1x}, \tag{23}$$

where the subscript x denotes the x component of the vector. When substituted together with (20) into the equations for the y and z components of (17), this yields two scalar design equations. Taken together with (22) we now again have a set of three design equations for each ISA specification. Now, however, we only have ten scalar unknowns (since $\mathbf{Q}_1$ no longer matters), and we are at liberty to specify only three arbitrary ISAs. The result is then nine equations in terms of ten unknowns, so we have one free parameter. In a similar way we could analyze the cases with two and three prismatic joints.

For the four-revolute chain, RRRR, we can proceed as we did above in obtaining equation (20), but instead of (20) we shall now have

$$\begin{aligned}
\theta_3 &= \{U(\mathbf{F} \times \mathbf{M}_1)\cdot\mathbf{U} - \theta_4(\mathbf{F} \times \mathbf{M}_1)\cdot\mathbf{M}_3\}/[(\mathbf{F} \times \mathbf{M}_1)\cdot\mathbf{M}_2],\\
\theta_2 &= \{U(\mathbf{F} \times \mathbf{M}_2)\cdot\mathbf{U} - \theta_4(\mathbf{F} \times \mathbf{M}_2)\cdot\mathbf{M}_3\}/[(\mathbf{F} \times \mathbf{M}_2)\cdot\mathbf{M}_1],\\
\theta_1 &= \{U(\mathbf{M}_1 \times \mathbf{M}_2)\cdot\mathbf{U} - \theta_4(\mathbf{M}_1 \times \mathbf{M}_2)\cdot\mathbf{M}_3\}/[(\mathbf{M}_1 \times \mathbf{M}_2)\cdot\mathbf{F}].
\end{aligned} \tag{24}$$

Substituting these into (17) yields

$$\begin{aligned}
p\mathbf{U} = {} & \{[(\mathbf{M}_1 \times \mathbf{M}_2)\cdot\mathbf{U} - \theta_4/U(\mathbf{M}_1 \times \mathbf{M}_2)\cdot\mathbf{M}_3]\\
& /[(\mathbf{M}_1 \times \mathbf{M}_2)\cdot\mathbf{F}]\}\mathbf{F} \times (\mathbf{A} - \mathbf{G}) + \{[(\mathbf{F} \times \mathbf{M}_2)\cdot\mathbf{U} - \theta_4\\
& /U(\mathbf{F} \times \mathbf{M}_2)\cdot\mathbf{M}_3]/[(\mathbf{F} \times \mathbf{M}_2)\cdot\mathbf{M}_1]\}\mathbf{M}_1 \times (\mathbf{A} - \mathbf{Q}_1)\\
& + \{[(\mathbf{F} \times \mathbf{M}_1)\cdot\mathbf{U} - \theta_4/U(\mathbf{F} \times \mathbf{M}_1)\cdot\mathbf{M}_3]\\
& /[(\mathbf{F} \times \mathbf{M}_1)\cdot\mathbf{M}_2]\}\mathbf{M}_2 \times (\mathbf{A} - \mathbf{Q}_2).
\end{aligned} \tag{25}$$

Equation (25) can be written as three scalar equations. From the first of these we solve for the scalar ratio θ_4/U and substitute this value into the remaining two equations. These then are our design equations. For each specification of an ISA (i.e., p, $\mathbf{U}$, and $\mathbf{A}$) these two scalar equations contain the same sixteen unknowns (two each for $\mathbf{F}$, $\mathbf{M}_1$, $\mathbf{M}_2$, $\mathbf{M}_3$, $\mathbf{G}$, $\mathbf{Q}_1$, $\mathbf{Q}_2$, and $\mathbf{Q}_3$.) Therefore we can specify up to eight ISAs; at eight we would have a balance of sixteen equations and sixteen unknowns.

If one or more of the joints are prismatic, those joints' angular velocities must be set to zero in (24). This means that if there is one prismatic joint, all the joint angular velocities can be expressed in terms of the joint directions by (24). If there is more than one prismatic joint, (24) gives us one additional geometric design equation for each prismatic joint beyond the first. The linear velocities of the prismatic joints can be determined from (25) (after it has been appropriately modified for prismatic joints). This is accomplished by writing the modified equation (25) as three scalar equations and then solving for the prismatic joints' linear velocity from the first one or two (depending upon if we have one or two prismatic joints) of these. Substituting the expressions for these linear velocities into the remaining (one or two) scalar equations from the modified equation (25) yields the design equations. Since we lose two design parameters for each prismatic substitution for a revolute, the maximum number of possible ISA specifications will be less than (the eight) for the four-revolute case. We have seven when there is one prismatic joint and six when there are two prismatic joints.

For five revolute joints, an RRRRR, we again proceed as we did in obtaining equation (20), but instead of (20) we obtain

$$\begin{aligned}
\theta_3 &= \{U(\mathbf{F} \times \mathbf{M}_1)\cdot\mathbf{U} - \theta_4(\mathbf{F} \times \mathbf{M}_1)\cdot\mathbf{M}_3 - \theta_5(\mathbf{F} \times \mathbf{M}_1)\cdot\mathbf{M}_4\} \\
&\quad /[(\mathbf{F} \times \mathbf{M}_1)\cdot\mathbf{M}_2], \\
\theta_2 &= \{U(\mathbf{F} \times \mathbf{M}_2)\cdot\mathbf{U} - \theta_4(\mathbf{F} \times \mathbf{M}_2)\cdot\mathbf{M}_3 - \theta_5(\mathbf{F} \times \mathbf{M}_2)\cdot\mathbf{M}_4\} \\
&\quad /[(\mathbf{F} \times \mathbf{M}_2)\cdot\mathbf{M}_1], \\
\theta_1 &= \{U(\mathbf{M}_1 \times \mathbf{M}_2)\cdot\mathbf{U} - \theta_4(\mathbf{M}_1 \times \mathbf{M}_2)\cdot\mathbf{M}_3 - \theta_5(\mathbf{M}_1 \times \mathbf{M}_2)\cdot\mathbf{M}_4\} \\
&\quad /[(\mathbf{M}_1 \times \mathbf{M}_2)\cdot\mathbf{F}].
\end{aligned} \tag{26}$$

Substituting these into (17) yields

$$\begin{aligned}
p\mathbf{U} &= \{[\mathbf{M}_1 \times \mathbf{M}_2)\cdot\mathbf{U} - \theta_4/U(\mathbf{M}_1 \times \mathbf{M}_2)\cdot\mathbf{M}_3 - \theta_5/U(\mathbf{M}_1 \times \mathbf{M}_2)\cdot\mathbf{M}_4] \\
&\quad /[(\mathbf{M}_1 \times \mathbf{M}_2)\cdot\mathbf{F}]\}\mathbf{F} \times (\mathbf{A} - \mathbf{G}) + \{[\mathbf{F} \times \mathbf{M}_2]\cdot\mathbf{U} - \theta_4 \\
&\quad /U(\mathbf{F} \times \mathbf{M}_2)\cdot\mathbf{M}_3 - \theta_5/U(\mathbf{F} \times \mathbf{M}_2)\cdot\mathbf{M}_4] \\
&\quad /[(\mathbf{F} \times \mathbf{M}_2)\cdot\mathbf{M}_1]\}\mathbf{M}_1 \times (\mathbf{A} - \mathbf{Q}_1) + \{[(\mathbf{F} \times \mathbf{M}_1)\cdot\mathbf{U} - \theta_4 \\
&\quad /U(\mathbf{F} \times \mathbf{M}_1)\cdot\mathbf{M}_3 - \theta_5/U(\mathbf{F} \times \mathbf{M}_1)\cdot\mathbf{M}_4] \\
&\quad /[(\mathbf{F} \times \mathbf{M}_1)\cdot\mathbf{M}_2]\}\mathbf{M}_2 \times (\mathbf{A} - \mathbf{Q}_2),
\end{aligned} \tag{27}$$

writing (27) as three scalar equations and solving the first two for the scalar ratios $\dot{\theta}_4/U$ and $\dot{\theta}_5/U$. These expressions can be substituted into the third equation to yield a single scalar design equation in terms of the 20 geometric design variables associated with the five revolute axes. Since we have only one scalar design equation, it is clear we can specify as many as 20 different ISAs. If we have one or more prismatic joints we use (26) and (27) in a manner analogous to that outlined for (24) and (25) for the case of four axes.

As expected, if we apply this method to six or more revolute joints, in a series chain, we find that there are no design equations analogous to (27). This implies that an unlimited number of specified ISAs may be attained by the last link—which is of course why six-degrees-of-freedom manipulators are so common.

If, in any of the foregoing cases, one or more of the ISAs refer to the motion of the last link when it is in a position other than the original reference position, all the moving-link axes need to be expressed as linear functions of their values in the reference position. This requires the introduction of the parameters $\Delta\theta_i$ as discussed in the time-based cases analyzed earlier in this chapter.

12.6 Synthesis for Force and Moment Design Specifications

For a force-based design, equation (21) represents the design equation for a parallel $\bar{P}\bar{P}\bar{P}$ configuration, equation (25) represents the design equation for a parallel $\bar{P}\bar{P}\bar{P}\bar{P}$ configuration, and (27) represents a $\bar{P}\bar{P}\bar{P}\bar{P}\bar{P}$ configuration of the type shown in figure 12.5 (We remind the reader that each crank's grounded spherical joint can be replaced by two orthogonal revolute joints in a plane perpendicular to the prismatic joint's line of slide.) If we

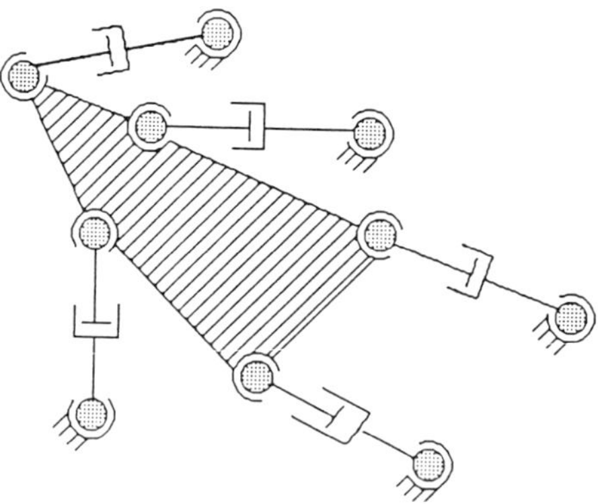

Figure 12.5

introduce prismatic joints in the series chain, then the analogous parallel chain has pure moment-generating revolutes as duals to these joints. So, for example, our discussion of the series chain *RPR* yields exactly the design equations for the parallel PRP shown in figure 12.6. The symbol *R* denotes a pure moment generator composed of one active revolute joint and one passive prismatic and two passive cylindrical joints (Waldron and Hunt, 1988). As shown in figure 12.6, the prismatic joint is in-line with the revolute and the cylindrical joints are normal to it.

For force based design of series revolute jointed chains we need to use equations analogous to (14) and (15). For a series chain, with j revolute joints, the torques at the joints are given by

$$t_{i+1} = \mathbf{M}_i \cdot (n\mathbf{U} + (\mathbf{A} - \mathbf{Q}_i) \times f\mathbf{U}), \qquad i = 1, 2, \ldots, j-1, \tag{28}$$

and

$$t_1 = \mathbf{F} \cdot (n\mathbf{U} + (\mathbf{A} - \mathbf{G}) \times f\mathbf{U}), \tag{29}$$

where t_1 and t_{i+1} are, respectively, the torques applied to joint axes 1 and $i + 1$. n and f are the torque and force on the last link (the link that rotates about axis j) due to its contact with the environment; these are taken in the form of a wrench with intensity f acting on an ISA with direction $\mathbf{U}$ through point $\mathbf{A}$. The pitch of the ISA is p, and therefore the moment is $fp = n$.

In order to use (28) and (29) for design we need to specify the wrench applied by the environment on the last link (i.e., n, p, $\mathbf{U}$, and $\mathbf{A}$) and also the joint torques. Under these circumstances, these j scalar equations contain as unknowns the $4j$ scalars that define the chain. Hence if we define

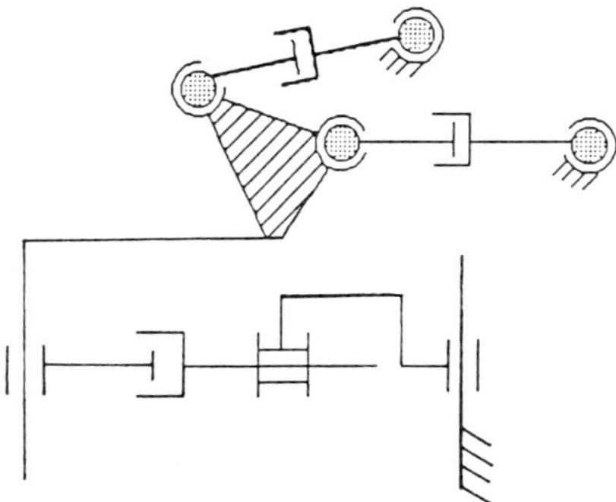

Figure 12.6

k different wrenches and k corresponding sets of joint torques—these torques could if we wished all be equal—we would have kj equations in terms of $4j$ unknowns. This means that the balance between unknowns and equations is independent of the number of joints and always occurs at $k = 4$.

Instead of specifying all the wrenches at one design position, we can choose different positions. This simply requires that $\mathbf{M}_i$ and $\mathbf{Q}_i$ in (28) be written as linear functions of their values in the original position. We would proceed as in the velocity case, and the result would be the introduction of a set of rotation angles $\Delta\theta_i$ $(i = 1, 2, \ldots, j - 1)$ for each new design position. If these angles are specified, we can choose at most four positions since we must remain with at most $4j$ equations in the $4j$ unknowns. However, even if the rotation angles are left as design parameters, if all t_1 are specified, we still can only write the set (29) four times.

If any of the joints is prismatic, the corresponding equations in (28) or (29) need to be replaced by an equation of the form

$$f_i = f\mathbf{U} \cdot \mathbf{M}_i.$$

The result is that there are the same number of equations but two less unknowns than when we had a revolute. Hence prismatic joints reduce the number of arbitrary wrenches that can be specified.

12.7 Conclusions

This chapter has outlined the use of constraint equations for determining the dimensions of the types of structures commonly used in manipulators, dexterous hands, and walking machines. The approach here has been to derive simple sets of equations that express the relationships between the entities that describe the desired capabilities of the devices—such as a point's velocity, a body's velocity, or a set of applied forces—and the structural dimensions that need to be determined.

This chapter is part of a set of studies that involve the basic kinematics and mechanics of robotic devices. Our overall aim is to study the basic kinematic properties of multi-degree-of-freedom systems. This paper deals mainly with the dimensional design of such devices, and as such is most directly related to Roth [1986] and Roth [1987]. Some of our recent works on instantaneous properties may also be of interest to the reader since they also deal with the force and velocity aspects of serial link chains. These treat general motions [Ghosal and Roth, 1987a, b], direct contact between bodies [Cai and Roth, 1987a, b], and the special problems of dexterous hands [Ji and Roth, 1987]. Finally, we mention that hardware development is currently underway to assist in the understanding and study oı the properties of force controlled systems.

Bibliography

Cai, C.-S., and Roth, B. 1987. "On the planar motion of rigid bodies with point contact," *Mechanism and Machine Theory*, Vol. 21, pp. 453–465.

Cai, C.-S., and Roth, B. 1987. "On the spatial motion of a rigid body with point contact," *Proceedings of IEEE International Conference on Robotics and Automation*, pp. 686–695.

Ghosal, A., and Roth, B. 1987. "Instantaneous properties of multi-degrees-of-freedom motions-point trajectories," *Trans. ASME, Journal of Mechanisms, Transmissions, and Automation in Design*, Vol. 109, pp. 107–115.

Ghosal, A., and Roth, B. 1987. "Instantaneous properties of multi-degrees-of-freedom motions-line trajectories", *ibid.*, pp. 116–125.

Ji, Z., and Roth, B. 1987. "Contact force in grasping and kinematic constraints," *Proceedings Seventh IFToMM World Congress, The Theory of Machines and Mechanisms*, Pergamon Press, pp. 1219–1222.

Roth, B. 1986. "Analytic design of open chains," *Proceedings of the Third International Symposium on Robotics Research*, MIT Press, pp. 281–288.

Roth, B. 1987. "Analytic design of two-revolute open chains," *Proceedings of the Sixth CISM-IFToMM Symposium on Theory and Practice of Robots and Manipulators*, MIT Press.

Tsai, L.-W. 1972. *Design of Open Loop Chains for Rigid Body Guidance*, Ph.D. Dissertation, Mech. Eng. Dept., Stanford University.

Waldron, J. K., and Hunt, K. H. 1988. Series—parallel dualities in actively coordinated mechanisms," *Proceedings of the Fourth International Symposium on Robotics Research*, MIT Press.

13 Arm Design

Haruhiko Asada

13.1 Introduction

A majority of today's manipulator arms are open-kinematic chains in mechanical construction. They are basically composed of cantilevered beams connected by hinged joints. This unique construction allows the robot to work in a large reachable range, access crowded places, and perform dexterous motions, all of which are desired features for manipulator arms. Compared with an x-y table, for example, which consists of all pristimatic joints, the manipulator arm is capable of highly flexible motions that result from the unique mechanical construction [1].

The manipulator arm design significantly departs from traditional machine design in many ways. The mechanical stiffness and accuracy of the arm structure are inherently poor, while traditional machines such as machine tools are designed for maximum stiffness and precision. In the serial sequence of servoed joints, errors accumulate quickly along the arm linkage. As the length of arm links gets longer, even a small error at the shoulder joint results in a significantly large error at the arm tip. The stiffness of the arm structure decreases rapidly when the arm length gets longer. In order to exploit fully the high mobility and dexterity features, these inherent difficulties to the articulated joint structure must be overcome by advanced design techniques.

In this chapter we look at some of the efficient design methods recently developed. The unique construction of manipulator arms needs powerful tools for design. Kinematic and dynamic behaviors are highly complex and nonlinear. Efficient methods for analyzing the mechanical behavior and evaluating arm performances are necessary in the arm design.

13.2 Kinematic Design Considerations

A manipulator arm performs three-dimensional motions in space. Spatial kinematics of such multi-degrees-of-freedom mechanisms are intricate, creating unique problems in mechanism design.

13.2.1 Workspace

B. Roth [2] first addressed the workspace problem: to find the reachable range of a manipulator arm, provided a kinematic structure of the arm linkage. Since manipulator arms are required to possess a large workspace, the volume of the workspace is an important performance index in the arm design. It is, however, not a simple problem to identify the workspace and to compute the volume, since the arm kinematics are highly nonlinear and complex. Efficient methods to obtain or evaluate the workspace were then addressed by Shimano [3], Sugimoto and Duffy [4], and Kumar and Waldron [5]. The synthesis of arm linkage was also addressed on the basis of the workspace evaluation. The problem is to find an optimal kinematic structure that maximizes the workspace volume or other indices on workspace evaluations. Gupta and Roth[6], for example, have addressed the synthesis of arm linkage when the dexterous workspace [5] is specified.

13.2.2 Solvability

To locate an end effector at a specified position and orientation, it is necessary to obtain joint displacements of the manipulator arm that lead the end effector to the specified location. To this end, kinematic equations must be solved for joint displacements. This problem is referred to as the inverse kinematics problem—one of the most fundamental problems in robot mechanics. The issue in the inverse kinematics is that closed-form solutions to the kinematic equation do not always exist—the equation is not solvable in general [1]. As a result, one needs to apply numerical methods to solving the nonlinear simultaneous equations, which involve heavy computation. A way of avoiding the heavy computation is to design the kinematic structure of the arm so that there exist closed-form analytic solutions. Pieper [7] first addressed the solvability of the kinematic equations and has found a sufficient condition for the kinematic structure to be solvable: closed-form solutions exist if three axes of revolute joints intersect at a single point for all arm configurations.

Recently, efficient numerical methods for solving the inverse kinematics were developed. In consequence, the kinematic solvability is now less significant. Nevertheless, all of the commercialized robots on the market still possess solvable structures, which significantly reduce computational complexity.

13.2.3 Singularity

Robot manipulators may have a singular point at which the end effector cannot be moved in a particular direction. In the vicinity of the singular point, the mobility of the arm linkage becomes extremely poor as well. It is therefore a critical issue in the arm design to avoid singularity and achieve high mobility inside the workspace. Paul and Stevenson [8] analyzed singularity for 3-dof wrist joints, and Hollerbach [9] addressed the design of 7-dof manipulators, in which a redundant degree of freedom is added in order to eliminate singular points inside the workspace.

13.2.4 Kinematics Evaluations Using Jacobian Matrices

Singularity is a problem associated with differential motions of a manipulator arm, which are described by the differential form

$$d\mathbf{p} = \mathbf{J}\, d\mathbf{q}, \tag{1}$$

where $\mathbf{J}$ is referred to as the Jacobian matrix associated with the kinematic relation between joint displacements $\mathbf{q}$ and the end effector position and orientation $\mathbf{p}$. The Jacobian matrix also relates the force and moment to be exerted at the end effector to the joint torques τ that are generated by actuators in the form

$$\tau = \mathbf{J}^T \mathbf{F}, \tag{2}$$

where $\mathbf{F}$ represents, collectively, the force and moment at the end effector [1].

Investigations about the Jacobian matrix provide useful information for the arm design. Many researchers, including Uchiyama et al. [10], Yoshikawa [11], Asada and Youcef-Toumi [12], [13], Salisbury and Craig [14], and Asada and Cro Granito [15] have addressed the kinematic and static evaluations based upon the Jacobian matrix. A variety of indices characterizing the Jacobian matrix have been proposed. The determinant of the matrix, det $\mathbf{J}$, is a measure to evaluate the mobility of the manipulator arm as well as to indicate how close the arm configuration is to a singular point. Singular values of the Jacobian matrix and eigenvalues of the squared matrix $\mathbf{J}^T\mathbf{J}$ represent the mobility in each direction of principal axes. The condition number of the Jacobian matrix indicates the isotropy of the mobility performance. The instantaneous kinematics and statics evaluations have been utilized in arm linkage design, as we discuss later in this chapter.

13.3 Arm Design Using Generalized Inertia Ellipsoids

Dynamic behavior of a manipulator arm is complex and intricate, as is the kinematic behavior. Multiple joints are dynamically coupled, having interaction torques acting on each other. Arm inertia reflected to each joint varies widely depending on the arm configuration. Coriolis and centrifugal effects are prominent as the speed of the arm motion increases. Thus the dynamics complexity creates unique problems in robots.

Advanced control techniques have been applied in order to overcome the complex dynamics and improve control performance, as discussed in control chapters. The techniques we address in this chapter are by mechanism design rather than by control. Arm design can be an alternative approach to the improvement of control performances. Here, the mechanical construction is to be devised along with the controller so that the total control performance may be further improved. For the improved hardware, the controller can be more effective. Followings this section we discuss arm design issues based on dynamics considerations. To aid the designer of a manipulator arm, analytic tools of the manipulator dynamics need be provided. It is critically important for the designer to comprehend dynamic behavior and evaluate dynamic performance, which are reflected in the total control performance.

The dynamics of a manipulator arm are primarily dependent upon mass properties of each member involved in the arm linkage as well as its kinematic structure. The inertia matrix associated with equations of motion reflect both mass properties and the kinematic structure collectively, and governs the dynamic behavior. In Lagrangian formulation, the inertia matrix defines the kinematic energy given by

$$\mathbf{T} = \tfrac{1}{2}\dot{\Theta}^T \mathbf{H}(\theta)\dot{\Theta}, \tag{3}$$

where $\mathbf{H}$ is the $n \times n$ inertia matrix represented in joint space $\Theta = [\theta_1, \ldots, \theta_n]^T$. Note that matrix $\mathbf{H}$ is space-dependent, hence a function of joint displacements. Matrix $\mathbf{H}$ is real, symmetric and, positive-definite, hence invertible.

Let $\mathbf{h}$ be momentum of the manipulator arm given by

$$\mathbf{h} = \mathbf{H}\dot{\Theta}. \tag{4}$$

Let us also consider the momentum with respect to a Cartesian coordinare system fixed in space, which is sometimes referred to as world coordinates representing the end effector motion. Denoting the momentum reflected to the end effector by $\mathbf{p}$, we can derive the coordinate transformation of the momenta so that

$$\mathbf{h} = \mathbf{J}^T \mathbf{p} \tag{5}$$

where $\mathbf{J}$ is the Jacobian matrix. Substituting (4) and (5) into (3), we can derive another expression for the kinetic energy:

$$\mathbf{T} = \tfrac{1}{2}\mathbf{p}^T\hat{\mathbf{G}}\mathbf{p}, \tag{6}$$

where

$$\hat{\mathbf{G}} = \mathbf{J}\mathbf{H}^{-1}\mathbf{J}^T. \tag{7}$$

The kinetic energy expression given by (6) is sometimes referred to as the kinematic coenergy [16], which is the same as the original energy if the momentum is a linear function of velocities. Note, however, that the matrix $\hat{\mathbf{G}}$ is 6×6 rather than $n \times n$, representing the inertial characteristics reflected in the end effector motion. We call $\hat{\mathbf{G}}$ the inverse inertia matrix in the end effector space. Since $\mathbf{H}$ is invertible, the inverse inertia matrix $\hat{\mathbf{G}}$ always exist no matter what the dimension and singularity of the Jacobian matrix are.

A geometrical interpretation has been given to the inertia matrices derived above. The characteristics of manipulator dynamics are thus visualized with use of the geometrical representation, which provides useful information on the arm design in evaluating the performance.

Associated with the inertia matrix $\mathbf{H}$, there is a quadratic form and a quadratic surface defined by

$$\mathbf{u}^T\mathbf{H}\mathbf{u} = 1. \tag{8}$$

Since the matrix $\mathbf{H}$ is positive-definite, the above quadratic surface is that of an ellipsoid. In the case of a single rigid body, the ellipsoid is known as the inertia ellipsoid [17]. We have extended the inertia ellipsoid of a single rigid body to the one for a collection of multiple bodies. The new ellipsoid is thus referred to as Generalized Inertia Ellipsoid (GIE) [18].

Similarly, we can define an ellipsoid associated with the inverse inertia matrix $\hat{\mathbf{G}}$, which is interpreted as the effective inertia reflected to the end point. We call the GIE for the inverse inertia matrix the inverse GIE. If $\hat{\mathbf{G}}$ is nonsingular, then the original GIE as well as the inverse GIE can be defined.

The GIE has principal axes along which the inertia matrix is diagonal. The principal axes of the GIE are aligned with the eigenvectors of the matrix, and the length of each principal axis is the reciprocal of the square root of the corresponding eigenvalue. Figure 13.1 shows an example, in which the GIE of a 2-dof manipulator is illustrated in space. In most cases, we are interested in the motion of an end effector mounted at the tip of the

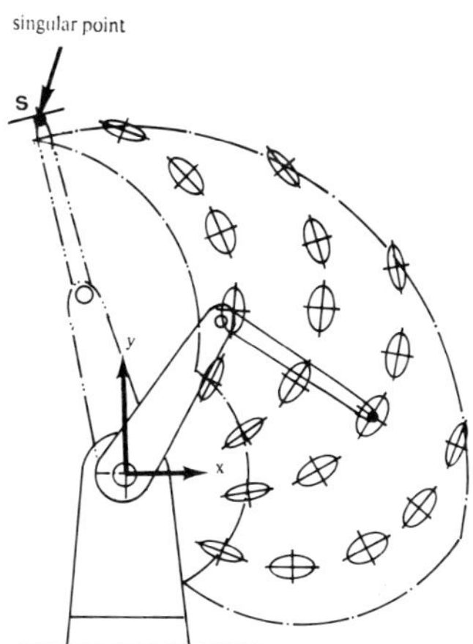

Figure 13.1
Generalized force-speed characteristics of multiple actuators.

arm. Therefore we investigate the manipulator dynamics with respect to the tip motion being referred to a Cartesian coordinate system fixed in space. Computing the eigenvalues and eigenvectors of the matrix, we draw a GIE locating its center at the location of the arm tip in space, as shown in figure 13.1 Unlike the inertia ellipsoid of a single rigid body, the GIE varies its configuration depending on the location in the generalized coordinate system. We draw the GIE at each point in a reachable region so that the global characteristics of the arm inertia can be represented in the whole.

Since the largest eigenvalue of the inertia matrix corresponds to the minor axis of the GIE, the moment of inertia is maximum along that axis. For motions with the same kinetic energy the velocity is minimum if it is in the direction of the minor axis. On the other hand, the moment of inertia in the direction of the major axis is the smallest; therefore, the speed is fastest in that direction.

If the lengths of the principal axes are the same—that is, the GIE is a pure sphere—the resultant inertia is isotropic. The difference between the lengths of the major and minor axes stands for the anisotropy of the resultant inertia. The shape of the GIE, in figure 13.1, is long and narrow at the peripheries of the reachable region, and thick and round at the center. Therefore, the resultant inertia is more isotropic at the center than at the peripheries. At point S in the figure, the arm cannot move in the radial direction. The point S is known as a singular point. The GIE becomes thinner and converges to a line when the tip of the arm approaches the singular point. The width of the GIE represents the degree of singularity.

Using the GIE diagram, we can also analyze the nonlinearity of a manipulator arm. The manipulator dynamics are characterized by such nonlinear effects as Coriolis and centrifugal forces, which are caused by the space-dependence of the inertia matrix. In fact, in Lagrangian formulation, the nonlinear terms are given by partial derivatives of the inertia matrix;

$$\sum_j \sum_k \left(\frac{\partial H_{ij}}{\partial \theta_k} - \frac{1}{2} \frac{\partial H_{jk}}{\partial \theta_i} \right) \dot{\theta}_j \dot{\theta}_k. \tag{9}$$

Thus the spatial changes of the inertia matrix yield the nonlinear terms. The spatial changes have been visualized by the GIE diagram. In consequence, we can analyze the nonlinear forces by investigating the geometrical changes in the GIE configuration. Figure 13.2 shows the two types of the geometrical changes in GIE; one is in its shape and the other is in the orientation. In the former case, it can be shown that the nonlinear force is exerted along the pricipal axis when the shape changes as shown in figure 13.2a). When the principal axis rotates, as in figure (13.2b), the nonlinear

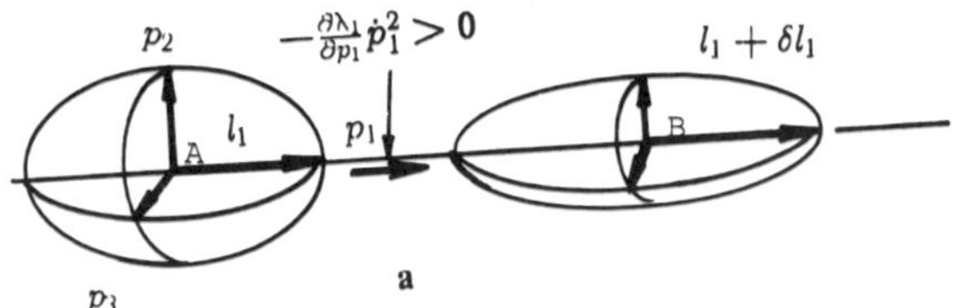

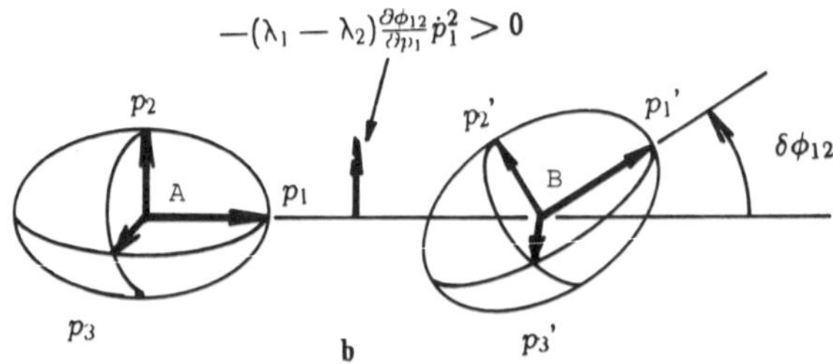

Figure 13.2
Change of GIE configuration.

force perpendicular to the principal axis is generated. Thus, observing the changes in shape and orientation, one can estimate the nonlinear forces and nonlinear effects.

The GIE diagram depicts the global characteristics of manipulator dynamics as a whole. From the graphical representation, one can understand the inertial effect and nonlinearities of mechanical arms. This technique has been applied to the design of a mechanical arm. The inertial effects and nonlinearities depicted by the GIE diagram provide useful data for designing the structure of the arm.

We discuss the arm shown in figure 13.3, where the dimensions and mass distributions are described by the following parameters:

l_1, l_2 = link lengths of bodies 1 and 2,
g_1, g_2 = distances between the centers of mass and the tips of links,
I_1, I_2 = moments of inertia, and
m_1, m_2 = mass.

Figure 13.1 shows the GIE diagram when the arm has the same link length for the upper and lower arms along with the parameters listed in the

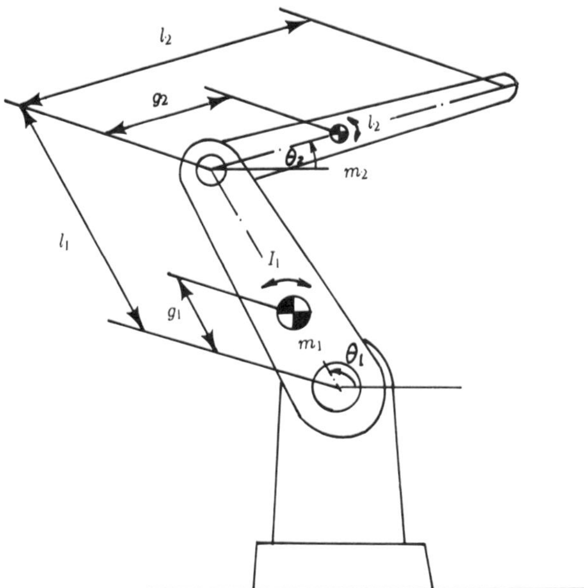

Figure 13.3
Dimensions and mass distribution of 2-dof arm.

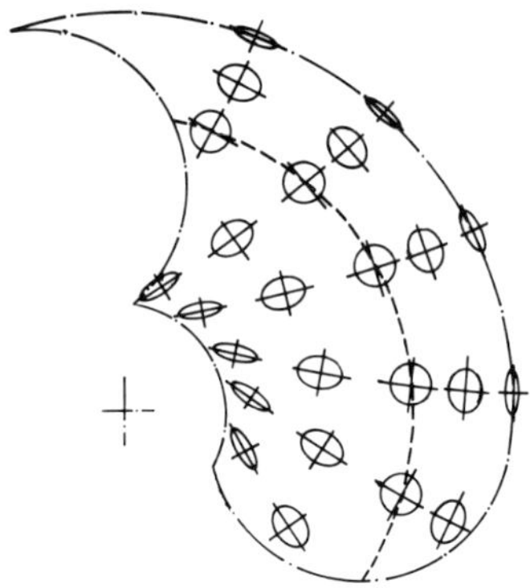

Figure 13.4
Modified GIE configuration I (isotropic on center line): $l_1 = 0.6$ m; $l_2 = 0.4$ m, $g_1 = 0$; $g_2 = 0.28$ m; $m_1 = 50$ kg; $m_2 = 30$ kg; $I_1 = 3.492$ kg m^2; $I_2 = 4$ kg m^2.

figure. From this diagram, it follows that the arm dynamics have the following problems.

1. The large difference in axial lengths between the major and minor axes shows that the moment of inertia at the tip of the arm varies significantly depending on the direction of motion.
2. Since the change of the GIE configuration, both in shape and in orientation, is significantly large, the arm dynamics involve large nonlinear forces.

By modifying the link lengths and the distribution of mass, we try to improve the arm dynamics so that the moment of inertia is uniform in any direction over a wide range of a reachable region and the nonlinear forces are reduced as well. If the GIE is a pure circle at any point in the reachable region, the arm is isotropic and has no nonlinearity. This uniform and isotropic configuration reduces the complexity in controlling the arm. Therefore the improvement of control performance can be expected.

Figure 13.4 shows a modified design, where the GIE is isotropic on the broken line on which the bending angle of the second joint is 90°. The configuration of the GIE, however, changes rapidly from the pure circles to long and narrow ellipses when the arm moves from the center to the peripheries. The rapid change of the GIE configuration leads to large nonlinear forces and unbalanced inertia characteristics. The dynamic characteristic of the arm can be improved by enlarging the region of isotropic GIE to include

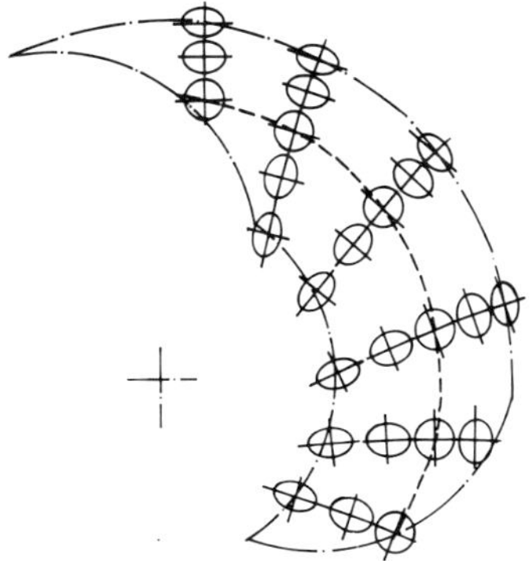

Figure 13.5
Modified GIE configuration II (isotropic and uniform): $l_1 = 0.4$ m; $l_2 = 0.6$ m; $g_1 = 0$; $g_2 = 0.42$ m; $m_1 = 50$ kg; $m_2 = 30$ kg; $I_1 = 1.552$ kg m^2; $I_2 = 9$ kg m^2.

more of the reachable region. As the angular difference of the second joint from 90° becomes larger, the GIE becomes slender. Therefore, we limit the bending angle to a smaller range where the GIE is almost isotropic. At the same time, we extend the link length of body 2 so that it covers a larger region with the small bending angle. Figure 13.5 shows the modified configuration, where the GIE is almost isotropic and uniform over the wide range of the reachable region. The improvement is noticeable.

13.4 Arm Design for Decoupled and/or Configuration Invariant Inertia

The arm design discussed in the previous section is basically a trial-and-error optimization, which is not systematic. We now consider a systematic and theoretical approach to the arm design in which the kinematic structure and mass distributions are determined. A few particular forms of desired dynamics are first presented and we develop design theory to accomplish the desired dynamics [19].

Let us write the equation of motion again,

$$\tau_i = H_{ii}\ddot{\theta}_i + \sum_{j \neq i} H_{ij}\ddot{\theta}_j + \sum_j \sum_k \left(\frac{\partial H_{ij}}{\partial \theta_k} - \frac{1}{2}\frac{\partial H_{jk}}{\partial \theta_i} \right) \dot{\theta}_j \dot{\theta}_k + \tau_{gi}, \tag{10}$$

where H_{ij} is the $i - j$ element of the inertia matrix, and τ_{gi} is the torque due to gravity. The first term on the right-hand side represents the inertia torque generated by the acceleration of the ith joint, while the second term is the interactive inertia torque caused by accelerations of the other joints. The interactive inertia torque is linearly proportional to acceleration. The third term is nonlinear velocity torques due to Coriolis and centrifugal effects.

The inertia matrix is determined by the kinematic structure of the manipulator arm and mass properties of individual links. The design problem we discuss in this section is to modify both the kinematic structure and mass properties of the arm linkage so that the inertia matrix reduces to particular forms.

Consider the inertia matrix that reduces to a diagonal matrix for an arbitrary arm configuration; then the second term in (10) vanishes and consequently no interactive inertia torques appear. The manipulator inertia matrix in this case is referred to as a decoupled inertia matrix. The significance of the decoupled inertia matrix is that the control system can be treated as a set of single-input, single-output subsystems associated with individual joint motions. The equation of motion reduces to

$$\tau_i = H_{ii}\ddot{\theta}_i + \sum_k \left(\frac{\partial H_{ii}}{\partial \theta_k} \dot{\theta}_i \dot{\theta}_k - \frac{1}{2} \frac{\partial H_{kk}}{\partial \theta_i} \dot{\theta}_k^2 \right) + \tau_{gi}, \tag{11}$$

where the second term represents the nonlinear velocity torques resulting from the spatial dependency of the diagonal elements of the inertia matrix. Note that the number of terms involved in (11) is much smaller than the original nonlinear velocity torques, because all the off-diagonal elements are zero for all $\theta_1, \ldots, \theta_n$. This reduces the computational complexity of the nonlinear torques.

Another significant form of the inertia matrix that reduces the dynamic complexity is a configuration-invariant form. The inertia matrix in this form does not vary for an arbitrary arm configuration. In other words, the matrix is independent of joint displacements; hence the third term in (10) vanishes and the equation of motion thus reduces to

$$\tau_i = H_{ii}\ddot{\theta}_i + \sum_j H_{ij}\ddot{\theta}_j + \tau_{gi}. \tag{12}$$

Note that the coefficients H_{ii} and H_{ij} are constant for all the arm configurations. Thus the equation is linear except for the last term, that is, the gravity torque. The inertia matrix in this form is referred to as an invariant inertia matrix. The significance of this form is that linear control schemes can be adopted, which are much simpler and easier to implement.

When the inertia matrix is both decoupled and configuration invariant, the equation of motion reduces to

$$\tau_i = H_{ii}\dot{\theta}_i + \tau_{gi}. \tag{13}$$

The system is completely decoupled and linearized except for the gravity term. We can treat the system as single-input, single-output systems with constant parameters.

We now discuss how the decoupled and/or configuration-invariant dynamics can be accomplished by design. The goal here is to find conditions on the kinematic structure and mass distributions that must be satisfied to achieve the decoupled and/or configuration-invariant dynamics. It is particularly important to derive necessary conditions, which will determine a group of all possible arms that meet the goal. In this section we consider open-loop manipulator arms, while in the following section some closed-loop mechanisms are to be discussed.

First we investigate the inertia matrix **H** and derive the necessary conditions for each element of **H** to be zero or invariant for all the arm configurations. To this end, let us first immobilize the last $(n - l + 1)$ links as shown in figure 13.6 and derive the *i-l* element of **H**, which accounts for

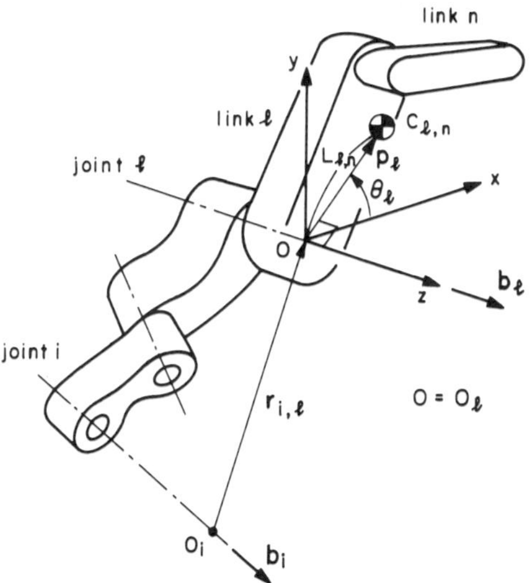

Figure 13.6
Coordinate frame used in the derivation of necessary conditions, where joint *i* is not orthogonal to joint *l*.

interactions between joints i and l:

$$H_{il} = (ML^2 + N_{zz})b_z + (b_x N_{xz} + b_y N_{yz} + MLb_z r_x)\cos\theta_l + (b_y N_{xz} - b_x N_{yz} + MLb_z r_y)\sin\theta_l, \tag{14}$$

where M is the mass of the last $(n - l + 1)$ links; L is the distance between the centroid of the immobilized part; N_{zz}, N_{xz}, etc., are components of the centroidal inertia tensor with respect to the O-xyz coordinate system as shown in figure 13.6; b_x, b_y, and b_z are direction cosines of the joint axis i; and θ_l is the displacement of joint l defined in the figure.

From (14) we can derive necessary conditions for the off-diagonal element of the inertia matrix **H** to be invariant for an arbitrary joint displacements θ_l:

$$b_x N_{xz} + b_y N_{yz} + MLb_z r_x = 0 \tag{15}$$

and

$$b_y N_{xz} - b_x N_{yz} + MLb_z r_y = 0. \tag{16}$$

In addition to the above conditions, the first term in (14) must be zero when the decoupled dynamics are required. However, the first term does not vanish unless $b_z = 0$, since the diagonal inertia tensor element N_{zz} is always positive. The condition that $b_z = 0$ implies that the two joint axes are perpendicular to each other. Therefore, an open-loop manipulator arm cannot possess the decoupled dynamics unless the joint axes are orthogonal to each other.

Similary, we can derive necessary conditions for diagonal elements of the inertia matrix, say H_{ii}, to be invariant for an arbitrary joint displacement. From these conditions on the diagonal and off-diagonal elements, we obtain the following.

THEOREM 1 An open-kinematic-chain manipulator possesses a decoupled inertia matrix for an arbitrary arm configuration if, and only if, its kinematic structure and mass properties satisfy the following conditions:

structure: 2-dof with orthogonal joint axes;
mass properties:

$$m_2 Lr_z = N_{xz}, \tag{17}$$

$$N_{yz} = 0. \tag{18}$$

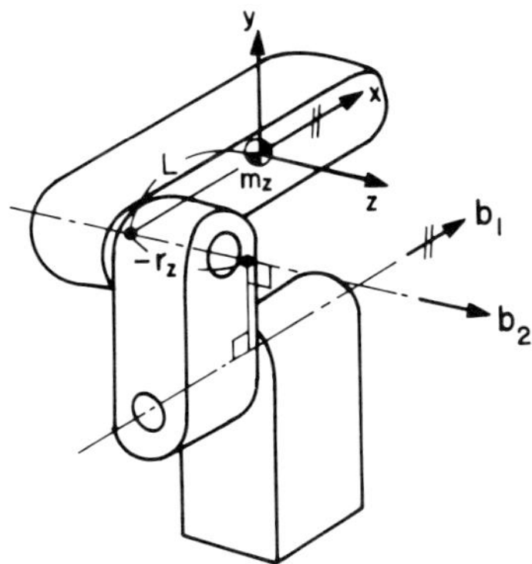

Figure 13.7
Manipulator arm with decoupled inertia.

The corresponding arm structures are shown in figure 13.7. The sufficiency of the conditions can be verified simply by evaluating the inertia matrix **H**.

When only decoupled dynamics are required, the above conditions reduce to the following.

THEOREM 2 The necessary and sufficient conditions for an open-kinematic-chain manipulator arm to possess a decoupled and configuration-invariant inertia matrix are given by:

structure: 2-dof arm with orthogonal joint axes;
mass properties: one of the following two sets of conditions must hold:

i. $L = 0,$ (19)

$$N_{xx} = N_{yy}, \tag{20}$$

$$N_{xy} = N_{yz} = N_{xz} = 0; \tag{21}$$

ii. $r_y = 0,$ (22)

$$N_{xx} = N_{yy} + ML^2, \tag{23}$$

$$N_{xy} = N_{yz} = 0, \tag{24}$$

$$m_2 L r_z = N_{xz}. \tag{25}$$

The arm designs corresponding to the above two case are shown in figure 13.8.

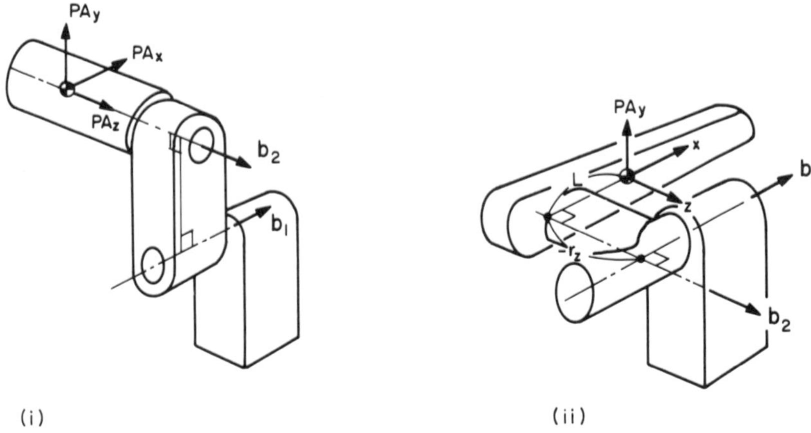

Figure 13.8
Manipulator arms with decoupled and invariant inertia. PA_x, PA_y, and PA_z are principal axes: (i) DI − 1, $L = 0$; (ii) DI − 2, $L \neq 0$.

If only the invariance of inertia matrix is required, the conditions described in theorem 1 are relaxed: all the possible 2-dof arm designs with the configuration-invariant dynamics are extended to the following four types. Two of them are shown in figure 13.9, and the other two, shown in figure 13.8, possess decoupled and invariant dynamics. The configuration invariant inertia can be accomplished not only for 2-dof manipulator arms but also for arms with three degrees of freedom or more. However, the possible designs are limited to combinations of the above four types of 2-dof manipulator arms. Figure 13.9 shows an example of a 3-dof arm with the configuration-invariant inertia, where the arm structure consists of the first type in figure 13.9 and the first one in figure 13.7.

Thus we have obtained the conditions for the decoupled and/or configuration-invariant inertia matrices. These conditions provide useful guidelines for arm design—what the kinematic structure and mass distributions should be in order to reduce dynamics complexity. In the following section, we shall apply the design theory to high-speed manipulator arms with direct-drive actuators.

13.5 Design of Direct-Drive Arms

In this section we consider arm design problems for direct-drive robots, high-performance manipulator arms that have been recently developed

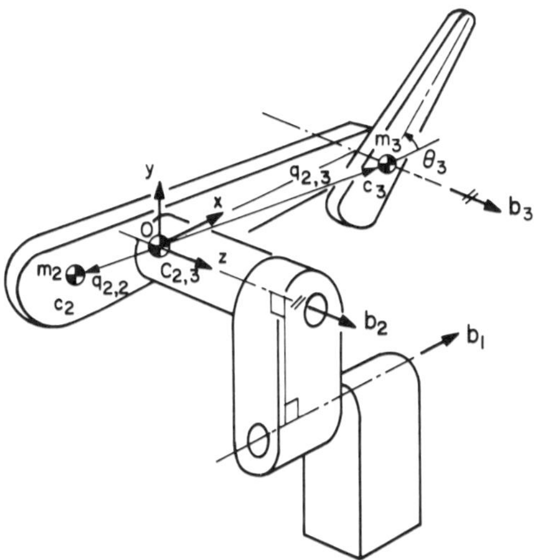

Figure 13.9
3-dof manipulator arm with invariant inertia consisting of DI − 1 and $I - 1$.

[20]. As required speed and accuracy increase, dynamic effects of the arm links become more prominent and critical in the arm design. In addition, the direct-drive method has some drawbacks as well as a number of desired features for manipulator arm drive. The arm mechanism design must be done in such a way that the features are fully exploited, while the drawbacks are overcome. The dynamic analysis methods we developed previously are now used in the arm mechanism design by considering the effects upon the actuators.

A majority of today's manipulator arms are driven by electromechanical drives. Electromechanical drives need gearing mechanisms to match the impedance of the motors with corresponding loads. These mechanisms sometimes cause serious problems that degrade the control performance of the robots. Harmonic drives, for example, which are widely used in existing robots, have insufficient mechanical stiffness and significant fluctuations in transmitted torque. Other robots have gear trains for which backlash is the main problem [21]. A common technique for eliminating backlash is the preloading of gears. A large preload, however, leads to large friction, which causes gears to wear quickly. These gears must be adjusted regularly to keep the backlash minimal.

Another problem is that conventional electromechanical robots are not suitable for torque control. Sophisticated control schemes recently developed, such as compliant motion control [22], generalized springs and dampers [23], impedance control [24], and hybrid/force control [25], are all dependent on the robot's ability to control torque, directly or indirectly. These control schemes were implemented mostly on conventional electromechanical robots, which were not designed as torque-controlled devices. The friction and deflection that exist at the gearings limited the performance under torque control.

A direct-drive arm is an innovative mechanical arm in which high-torque, low-speed motors are directly coupled to their loads [26, 27] (see figure 13.10). Since gearings are eliminated, the robots have no backlash, low friction, and high mechanical stiffness. Therefore the direct-drive arm is capable of fast and accurate positioning as well as precise torque control.

We now consider drawbacks of the direct-drive method. First, since the drive system cannot amplify the motor output torque mechanically, the robot would not be able to exert a large force for long without overheating the motors. Second, as the motor is mounted at each joint of a serial arm linkage, the weight of the motor itself is a load for the next motor down the

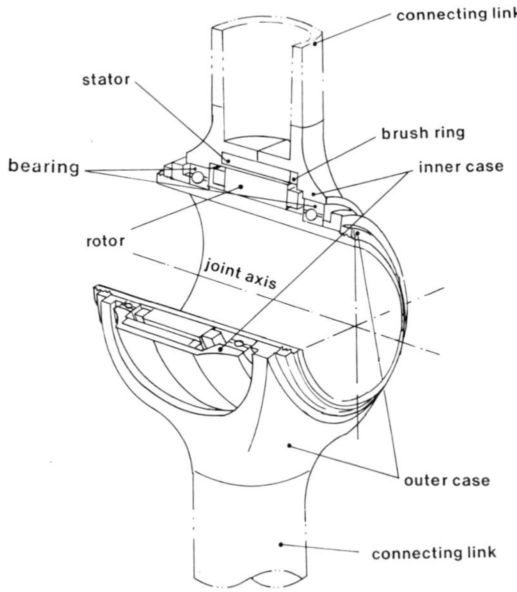

Figure 13.10
A direct-drive joint.

serial linkage. Therefore the drive torque required for each motor increases rapidly from the distal joint to the proximal joint. The rapid increase of motor sizes results in an impractical arm design with heavy arm weight and relatively small load capacity [26]. The third drawback is that direct-drive motors are too sensitive to load changes or disturbances.

If the ith joint of the arm is driven through gearing with gear ratio k_i, the dynamic equation in terms of the motor rotor is given by

$$\left(h_{ri} + \frac{h_{ai}}{k_i^2}\right)\ddot{\theta}_i + \frac{\tau_{ci} + \tau_{ni}}{k_i} = \tau_i, \tag{26}$$

where h_{ri} is the inertia of the motor rotor including the gearing, which is invariant; h_{ai} is the arm's inertia reflected to the ith joint axis, which varies depending on the arm configuration; and τ_{ci} and τ_{ni} are the interactive inertia torque and nonlinear torque reflected to the ith joint axis. Although h_{ai} is configuration-dependent, the effect of the varying inertia is attenuated by a factor of k_i^2 when reflected to the motor axis, and the rest of the inertial load—that is, h_{ri}—is invariant. Also the interactive and nonlinear torques at the joint axis are reduced by a factor of k_i. The complicated arm dynamics are less significant when the actuator has a large gear ratio. The direct-drive arm, on other hand, is most affected by the complicated dynamics. The varying inertia and the interactive and nonlinear forces are reflected directly to the motor axes and make the direct drive most sensitive to arm dynamics [27]. This drawback in the direct-drive method complicates the control problem.

In this section, arm design for overcoming the difficulties inherent to the direct-drive method is discussed. First, locations of motors to avoid the rapid increase of the arm weight are presented. Second, an arm mass distribution that reduces the interactions and nonlinearities is analyzed. Third, an arm structure that causes its actuators to have lower power dissipation, while the arm tip bears a load, is discussed. Finally, a prototype robot designed on the basis of analysis is described and evaluated.

Figure 13.11 shows typical construction of a serial drive arm mechanism, in which the lower link is driven by a motor fixed on the base and the upper link is driven by a motor attached at the tip of the lower link. In this serial configuration, the weight of the second motor is a load to the first motor. Moreover, the reaction torque of the second motor acts on the first motor. When the second motor accelerates in the clockwise direction, a counterclockwise torque acts on the first motor, and vice versa. Thus the

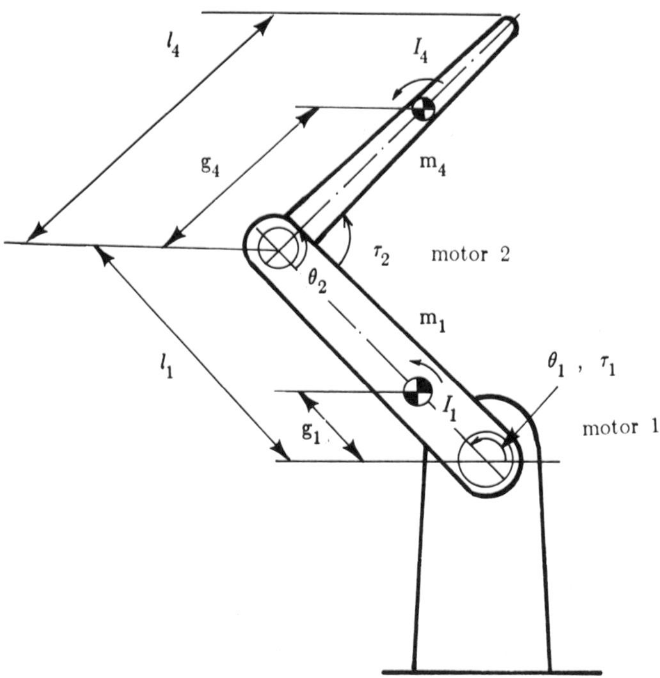

Figure 13.11
Dimensions and mass properties of a 2-dof serial drive arm mechanism.

two motors have significant interactions. An arm mechanism in which each motor is located at a joint of a serial linkage between the adjacent links is referred to as a serial drive mechanism.

Figure 13.12 shows an alternative arm construction consisting of a five-bar-link mechanism. Two motors fixed on the base drive the two input links and cause a two-dimensional motion at the tip of the arm. The weight of one motor is not a load on the other. Also the reaction torque of one motor does not act directly upon the other, because the motors are fixed on the base. An arm mechanism in which motors are mounted on a fixture and the weight and reaction torque of one motor does not affect the other motors directly is referred to as a parallel-drive mechanism. The parallel-drive mechanism has been used in several models of commercial robots that have conventional electromechanical drives with gearings. The five-bar-link mechanism, however, is particularly useful for overcoming the inherent difficulties of the direct drive method. The advantages of the parallel-drive

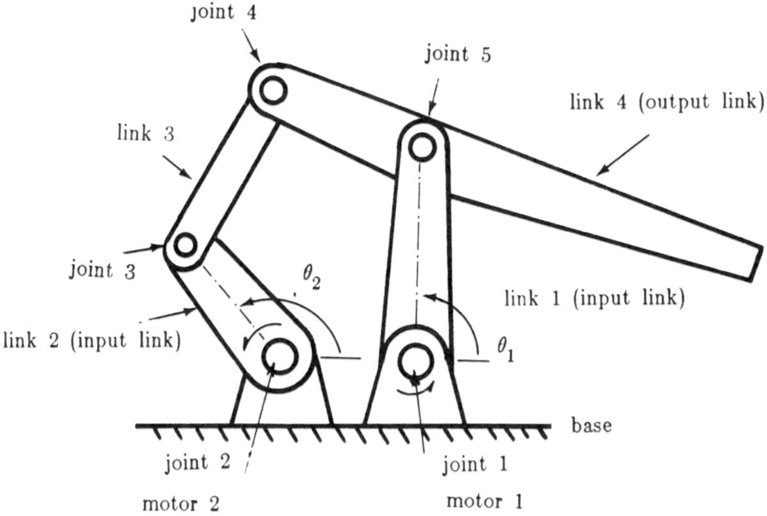

Figure 13.12
Parallel-drive fiver-bar-link mechanism.

mechanism are (i) light weight, (ii) lower interaction and nonlinearity, and (iii) lower power dissipation.

In the previous section, we discussed the arm design for decoupled and configuration-invariant inertia. For open-loop manipulator arms with serial drive mechanisms, the decoupling cannot be achieved unless joint axes are orthogonal to each other. This is the reason why only 2-dof manipulator arms may possess the decoupled inertia matrix. The orthogonality condition, however, is not required for the parallel drive mechanism, which forms a closed loop in the arm linkage. With a certain mass distribution the parallel-drive mechanism allows the inertia matrix to be diagonal even if the joint axes are parallel. Figure 13.13 shows a special case of the parallel-drive five-bar-link mechanism in which the distance between the two motors is zero and the two pairs of opposite links are parallel. First we immobilize motor 2 and compute the moment of inertia about the axis of motor 1, which corresponds to H_{11} in the inertia matrix. Inertia H_{11} is the resultant inertia of link 1, link 3, and link 4 about joint 3. By using the notation shown in figure 13.13, the inertia of link 1 about the first motor axis is given by $I_1 + m_1 g_1^2$, which is invariant. Since link 3 rotates about joint 3 when link 1 rotates, the resultant inertia of link 3 involved in H_{11} is

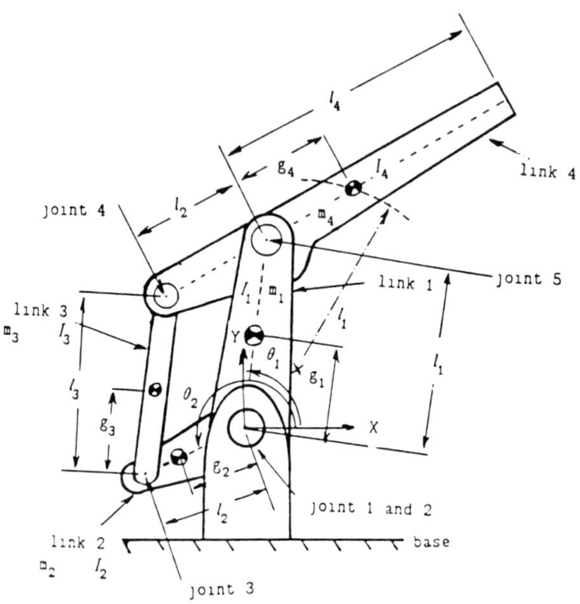

Figure 13.13
Parallelogram arm mechanism.

$I_3 + m_3 g_3^2$, which is also invariant for all the arm configuration. Similarly the mass center of link 4 moves along the circle of radius l_1, keeping constant attitude. Therefore, the resultant inertia of link 4 appears as $m_4 l_1^2$ in H_{11}, which is also invariant and configuration independent. Thus the overall inertia about the first motor axis is invariant. The moment of inertia about the second motor axis, when motor 1 is immobilized, is also invariant by the same reason. The inertia matrix of the arm mechanism is then given by

$$\mathbf{H} = \begin{bmatrix} H_{11} & H_{12} \\ H_{12} & H_{22} \end{bmatrix},$$

where

$$H_{11} = I_1 + m_1 g_1^2 + I_3 + m_3 g_3^2 + m_4 l_1^2,$$

$$H_{22} = I_4 + m_4 g_4^2 + I_2 + m_2 g_2^2 + m_3 l_2^2, \tag{27}$$

$$H_{12} = (m_3 l_2 g_3 - m_4 l_1 g_4)\cos(\theta_2 - \theta_1).$$

Off-diagonal element H_{12} governs the interaction between the two motors. It is possible to eliminate the interaction completely for all configurations by modifying the dimensions and mass properties of the arm. In the above expression for H_{12}, we can modify the mass ratio of link 3 and link 4, m_4/m_3, the ratio of mass center distances of the two links, g_4/g_3, and/or the length l_2 so that the coefficient $(m_3 l_2 g_3 - m_4 l_1 g_4)$ becomes zero. The condition for which the arm has no interaction for all configuration is given by

$$\frac{m_4}{m_3}\frac{g_4}{g_3} = \frac{l_2}{l_1}. \tag{28}$$

The resultant inertia matrix that satisfies the above condition reduces to

$$\mathbf{H} = \begin{bmatrix} I_1 + I_3 + m_1 g_1^2 + m_4 l_1^2\left(1 + \dfrac{g_3 g_4}{l_1 l_2}\right) & 0 \\ 0 & I_2 + I_4 + m_3 l_2^2 + m_4 g_4^2\left(1 + \dfrac{l_1 l_2}{g_3 g_4}\right) \end{bmatrix}. \tag{29}$$

The inertia matrix is invariant and completely decoupled. Therefore, there are no variations in motor load, no interactions, and no Coriolis and centrifugal forces.

Another problem with the direct drive is that the robot cannot exert a large force. Because of the direct coupling of motors to their loads, the output torques of motors are not amplified through reducers. In consequence, the motors have to bear the loads directly and might overheat (particularly in a stall condition). In what follows, we analyze the power dissipated in the motors when a force is exerted at the tip of the arm. We also compare the five-bar-link parallel-drive mechanism with the serial-drive mechanism, and show another significant advantage of the parallel drive over the serial drive regarding the power dissipation.

An important specification of a torque motor is the motor constant, that is, the ratio of output torque τ to the square root of power dissipated in the motor P [28]:

$$k_m = \frac{\tau}{\sqrt{P}}. \tag{30}$$

The motor constant represents how efficiently the motor converts its input power into output torque. If k_m is small, the motor dissipates a large amount of power to exert torque.

When a robot bears a load at the arm tip, the load is distributed to the multiple motors connected to the arm mechanism. In addition to the individual motor characteristics, the efficiency of exerting a force at the arm tip also depends on the kinematic structure of the arm mechanism.

Let P be the total power dissipated in the n motors driving the arm mechanism and k_{mi} be the motor constant of the ith motor; then the total power dissipation P is given by the following quadratic form of end point force $\mathbf{F}$:

$$P = \sum_{i=1}^{n} \frac{\tau_i^2}{k_{mi}^2} = \mathbf{F}^T \mathbf{L}_m \mathbf{F}, \tag{31}$$

where $\mathbf{L}_m$ is given by

$$\mathbf{L}_m = \mathbf{J} \begin{bmatrix} 1/k_{m1}^2 & & 0 \\ & \ddots & \\ 0 & & 1/k_{mn}^2 \end{bmatrix} \mathbf{J}^T. \tag{32}$$

The matrix $\mathbf{L}_m$ has all real, positive eigenvalues, $\lambda_i > 0$, and the quadratic form can be rewritten as

$$P = \sum_{i=1}^{n} \lambda_i F_i^2, \tag{33}$$

where F_i is the component of end point force along the eigenvector corresponding to λ_i. If the eigenvalue λ_i is large, a large amount of power is dissipated to exert force F_i in the direction of the corresponding eigenvector. For smaller eigenvalues, the arm can exert an end point force more efficiently. The force characteristics of this arm have been analyzed in the previous section. To evaluate the overall efficiency of the arm-motor combination, we define the mean power dissipation ratio, λ_{m}, which is the mean of the eigenvalues and is given by

$$\lambda_{\mathrm{m}} = \frac{1}{n} \operatorname{trace}(\mathbf{L}_{\mathrm{m}}). \tag{34}$$

The units of λ_{m} are watts per square newton.

Let us compare the parallel-drive mechanism with the serial drive in terms of the value of λ_{m}. The mean power dissipation ratio of the serial drive mechanism is larger than the parallel drive as shown in figure 13.14. Thus, the parallel-drive mechanism has a better efficiency for most arm configurations. This is another important advantage of the parallel-drive

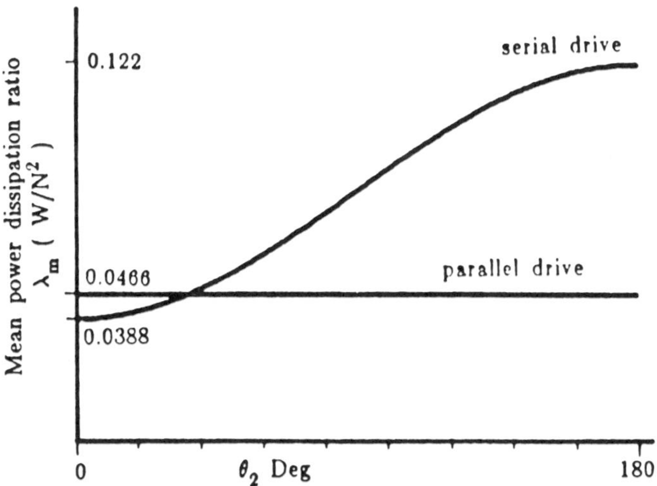

Figure 13.14
Mean power dissipation ratio.

mechanism over the serial drive. Having the same motor constants and the same link lengths covering the same workspace, the parallel-drive mechanism can more efficiently exert force at the arm tip.

Figure 13.15 shows the prototype direct-drive arm developed at the Massachusetts Institute of Technology. The arm structure is basically the five-bar-link parallel-drive mechanism discussed previously. In addition to the five-bar-link arm mechanism, the arm has another rotational joint at the base part to permit three-dimensional motions. The upper parallelogram mechanism and the lower base joint are designed in such a way that the conditions in theorem 1 for an open-loop arm to be decoupled are satisfied; thus no interactive inertia torques are generated.

The motors used for the direct-drive arm are brushless dc motors specially developed for the arm. Brushless motors have better heat dissipation characteristics than do dc motors. No spark arises at the commutator when large currents are delivered to the windings. The motor stator consisting of windings is housed in an aluminum case through which heat is dissipated, and the rotor having many poles consisting of permanent magnets on its surface is coupled directly to the joint axis. The permanent magnet is a new type of samarium cobalt magnet whose maximum magnetic energy is 26 MGOe. The motor used for the base joint has 660-Nm peak torque with

35-cm outer diameter. In comparison with the largest motor used for the direct-drive robot in reference [26], which is a dc torque with alnico magnet, the newly developed one has over three times the peak torque and is smaller by 60% in outer diameter. The motor constant is 7.5 Nm/$\sqrt{\mathrm{W}}$, and has a peak torque of 230 Nm. By combining the motor specification with the link's dimensions, the mean power dissipation ratio λ_m in (32) is computed. The mean is 0.0466 W/N^2 for all the arm configurations. This implies, for example, that the two motors dissipate 466 W of power on exerting an end point force of 100 N. In the case of a serial drive mechanism, the mean power dissipation ratio for the same motor constant and the same link lengths as the parallel drive is 0.0804 W/N^2 when $\theta_2 = 90°$, which is 73% inefficient.

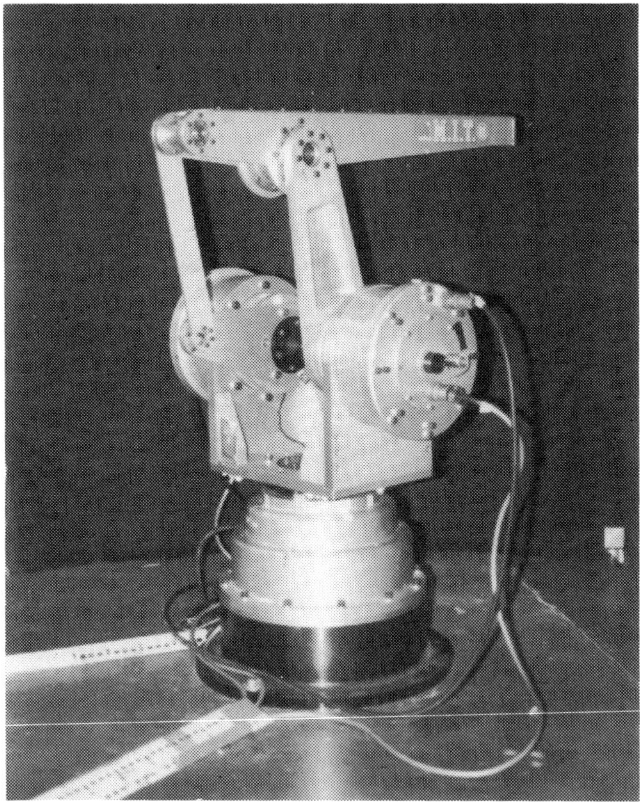

Figure 13.15
MIT direct-drive arm.

Figure 13.16 shows a velocity profile when the base joint traveled from -90° to $+90^\circ$. The speed indicated is that of the end point. The end point was accelerated very rapidly and reached the top speed. The maximum acceleration determined by the slope is 52 m/sec^2, which is more than 5 G. Since most industrial robots currently used have at most 0.1–1 G, the developed direct-drive arm is an order of magnitude faster in acceleration. The top speed measured was 9.36 m/sec, which is also much faster than existing robots.

13.6 Conclusion

In this chapter we have discussed the design problem of manipulator arms. Kinematic design issues were first presented. Previous papers on workspace, solvability, and singularity were surveyed, and mobility evaluations based on Jacobian matrices were described briefly. Then the dynamic problems were discussed. The analysis method using the Generalized Inertia Ellipsoid was described along with the application to the design of a 2-dof manipulator arm. Decoupled and/or configuration-invariant arm dynamics have been addressed, and the necessary and sufficient conditions on the kinematic structure and mass distributions were derived in order to accomplish the decoupled and/or configuration-invariant dynamics.

The analysis tools were then applied to the design of direct-drive arms. The direct-drive arms are high-speed, high-accuracy robots, while the drive

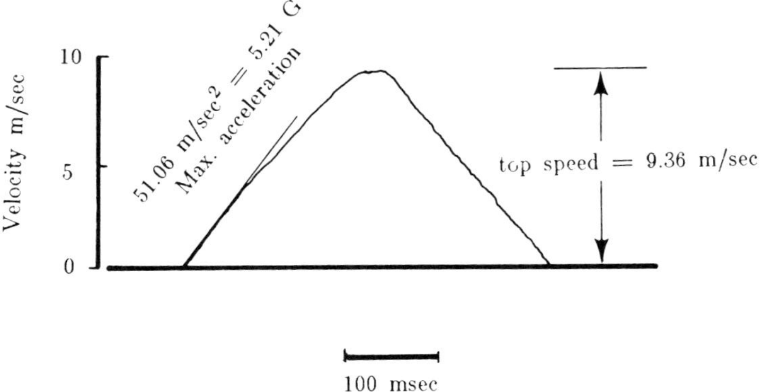

Figure 13.16
Velocity profile of base motor: arm stretched 0.75 m from base center; no load at tip of the arm; distance traveled from -90° to $+90^\circ$.

systems are difficult to control because of the high sensitivity to load changes. An efficient arm design was then introduced so as to exploit fully the desired features of the direct-drive arm. Direct-drive arms are currently studied at several companies and universities [29–32]. Takase et al. [29], for example, developed a 6-dof direct-drive manipulator in which the kinematic structure is devised so that static mass balancing can be achieved at each joint. Kuwahara et al. [31] developed a PUMA-type direct-drive arm with hybrid variable reluctance motors, which have been commercialized. The first commercialized direct-drive robot, developed at ADEPT Technology Inc. [30], has been used extensively in industries.

There are many other manipulator arms that have unique features in mechanical design. The micromanipulator by Sharon [33], for example, is capable of fine motion even under heavy loads. The arm is mounted at the tip of a large arm, which is appropriate for gross motion but is not suited for accurate positioning. The micromanipulator is then used to compensate for errors at the macromanipulator. Also the micromanipulator has a wider bandwidth in the servo system than in the macromanipulator, so that high-speed, high-frequency motions can be achieved. For ultrahigh precision manipulation, Hollis et al. at IBM [35] developed a 6-dof magnetically levitated wrist, which is mounted at an arm tip for fine manipulation. This is basically a direct-drive system where the moving part is directly driven by magnetic forces having no mechanical contact with the arm tip. It is expected to perform extremely high-precision tasks such as in microelectronics assembly. To bear a large load, macromanipulators having cantilevered beam structures are not suited, as mentioned before. A way of increasing the mechanical stiffness is to brace the arm against the environment near the arm tip. Asada and West [36] developed a braced manipulator in which the wrist of the arm is pressed against the workpiece surface through a bracing mechanism. Heavy-duty tasks such as metal removing are accomplished with the braced manipulator.

Thus, arm design problems are closely related to end effector design problems in many ways. To increase accuracy, stiffness, and load-bearing capability, it is efficient to construct the arm by incorporating microactuation devices and other end effectors. This is a promising and important approach to advanced robot design. Currently, several research groups are working on innovative arm design with unique end effectors—an example is the hybrid serial-parallel manipulator by Waldron and Roth at Stanford University [34].

References

[1] Asada, H., and Slotine, J.-J. E., *Robot Analysis and Control*, John Wiley and Sons (1985).

[2] Roth, B., "Performance Evaluation of Manipulators from a Kinematic Viewpoint," NSF Special Publication: Performance Evaluation of Programmable Robots and Manipulators (1975).

[3] Shimano, B., "The Kinematic Design and Force Control of Computer Controlled Manipulators," Ph.D. Thesis, Department of Mechanical Engineering, Stanford University, (1978).

[4] Sugimoto, K., and Duffy, J., "Determination of Extreme Distances of Robot Hands and Arms with Special Geometry," *ASME J. Mechanical Design*, 103–4 (1981).

[5] Kumar, A., and Waldron, K., "The Workspace of a Mechanical Manipulator," *ASME J. Mechanical Design* 103–3 (1981).

[6] Gupta, K. C., and Roth, B., "Design Considerations for Manipulator Workspace," *ASME J. Mechanical Design*, 104–4 (1982).

[7] Pieper, D. L., and Roth, B., "The Kinematics of Manipulator under Computer Control," "*Proc. 2nd Int. Conf. Theory of Machines and Mechanisms* (1969).

[8] Paul, R. P., and Stevenson, C. N., "Kinematics of Robot Wrists," *Int. J. of Robotics Research*, 2-1, 31/38 (1983).

[9] Hollerbach, J. M., "Optimal Kinematic Design for a Seven Degree of Freedom Manipulator," *Proc. 2nd Int. Symp. on Robotics Research*, 349/356 (1984).

[10] Uchiyama, M., Shinizu, K., and Hakomori, K., "Performance Evaluation of Manipulators Using the Jacobian and Its Application to Trajectory Planning," *Proc. 2nd Int. Symp. on Robotics Research*, 447/454 (1984).

[11] Yosikawa, T., "Manipulability of Robotic Mechanisms," *ibid.*, 439/446 (1984).

[12] Asada, H., and Yousef-Toumi, K., "Development of a Direct-Drive Arm Using High Torque Brushless Motors," *Proc. 1st Int. Symp. on Robotics Research, New Hampshire* (1983).

[13] Asada, H., and Yocef-Toumi, K., "Analysis of Multi-Degree-of-Freedom Actuator Systems for Robot Arm Design," *Proc. 1983 ASME Winter Annual Meeting, Control of Manufacturing Processes and Robot Systems* (1983).

[14] Salisbury, J. K., and Craig, J. T., "Articulated Hands: Force Control and Kinematic Issues," *Int. J. of Robotics Research*, 1-2, 4/17 (1982).

[15] Asada, H., and Cro Granito, J. A., "Kinematic and Static Characterization of Wrist Joints and Their Optimal Design," *Proc. 1985 IEEE Int. Conf. on Robotics and Automation*, 244/250 (1985).

[16] Crandall, S. H., et al., *Dynamics of Mechanical and Electromechanical Systems*, Chapter 2, Mc Graw-Hill (1968).

[17] Goldstein, H., *Classical Mechanics*, 2nd Edition, chapter 5, Addison-Welsey (1980).

[18] Asada, H., "A Geometrical Representation of Manipulator Dynamics and Its Application to Arm Design," *ASME J. Dynamic Systems, Measurement and Controls*, 105–3 (1983).

[19] Youcef-Toumi, K., and Asada, H., "The design of Open-Loop Manipulator Arms with Decoupled and Configuration-Invariant Inertia Tensors," *Proc. 1986 IEEE Int. Conf. on Robotics and Automation*, Vol. 3, 2018/2026 (1986).

[20] Asada, H., and Kanade, T., "Design of Direct-Drive Mechanical Arms," *ASME J. of Vibration, Stress, and Reliability in Design*, 105–3, 312/316 (1983).

[21] *Unimate PUMA Robot*, Technical Manual, Unimation Inc., Apr. 1980.

[22] Mason, M., "Compliance and Force Control for Computer Controlled Manipulators," *IEEE Transactions on Systems, Man, and Cybernetics*, Vol. SMC-11, No. 6, June 1981.

[23] Nevins, J. L., and Whitney, D. E., "The Force Vector Assembler Concept, presented at First IFToMM Symp. Theory, Practice of Robots, Manipulators, 1974.

[24] Hogan, N., "Impedance Control of a Robotic Manipulator," *Proc. of 1981 ASME, Winter Annual Meeting*, Washington, D. C., Nov. 1981.

[25] Raibert, M. H., and Craig, J. J., "Hybrid Position/Force Control of Manipulators," *Dynamic Systems and Control Division of ASME, Winter Annual Meeting*, Nov. 1980.

[26] Asada, H., Kanade, T., and Takeyama, I., "Control of a Direct-Drive Arm," *ARME J. of Dynamic Systems, Measurement, and Control*, 105-3, 136/142 (1983).

[27] Asada, H., and Youcef-Toumi, K., "Analysis and Design of a Direct-Drive Arm with a Five-Bar-Link Parallel Drive Mechanism," *ASME J. of Dynamic Systems, Measurement and Control*, 106–3, 225/230 (1984).

[28] *Electro-Craft, DC Motors, Speed Controls, Servo-Systems*, Engineering Handbook by Electro-Craft Corporation, Minn.

[29] Suehiro, T., and Takase, K., "Development of a Direct-Drive Manipulator ETA-3 and Enhancement of Servo Stiffness by a Second-Order Digital Filter," *Proc. of 15th Int. Symp. on Industrial Robots*, Tokyo 1985.

[30] Curran, R., and Mayer, G., "The Architecture of the ADEPT One Direct-Drive Robot, *Proc. of 1985 American Control Conference*, 716/721, Boston 1985.

[31] Kuwahara, H., et al., "A Precision Direct-Drive Robot Arm," *Proc. of 1985 American Control Conference*, 722/724, Boston 1985.

[32] Asada, H., and Youcef-Toumi, K., *Direct-Drive Robots*, MIT Press, 1987.

[33] Sharon, A., Hogan, N., and Hardt, D., "More Analysis and Experimentation on a Macro/Micro Manipulator System," *Proc. of 1987 ASNE Winter Annual Meeting, Modeling and Control of Robotic Manipulators*, 417/422, 1987.

[34] Waldron, K. J., Raghavan, M., and Roth, B., "Kinematics of a Hybrid Serial-Parallel Manipulator System, Parts 1 and 2, *Proc. of 1987 ASME Winter Annual Meeting, Modeling and Control Robotic Manipulators*, 127/148, 1987.

[35] Hollis, R. L., Allan, A. P., and Salcudean, S., "A Six Degree-of-Freedom Magnetically Levitated Variable Compliance Fine Motion Wrist, "*Proc. of 4th Int. Symp. of Robotics Research*, Santa Cruz, 1987.

[36] Asada, H., and West, H., "Design and Analysis of Braced Manipulators for Improved Stiffness," *Proc. of 3rd Int. Symp. of Robotics Research*, 62/67, 1985.

14 Behavior-Based Design of Robot Effectors

Stephen C. Jacobsen, Craig C. Smith, Klaus B. Biggers, and Edwin K. Iversen

14.1 Introduction

14.1.1 Background

Over the past decade, a major focus of the Center for Engineering Design (CED) at the University of Utah has been the design of actuation systems for robots, teleoperators, prosthetic limbs, and medical devices. Each project has required the development of systems with different combinations of behavioral characteristics such as strength, speed, precision, grace, efficiency, size and weight. Necessarily, the projects have included consideration of the six subsystems of a robot shown in figure 14.1.

Block 3 represents the effector, the *distinguishing feature* that separates a robot from other smart machines. The effector includes actuators, structures, a servo-level controller, and internal sensors. The actuators, which are energy transforming devices, can be electric, hydraulic, or pneumatic systems controlled by amplifiers and/or valves. Actuators may also include other mechanisms such as reducers and drives.

Block 4 represents the environment, which can impose external loads on the effector by touching or pushing. Block 2 represents the mid- and high-level controller, which plans actions and tasks. Block 5 includes external sensors that observe the environment, such as vision systems, tactile sensors, and acoustic systems. Block 1 represents the command source, and Block 6 is the model, which provides information to the higher level controller and the command source. The model can be as simple as a group of potentiometer settings in an analog electronic system, or as complex as a complete representation of the dynamics of the system.

14.1.2 Problems with Effector Development

Successful effector systems have been developed by the CED for the Utah Arm, the Utah/MIT Dextrous Hand, entertainment robots, and high performance servo-pumps for use in medical devices. The CED is a design-oriented group that has believed itself quite good at developing effectors. This belief was somewhat shaken when, during a workshop, a sponsor asked the question, "What is a good actuator and how is one designed?" (Kelly, 1985). No one responded very well and attempted answers were fuzzy and circular. The discussion demonstrated clearly that it is not yet

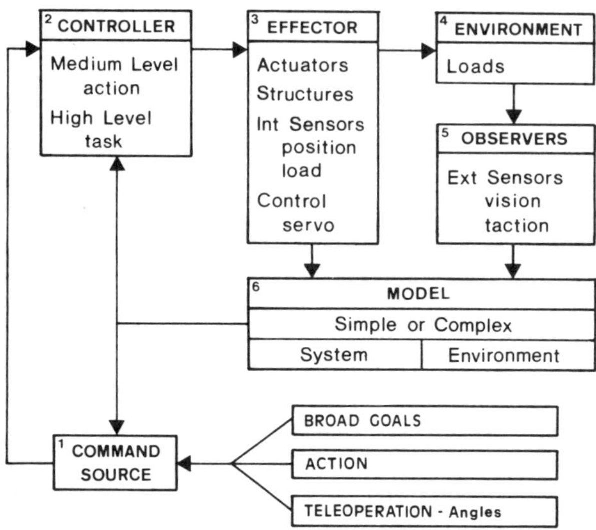

Figure 14.1
The six subsystems of a robot.

easy to talk about effectors in a comprehensive way. As a result of the discussions, it became clear that the reasons for lack of progress in the development of effectors stem from (1) lack of *need*, (2) problems with *complexity*, and (3) a general *underestimation* of the importance of effecting systems.

(1) Need: The performance demands on effectors have only recently begun to stretch the limits of available physical systems. Thus, it has been possible to consider effectors as merely simplified elements in block diagrams. In short, effectors have been neglected because neglecting them has been possible. **(2) Complexity:** Effectors have also been neglected because their characteristics are difficult to understand and represent. Effectors are complex thermodynamic machines, which are typically only one interactive element of an even larger, more complex system. **(3) Underestimation:** Finally, and most significantly, the importance of effectors, and the limitations they will impose on future robots, has not been clearly articulated by researchers interested in this area.

14.1.3 The Real Question

It was interesting that during the Kelly actuator discussions the question "What is a good actuator?" raised a set of concerns and antagonisms that have been growing over the past decade. Typical comments on this issue have been "Why all this talk from engineers about problems with machinery that moves? Surely, problems with actuators can't represent a major obstacle to progress in robotics." "Clearly, actuators rank in difficulty with Artificial Intelligence, vision, and other topics in robotics." "Motors just can't be a problem, they're just motors." "Where's the fundamental content, anyway?" "When we want them, we'll simply build them." "There has been too much talk and no clear answers. This problem should be solved or go away."

Out of these discussions, three important points emerged:

1. With sufficient effort, the actuator problem can be completely understood, certainly for linear systems. The relationships between effector configuration and behavior can be rigorously defined. This is the focus of this chapter.
2. The development of new effector systems, appropriate for future robots, does rank in importance and difficulty with other major engineering problems. Barriers to progress range from the development of better approaches to the dynamic control of energy conversion to basic issues in machine design (i.e., packaging, heat transfer, friction, wear, efficiency, etc.).
3. Systems of the future must exhibit extraordinary behavioral capabilities while simultaneously being reliable, efficient, and economic. If these problems are not addressed now, in a fundamental way, the effector problem will become a major obstacle to future progress in the area of robotics.

14.1.4 Objectives

The goal of this chapter is to explore issues important to the systematic design of effectors. Of specific importance is how to assemble effector system elements into machinery that will produce desired behaviors.

The chapter addresses the effector design problem using no new approaches, only an examination of some fundamental concepts via an application of performance criteria to an effector model. The objective is to improve our understanding of concepts and to expand the vocabulary necessary to discuss effectors systematically.

Results are not expressed in a particularly pretty or comprehensive form. Generalized representations such as nondimensional groups and universal graphs are deferred for later work. The system model is not dismembered into elements in an attempt to understand overall behaviors in terms of component characteristics such as internal impedances. It is the behavior of the *entire system, when assembled*, that is of primary importance.

14.2 Approach

14.2.1 A Simple Effector System

Our goal is to understand the behavioral characteristics of an effector system in the simplest possible way. Of importance is understanding how to trade off component and controller parameters in order to achieve desired performance characteristics. The following review serves to acquaint the reader with a typical servo experiment.

The usual approach is to assemble a servo-system such as that shown in figure 14.2. The system includes a one-degree-of-freedom arm with motor, amplifier, reducer, sensors (colocated with the reducer output), a compliant arm, and a controller. Alternatively, the example could utilize a hydraulic or pneumatic actuator in place of the electric motor.

A process of optimization (compromise) is then undertaken to determine acceptable system parameter settings, i.e., via knob twisting. As variables are changed, a number of behaviors can result as depicted in figure 14.3. Figure 14.3 is a simplified schematic that shows the load, the flexible arm structure, and the armature mass as elements that under various conditions combine to exhibit sluggish, rapid, oscillatory, or unstable behaviors. Note that the armature mass M_A is actuated by the feedback system, which

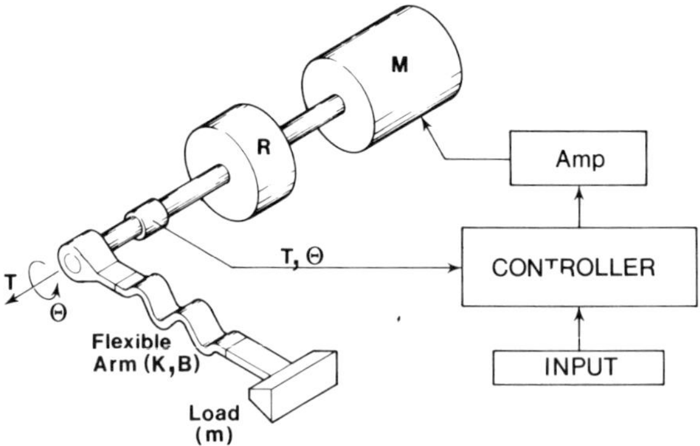

Figure 14.2
Simple one-dimensional model of a one-link effector arm.

includes position and force sensors (shown as the eye) and a motor (shown as the coil).

First, the position feedback loop gain, K_p, is increased from zero. If the position loop gain is too low the system will be soft, slow, and positional accuracy will be low (case A in figure 14.3). As K_p is increased the system exhibits sufficient stiffness to improve positional accuracy while remaining stable. Typically, the resulting stiff system does not interact well with touched objects and can encounter combinations of servo and structural oscillations when speed demands are high enough for actuator loadings to deform structures (case B, figure 14.3). If K_p is further increased, the system becomes unstable and the actuator mass M_A oscillates against the load mass M_L through the structural impedance Z_s (case C, figure 14.3).

In an attempt to correct the problem, a force feedback loop is usually added in order to modify passive qualities of the system's response. As the

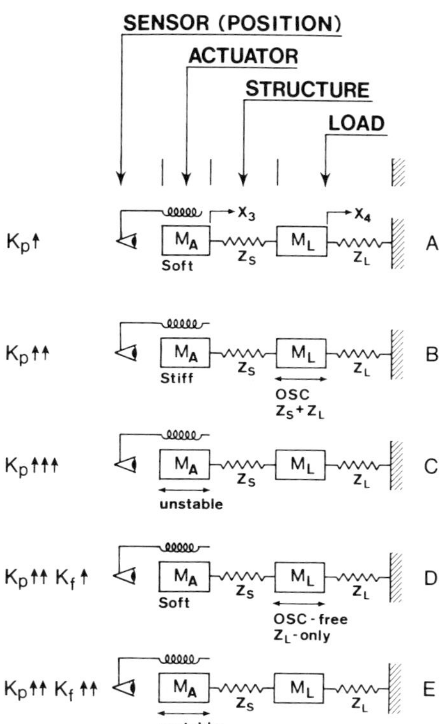

Figure 14.3
Simple servo-system behaviors.

force feedback gain, K_f, is increased, the system becomes increasingly softer, more graceful, and the fundamental oscillation frequency of the load decreases and approaches that of the free load without an actuator. Now the system interacts more successfully with external objects but steady state position error is increased (case D, figure 14.3). If K_f is further increased, problems with high frequency structural oscillations and instability emerge (case E, figure 14.3).

Root locus plots, shown later in the chapter, further discuss these behaviors. Note that these behavioral trade-offs strongly suggest the development of nonlinear adaptive controllers in which the gains of selected feedback loops are dynamically adjusted according to system behavioral demands (Jacobsen, 1973; Jacobsen, 1986; Meek, 1986; Fullmer, 1986; Seamons, 1986; Hooker, 1987).

14.2.2 A Linear Effector Model

A general linear model of the system in figure 14.2 is shown in figure 14.4. The model contains ten elements: a motor, amplifier, reducer, drive, sensors, structure, load, touch load, and a linear feedback control system.

This model has sufficient complexity to exhibit behaviors of interest while also being relatively simple. In the following sections, the system is separated into second- and third-order approximate models in order to examine sequentially the influence of actuators, structures, loads, and controllers on behavior.

14.2.3 Design Sequence

Of importance here is that this problem is not merely one of addressing an existing plant to be controlled. The entire problem must be impartially reexplored not only in the area of control strategy but also in the design of actuators, structures, and load.

The design approach includes five steps:

First Identify, in general terms, the desired system behavior versus anticipated types of inputs (including externally produced disturbances). Important behaviors can be characterized by terms such as quickness, gracefulness, trajectory quality, ability to maintain stability while touching objects, and absence of structural oscillations during operation, as well as others.

Second Define quantitative performance criteria (QPCs), which, if satisfied, will constrain the system to exhibit desired behaviors. Typical perfor-

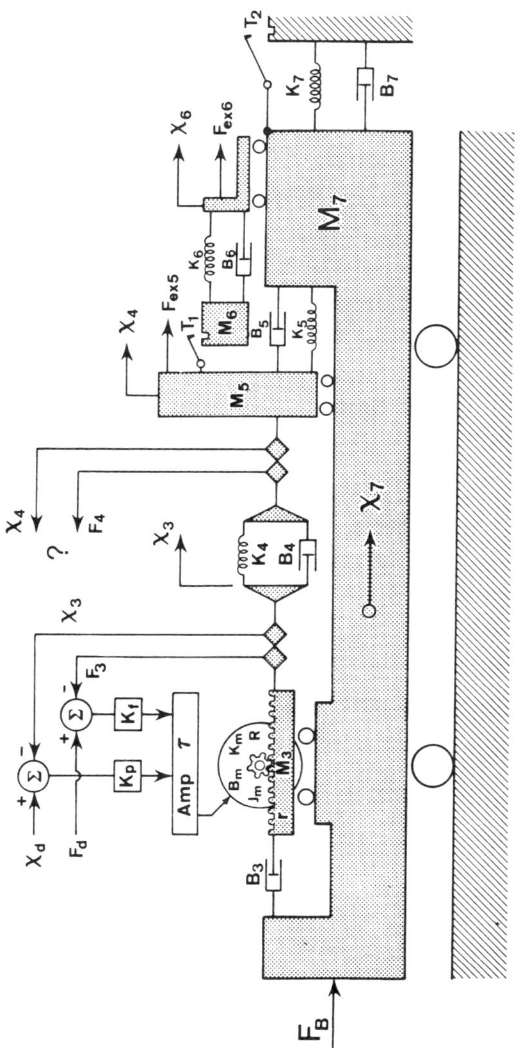

Figure 14.4
One-dimensional linear effector model.

mance criteria include rise-time, overshoot, steady state error, relative stability, and others.

Third Formulate a model that is complex enough to generate behaviors of interest and exhibit the relationships between the system and these behaviors, yet that is simple enough to allow solutions to be obtained systematically.

Fourth Combine results of steps 1–3 to define a set of parameter constraint equations that can then be used in different design scenarios.

Fifth Explore design scenarios, such as

A. Given structure, load, and desired system behavioral characteristics, what is a good actuator?

B. What behavior is possible given the characteristics of the actuator, load, controller, and structure?

Or, in more simple terms: Use the desired system behavior (1) to generate quantitative performance criteria (2), which when used with the system model (3) can generate a set of constraint equations (4), which are useful in the solution of various design scenarios (5).

14.3 A Linear Effector Model

14.3.1 Introduction

Figure 14.4 shows a simple linear model of the system in figure 14.2, which includes one external degree of freedom. The model includes no stiction, backlash, saturation, drive compliance (in the shaft between the motor armature and the pinion gear), or other complications.

The actuator, which consists of the motor, reducer, sensors, and drive, is represented in the model as a second-order system with a torque delay caused by the motor inductance. Additional delays could be caused via the amplifier, drive compliance, and backlash, but they are not considered here.

The motor is a permanent magnet, dc electric motor (see table 14.1) connected to a reducer and drive (rack and pinion gears), which includes inertial and damping characteristics (M_3, B_3). The diamond-shaped elements interposed between the rack and structure, and between the structure and load mass, are transducers which measure force and position. The inverse of the radius of the pinion gear, r, is defined as the reduction parameter G.

Table 14.1
Model parameters: ten system elements are defined by twenty-six parameters

K_m	53.2×10^{-3} N·m/W$^{1/2}$	Motor[a]
R	.631 Ω	Motor[a]
J_m	38.1×10^{-6} kg·m^2	Motor[a]
B_m	17×10^{-6} N·m·s/rad	Motor[a]
L	.0975 mH	Motor[a]
τ	0	Amplifier
G	$108.7\ \mathrm{m}^{-1}\left(\frac{1}{\mathrm{r}}\right)$	Reducer
B_3	3.46 N·s/m	Drive
M_3	0	Drive
B_4	3.464 N·s/m	Structure
K_4	2.83×10^5 N/m	Structure
M_5	2.65 kg	Load
B_5	17.32 N·s/m	Load
K_5	113.2 N/m	Load
M_6	0	Touch load
B_6	0	Touch load
K_6	0	Touch load
T_1	0	Touch load
K_p	Variable	Controller
K_f	Variable	Controller
S_f, S_p	Unity	Sensors
M_7, B_7, K_7, T_2	Fixed	Base

a. Pitman Corp Motor, Model 5113.

The structure is represented by parallel spring and damping elements (B_4, K_4), which permit the representation of structural oscillations between the actuator and load. The load is a second-order parallel combination of the mass, spring, and damper (M_5, B_5, K_5), and it can be contacted by a second-order touch impedance defined by a mass, spring, and damper (M_6, B_6, K_6). Note that touch is engaged by the binary parameter T_1 ($T_1 = 1$, connected) and pushing (with impedance) via movement of the variable X_6. An external force (with zero impedance) can also be applied via F_{ex5}.

Although the position of the base X_7 is assumed fixed in this case (parameter $T_2 = 1$), for later studies it can be freed to include base dynamics (M_7, B_7, K_7) and the application of external force F_B. Observe that, in future work, the model can be cascaded to represent higher complexity systems by connecting the load mass (M_5) in one system to the base mass (M_7) of the next system. Of course, successive systems would typically be free with their ground impedances (K_7, B_7) set at zero.

The simple linear control structure includes position and force loops adjusted by gains K_p and K_f. The system input is X_d and the output for this paper is considered as X_3. If the output is considered as X_4 and information from noncolocated sensors X_4 and F_4 is used, then more complicated stability issues arise (Cannon, 1984).

Of course, initial conditions can be specified in order to simulate specific effector experiments. For example, with an appropriate initial velocity on X_4 and suitable load impedance, the model can be used to simulate system response to external insults.

Case I—The Total System Case I is the total fifth-order system obtained by using parameter values listed in table 14.1. Equations representing system elements are described by equations (1a)–(1l).

Third-order (case II) and second-order (cases III and IV) approximations of case I will be used to generate eleven constraint equations.

Motor:

$$P_1 \frac{d\theta}{dt} = \frac{K_m}{R} V_{in} - P_2 T_{dr}, \tag{1a}$$

where V_{in} is the voltage applied to the motor;
Reducer:

$$T_{dr} = \frac{1}{G} F_{dr}, \tag{1b}$$

where F_{dr} is the force applied to the rack by the pinion gear and

$$\theta = GX_3; \tag{1c}$$

Drive:

$$F_{dr} = P_3 X_3 + F_{st}, \tag{1d}$$

where F_{st} is the force applied to the structure;
Structure:

$$F_{st} = P_4(X_3 - X_4); \tag{1e}$$

Load:

$$F_L = P_5 X_4 - F_{ex5}; \tag{1f}$$

Controller:

$$V_{in} = K_p(X_d - X_3) + K_f(F_d - F_{st}). \tag{1g}$$

Here

$$P_1 = \left[\frac{L}{R} J_m\right] s^2 + \left[J_m + \frac{L}{R} B_m\right] s + \left[B_m + \frac{K_m^2}{R}\right], \tag{1h}$$

$$P_2 = \left[\frac{L}{R}\right] s + 1, \tag{1i}$$

$$P_3 = [M_3]s^2 + [B_3]s, \tag{1j}$$

$$P_4 = [B_4]s + K_4, \tag{1k}$$

$$P_5 = [M_5]s^2 + [B_5]s + K_5. \tag{1l}$$

The case I equations are combined below as equations (2a)–(2j):

$$(D_5 s^5 + \cdots + D_1 s + D_0) X_3 = \varphi_p (g s^2 + b s + a) X_d. \tag{2a}$$

Here

$$\begin{bmatrix} D_0 \\ \cdot \\ \cdot \\ \cdot \\ \cdot \\ D_5 \end{bmatrix} = \tilde{\beta} \begin{bmatrix} M_A \\ B_A \\ B_L \\ \varphi_p \\ \varphi_F \\ 1 \end{bmatrix} \tag{2b}$$

and

$$M_A = G^2 J_m + M_3, \tag{2c}$$

$$B_A = G^2 B_m + B_3, \tag{2d}$$

$$B_L = G^2 \left(\frac{(K_m)^2}{R}\right), \tag{2e}$$

$$\varphi_F = \left(1 + G \frac{K_m}{R} K_f\right), \tag{2f}$$

$$\varphi_{\mathrm{p}} = \left(G \frac{K_{\mathrm{m}}}{R} K_{\mathrm{p}} \right), \tag{2g}$$

$$G = 1/r, \tag{2h}$$

and

$$\tilde{\beta} = \begin{bmatrix} 0 & 0 & 0 & a & e & 0 \\ 0 & a & a & b & c & \left(\frac{L}{R}\right)e \\ a & b + \left(\frac{L}{R}\right)a & b & g & d & \left(\frac{L}{R}\right)c \\ b + \left(\frac{L}{R}\right)a & g + \left(\frac{L}{R}\right)b & g & 0 & f & \left(\frac{L}{R}\right)d \\ g + \left(\frac{L}{R}\right)b & \left(\frac{L}{R}\right)g & 0 & 0 & 0 & \left(\frac{L}{R}\right)f \\ \left(\frac{L}{R}\right)g & 0 & 0 & 0 & 0 & 0 \end{bmatrix}, \tag{2i}$$

$$\begin{bmatrix} a \\ b \\ c \\ d \\ e \\ f \\ g \end{bmatrix} = \begin{bmatrix} K_4 + K_5 \\ B_4 + B_5 \\ B_4 K_5 + B_5 K_4 \\ B_4 B_5 + K_4 M_5 \\ K_4 K_5 \\ M_5 B_4 \\ M_5 \end{bmatrix}. \tag{2j}$$

The D's are coefficients of the system's characteristic equation and s is the Laplace transform operator. φ's are nondimensional controller gains and a through g are constants composed of the lumped parameter coefficients of the structure and load. M_{A}, B_{A}, and B_{L} are motor, reducer, and drive system characteristics.

14.3.2 Root Loci Illustrating Case I Behavior

In order to examine the behavior of the fifth-order system in a qualitative way, a number of root loci are plotted for a typical system whose parameters are quantitatively defined in table 14.1. (Note that for illustration

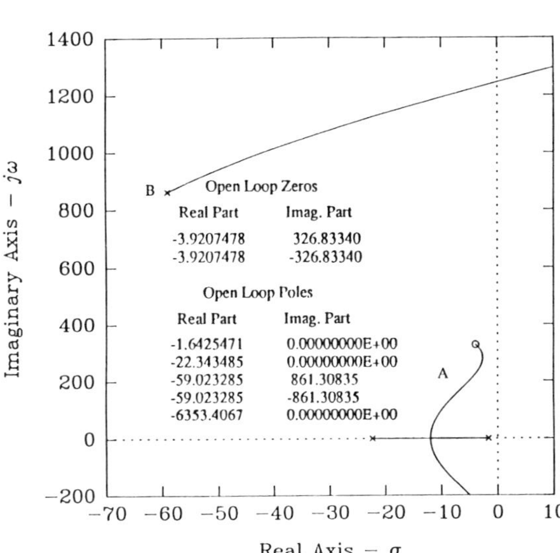

Figure 14.5
Root locus of case I system (fifth-order system with position feedback only).

purposes all root locus plots are distorted by an expansion of the real axis and some poles and/or zeros are off the figure.)

Figure 14.5 contains a root locus plot for the case I system with only the position feedback loop in operation ($K_f = 0$). Observe that as the gain of the position loop, K_p, is increased from 0 to ∞, five poles migrate to two finite zeros and three infinite zeros.

The two finite zeros marked A represent an oscillation of the load together with the actuator, which occurs at relatively low frequencies. The complex poles marked B are the result of oscillatory interactions between the actuator and structure, which ultimately result in unstable operation as those poles cross the imaginary axis. The fifth pole migrates out the negative real axis (recall figure 14.3, cases A, B, and C).

Note that the behavior of poles and zeros on the negative real axis can be important, since they shift the intersection of asymptotes with the real axis.

Figure 14.6 shows a root locus for case I with $K_p = 30{,}000$ V/m and the force feedback gain K_f varying from 0 to ∞ V/N. Observe again two types of oscillation. The low magnitude poles labelled A represent a load oscillation that becomes progressively more free (slower and less damped than

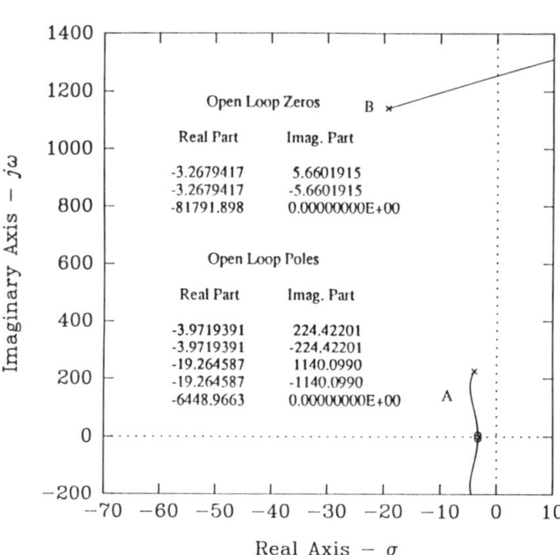

Figure 14.6
Root locus of case I system for $K_p = 30{,}000$ V/m and K_f varying from 0 to ∞ V/N.

the case in figure 14.5 where only K_p is a variable) as the force feedback gain is increased. These poles approach the eigenvalues of the load with the actuator unattached (which show up as zeros on the plot). The higher frequency poles labeled B represent oscillations that occur as a result of interactions between the actuator and the load through the compliant structure. As K_f increases, the B poles continue out toward the imaginary axis and can, if the system is not properly designed, produce undesirable instabilities before the desired level of softness is achieved (recall figure 14.3, cases D and E).

Note again that the motivation for adding force feedback is to soften the system's behavior. This in fact occurs as K_f is turned up and the A poles, which represent the load (swinging against gravity in figure 14.2), decrease in magnitude (frequency) and remain less damped than without force feedback.

14.3.3 Use of Reduced Models

Generating quantitative relationships between system parameters and behavior is difficult using the fifth-order model. Therefore, the fifth-order case is separated into three simpler cases.

Case II is a third-order system that focuses on oscillations between the actuator and structure. Case III is a second-order system that addresses behavior of the load assuming a rigid structure. Case IV is a second-order system used to examine system response to insults to the structure and actuator assuming the motor is fast and the load is slow. These reduced equations are than used, together with quantitative performance criteria (QPCs), to develop system parameter constraint equations.

Case II—Compliant Structure with a Slow Load This case is generated by assuming that the load mass (M_5) is large such that its motions are slow (assumed fixed) compared to the dynamics of the actuator and structure. Case II is described by equations (3a)–(3f) and (4a)–(4f):

Actuator:

$$\{[P_6]s^3 + [P_7]s^2 + [P_8]s + [P_9]\}X_3 = G\frac{K_m}{R}V_{in}; \tag{3a}$$

Controller with $F_d = 0$:

$$V_{in} = K_p X_d - \{[K_f B_4]s + [K_p + K_f K_4]\}X_3. \tag{3b}$$

Here

$$P_6 = \frac{L}{R}[M_A], \tag{3c}$$

$$P_7 = M_A + \frac{L}{R}[B_A + B_4], \tag{3d}$$

$$P_8 = B_A + B_L + B_4 + \frac{L}{R}[K_4], \tag{3e}$$

$$P_9 = K_4. \tag{3f}$$

For later use, combining equations (3a) and (3b) yields equation (4a):

$$[A_3 s^3 + A_2 s^2 + A_1 s + A_0]X_3 = \varphi_P X_d + G\frac{K_m}{R}K_f F_d. \tag{4a}$$

Here

$$A_3 = \frac{L}{R}[M_A], \tag{4b}$$

$$A_2 = M_A + \frac{L}{R}[B_A + B_4], \tag{4c}$$

$$A_1 = B_A + B_L + \varphi_F B_4 + \frac{L}{R}[K_4], \tag{4d}$$

$$A_0 = \varphi_F K_4 + \varphi_P, \tag{4e}$$

$$T(s) = \frac{X_3}{X_d}; \tag{4f}$$

A_n represents the coefficients of the characteristic equation, and $T(s)$ is the third-order system's closed loop transfer function.

Case III—Rigid Structure with Moving Load This case, described by equations (5a)–(5f) and (6a)–(6d), is generated by assuming that L is very small and K_4 is very large, i.e., a rigid structure:

Actuator:

$$\{[P_{10}]s^2 + [P_{11}]s + [P_{12}]\}X_3 = G\frac{K_m}{R}V_{in} + F_{ex5}; \tag{5a}$$

Controller:

$$V_{in} = K_p X_d + K_f F_d + K_f F_{ex5} - P_{13} X_3. \tag{5b}$$

Here

$$P_{10} = M_A + M_5, \tag{5c}$$

$$P_{11} = B_A + B_L + B_5, \tag{5d}$$

$$P_{12} = K_5, \tag{5e}$$

$$P_{13} = [K_f M_5]s^2 + [K_f B_5]s + [K_f K_5 + K_p]. \tag{5f}$$

For later use, combining equations (5a) and (5b) yields equation (6a):

$$\{[M_1]s^2 + [B_1]s + [K_1]\}X_3 = \varphi_P X_d + G\frac{K_m}{R}K_f[F_d + F_{ex5}]. \tag{6a}$$

Here

$$M_1 = M_A + \varphi_F M_5, \tag{6b}$$

$$B_1 = B_A + B_L + \varphi_F B_5, \tag{6c}$$

$$K_1 = \varphi_F K_5 + \varphi_P. \tag{6d}$$

Figure 14.7 shows a typical root locus of case III, the second-order system, with K_p fixed. As in the previous plot, note that the load oscillations become more free as K_f is increased. At very large values of K_f the A poles approach the load zeros and the actuator is functionally absent; i.e., the system behaves as the free load.

Figure 14.8 shows a typical root locus of case II, the third-order system used to represent structural oscillations. Note that as K_f is increased, the complex poles (B poles) approach the imaginary axis.

Figure 14.9 is a root locus plot in which the second- and third-order models have been overlayed onto the fifth-order model with K_p fixed and K_f varying. The objective of this figure is to illustrate the validity of the assumptions that allow the separation of the fifth-order system into the second- and third-order models. Note that the large magnitude B poles overlay closely, indicating that using the third-order structural model is acceptable for the system under consideration. In this case, the second-order model does not agree quite as well for low values of K_f. For higher

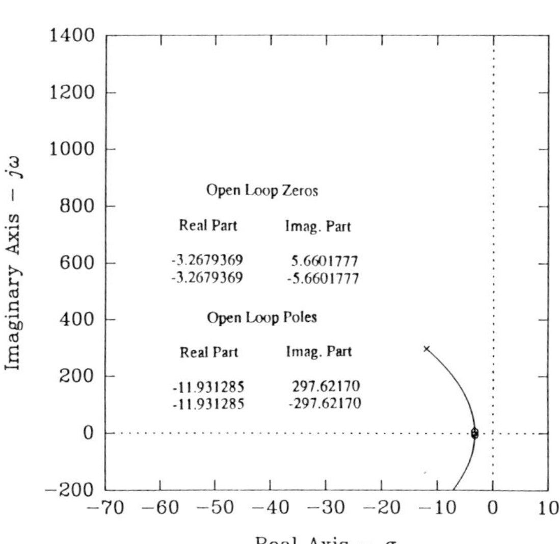

Figure 14.7
Root locus for case III with $K_p = 30{,}000$ V/m and K_f varying.

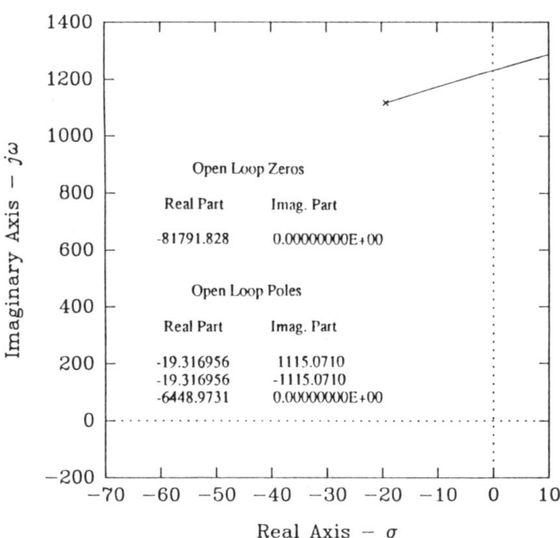

Figure 14.8
Root locus for case II with $K_p = 30{,}000$ V/m and K_f varying.

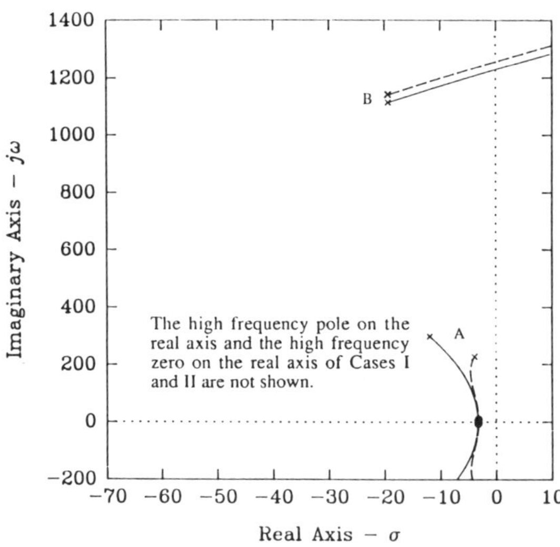

Figure 14.9
Cases II and III (solid lines) superimposed on case I (dotted lines).

K_f the agreement is quite good. Increasing K_4 and B_4 would move the A poles of figure 14.7 closer to the A poles of figure 14.6. Then, however, the agreement for intermediate values of K_f would not be as accurate.

Case IV—Compliant Structure with Slow Load and Fast Motor This case is generated by assuming that the motor inductance (L) is zero and is described by equations (7a)–(7d):

$$\{[M_2]s^2 + [B_2]s + [K_2]\}X_3 = P_4\varphi_F X_4 + \varphi_P X_d + \left[G\frac{K_m}{R}K_f\right]F_d, \quad (7a)$$

$$M_2 = M_A, \quad (7b)$$

$$B_2 = B_A + B_L + \varphi_F B_4, \quad (7c)$$

$$K_2 = \varphi_P + \varphi_F K_4. \quad (7d)$$

14.4 Characterizing System Behavior

14.4.1 Behaviors

Clearly, system behavioral requirements, rather than the convenience of the designer, should drive design decisions. Most often, however, quantitative performance criteria (QPCs) are favored as a starting point for design. QPCs, which are intentionally structured for convenient use in optimization procedures, can be simple, such as rise time and bandwidth, or more comprehensive, such as quadratic forms on integral functions of error and time. Using well-defined QPCs without reflection can divert focus from the main objective of any design procedure, which is the production of a system which achieves desired behavioral characteristics.

Approaching effector design from a behavior-based viewpoint can be extremely valuable, since it alters the designer's mentality by forcing acceptance of nonlinear issues and by encouraging the inclusion of experimental validation procedures into development methodologies. Finally, a behavior-based viewpoint makes clear the advantages of designing systems that behave appropriately as a result of intrinsic physics rather than as a result of compensation via ad hoc feedback loops.

14.4.2 Definition of Behaviors

Certain behavioral characteristics, such as speed and strength, are easily defined. However, many are hard to think about, being multidimensional, nonlinear, and difficult to quantify for convenient use in analytical optimization procedures. Examples of behavioral characteristics that are hard to define include quickness, grace, trajectory execution precision, ability to maintain stability when touching objects, response to external loading insults, saturation avoidance, and functioning without structural oscillations.

Some behaviors can be characterized as active in that they are a result of control inputs, and others can be considered as passive since they are responses to environmental insults or influences. It is important to understand the difference, since passive responses, many times ignored, may actually have much higher demands in terms of bandwidth and power and thus emerge as fundamental determinants of system success.

Consider desired behaviors for the one-degree-of-freedom effector in figure 14.2. A typical goal statement for the one-link arm would be that the system should have appropriate active responses to inputs being strong, fast, accurate, and graceful. The system should also exhibit minimal high-frequency oscillation (hum) between the actuator and structure and should not oscillate (shake) while in contact with objects.

The system should be compliant (appropriately passive) and experience minimal internal loads when pushed out of the way. Finally, system elements should not saturate or overheat in operation.

These are in fact complex goals that involve the simultaneous satisfaction of multiple conflicting constraints (SSMCC). The really difficult problem is to translate these behaviors into quantitative performance criteria so that they can be applied to the system model.

14.5 Formulation of Design Equations

14.5.1 Comments

Previous sections introduced a comprehensive model and a number of individual cases useful as a basis for *design*. This section introduces a number of quantitative performance criteria that can be used to generate *design equations*. The final section, "Design Scenarios," briefly introduces the application of design equations to specific situations. This final section, which is rough and conceptual, is by no means a complete treatment of all issues. It serves to introduce an enormous amount of work that remains to be completed before the design of effector systems can be a truly rigorous procedure.

14.5.2 Basis for Design Equations

The model includes ten system elements with twenty-six parameters as shown in table 14.1. In this chapter, position variables include the angular position of the motor (θ), the deflection of the drive (X_3), and the deflection of the load (X_4). Inputs are the desired position (X_d), the desired force (F_d), the externally applied push forces (F_{ex5} and F_{ex6}), the externally applied base force (F_B), the binary touch parameter (T_1—which engages the contact between the load and the touch impedances), the shove variable (X_6), and the base shaking variable (X_7 together with T_2).

14.5.3 Desired Effector Behaviors

For the model shown in figure 14.4, eleven desired behaviors are considered along with the quantitative performance criteria they generate. Note that for each of the behaviors, many different quantitative performance criteria can be defined. Each QPC will succeed, to various degrees, in imposing desirable behaviors on the system. In this paper we select the simplest possible QPCs. In later papers, more comprehensive but less mathematically convenient QPCs will be applied.

QPC 1 Structural smoothness, ζ_2, is defined here as the absence of high-frequency structural oscillation or instability. These oscillations are caused by interactions between the actuator and load through the structural impedance. A smoothness constraint equation can be generated based on a number of approaches. However, to maintain simplicity in this case, smoothness will be characterized by the damping ratio, ζ_2, of the complex poles in the third order model of case II.

From equation (4a) the system is represented by equation (8):

$$\left\{s^3 + \left[\frac{A_2}{A_3}\right]s^2 + \left[\frac{A_1}{A_3}\right]s + \left[\frac{A_0}{A_3}\right]\right\}X_3 = \left[\frac{\varphi_P}{A_3}\right]X_d. \tag{8}$$

Equation (9) shows a form that allows a convenient (traditional) definition of pole locations:

$$[(\tau_2)s + 1]\{s^2 + [2\zeta_2\omega_{n2}]s + [\omega_{n2}]^2\} = \tau_2 Z_9 X_d. \tag{9}$$

Expanding the terms of equation (9) yields equation (10):

$$\{s^3 + [Z_6]s^2 + [Z_7]s + [Z_8]\}X_3 = [Z_9]X_d. \tag{10}$$

Here

$$Z_6 = \frac{2\zeta_2(\omega_{n2})\tau_2 + 1}{\tau_2}, \tag{11a}$$

$$Z_7 = \frac{\tau_2(\omega_{n2})^2 + 2\zeta_2(\omega_{n2})}{\tau_2}, \tag{11b}$$

$$Z_8 = \frac{(\omega_{n2})^2}{\tau_2}. \tag{11c}$$

Equating like terms yields equations (12a)–(12c):

$$Z_6 = \frac{A_2}{A_3} = \frac{M_A + (L/R)[B_A + B_4]}{(L/R)[M_A]}, \tag{12a}$$

$$Z_7 = \frac{A_1}{A_3} = \frac{B_A + B_L + \varphi_F B_4 + (L/R)[K_4]}{(L/R)[M_A]}, \tag{12b}$$

$$Z_8 = \frac{A_0}{A_3} = \frac{\varphi_F K_4 + \varphi_P}{(L/R)[M_A]}. \tag{12c}$$

If one chooses marginal stability for the structural oscillations, then $\zeta_2 = 0$ and

$$\frac{A_2}{A_3} = \frac{1}{\tau_2}, \tag{13a}$$

$$\frac{A_1}{A_3} = (\omega_{n2})^2, \tag{13b}$$

$$\frac{A_0}{A_3} = \frac{(\omega_{n2})^2}{\tau_2}. \tag{13c}$$

The results in equations (13a)–(13c) are the same as would be obtained using a Routh array, as shown in equation (14):

$$A_2 A_1 = A_0 A_3. \tag{14}$$

One can set ω_{n2} and τ_2 in equations (13a)–(13c) as desired to further constrain system variables.

QPC 2 Structure/actuator stability assurance, AI, is an alternative to QPC 1 and applies to case II as shown in figure 14.10. By placement of the zero (via adjustment of K_4 and B_4), the asymptote intersection (AI) with the real axis can be moved into the left half plane thereby indemnifying the system against instability regardless of the value of K_f.

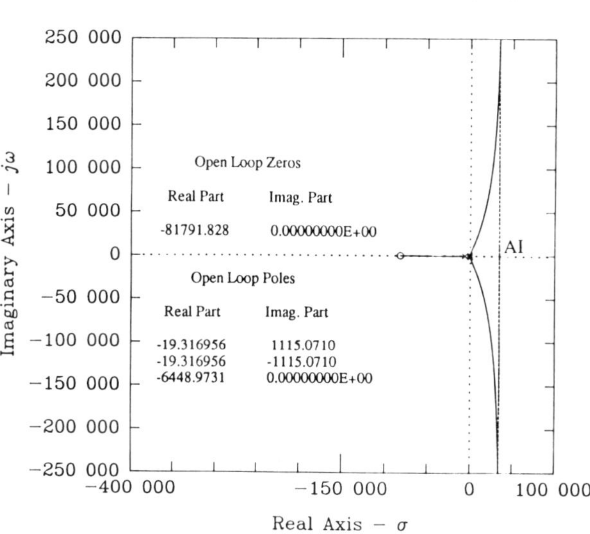

Figure 14.10
Schematic of root locus plot for case II showing the location of intersection of asymptotes with the real axis.

Recall that the closed loop transfer function, $T(s)$ in equation (4f), can be written in the form

$$T(s) = \frac{K_{10}G_1(s)}{1 + K_{10}G_1(s)}. \tag{15}$$

Then the open loop transfer function can be shown for K_p fixed and K_f varying as equations (16a)–(16c):

$$K_{10}G_1(s) = \frac{[G(K_m/R)K_f B_4(1/A_3)][s + (K_4/B_4)]}{[s^3 + (A_2/A_3)s^2 + (A_4/A_3)s + (A_5/A_3)]}. \tag{16a}$$

Here

$$A_4 = B_A + B_L + B_4 + \frac{L}{R}[K_4], \tag{16b}$$

$$A_5 = \varphi_P + K_4. \tag{16c}$$

The intersection can then be determined by selecting the second terms in the numerator and denominator polynomials defining sums of poles and zeros as shown by equations (17a)–(17b):

$$\mathrm{AI} = \frac{-(A_2/A_3) + (K_4/B_4)}{2} \tag{17a}$$

or, in expanded terms,

$$B_4\left[M_A + \frac{L}{R}(B_A + B_4)\right] = K_4\frac{L}{R}[M_A]. \tag{17b}$$

QPC 3 Static force accuracy, $\mathbf{A_{SF}}$, is derived from equations (4) with $X_d = 0$. This yields equations (18) and (19):

$$A_0 X_3 = G\frac{K_m}{R}K_f F_d, \tag{18}$$

$$F_{st} = K_4 X_3 = F_a. \tag{19}$$

Combining these, equation (20) is derived:

$$\frac{F_a}{F_d} = \frac{K_f K_4}{K_f K_4 + K_p + (RK_4/GK_m)}. \tag{20}$$

Equation (20) shows that a system with a high reduction and no position feedback ($K_p = 0$) will be statically accurate in terms of force generation.

QPC 4 Quickness, $\mathbf{Q_L}$, is probably best represented by the inverse of the time required for the system to transfer the load between two end states. A more simple indicator of quickness used in this chapter is the inverse of rise time, as shown by equation (21):

$$\frac{1}{RT} = Q_L = \frac{\omega_{nL}}{\alpha}. \tag{21}$$

The term alpha is defined by equation (22):

$$\alpha = [0.8 + 2.5\zeta_L]. \tag{22}$$

The natural undamped frequency is defined by equation (23):

$$\omega_{nL} = \sqrt{K_1/M_1}. \tag{23}$$

The quickness equation can then be written as equation (24):

$$(Q_L)^2\alpha^2 M_1 = K_1. \tag{24}$$

QPC 5 Positional accuracy, A_p, is an indicator of a system's ability to position itself in its working space within a specified tolerance. Here, accuracy will be defined by equation (25) as one minus the steady state error:

$$A_p = \left[\frac{X_3}{X_d}\right]_{\text{Steady State}} = \frac{\varphi_P}{K_1}. \tag{25}$$

Rearranging and combining with equations (6) produces the constraint equation (26):

$$K_5 \varphi_F A_p = \varphi_P (1 - A_p). \tag{26}$$

QPC 6 Load movement grace, ζ_L, depends on the specific application of the system. For example, it might be desirable to achieve a good step response while positioning a load, or, as in the case of an entertainment robot, it might be desirable to have the system swing freely or "dangle" well. For this case, a reasonable indicator of grace will be the damping ratio (ζ_L) of the load in case III.

The constraint is shown below as equation (27):

$$4\zeta_L^2 M_1 K_1 = B_1^2, \tag{27}$$

where $\zeta_L = 0.707$ produces a good step response and ζ_L less than 0.1 yields a system with good dangle.

QPC 7 Strength, S_{LM}, indicates the capability of the system to move its load through an excursion of Δ without exceeding power requirement P.

From equations (5a)–(5f) the static deflection versus input voltage is shown in equation (28):

$$P_{12} X_3 = G \frac{K_m}{R} V_{in}. \tag{28}$$

If the input voltage is V_m and the maximum deflection is Δ (also called strength), then equation (29) results as the constraint:

$$S_{LM} = \Delta = \frac{G K_m}{\sqrt{R K_5}} \sqrt{P}, \qquad P = V_{in} i_m, \tag{29}$$

where i_m is the armature current and power is the product of motor current and voltage.

QPC 8 Saturation avoidance, $\mathbf{S_A}$, constrains the system to stay within the voltage of V_m when the system undergoes a maximal step input of Δ.

From equations (5a)–(5f), the controller relationship together with $V_{in} = V_m$, and $X_d = \Delta$, equation (30) results:

$$V_m = K_p\Delta - X_3[K_f M_5]. \tag{30}$$

The acceleration $\ddot{X}_3$ is shown by equation (31):

$$\ddot{X}_3 = \frac{G(K_m/R)V_m}{M_A + M_5}. \tag{31}$$

Then the constraint is as shown in equation (32):

$$S_A = V_m = K_p\Delta\left[\frac{M_A + M_5}{M_A + \varphi_F M_5}\right]. \tag{32}$$

QPC 9 Response of the *load* to insults, ζ_3, ω_{n3}, and Δ_3, is used to describe the response of the load to external disturbances. Thus QPC 9 is an indicator of the expected passive behavior. It can be used, for example, to determine the behavior of a leg contacting the ground during running or walking motions.

QPC 9 assumes that F_d, X_d, B_5, and K_5 are all zero, that M_5 is a vehicle mass, and that F_{ex5} is the vehicle weight ($g_c M_5$). This produces equation (33a)–(33d):

$$\{[M_3]s^2 + [B_3]s + [K_3]\}X_3 = G\frac{K_m}{R}K_f g_c M_5. \tag{33a}$$

Here

$$M_3 = M_A + \varphi_F M_5, \tag{33b}$$

$$B_3 = B_A + B_L, \tag{33c}$$

$$K_3 = \varphi_p. \tag{33d}$$

The load response can then be constrained by setting the damping ratio ζ_3, the natural frequency ω_{n3}, and the static deflection Δ_3 in equations (34a)–(34c):

$$4(\zeta_3)^2 M_3 K_3 = (B_3)^2, \tag{34a}$$

$$(\omega_{n3})^2 M_3 = K_3, \tag{34b}$$

$$K_p\Delta = gK_f M_5. \tag{34c}$$

QPC 10 Response of the *structure* to insults, ζ_5, is similar to QPC 9, but it describes the behavior of the structure to externally imposed disturbances. During encounters with external loadings, structural force can grow dramatically until the actuator catches up with the insult. The system can experience an insult from a load as represented by a ramp of variable X_4 as in equations (7). Equation (35) shows the response X_3 to input X_4 with φ_P, X_d, and F_d all set to zero:

$$X_3 = \frac{P_4 \varphi_F}{\{[M_2]s^2 + [B_2]s + [K_2]\}} X_4. \tag{35}$$

The structural force F_{st} can then be determined via equation (36).

$$F_{st} = P_4(X_3 - X_4). \tag{36}$$

Although not completed here, equation (36) can be used to control force overshoot in response to ramp inputs X_4 by setting its damping ratio ζ_5 at an appropriate value such as 0.707.

In this case, given the speed limitations of actuators (which simultaneously must also be strong), the structural parameters B_4 and K_4 can be chosen to soften (minimize) actuator loading during rapid transients. This improves passive performance at the expense of active speed, as shown in the next paragraph defining force generation quickness.

QPC 11 Force generating quickness, Q_4. In case IV, the actuator applies force to the load through the structure. If the structure has been softened by application of QPC 10, its force generation speed can be degraded [the actuator must store energy in the structure (B_4 and K_4) in order to transfer force to the load]. QPC 11 uses a quickness criterion Q_4 similar to QPC 4 to impose a speed constraint on the force generation capability of the system.

The actuator generates force on a static load through the structure by moving X_3. This situation is represented by equations (7) with X_4 and X_d equal to 0 and with F_d as the input. The quickness of force control is then shown by equations (37a)–(37b) (RT = rise time).

$$\frac{1}{RT} = Q_4 = \frac{\omega_{n4}}{\alpha_4}. \tag{37a}$$

Here

$$\alpha_4 = (0.8 + 2.5\zeta_4). \tag{37b}$$

Using the definition of natural undamped frequency produces the constraint equation (38), where Q_4 and α_4 must be specified:

$$Q_4^2\alpha_4^2M_2 = K_2. \tag{38}$$

14.5.4 Comments

In addition to the eleven behaviors mentioned above, many others can be used to generate QPCs and constraint equations. Their selection depends on the specific intended application of the effector. Just a few examples are

1. pushing capability, which can quantify the system's capability to generate external forces,
2. stability margin when touching objects, i.e., tolerance to changes in load impedance,
3. response to shaking of the base (variable X_7 on figure 14.4),
4. the system's tolerance to degrading functional nonlinearities such as backlash and stiction, and
5. various active and passive bandwidth measures.

Of course, other *static* system descriptors are also important. Examples of these are size, weight, geometry, kinematics, structural strength, acoustic noise generation, etc. It should also be remembered that performance measures are functions of more fundamental characteristics of the materials and components that comprise the system. For example, the following all have an effect on the final achievable performance of the overall system: (1) field strengths, which can be generated by magnetic materials, (2) magnetic characteristics of iron, (3) electrical properties of field generating coils, (4) power capacities and speed of commutation systems, (5) losses in bearings, and (6) inertial characteristics of armatures.

Ideally, as the relationships between fundamental characteristics and system model parameters are better defined, the design synthesis can proceed directly from a statement of desired behaviors into static characteristics such as size and weight. Actuators can then be more precisely designed to optimize the complete system in which they function.

14.5.5 Interesting Cases

Various combinations of equations (8)–(38), when combined with system equations, can be used to generate constraint equations. For example, using QPCs 1, 4, 5, 6, 7, and 8 [via equations (39) through (44)], structural stability, quickness, accuracy, grace, strength, and linearity can be constrained.

Assuming $\zeta_s = 0$ yields equation (39):

$$\left(\frac{L}{R}\right) M_A(\varphi_F K_4 + \varphi_P) = \left[M_A + \left(\frac{L}{R}\right)(B_A + B_4)\right]\left[(B_A + B_L) + \varphi_F B_4 + \left(\frac{L}{R}\right) K_4\right]. \tag{39}$$

Rearranging and combining equations (24), (26), and (27) with system equations produces the very interesting set of equations (40), (41), (42). Equation (40) is a mass equation, equation (41) is a damping equation, and equation (42) is a stiffness equation:

$$Z_1 \varphi_P - M_5 \varphi_F = M_A, \tag{40}$$

$$Z_2 \varphi_P - B_5 \varphi_F = B_A + B_L, \tag{41}$$

$$Z_3 \varphi_P - K_5 \varphi_F = 0. \tag{42}$$

Here the performance goals are embedded in the equalities,

$$Z_1 = \frac{1}{A_p (Q_L)^2 (\alpha)^2}, \tag{43a}$$

$$Z_2 = \frac{2\zeta_1}{A_p Q_L \alpha}, \tag{43b}$$

$$Z_3 = \left(\frac{1}{A_p}\right) - 1. \tag{43c}$$

The strength equation from QPC 7 yields equation (43d):

$$S_{LM} = (G K_m \sqrt{P})/(\sqrt{R} K_5). \tag{43d}$$

Finally, the saturation avoidance equation from QPC 8 yields equation (44):

$$V_{max}(M_A + \varphi_F M_5) = K_p \Delta (M_A + M_5). \tag{44}$$

14.6 Design Scenarios

14.6.1 Use of Constraints

Obviously, the generation of constraint equations is only a beginning that can, as shown above, become quite complicated. In undertaking a real design procedure, the constraint equations must be carefully selected in an attempt to produce desired behaviors. Then constraints can be used in an appropriate computational format to determine design limits. Later papers will address the selection of constraints and their quantitative implementation.

14.6.2 Objectives

Our objective is to develop the capability to execute systematic design scenarios in terms of system variables, which include

1. 26 model parameters,
2. 2 controller gains, and
3. 12 performance goals.

In each scenario, some of the variables are specified and other are unknown. The object in each scenario is to determine the range of values permissible for unknowns given the knowns. Then the selection of physical hardware can begin.

Prior to discussing the scenarios it is advantageous to reorganize constraint equations by eliminating φ_P and φ_F from equations (40), (41), and (42). The simple design number shown in equation (45) results, where desired performance is carried by parameters Z_1, Z_2, Z_3, the load is defined by M_5, K_5, B_5, and actuator characteristics are defined in terms of M_A, B_A, and B_L:

$$\frac{Z_3 M_5 - Z_1 K_5}{Z_3 B_5 - Z_2 K_5} = \frac{M_A}{B_A + B_L}. \tag{45}$$

With some effort, similar equations can be generated for combination with other sets of constraint equations (design rules). Equation (45) is particularly well structured since it so cleanly states the requirements for a system that obeys constraints on quickness, accuracy, and grace.

Two example scenarios are included below to introduce concepts briefly. In these discussions only six QPCs (1 and 4 through 8) are used. Later papers will expand the number of QPCs used.

Scenario A—"What's a Good Actuator?" In this scenario the goal is to determine those actuator characteristics required to produce the desired effector system behavior given the characteristics of structure, load, and controller. In this case, the sixteen known parameters are K_4, B_4, M_5, B_5, K_5, M_6, B_6, K_6, T_1, T_2, τ, M_7, B_7, K_7, S_p, S_f, and the six QPCs ζ_s, S_{LM}, Q_L,

A_p, ζ_L, and S_A. The ten unknown parameters are L, J_m, B_m, K_m, R, B_3, M_3, G, K_p, and K_f.

The constraint equations are used to limit possible values of the unknown parameters, thus defining limits on what the motor, reducer, drive, and controller can be. Note that if performance requirements are too aggressively selected the analysis might ask for a motor that cannot exist due to physical realities.

By pursuing scenario A in greater depth, general design rules can be generated which permit not only a better selection of commercially available system components, but a better understanding of the design of motors, drives, and controllers that are more appropriate for use in robotic applications.

Scenario B—"What Are the Limits of a Given System's Performance?" In this scenario the goal is to determine what performance is possible for various feedback gains given the system characteristics. In this case all parameters are defined except for the performance criteria ζ_s, S_{LM}, Q_L, A_p, ζ_L, and S_A, and the controller gains K_p and K_f. The objective is generating an understanding of the relationships between controller settings and the limits of system performance.

14.7 Conclusions

Being able to answer questions such as "What's a good actuator?" and "How well can a specific effector system perform?" will be one of the fundamental determinants of success in the design of future robots. Attempting to answer these complex questions motivated this chapter, which is the first of a number of related publications from the CED.

This chapter applies a relatively conventional design methodology to a carefully structured, fifth-order, one-degree-of-freedom study model so that important issues can be identified and understood. Using this approach, simple effector systems can be designed in a comprehensive way, beginning with a systematic understanding of system behavioral objectives that guide selection of system elements and their controller (i.e., servo science).

Clearly, the design procedures will be complex for more comprehensive systems, which include multiple degrees-of-freedom, nonideal elements, nonlinear kinematics, and other realities. However, in these more complicate cases, numerical methods will be applied, together with an expanded data base on actual subcomponents in order to facilitate the design of effectors.

14.7.1 Future Work

Work in the CED is now pursuing answers to questions in three areas:

1. Thoroughly understanding the linear, one-degree-of-freedom model in figure 14.4. Specifically
   a. *Response to shove*—How well can the output X_3 be maintained in the presence of external disturbances represented, for example, as a step in F_{ex5} or F_{ex6} or as an initial velocity of M_5? Should the response be characterized by the behavior of displacement or force?
   b. *Stability margin when touching external loads*—What is the effect on performance of changing the load impedance by touching?
   c. *Response to base shaking*—Using X_d as an input, how well can the effector maintain its output, X_3, with a disturbed base X_7?
   d. *Performance in operation with a compliant base*—With a soft footing ($T_2 = 0$ and M_7, B_7, K_7 "soft") how well can the output X_3 be controlled by the input X_d?
   e. *Operation with noncolocated sensors*—As sensors are separated from actuators across dynamic structures, how are resulting delays tolerated?
   f. *Intrinsic qualities*—Can they be used to design in desirable system characteristics? For example, can structural damping, B_4, as presented in QPC 2, be used to eliminate the instability caused by interactions between the actuator and structure?
   g. *Understanding the behavior of systems in specific simplified operating regimes*—For example, how will the system behave while (1) operating at high speed with small loads (i.e., as an indicator with small M_5, B_5, K_5); (2) operating massy loads (i.e., as a free arm with B_5 and K_5 small); and (3) running against viscous loads (i.e., as a speaker cone with B_5 large)?
2. The study of more complicated systems with
   a. *Higher-order models*—How will the system behave with more complicated motors, reducers, drives, structures, loads, sensors, and controllers?
   b. *Additional degrees of freedom*—How should a cascade of the model be controlled? (Cascade is when successive models are connected at base and load masses.) Can a cascade be balanced so that an intrinsic

passivity is created which causes masses to respond to disturbances in a coordinated fashion?

c. *Surviving nonlinearities*—How can saturation, backlash, stiction, and delays be tolerated? Can additional elements such as K_3 be used to counteract problems such as backlash?

d. *Adaptive control capabilities* used for both calibration maintenance and behavior modification in response to inputs. For example, can a variation of actuator and load damping be used to avoid instabilites?

e. *Alternate components*—Can a simple, fluid-based actuator model be inserted in order to allow the design of hydraulic or pneumatic systems?

3. Effects to make results more useful for general design procedures

a. Nondimensional groups, like those used in mechanics, will be defined to better represent system elements, such as motors, structures, and loads. Such normalized representations will simplify the understanding of system behavior, as well as form the basis for a more useful method of characterizing system elements supplied by manufacturers.

b. Quantitative performance criteria that can impose constraints which produce desired effector system behavior will be defined.

c. Actuator parameters, such as K_m, R, L, and G, will be related to static actuator characteristics, such as size and weight, for use in comprehensive design procedures.

Acknowledgments

The authors wish to acknowledge the support of the Systems Development Foundation, the Defense Advanced Research Projects Agency (DARPA/USAF, Wright-Patterson), and the Naval Ocean Systems Center of the Office of Naval Research.

The authors also thank the Artifical Intelligence Laboratory at the Massachusetts Institute of Technology for its collaborative efforts.

References

An, C. H., and Hollerbach, J. M.: "Dynamic Stability Issues in Force Control of Manipulation," Proc. I.E.E.E. Int. Conf. Robotics and Automation, Raleigh, NC, March 30–April 3, 1987.

Biggers, K. B., Jacobsen, S. C., and Gerpheide, G. E.: "Low-Level Control of the Utah/M.I.T. Dextrous Hand," Proc. I.E.E.E. Int. Conf. Robotics and Automation, San Francisco, CA, April 7–10, 1986.

Book, W. J., Maizza-Neto, O., and Whitney, D. E.: "Feedback Control of Two Beam, Two Joint Systems with Distributed Flexibility," Trans. A.S.M.E. J. Dyn. Sys. Meas. and Control, pp. 421–424, December 1975.

Book, W. J.: "Recursive Lagrangian Dynamics of Flexible Manipulator Arm Via Transformation Matrices," I.F.A.C. Symp. CAD Multivariable Technological Sys., W. Lafayette, IN, Vol. 5, No. 17 (1983).

Book, W. J.: "Recursive Lagrangian Dynamics of Flexible Manipulator Arms," 2nd Sym. of Robotics Research, Kyoto, 1984.

Cannon, D. H., and Schmitz, E.: "Initial Experiments on the End-Point Control of a Flexible One-Link Robot," International Journal of Robotics Research, Vol. 3, No. 3, pp. 62–75, Fall 1984.

Coy, J. R.: "Dynamic Performance and Control of an Electromechanical Actuator," M.S. Thesis, Department of Mechanical Engineering, Brigham Young University, Provo, UT, 1987.

Fullmer, R. R., Meek, S. G., and Jacobsen, S. C.: "Generation of the 7 Degree-of-Freedom Controller Equations for a Prosthetic Arm," Presented at the Conference on Applied Motion Control, University of Minnesota, Minneapolis, MN, 11–13 June 1985.

Hollerbach, J. M.: "A Recursive Formulation of Lagrangian Manipulator Dynamics," I.E.E.E., Trans. Systems, Man, Cybernetics, Vol. 10, No. 11 (1980).

Hollerbach, J. M.: "Dynamic Scaling of Manipulator Trajectories," A.S.M.E. J. Dyn. Syst. Meas. and Control, Vol. 106, No. 1 (1984-a).

Hooker, D. J.: "Robot and Payload Weight Compensation for a Compliant Robot in Free Space," M.S. Thesis, Department of Mechanical Engineering, Brigham Young University, Provo, UT, July 1987.

Hooker, D. J., Seamons, M. B., and Smith, C. C.: "Position and Orientation Control of a Compliant Robot with Weight Compensation," Society of Engineering Science, 24th Annual Meeting, Salt Lake City, UT, September 1987.

Jacobsen, S. C.: "Control Systems for Artificial Limbs," Ph.D. Thesis, Massachusetts Institute of Technology, Cambridge, MA, 1973.

Jacobsen, S. C., Meeks, S. G., and Fullmer, R. R.: "An Adaptive Myoelectric Filter," 6th Conference I.E.E.E./Eng. in Med. and Biol. Soc., 1984.

Jacobsen, S. C., Iversen, E. K., Knutti, D. F., Johnson, R. T., and Biggers, K. B.: "Design of the Utah/MIT Dextrous Hand," I.E.E.E. International Conference on Robotics and Automation, San Francisco, CA, March 1986.

Johnson, D. A.: "A Simulation of Force Feedback Control of Robotic Assembly," M.S. Thesis, Department of Mechanical Engineering, Brigham Young University, Provo, UT, December 1984.

Kelley, C., III: Personal communication, 1985.

Meeks, S. G.: "An Adaptive Myoelectric Filter," M.S. Thesis, University of Utah, Salt Lake City, UT, 1980.

Reese, B. S.: "Excitation of Rotary Mechanical Systems for System Identification," M.S. Thesis, Department of Mechanical Engineering, Brigham Young University, Provo, UT, December 1984.

Seamons, M. B.: "Compliant Position Control for a Robot Manipulator in Free Space," M.S. Thesis, Department of Mechanical Engineering, Brigham Young University, Provo, UT, March 1986.

Seering, W. P., and Eppinger, S. D.: "Understanding Bandwidth Limitations in Robot Force Control," Proc. I.E.E.E. Int. Conf. on Robotics and Automation, Raleigh, N.C., March 30–April 3, 1987.

Smith, C. C.: "Robot Control for Assembly," Alliance with Industry Conference, Brigham Young University, Provo, UT, November 1985.

Whitney, D. E.: "Resolved Motion Rate Control of Manipulators and Human Prostheses," I.E.E.E. Trans. Man-Machine Systems, 10, 1969-a.

Whitney, D. E.: "Force Feedback Control of Manipulator Fine Motions," 1976 Fine Automatic Control Conf., San Francisco, 1976.

Whitney, D. E.: "Force Feedback Control of Manipulator Fine Motions," A.S.M.E. J. Dyn. Sys. Meas. and Control, 1977.

15 Using an Articulated Hand to Manipulate Objects

Kenneth Salisbury, David Brock, and Patrick O'Donnell

15.1 Introduction

We manipulate objects with our own hands by a complex and synergistic combination of motion and sensing. Our skillfulness results not only from the quality with which we can impose motion and force, but also from well-developed sensory and cognitive capabilities. The "dexterity" of an articulated robot hand is similarly a function of the quality of the hand's sensory and motor skills and the quality of high-level control imposed on it.

Interest in replicating human hands, in both form and function, goes back centuries. Early prosthetic implements were most likely simple static devices used to provide some pushing and pulling capability. By the early 1500s there existed examples of relatively complex mechanisms used as lower arm prosthesis. One such hand employed articulated fingers that could be manually locked into position to provide a variety of static grasps [Childress]. As electric and other portable types of actuators were developed, prosthetic hands began to employ nonhuman power to perform their functions. Whether or not through conscious intent by the designers, these hands began to integrate simple grasping *strategies* such as compliance (compliant transmissions or coverings) and coordination (gear coupled motions). Durability and ease of control seem to have been driving forces in prosthetic development and only the simplest of prostheses have persisted in wide use for functional application. Perhaps because of the difficulty in directly communicating human intent to a multi-degree-of-freedom mechanical hand, prosthetic devices have tended to employ simple single-degree-of-freedom control inputs.

With the advent of sophisticated computer systems we are now able to assume a degree of "intelligent" control over much more complicated mechanical systems than ever before. Concurrently, we have begun to see new and more complex types of hand mechanisms with the potential for performing highly dexterous manipulation. The marriage of high-performance computational systems with sophisticated mechanical hand systems promises to provide increased dexterity in prosthetic, robotic and teleoperation applications. In addition, the experience we gain in trying to understand and achieve machine dexterity may give us insight into some of the mysteries of human dexterity.

It is important to observe that, despite the current availability of relatively sophisticated mechanical hands (most notably the Salisbury hand and the Utah/MIT hand), we have yet to see application of these devices outside the laboratory. While a high degree of mechanical sophistication is necessary for achieving dexterity with a robot hand, complexity alone will not make the hand dexterous. Without diminishing the importance of good mechanical design in hands (and indeed, good hands are hard to build), it appears that the real problem lies in integrating the control, sensory, and perceptual components of manipulation. Adding more fingers and motion freedoms to a robot does not free us from having to address the problems of planning and perception in the face of uncertainty; if anything, it forces us to face them more directly. The very nature of manipulation with cooperating fingers implies a degree of unpredictable behavior that requires more sophisticated sensing, feedback, and planning than found in existing robots.

In this chapter, then, we shall try to present our view of the steps that must be taken to achieve dexterity with a robot hand. Some of these, such as mechanical design, low-level control, and simple languages, have already been begun. Others, such as comprehensive sensory integration and high-level planning, are still in their formative stages. Our experience primarily has been with the Salisbury hand experimental environment at MIT and this will be used to illustrate our perspective.

15.2 Hand Functions

The functions that hands are required to perform can be divided into several areas: prehensile, manipulative, and sensory functions are perhaps the most obvious. In human experience they often occur in combination with each other in a mixture of exploratory and purposeful motions. In performing complex tasks humans simultaneously employ all three functions to verify, guide, and plan the sequence of operations performed. In the robotic world we are not yet so lucky as to be able to integrate these complex functions smoothly into coherent actions; current robotic systems employ a rather loose coupling of their prehensile, manipulative, and sensory functions, both structurally and operationally.

Prehensile capability implies the capacity to hold an object in some controlled state relative to the hand. This requires that the grasp allow exertion of sufficient restoring forces on an object to hold it against arbitrary disturbance forces. Secure grasping requires at least that there be sufficient dimension to the collective effect of forces acting through all the contact points to span the space of disturbance forces that may act on the object. For spatial motion of rigid objects this is six. The constraints imposed by contacts can be divided into structural constraint and frictional constraint. A grasp that employs more structural constraints will be more secure against large disturbance forces; frictional constraints will cease to be active if disturbance forces become too large in certain directions. The dimension of net forces acting on the grasped object can be no greater than six; therefore the extra freedoms in force exertion by the hand result in the freedom to adjust the internal forces. These internal forces may be varied without changing the net force acting on the body. They do permit, within certain limits, changing the direction and magnitude of forces acting through individual contact points. This freedom in allocation of grasp forces permits a degree of optimization of grasp force distribution. It also raises the possibility of modulating the frictional restraints so as to introduce partial freedom of motion within the hand (controlled slipping).

Manipulation of objects with a hand can take several forms. It may simply hold an object securely relative to the wrist and rely on the rest of the arm for imparting motion. This is typical of most robots and of humans moving relatively large objects. Motion of smaller objects can be achieved with the fingers alone. Objects held in the fingertips can be moved and rotated short distances without breaking or sliding finger contact. Larger motions within the hand require controlled slipping and repositioning of fingers. In manipulating very large objects the hand may recursively act as a large fingertip in concert with other cooperating hands.

Sensing includes a number of functions, ranging from simple measurement of the mechanism's position, velocity, acceleration, and force states, to detecting the characteristics of contacts with objects such as location and force distribution, to the measurement of manipulated object characteristics such as curvature, texture, and temperature. Other sensory modes, such as visual and auditory, may be brought into play to provide higher-level feedback.

15.3 Dexterity

Dexterity then, is the integration of all these and more capacities into a higher level of competence. It is a quality of manipulative capability that implies a degree of skill and deftness of operation. These in turn, depend on the quality of motion, sensing, and control. If we look at examples of increasingly dexterous motion, the distinction between prehension, manipulation, and sensing becomes less obvious. Sensing the shape of an object may require manipulation of it to detect its contours. Holding an object may require adjusting to a more favorable grasp. And moving an object may require a sequence of regrasping. Humans easily integrate these functions to perform wonderfully complex operations. High degrees of mechanical functionality, rich sensory input, and intellectual capacity combine to produce human dexterity. Dexterity in the machine sense is still in its infancy and suffers from limited functionality in all three areas.

True *machine dexterity* necessitates that a high degree of mechanical functionality be available. In the same way that the quality of sound from a stereo radio can be no greater than that allowed by the speakers, the net dexterity of a robot hand can be no greater than is allowed by the mechanical and sensory performance of the system. Yet, even with the best possible mechanism, the dexterity of the hand will be limited by the quality of commands or "intelligence" used to control the system.

From a mechanical point view, the potential dexterity of a hand is dependent upon a number of elements including its kinematic, dynamic, and sensory qualities. The obvious kinematic description of a hand refers to the number of joints and links composing the hand, their placement relative to each other, and the ranges of motion possible. The more subtle aspects of a hand's kinematics deal with the shapes of the finger links and their ability to present suitable working surfaces to grasped objects for constraining and manipulating them. The dynamic performance of a hand is affected by the distribution of mass and compliance within the hand, the forces that may be exerted, and the speeds and accelerations that may be obtained. In addition, the resolution of force and motion affect the precision with which actions may be performed. Finally the type, accuracy, sensitivity, and resolution of sensing must be adequate to provide required perception of hand states and object interactions. The kinematics and dynamics of a hand limit the controllability of states of the manipulated objects. The sensory system limits the observability of task states. A mechanism lacking certain degrees of freedom may not be able to effect certain motions of a manipulated object; a mechanism lacking certain state information may not be able to detect particular motions. Thus the controllable and observable states permitted by a hand must overlap in the dimensions in which we want to achieve closed loop control. Perhaps the most difficult aspects of machine dexterity are the closely coupled intellectual and perceptual components. It is in performing complex tasks that require a high degree of dexterity that the demand for high-level "understanding" of task goals and progress is greatest. In prosthetic or teleoperation modes these high-level functions can be subsumed by the human. In the autonomous system either they must be explicitly encoded in the program or embodied in a more comprehensive world model from which motions are planned. This is one area where experience and experimentation with hand mechanisms will be of significant value in inspiring and guiding our thinking.

It is also important to realize that the skills we observe when several fingers cooperate to manipulate an object are but one example of a more general capacity to understand motions of objects subjected to forces and constraints. These same skills come into play when we clutch an object to our side with an elbow or when several hands cooperate to move a large object. Thus our research into dexterous hands is really a metaphor for a much broader class of manipulative skill that we may one day employ in robotic systems, including multiarm and multirobot operation.

In what follows we describe research with the 3-finger Salisbury hand at MIT. The work is aimed at gaining experience with articulated hand operation and developing strategies for bringing this type of hand into useful application. In addition, we are interested in determining what are the minimal sensor/hand mechanism requirements for performing reasonable and reliable grasping and manipulation with an articulated gripper. Our approach has been to develop an experimental environment integrating an articulated hand with a variety of control systems, sensors, and command language constructs.

15.4 Control System

The Salisbury hand (which was originally called the Stanford/JPL hand) is comprised of 3 fingers, each with three degrees of freedom (see figure 15.1). This particular choice of kinematics was made because it is the minimal configuration capable of imparting arbitrarily directed forces (including torques) and small motions (including rotations) to objects held in the fingertips. At the same time it permits the imposing of internal forces required for stable grasping with friction contacts [Mason and Salisbury]. In order to actuate the fingers, flexible-Teflon coated steel cables

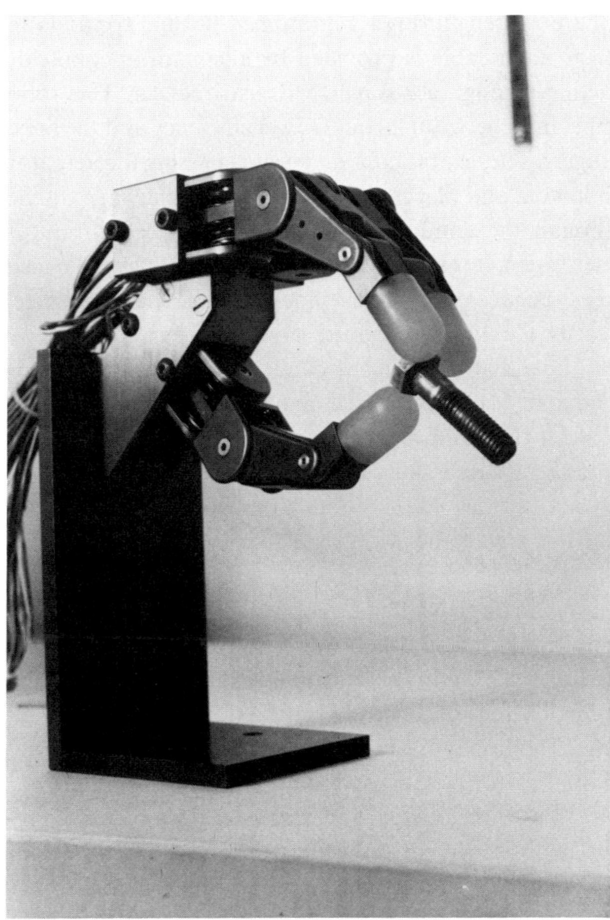

Figure 15.1
The Salisbury 3-finger hand (photographs courtesy of David Lampe, MIT).

(7 × 7 construction) are routed through Teflon-lined flexible conduits to each finger. Tension to each cable is provided by a samarium-cobolt dc brush-type motor acting through a 2-stage 25 : 1 gear reducer. This ratio was chosen as a compromise between torque/power efficiency and the back drive inertia of the fingers. The flexible conduit permits passing the actuator cables around a robot wrist and placement of the actuator package on the robot forearm. Although the conduit/cable sliding interface introduces some friction into the system, it is significantly less than the reflected brush friction in the motors. Because the cables slide over a Teflon interface, the friction is primarily Coulomb in nature and can be minimized by limiting the amount of bend allowed in the conduit. While initial experimentation with the hand at MIT has focused on hand operation alone and has been performed with the hand located on a manually repositionable support, the device has been successfully mounted on a number of PUMA 560 robots at other laboratories. [Loucks et al.] have demonstrated coordinated control of this hand and an arm in the grasping of moving objects.

The initial control system implemented on this hand (actually one finger) was at Stanford University and is described in [Salisbury and Craig]. The system consisted of a hierarchy of controllers (implemented on one PDP 11/45) with a tendon tension controller at the lowest level. Corrections to errors in desired tendon tensions were made by altering the torque (current) command applied to the corresponding motor (through appropriate compensation). "Position" control with this system was achieved by computing the Cartesian position error at the fingertip and using a Cartesian stiffness and damping matrix to control the endpoint impedance. Corrective forces to fingertip displacement were then mapped into joint space and then into tendon space to establish the tendon tension controller setpoints. Because each three-degrees-of-freedom finger is actuated by four tendons (the minimum possible), the extra freedom in tension selection was used to minimize the tensions required subject to the constraint that they all be positive. While the original Stanford implementation worked, it suffered from a number of problems. The Cartesian nature of the controller prevented explicit positioning of joints. This allowed multiple configurations of the fingertip to satisfy a given position command (i.e., knuckle up or knuckle down). The configuration to which the finger settled was dependent upon the path of the finger. Worse yet, when attempting to move the finger to singularities (such as fully straightening it), the finger would exhibit large position errors. At these positions the small joint friction torques overwhelmed the small corrective torques mapped from the Cartesian corrective forces through a nearly singular Jacobian. Finally the force control was not particularly robust to large disturbance forces. The motor brush friction effectively introduced hysteresis into the transmission of force from the motor to the tension sensor, resulting in limit cycles for only moderate gains.

The second control system was implemented at MIT and is described in detail in [Salisbury, Brock and Chiu]. This much improved system (called HAND) eliminated the motor brush friction problem by closing a high-gain velocity servo around each motor using a shaft-mounted encoder for feedback and a microprocessor operating at servo rate of 1,000 Hz. This allows us to assume velocity control over the proximal end of each tendon. This provides a stable "platform" from which force control may be effected. In its basic form the force control servo commands tendon velocities proportional to tendon tension error and the closed loop behavior approximates a first-order system response up to about 5 Hz. The behavior of this type of force controller in the presence of Coulomb friction and stiction has been analyzed in detail by [Townsend and Salisbury] and is stabilizable in the presence of Coulomb friction. Stiction does introduce limit cycles, the amplitude of which can be minimized by keeping the transmission stiffness high between the actuator and source of stiction.

The HAND system closes the force control loop at 50 Hz servo rate using a VAX-750 to command motor velocities and process tension sensor readings. As described in [Salisbury, Brock and Chiu] the controller running on the VAX specifies motions of the hand expressed in joint space. A joint space stiffness specification can be utilized to modify the local compliant behavior of the fingers. Trajectories in joint space can be flexibly constructed through a simple editor and used to command the hand motions. While being simple and computationally efficient, the joint space stiffness formulation permits a wide variety of compliant behaviors to be constructed including controlled stiffness in Cartesian space as well as the simple grasping "reflexes." It is also possible to command motions strictly in position control mode. The resulting maximum stiffness in this mode is dominated by the tendon stiffness and is roughly twice that of the maximum stable stiffness attainable under active stiffness control. This is the most precise mode of positioning and is often used for commanding motion of the fingers when they are not expected to contact an object. With control loop closure in joint space there are no problems with moving the fingers through singularities in the work space. The upper limit on actively controlled stiffness is due to a combination of limited servo rate and unmodeled dynamic modes in the system.

At the highest level, a Symbolics 3600 Lisp Machine communicates with the joint space controller to provide a flexible user interface for commanding motions, reacting to events and constructing programs. The OOLAH system developed by [Chiu] permits the user to construct Cartesian-based finger and object motions with the option of superimposing controlled stiffnesses and bias forces. It provides a means for reacting to user-defined force conditions. It also provides a common interface for coupling the hand to higher-level sensory, grasp, and trajectory planning modules currently under development. Communication between the Lisp-based OOLAH system and the VAX-based controller is achieved by an asynchronous message passing system described in the following section.

15.5 Object Manipulation

When we wish to move an object with a traditional manipulator we have little choice but to grasp it firmly with the (simple) gripper and command the necessary arm joint motions to effect the movement. Even a simple reorientation may require large motions of massive links to achieve it. The involvement of the whole arm in even small motions is one of the fundamental limitations we find in current manipulation practice. The power and complex control required to move massive links quickly and precisely place severe constraints on the quality of manipulation achievable. Precise control of forces is limited even more by these massive links. It is these shortcomings that provide one of the prime motivations for developing articulated hands. By placing low-mass, high-speed links (fingers) near the object to be manipulated, we provide for a whole new range of manipulation modes. Instead of using articulated handlike mechanisms, we could simply place multi-degree-of-freedom small motion devices between the gripper and wrist to gain some of the benefits of local small motions control. However, it is only by using a number of multi-degree-of-freedom fingers that we gain the additional benefits of reconfigurable and adaptable grasping.

Humans employ a wide range of grasps that permit varying degrees of dexterity and security in hand-based manipulation. Most of our experience with the Salisbury hand has been in the mode of grasping called *fingertip prehension*. In this mode an object is held in the fingertips only and the fingers are commanded to move simultaneously so as to impart the desired net motion or force to the object. If we make the simplifying assumption that each fingertip contacts the object at a single point, fixed in the object, with three degrees of freedom (i.e. a point contact with friction that is instantaneously equivalent to a "ball and socket" joint), then the kinematic and force relationships are easy to compute. Such a grasp defines a rigid grasping triangle. To achieve a desired motion of the object we simply compute the corresponding rigid body displacement of the triangle and the corresponding fingertip motions necessary to keep them in contact with the corners of the triangle. There are two subtle aspects to this process which must be dealt with.

15.5.1 Quasi-Static Force Relationships

The first is control of the internal forces. Since there are more than six degrees of freedom in the hand motion we must impose additional constraints to completely define the finger motions. Specifying that the grasping triangle shall not deform implicitly imposes these constraints but does not address the need to control the internal or grasping forces. With the three-finger Salisbury hand there are three such grasping forces that may be controlled. These forces *must* be imposed so as to present sufficient normal force at each contact to keep the tangential frictional forces large enough to constrain the object. The easiest way to do this is to rely on the inherent compliance of the mechanism (servo, transmission, and finger covering) to maintain sufficient internal forces after a grasp has been achieved. This has been repeatedly demonstrated to work well for a wide range of objects in our experimentation. More fragile objects and delicate operations require that we be able to assume more precise control over the force interaction between the hand and grasped objects.

Using active force and stiffness control [Chiu] we can specify more precisely the internal grasping forces. This mode is particularly useful for handling fragile objects where "grasping as hard as you can" is not appropriate. The transformation required for determining the fingertip forces necessary to achieve a desired set of internal and external forces on a grasped object is called the *grip transform*, G [Mason and Salisbury]. Because forces are applied by several contacts acting in parallel, we cannot compute this transform directly from the grasp geometry. We compute the grip transform by first constructing a matrix relating forces applied at the fingertips to the net force and torque on the object. We define a (6×9) $\mathscr{W}$ matrix to have each of its nine columns correspond to a component of net force that may be exerted upon the grasped object by finger action. This matrix will have rank six unless we have a degenerate grasp (3 collinear or 2 coincident grasp points). The first three columns will correspond to forces exertable by finger 1, the second three columns to those exertable by finger two, and the last three to forces exertable by finger three. If we also define a 9-vector of finger force intensities, $\mathbf{F} = [c_1, \ldots, c_9]^T$, then the net (external) force acting on the body is

$$\mathscr{W}\mathbf{F}. \tag{1}$$

In order to parameterize and assume control over the internal forces on a grasped object we first define a net generalized force acting on the object,

$$\mathscr{F} = [f_x, f_y, f_z, m_x, m_y, m_z, \lambda_1, \lambda_2, \lambda_3]^T. \tag{2}$$

The first six components are the familiar external forces on the object and the last three are the intensities of the internal forces (in a coordinate system yet to be chosen). For reasons that will be obvious in a moment we next form the (9×9) matrix

$$G^{-T} = \begin{bmatrix} \mathscr{W} \\ \text{------} \\ \mathbf{c}_{1,h}^T \\ \vdots \\ \mathbf{c}_{3,h}^T \end{bmatrix} \tag{3}$$

by augmenting the $\mathscr{W}$ matrix with three linearly independent row vectors that are orthogonal to the rows of $\mathscr{W}$ (e.g., in its null space). The product

$$\mathscr{F} = G^{-T}\mathbf{F} \tag{4}$$

then gives the net generalized forces acting on a grasped object when a given set of fingertip forces are exerted. The equation's inverse,

$$\mathbf{F} = G^T\mathscr{F}, \tag{5}$$

indicates what forces the fingers must be commanded to exert in order to apply a desired set of external and internal forces to an object. This equation is of fundamental importance in hand control and, as will be shown below, it can be used to construct a fairly general stiffness specification. It should be noted that the parameterization of the null space here is not specified. The choice of the best set of basis vectors for the null space ($c_{i,h}$s) is an open research topic; we suggest one useful choice below. It should also be noted that having rank($\mathscr{W}$) = 6 is not sufficient alone for achieving a stable grasp. The orientation of the uni-sense contact forces (i.e., normal contact forces) must be such that their forces can always be kept positive (and in contact) by readjustment of internal forces.

Via energy conservation principles we can also derive the grip velocity relationships. First we must clarify the meaning of input and output power in grip systems. We define **V** to be a vector of contact velocities occurring at the fingertips, expressed in the same coordinate systems as the contact forces, **F**,

$$\mathbf{V} = [d_1, \ldots, d_9]^T. \tag{6}$$

Since the elements of **V** are the intensities of motion along the axes of the corresponding contact forces,

$$\mathbf{F}^T\mathbf{V} = \sum_{i=1}^{n} c_i d_i \tag{7}$$

is the power being input to the object. It is the sum of the power being input at the contact points as a result of motion along each axis of force application.

We define $\mathscr{V}$ to be a 6-vector, the first six components of which represent the body's linear and angular velocity and the remaining 3 γ_is are its (virtual) velocities resulting from deformation of the body (grasp triangle). Thus

$$\mathscr{V} = [v_x, v_y, v_z, \omega_x, \omega_y, \omega_z, \gamma_1, \ldots, \gamma_3]^T \tag{8}$$

and

$$\mathscr{F}^T\mathscr{V} = f_x v_x + f_y v_y + f_z v_z + m_x\omega_x + m_y\omega_y + m_z\omega_z + \lambda_1\gamma_1 + \cdots + \lambda_3\gamma_3 \tag{9}$$

is the sum of power that is used to do work on the environment and to store energy as the object deforms. However, since we consider objects rigid and massless no energy can be stored. The internal forces, λ_i, may be nonzero, but they can do no work without violating our assumptions. Thus we expect the γ_is to be zero, and if we do encounter nonzero values for these internal displacements it will mean that our grip has slipped and/or our assumptions about contact constraints have been violated.

Equating power input with power output and rate of storage [(7) and (9)] and using (4), we have

$$\mathbf{F}^T\mathbf{V} = \mathscr{F}^T\mathscr{V} = (\mathbf{F}^T G^{-1})\mathscr{V}, \tag{10}$$

from which it follows that if $\mathbf{F} \neq 0$

$$\mathbf{V} = G^{-1}\mathscr{V} \tag{11}$$

and finally, by inversion, that

$$\mathscr{V} = G\mathbf{V}. \tag{12}$$

Equation (11) allows us to take a desired set of body velocities (with the γ_is $= 0$) and determine what set of finger velocities are required. Equation (12) permits us to determine the body's velocity (as well as check for a slipping grip) from a set of finger velocity measurements.

The grip transform, G, is an integral part of the stiffness control strategy which we have implemented in the HAND and OOLAH systems. For example, it can be used to take a Cartesian stiffness matrix, K_C, which represents the desired local compliant behavior of a grasped object, and transform it to an efficient joint space representation,

$$K = \mathscr{J}^T G^T K_C G \mathscr{J}, \tag{13}$$

where $\mathscr{J}$ is a block diagonal matrix composed of the three (3×3) finger Jacobian matrices. The stiffness control system was originally designed to impart controlled stiffness to grasped objects as might be required to place a peg in a hole or push an object into alignment. We have found that it also permits construction of stiffnesses that aid in the acquisition of objects. One particularly useful and simple construct we have employed to build joint space stiffness matrices makes use of the fact that a matrix of rank n may be spectrally decomposed into a weighted sum of n unitary matrices. For example, if K is a matrix of rank 9 with nine linearly independent unit eigenvectors, k_i, and eigenvalues, λ_i, then

$$K = \sum k_i * k_i^T * \lambda_i. \tag{14}$$

If K is our joint space stiffness matrix, then the elements of K represent the rate at which the torque increases on the ith joint for a given deflection in the jth joint. The principal stiffness "directions" are defined by each k_i.

In this case the direction may be simply the motion of a particular joint (k_i has a nonzero element only in the ith position) or it may be a combination of simultaneous joint motions. By choosing directions along which we wish to resist motion, k_i, and magnitudes of stiffness along those directions, λ_i, we can construct K from the above relationship. If a set of directions is chosen that do not span the full space of motion, there will not be resistance to motion in the unspanned directions.

It is also possible to use the reverse of this process to remove certain stiffnesses from a grasp. For example, to permit the two fingers to balance the forces on them against the thumb, we would want to remove stiffness for differential motion of the fingers (here we mean moving finger 1 up while finger 2 moves down). By taking a nominal joint space stiffness, k, of full rank (for example, a diagonal matrix) and subtracting the matrix $kx_1x_1^t$ from it, we reduce the rank of k by one. (This is strictly true only if x_1 is an eigenvector of k, but is often close enough.) If x_1 is a vector representing the desired differential motion of the fingers, then the reduction of k's rank corresponds to the freedom introduced between finger 1 and 2. In general we can construct a new reduced rank stiffness matrix by removing the stiffnesses from the "directions" in which we require compliance:

$$K_{\text{new}} = K - Kx_1x_1^T - Kx_2x_2^T - \cdots. \tag{15}$$

15.5.2 Other Types of Contact

A second subtle aspect of moving and exerting forces upon objects held in fingertip prehension stems from the fact that the contact points with the object are not really fixed on its surface. While the ball and socket contact assumption is reasonably accurate for small motions of large objects, it is not precise for very small objects or very large displacements. Clearly when rolling a pea in our fingertips consideration must be made for the change in contact location on both the finger and pea as the motion proceeds. While it seems possible actually to derive kinematic relationships for this type of rolling contact, it is not clear that they would be of much use. Slipping in both the contact plane and about the contact normal will introduce inaccuracies that will be impossible to model and predict. Thus, as we press our hand to extreme operating conditions it appears that we will have to rely upon more novel manipulation, sensing, and feedback strategies.

Controlled slip manipulation is one such class of manipulation that we have begun to explore at MIT. [Brock87] has analyzed a broad class of controlled slip manipulation and demonstrated a number of specific examples with the Salisbury hand. When humans manipulate objects, our fingers are not always fixed securely to the object's surface. Many times we allow objects to slide or rotate on our fingertips, consciously controlling the motion of the object rather than the motion of our fingers. This controlled slipping technique of manipulation may in fact be the dominant form of dexterous human manipulation. For example, try putting a lid on a jar, but start with the lid top down on a table. Without thinking, we pick up the lid, spin it around between our two fingers using the edge of the jar or another finger, and screw it on the jar. The use of a pencil eraser is another example of controlled slipping. We stop writing, flip the pencil over, push the pencil through our finger to gain leverage, and erase. In both these motions and in many others, objects are allowed to slide and rotate at the points of contact with the fingers.

To employ this type of manipulation fully we need to understand a whole new class of "grasps" that do not fully constrain the object. While the design of the Salisbury hand was aimed a obtaining secure (force closure) grasps, it is possible to modify the internal grasping forces to permit some freedom of motion within the grasp. Secure constraint is really a matter of degree. While a given grasp may completely restrain an object for small disturbance forces, it may allow the object to slip for large disturbance forces. Thus, by carefully modulating the constraint state and applying disturbance forces, we may cause the object to slip controllably. The direction of slip will depend upon the grasp's constraint state and the direction of applied forces. These intentionally applied disturbance forces may arise from a number of sources, including gravity, acceleration, and contact with external objects or other fingers. An obvious and familiar example can be demonstrated by holding one end of a pencil horizontally between two fingers. If you then slightly relax the grip and/or accelerate your hand, the orientation of the pencil in the grasp can be altered. Actually to plan this type of motion for a robot hand we can construct a *constraint state map* by identifying the range of motions permitted for each combination of internal forces that may be imposed. For each point in this three-dimensional space of grasp forces there is an associated set of permissible motions (or *twists*). The goal then is to find the set of internal forces (if any) that will permit freedom of motion in the desired direction (with no more freedom than necessary) and then apply a disturbance force that will tend to move the object in the desired direction. To simplify this process we have parameterized the

3-space of internal forces in a geometrically appealing manner. We first choose a grasping force focus point (which lies in the plane of the three contacts). This requires 2 parameters and defines the direction and relative magnitudes of the internal components of the forces applied at the fingertips. The third parameter is the magnitude or *intensity* of the grasping force. The map of permissible motions can then be constructed by examining each possible constraint states and determining the associated freedom of motion. The choice of the grasp forces is then made by finding the set of permissible motions that most closely align with the desired motion and using the associated grasp forces. Finally, we apply the required disturbance force to make the object move. Obviously, controlled slip manipulation carries with it a degree of indeterminacy in many cases and will require more sophisticated touch or visual feedback to make it useful. There are, however, many cases where the motion is quite well determined. [Brock87] has demonstrated repeated reorientation of a rectangular solid by a two-step process. First, the block is held in two fingers and pressed against a fixed surface. As the block stops against and aligns with the surface the fingers slip down the block to its end. This brings the fingers to a known position on the block and the block to a known orientation. Then the fingers squeeze more tightly, lift the block away from the surface, and then slightly loosen the grip to allow the block to rotate nearly 180° due to the action of gravity but not fall. This places the block back in the original state (though rotated) and the whole process can be repeated again and again.

15.6 Sensing

In anticipation of harsh operating environments and to reduce complexity we have only employed as many sensors as can be justified by anticipated control and manipulation strategies. High-bandwidth control of the motors required direct measure of motor shaft position and this was achieved with incremental encoders. Given the need to control forces in the presence of some mechanism friction, force-sensing elements were also required. Strain gauge tendon tension sensors were used on each tendon to minimize the transformations required in the servo loop. The sensors were located at the base of the fingers as a compromise between putting them at the motors (where higher bandwidth might be attainable) and putting them on the joints or fingertips (where maximum accuracy would be attainable). The resulting design is sturdy and unencumbered by excessive wiring. Although direct measurement of finger joint positions and velocities could improve positioning accuracy and bandwidth with a higher-order controller, the additional complexity and mechanical vulnerability was not considered warranted.

It is less obvious what type of contact or tactile sensor is required to perform dexterous manipulation. The fact that many motions of objects in the hand are unobservable with traditional position and force measurements indicates that some type of touch information is required. While humans make good use of tactile sensing (measurement of pressure distribution) in manipulation, it is not yet clear how to implement, let alone use, this capability in a robot hand (despite vigorous research into tactile sensor design). Our approach has been to develop a simple sensor that can locate and track the effective point of contact with the fingertip. This has been achieved by building a fingertip that is supported on a miniature 6-axis force sensor. As described in [Salisbury] and [Brock84] (figure 15.2), these 6 measurements permit determination of the location of contact with the fingertip, orientation of the contact plane, and resolution of the contact forces into normal and tangential components. Though object features such as texture and local curvature are not statically observable with this type of contact-resolving sensor, it appears that by tracking the frequency

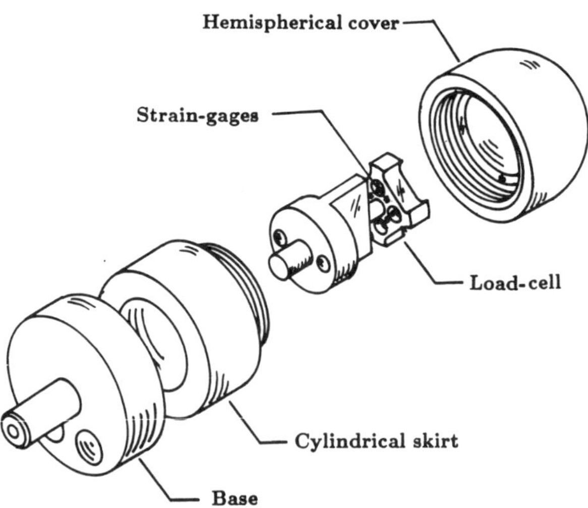

Figure 15.2
The contact-resolving fingertip.

content and time histories of the contact forces during exploratory motion we may, in fact, be able to observe texture and curvature. To date, this sensor has been used only to probe unknown object shapes for contact location and orientation, but has demonstrated good accuracy and tolerance to abuse. Two clear advantages of this type of sensor are the minimal wiring required (compared to tactile sensors) and the ability to measure tangential forces. The ability to measure normal and tangential forces at the contacts permits anticipating incipient slip and should guide us in readjusting the grasp to prevent slip.

15.7 Interprocessor Communication

In the current implementation of our control system and experimental environment, communication between a Lisp Machine and a VAX takes place in the form of a stream of packets. Whenever the Lisp Machine needs to send a command to the VAX, it sends a command packet. When the VAX needs to send data back to the Lisp Machine, it sends a response packet. These packets are transmitted over a dedicated 16-bit parallel connection. Each packet sent from the Lisp Machine to the VAX contains one command and its associated data; each packet sent from the VAX to the Lisp Machine contains information such as current joint position, joint torques, or finger forces.

The protocol used to transfer information between the two computers is very simple. There is minimal error checking, but there is also very little overhead for either computer to maintain the data connection or to implement failure recovery. This keeps the computational overhead on the Lisp Machine and the VAX to a minimum. Each packet sent from the Lisp Machine to the VAX contains one command and whatever data the command requires. For example, the command to move to a new set of angles contains twelve floating point numbers. Similarly, the data sent from the VAX to the Lisp Machine contains one block of information, such as the current sensed joint angles, or the current joint torques. The command and the type of response is indicated by an opcode in the packet. There is no indication within the packet itself of the type of data accompanying the command; it is assumed that each side of the connection understands how to interpret the bits being received. A sequence number begins each packet, followed by a count of the 16-bit words in the packet. Following the command opcode and the command data, the sequence number is repeated. These three words of control information enable the receiver of the packet to determine the limits of the packet.

The protocol described here is not dependent on any particular communication hardware; in fact, it mildly resembles a protocol also developed at MIT to communicate over a local area network to control a Unimation Puma robot. The communication could also be carried out over a serial line (with a substantial degradation in performance, of course). Although the control overhead is minimal, and the error checking is almost nonexistent the protocol is still flexible enough to adapt to varying circumstances.

On the Lisp Machine, a background process monitors the data connection, and sends and receives all the packets. The trajectory generator prepares the commands and data and stores them into buffers, which are then queued for delivery to the VAX by the background process. Incoming data from the VAX are separated into packets and queued for use by the higher-level facilities of the system. From the point of view of the background process, the data being transferred are 16-bit words. No interpretation is placed on the contents of the packets being sent or received. The background process exists to free the higher-level tasks on the Lisp Machine from the mundane communication details, and also allows the data transfer to proceed asynchronously with the VAX. Similarly, the higher-level tasks are relieved of the necessity of concerning themselves with checking for incoming words from the parallel port or separating these words into meaningful packets.

Since the Lisp Machine does not come equipped with a parallel port, and since we do not have access to the low-level hardware interrupt support, the background process must poll the hardware device or busy-wait while sending and receiving data. In the current implementation, when the process is idle, it periodically checks the hardware control lines for incoming data and the output queue for packets to be sent. While a transfer is actually in progress, however, the process assumes that the VAX will accept to send each word quickly, and so it continually tests the control lines until the VAX is ready for the next step. While this busy-waiting ties up the Lisp Machine processor during the transfer, the actual time spent doing the transfer is very small. The alternative, relinquishing the processor between each word, would cause the Lisp Machine's scheduler to switch contexts until the VAX had sent the next word, which would happen very quickly, but would consume even more processor time than busy-waiting. (In practice, the VAX can accept a word from the Lisp Machine in less time that it would take the Lisp Machine to check to see if it had.)

On the VAX side of the connection, the actual transfer of words to and from the parallel port is done by interrupt handler routines. These routines merely transfer the words to and from small buffers which are emptied or filled by the joint space hand control program, HAND.

There is no separate process on the VAX to handle the transmission and reception of packets, as there is on the Lisp Machine. Whenever HAND needs to send data to the Lisp Machine, it stores the "packet" in the buffer maintained by the output interrupt routine; it also periodically checks the buffer filled by the input interrupt routine for new words received. It takes the responsibility itself for determining the beginning and end of each packet, and when a complete packet has been read, HAND performs the operations indicated by the command opcode.

The ability to use the hardware interrupt mechanism on the VAX removed the need for busy-waiting in the time-critical control program. Also, by not using a separate process to handle the packet-level operations, there is no need for a change of process context.

If the Lisp Machine end could be made to be interrupt-driven as the VAX end is, then the reliability of the low-level communication could be drastically improved. It would also undoubtedly improve the throughput of the data transfer while relieving some of the computational burden, in the form of busy-waiting, of the background process. Table 15.1 summarizes the data transfer rates achieved with the system.

Times are calculated by measuring the time the Lisp Machine spends in its background process, and thus measuring the load on the Lisp Machine. A typical packet size would be around 26 words: 2 opcode words plus 12 floating point numbers. The time spent in the VAX interrupt routines is 102 μsec for each word when sending to the Lisp Machine, and 52 μsec for each word when receiving from the Lisp Machine.

Table 15.1
Average time for transfer of one packet (times are in milliseconds)

Number of words	Time to Lisp Machine	Time to VAX
26	27	14
100	77	47
250	160	116

In practice, the communication link, as implemented, has worked satisfactorily. Although it is subject to potential failure, such failure is not catastrophic, in that the control of the hand is not jeopardized. Failures in communication are very rare, in any event. The data transfer rate is satisfactory; it is much less than the time required to process the information that is being transferred.

15.8 Conclusion

This chapter has been aimed at giving our view of how to use an articulated hand to manipulate objects. It may be pretentious to say that we have achieved dexterous manipulation, but it is clear that we have extended the vocabulary of primitive manipulation and sensing elements required for truly dexterous manipulation. The boundary between simple preprogrammed motions and real dexterity is just as fuzzy as the boundary between simply clever and really intelligent programs.

There are a number of hard problems that must be addressed before dexterous articulated hands will be practical and useful to us. Among the unsolved problems are

• Hand Mechanics. How should we really design a hand to perform useful manipulation? Do we want more fingers and degrees of freedom? Do we want the fingers to spin and gyrate in ways that would make humans cringe? Why not? No person ever thought of rolling himself as a means of locomotion but it is certainty useful in a car. The debate over the choice of anthropomorphic and nonanthropomorphic designs probably will go on forever and there probably always will be an appropriate niche for each approach. Because of the difficulty in building, packaging, maintaining, and controlling these complex mechanisms, it is always important to rationalize the complexity in terms of potential performance. And yet we must not get stuck building hands for single applications and should keep sight of the benefits of designing a single device for many different operations.
• Actuators and Transmissions. Actuation and transmission of motion and force have been large stumbling blocks for hand designers. It is hard to build small, efficient actuators and transmit their actions through the robot arm. The headache of passing tendons (if indeed that is the answer) around or through a complex wrist has plagued every hand designer. Until we develop very powerful and compact actuators we probably will be compelled

to locate the actuators at some distance from the hand and the problem of efficient, stiff transmission of forces must be addressed.

• Hand/Arm Integration. It is wrong to consider the hand as an "add on" accessory to the arm. Both from the design and control points of view we must begin to consider these as components of a single larger system so that we may synergistically integrate their capabilities.

• Manipulation Modes. There exists whole range of manipulation modes that do not yet employ robots. The familiar deterministic modes of robot manipulation do not take advantage of most of the surface area on a robot. Can we extend our ideas of controlled slip to include all the links of the hand and all the surfaces of the arm? Are there novel dynamic modes of operation that humans are too slow or insensitive to employ?

• Sensing Modalities. What do we really need to know in order to manipulate an object reliably? Obviously force, position, and touch sensing are required, but until we have a clear plan of how to integrate and use the information it is hard to specify appropriate sensor performance. Determining what the useful bandwidth and sensitivity of these sensors should be would help focus the design efforts. Other sensory modes such as proximity, thermal, acoustic, and just about any part of the electromagnetic spectrum should be considered as candidates for augmenting the manipulative process.

Achieving dexterity with autonomous robots will ultimately depend upon integrating the ability to move and sense with the ability to recognize and plan. To close the high-level control loops needed for intelligent robot behavior will require a much broader and coherent understanding of the mechanics of object motion and interaction than we have yet achieved. Our work with robot hands has perhaps made a few of the steps in this direction but there is whole garden of delights yet to be discovered.

Acknowledgment

The authors gratefully acknowledge the financial support of the System Development Foundation.

References

Brock, D. L., "Strain Gauge Based Force and Tactile Sensors," Bachelor's Thesis, Department of Mechanical Engineering, MIT June 1984.

Brock, D. L., "Enhancing the Dexterity of a Multi-Fingered Hand through Controlled Slip Manipulation," Master's Thesis, Department of Mechanical Engineering, MIT June 1987.

Childress, D. S., "Artificial Hand Mechanisms," ASME Mechanisms Conference and International Symposium on Gearing and Transmissions, San Francisco, Oct. 1972.

Chiu, S. L., "Generating Compliant Motion of Objects with an Articulated Hand," Master's Thesis, Department of Mechanical Engineering, MIT May 1985.

Jacobsen, S. C., et al., "Design of the Utah/MIT Dextrous Hand," Proc. 1987 IEEE International Conference on Robotics and Automation, San Francisco, Apr. 1986.

Loucks, C. S., et al., "Modeling and Control of the Stanford/JPL Hand," Proc. 1987 IEEE International Conference on Robotics and Automation, Raleigh, North Carolina, April 1987.

Mason, M. T., and J. K. Salisbury, *Robot Hands and the Mechanics of Manipulation*, MIT Press, Cambridge, MA, 1985.

Salisbury, J. K., and J. J. Craig, "Articulated Hands: Force Control and Kinematic Issues," *International Journal of Robotics Research*, Vol. 1, No. 1, Spring 1982.

Salisbury, K., D. L. Brock, and S. L. Chiu, "Integration of Language, Sensing and Control for a Robot Hand," Proc. 3rd ISRR, Gouvieux France, Oct. 1985 (published by the MIT Press, 1986).

Salisbury, J. K., "Interpretation of Contact Geometries from Force Measurements," Proc. 1st ISRR Bretton Woods, NH, MIT Press, Sept. 1984.

Townsend, W. T., and J. K. Salisbury, "The Effect of Coulomb Friction and Stiction of Force Control," Proc. 1987 IEEE International Conference on Robotics and Automation, Raleigh, North Carolina, April 1987.

16 Legged Robots

Marc H. Raibert

16.1 Why Study Legged Machines?

Aside from the sheer thrill of creating machines that actually run, there are two serious reasons for exploring legged machines. One reason is mobility: There is a need for vehicles that can travel in difficult terrain, where existing vehicles cannot go. Wheels excel on prepared surfaces such as rails and roads, but perform poorly where the terrain is soft or uneven. Because of these limitations only about half the earth's landmass is accessible to existing wheeled and tracked vehicles, whereas a much greater area can be reached by animals on foot. It should be possible to build legged vehicles that can go to the places that animals can now reach.

One reason legs provide better mobility in rough terrain is that they can use isolated footholds that optimize support and traction, whereas a wheel requires a continuous path of suport. As a consequence, a legged system is free to choose among the best footholds in the reachable terrain whereas a wheel is forced to negotiate the worst terrain. A ladder illustrates this point: Rungs provide footholds that enable systems to climb, but the spaces between the rungs prevent the wheeled system from making progress.

Another advantage of legs is that they provide an active suspension that decouples the path of the body from the paths of the feet. The payload is free to travel smoothly despite pronounced variations in the terrain. A legged system can also step over obstacles. The performance of legged vehicles can, to a great extent, be independent of the detailed roughness of the ground.

The construction of useful legged vehicles depends on progress in several areas of engineering and science. Legged vehicles will need systems that control joint motions, sequence the use of legs, monitor and manipulate balance, generate motions to use known footholds, sense the terrain to find good footholds, and calculate negotiable foothold sequences. Most of these tasks are not well understood yet, but research is under way. If this research is successful, it will lead to the development of legged vehicles that travel efficiently and quickly in terrain where softness, grade, or obstacles make existing vehicles ineffective. Such vehicles may be useful in industrial, agricultural, and military applications.

A second reason for exploring legged machines is to understand how humans and animals use their legs for locomotion. A few instant replays on television will reveal the large variety and complexity of ways athletes can carry, swing, toss, glide, and otherwise propel their bodies through space, maintaining orientation, balance, and speed as they go. Such performance is not limited to professional athletes; behavior at the local playground is equally impressive from a mechanical engineering, sensory-motor integration, or computational point of view. Animals also demonstrate great mobility and agility. They use their legs to move quickly and reliably through forest, swamp, marsh, and jungle, and from tree to tree. They move with great speed and efficiency.

Despite the skill we apply in using our own legs for locomotion, we are still at a primitive stage in understanding the principles that underlie walking and running. What control mechanisms do animals use? The development of legged machines will lead to new ideas about animal locomotion. To the extent that an animal and a machine perform similar locomotion tasks, their control systems and mechanical structures must solve similar problems. Of course, results in biology will also help us to make progress with legged robots. This sort of interdisciplinary appoach has already become effective in other areas where biology and robotics have a common ground, such as vision, speech, and manipulation.

16.2 Research on Legged Machines

The scientific study of legged locomotion began just over a century ago when Leland Stanford, then governor of California, commissioned Eadweard Muybridge to find out whether or not a trotting horse left the ground with all four feet at the same time. (For milestones in the development of legged robots, see table 16.1.) Stanford had wagered that it never did. After Muybridge proved him wrong with a set of stop-motion photographs that appeared in *Scientific American* in 1878, Muybridge went on to document the walking and running behavior of over forty mammals, including humans (Muybridge 1955, 1957). Even after 100 years, his photographic data are of considerable value and beauty, and survive as a landmark in locomotion research.

The study of machines that walk also had its origin in Muybridge's time. An early walking model appeared in about 1870 (Lucas 1894). It used a linkage to move the body along a straight horizontal path while the feet moved up and down to exchange support during stepping (see figure 16.1). The linkage was originally designed by the famous Russian mathematician

Table 16.1
Milestones in the development of legged robots

1850	Chebyshev	Designs linkage used in early walking mechanism (Lucas 1894).
1872	Muybridge	Uses stop-motion photography to document running animals.
1893	Rygg	Patents human-powered mechanical horse.
1945	Wallace	Patents hopping tank with reaction wheels that provide stability.
1961	Space General	Eight-legged kinematic machine walks in outdoor terrain (Morrison 1968).
1963	Cannon, Higdon and Schaefer	Control system balances single, double, and limber inverted pendulums.
1968	Frank and McGhee	Simple digital logic controls walking of Phony Pony.
1968	Mosher	GE quadruped truck climbs railroad ties under control of human driver.
1969	Bucyrus-Erie Co.	Big Muskie, a 15,000-ton walking dragline is used for strip mining. It moves in soft terrain at a speed of 900 ft/hr (Sitek 1976).
1977	McGhee	Digital computer coordinates leg motions of hexapod walking machine.
1977	Gurfinkel	Hybrid computer controls hexapod walker in USSR.
1977	McMahon and Greene	Human runners set new speed records on *tuned track* at Harvard. Its compliance is adjusted to mechanics of human leg.
1980	Hirose and Umetani	Quadruped machine climbs stairs and climbs over obstacles using simple sensors. The leg mechanism simplifies control.
1980	Kato	Hydraulic biped walks with quasidynamic gait.
1980	Matsuoka	Mechanism balances in the plane while hopping on one leg.
1981	Miura and Shimoyama	Walking biped balances actively in three dimensional space.
1983	Sutherland	Hexapod carries human rider. Computer, hydraulics, and human share computing task.
1983	Odetics	Self-contained hexapod lifts and moves back end of pickup truck (Russell 1983).
1987	OSU	Three-ton self-contained hexapod carrying human driver travels at 5 mph and climbs over obstacle.

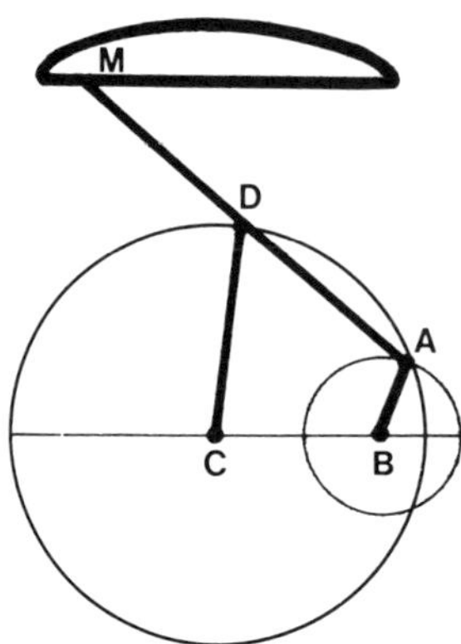

Figure 16.1
Linkage used in an early walking machine. When the input crank AB rotates, the output point M moves along a straight path during part of the cycle and an arched path during the other part of the cycle. Two identical linkages are arranged to operate out of phase so at least one provides a straight motion at all times. The body is always supported by the feet connected to the straight-moving linkage. $AB = 1$; $CD = AD = DM = (3 + \sqrt{7})/2$; $BC = (4 + \sqrt{7})/3$. After Lucas (1894).

Chebyshev some years earlier. During the eighty or ninety years that followed, workers viewed the task of building walking machines as the task of designing linkages that would generate suitable motions when driven by a source of power. Many designs were proposed (e.g., Rygg 1893; Nilson 1926; Ehrlich 1928; Kinch 1928; Snell 1947; Urschel 1949; Shigley 1957; Corson 1958; Bair 1959; Morrison 1968) (see figure 16.2), but the performance of such machines was limited by their fixed patterns of motion, since they could not adjust to variations in the terrain by placing the feet on the best footholds. By the late 1950s it had become clear that linkages providing fixed motion would not do the trick and that useful walking machines would need *control* (Liston 1970).

One approach to control was to harness a human. Ralph Mosher used this approach in building a four-legged walking truck at General Electric in the mid-1960s (Liston and Mosher 1968). The project was part of a decade-long campaign to build advanced teleoperators, capable of providing better dexterity through high-fidelity force feedback. The walking

Figure 16.2 ▶
Mechanical horse patented by Lewis A. Rygg in 1893. The stirrups double as pedals so the rider can power the stepping motions. The reins move the head and forelegs from side to side for steering. Apparently the machine was never built.

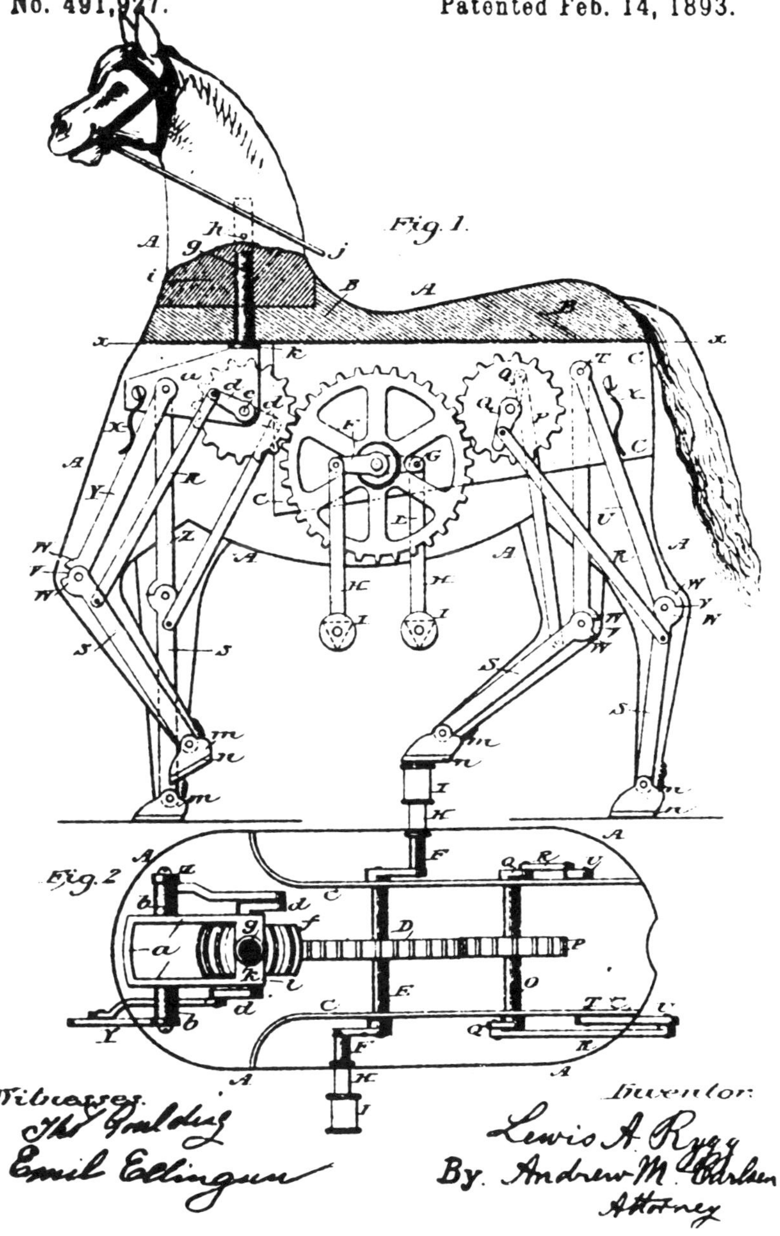
(No Model.)
L. A. RYGG.
MECHANICAL HORSE.
No. 491,927.
Patented Feb. 14, 1893.
Fig. 1.
Fig. 2
Witnesses.
Inventor:
Lewis A. Rygg
By. Andrew M. Carlsen
Attorney

machine Mosher built stood 11 feet tall, weighed 3,000 pounds, and was powered hydraulically. It is shown in figure 16.3. Each of the driver's limbs was connected to a handle or pedal that controlled one of the truck's four legs. Whenever the driver caused a truck leg to push against an obstacle, force feedback let the drive feel the obstacle as though it were his or her own arm or leg doing the pushing.

After about twenty hours of training Mosher was able to handle the machine with surprising agility. Films of the machine operating under his control show it ambling along at about 5 mph, climbing a stack of railroad ties, pushing a foundered jeep out of the mud, and maneuvering a large drum onto some hooks. Despite its dependence on a well-trained human for control, the GE Walking Truck was a milestone in legged technology.

An alternative to human control became feasible in the 1970s: use of a digital computer. Robert McGhee's group at the Ohio State University was the first to use this approach successfully (McGhee 1983). In 1977 they built an insectlike hexapod that would walk with a number of gaits, turn, walk sideways, and negotiate simple obstacles. The computer's primary task was to solve kinematic equations in order to coordinate the eighteen electric motors driving the legs. This coordination ensured that the machine's center of mass stayed over the polygon of support provided by the feet while allowing the legs to sequence through a gait (figure 16.4). The machine traveled quite slowly, covering several yards per minute. Force and visual sensing provided a measure of terrain accommodation in later developments. The hexapod provided McGhee with an experimental means of pursuing his earlier theoretical findings on the combinatorics and selection of gait (McGhee 1968; McGhee and Jain 1972; Koozekanani and McGhee 1973).

Gurfinkel and his coworkers in the USSR built a machine with characteristics and performance quite similar to McGhee's at about the same time (Gurfinkel et al. 1981). It used a hybrid computer for control, with analog computation aiding in kinematic calculations.

The group at Ohio State has recently built a much larger hexapod (figure 16.5), which they designed for self-contained operation on rough terrain (Waldron et al. 1984). It carries a gasoline engine for power, several computers and a human operator for control, and a laser range sensor for terrain preview. At the time of this writing this machine has walked at about 5 mph and negotiated simple obstacles.

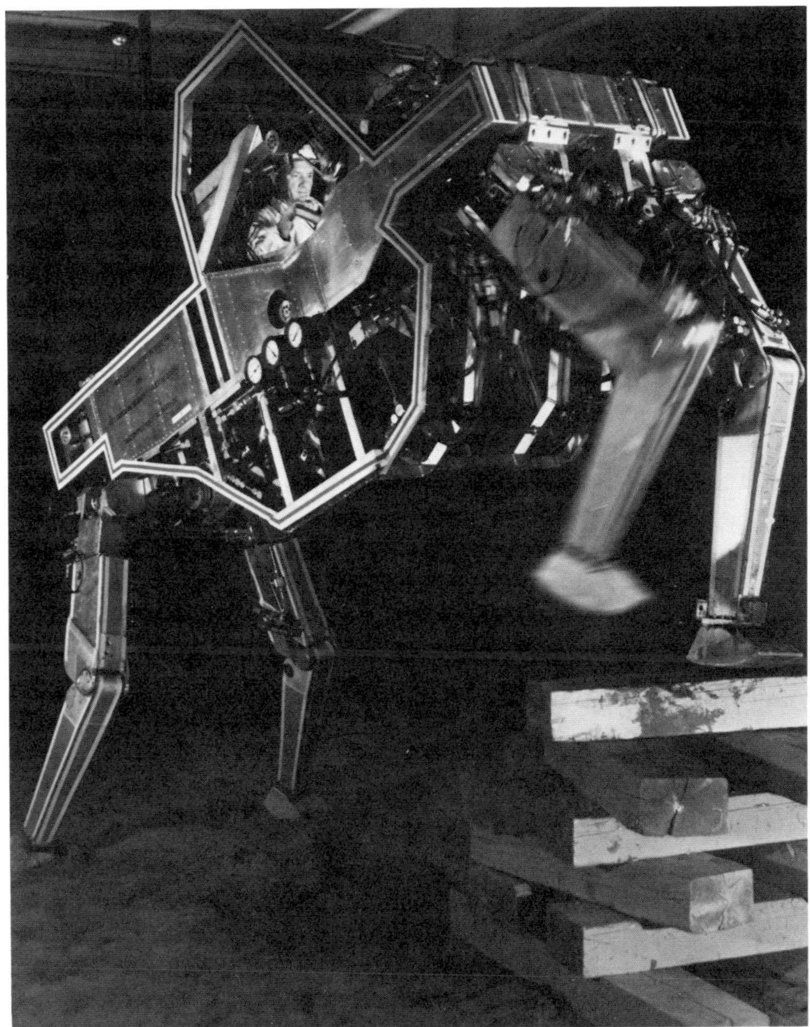

Figure 16.3
Walking truck developed by Ralph Mosher at General Electric in about 1968. The human driver controlled the machine with four handles and pedals that were connected to the four legs hydraulically. Photograph courtesy of General Electric Research and Development Center.

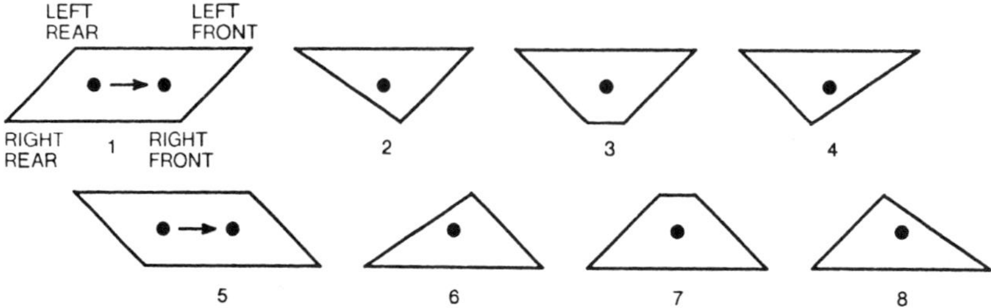

Figure 16.4
Statically stable gait. The diagrams shows the sequence of support patterns provided by the feet of a quadruped walking with a crawling gait. The body and legs move to keep the projection of the center of mass within the polygon defined by the feet. A supporting foot is located at each vertex. The dot indicates the projection of the center of mass. Adapted from McGhee and Frank (1968).

Hirose realized that linkage design and computer control were not mutually exclusive. His experience with clever and unusual mechanisms—he had built seven kinds of mechanical snake—led to his design of a special leg that simplified the control of locomotion and could improve efficiency (Hirose and Umetani 1980; Hirose et al. 1984). The leg was a three-dimensional pantograph that translated the motion of each actuator into a pure Cartesian translation of the foot. With the ability to generate x, y, and z translations of each foot by merely choosing an actuator, the control computer was freed from the arduous task of performing kinematic solutions. The mechanical linkage was helping to perform the calculations needed for locomotion. The linkage was efficient because the actuators performed only positive work in moving the body forward.

Hirose used this leg design to build a small quadruped, about 1 yard long. It was equipped with touch sensors on each foot and an oil-damped pendulum attached to the body. Simple algorithms used the sensors to control the actions of the feet. For instance, if a touch sensor indicated contact while the foot was moving forward, the leg would move backward a little bit, move upward a little bit, then resume its forward motion. If the foot had not yet cleared the obstacle, the cycle would repeat. The use of several simple algorithms like this one permitted Hirose's machine to climb up and down stairs and to negotiate other obstacles without human intervention (Hirose 1984).

These three walking machines, McGhee's Gurfinkel's, and Hirose's, represent a class called *static crawlers*. Each differs in the details of construction and in the computing technology used for control, but shares a common

Figure 16.5
The hexapod walking machine developed at Ohio State University. It stands about 10 feet tall, 15 feet long, and weighs 3 tons. A 90-horsepower motorcycle engine provides power to 18 variable displacement hydraulic pumps that drive the joints. The legs use pantographs linkages to improve energy efficiency. The operator normally provides steering and speed commands while computers control the stepping motions of the legs.

approach to balance and stability. They all keep enough feet on the ground to guarantee a broad base of support at all times, and the body legs move to keep the center of mass over this broad support base. The forward velocity is kept low enough so that stored kinetic energy can be ignored in the stability calculation. Each of these machines has been used to study rough terrain locomotion in the laboratory through experiments on terrain sensing, gait selection, and selection of foothold sequences. Several other machines that fall into this class have been studied in the intervening years, for example see (Russell 1983; Sutherland and Ullner 1984; Ooka et al. 1985).

16.3 Dynamics and Balance Improve Mobility

We now consider the study of dynamic legged machines that balance actively. These systems operate in a regime where the velocities and kinetic energies of the masses are important determinants of behavior. In order to predict and influence the behavior of a dynamic system, we must consider the energy stored in each mass and spring as well as the geometric structure and configuration of the mechanism. Geometry and configuration taken alone are not adequate to model a system that moves with substantial speed or has large mass. Consider, for example, a fast-moving vehicle that would tip over if it stopped suddenly with its center of mass too close to the front feet.

The exchange of energy among its various forms is also important in dynamic legged locomotion. For example, there is a cycle of activity in running that changes the form of the stored energy several times: the body's potential energy of elevation changes into kinetic energy during falling, then into strain energy when parts of the leg deform elastically during rebound with the ground, then into kinetic energy again as the body accelerates upward, and finally back into potential energy of elevation. This sort of dynamic exchange is central to an understanding of legged locomotion.

Dynamics also plays a role in giving legged systems the ability to balance actively. A statically balanced system avoids tipping and the ensuring horizontal accelerations by keeping its center of mass over the polygon of support formed by the feet. Animals sometimes use this sort of balance when they move slowly, but usually they balance actively.

A legged system that balances actively can tolerate departures from static equilibrium, tipping and accelerating for short periods of time. The control system manipulates body and leg motions to ensure that each tipping interval is brief and that each tipping motion in one direction is compensated by a tipping motion in the opposite direction. An effective base of support is thus maintained over time. A system that balances actively can also tolerate vertical acceleration, such as the ballistic flight and bouncing that occur during running.

The ability of an actively balanced system to depart from static equilibrium relaxes the rules govering how legs can be used for support, which in turn leads to improved mobility. For example, if a legged system can tolerate tipping, then it can position its feet far from the center of mass in order to use footholds that are widely separated or erratically placed. If it can remain upright with a small base of support, then it can travel where there are closely spaced obstructions or where is a narrow path of firm support. The ability to tolerate intermittent support also contributes to mobility. Intermittent support allows a system to move all its legs to new footholds at one time, to jump onto or over obstacles, and to use short periods of ballistic flight for increased speed. These abilities to use narrow base and intermittent support generally increase the types of terrain a legged system can negotiate. Animals routinely exploit active balance to travel quickly on difficult terrain; legged vehicles will have to balance actively, too, if they are to move with animal-like mobility and speed.

16.4 Research on Active Balance

The first machines that balanced actively were automatically controlled inverted pendulums. Everyone knows that a human can balance a broom on his finger with relative ease. Why not use automatic control to build a broom that can balance itself? Claude Shannon was probably the first to do so. In 1951 he used the parts from an erector set to build a machine that balanced an inverted pendulum atop a small powered truck. The truck drove back and forth in response to the tipping movements of the pendulum, as sensed by a pair of switches at its base. In order to move from one place to another, the truck first had to drive away from the destination to unbalance the pendulum, then proceed toward the destination. In order to balance again at the destination, the truck moved past the destination until the pendulum was again upright with no forward veocity, then moved back to the destination.

Figure 16.6
Cannon and his students built machines that balanced inverted pendulums on a moving cart. They balanced two pendulums side by side, one pendulum on top of another, and a long limber inverted pendulum. Only one input, the force driving the cart horizontally, was available for control. Adapted from Schaefer and Cannon (1966).

At Shannon's urging Robert Cannon and two of his students at Stanford University set about demonstrating controllers that balanced two pendulums at once. In one case the pendulums were mounted side by side on the cart, and in the other they were mounted one on top of the other (figure 16.6). Cannon's group was interested in the single-input multiple-output problem and in the limitations of achievable balance: how could they use the single force that drove the cart's motion to control the angles of two pendulums as well as the position of the cast? How far from balance could the system deviate before it was impossible to return to equilibrium, given such parameters of the mechanical system as the cart motor's strength and the pendular lengths?

Using analysis based on normal coordinates and optimal switching curves, Cannon's group expressed regions of controllability as explicit functions of the physical parameters of the system. Once these regions were found, their boundaries were used to find switching functions that provided control (Higdon and Cannon 1963; Higdon 1963). Later, they extended these techniques to provide balance for a flexible inverted pendulum (Schaefer 1965; Schaefer and Cannon 1966). These studies of balance for inverted pendulums were important precursors to later work on locomotion and the inverted pendulum model for walking would become the primary tool for studying balance in legged systems (e.g., Hemami and Weimer 1974; Vukobratovic and Stepaneko 1972; Vukobratovic 1973;

Hemami and Golliday 1977; Kato et al. 1983; Miura and Shimoyama 1984). It is unfortunate that no one has yet extended Cannon's elegant analytical results to the more complicated legged case.

The importance of active balance in legged locomotion had been recognized for some years (e.g., Manter 1938; McGhee and Kuhner 1969; Frank 1970; Vukobratovic 1973; Gubina, Hemami, and McGhee 1974; Beletskii 1975a), but progress in building physical legged systems that employ such principles was retarded by the perceived difficulty of the task. It was not until the late 1970s that experimental work on balance in legged systems got underway.

Kato and his coworkers built a biped that walked with a *quasi-dynamic* gait (Ogo et al. 1980; Kato et al. 1983). The machine had ten hydraulically powered degrees of freedom and two large feet. This machine was usually a static crawler, moving along a preplanned trajectory to keep the center of mass over the base of support provided by the large supporting foot. Once during each step, however, the machine temporarily destabilized itself to tip forward so that support would be transferred quickly from one foot to the other. Before the transfer took place on each step, the *catching* foot was positioned to return the machine to equilibrium passively. No active response was required. A modified inverted pendulum model was used to plan the tipping motion. In 1984 this machine walked with a quasi-dynamic gait, taking about a dozen 0.5-m steps per minute. The use of a dynamic transfer phase makes an important point: A legged system can exhibit complicated dynamic behavior without requiring a very complicated control system.

Miura and Shimoyama (1980, 1984) built the first walking machine that balanced itself actively. Their *stilt biped* was patterned after a human walking on stilts. Each foot provided only a point of support, and the machine had three actuators: one for each leg that moved the leg sideways and a third that separated the legs fore and aft. Because the legs did not change length, the hips were used to pick up the feet. This gave the machine a pronounced shuffling gait reminiscent of Charlie Chaplin's stiff-kneed walk.

Control for the stilt biped relied, once again, on the inverted pendulum model of its behavior. Each time a foot was placed on the floor, its position was chosen according to the tipping behavior expected from an inverted pendulum. Actually, the problem was broken down as though there were two planar pendulums, one in the pitching plane and one in the rolling plane. The choice of foot position along each axis took the current and desired state of the system into account. The control system used tabulated descriptions of planned leg motions together with linear feedback to perform the necessary calculations. Unlike Kato's machine, which came to static equilibrium before and after each dynamic transfer, the stilt biped tipped all the time.

Dynamic bipeds that balance are now being studied in several laboratories around the world. Miura (1986) has edited a videotape that reports work, including new machines by Kato, Arimoto, Masubuchi, and Furusho.

Matsuoka (Matsuoka 1979) was the first to build a machine that ran, where running is defined by a period when all feet are off the ground at one time. Matsuoka's goal was to model repetitive hopping in humans. He formulated a model with a body and one massless leg, and he simplified the problem by assuming that the duration of the support phase was short compared with the ballistic flight phase. This extreme form of running, in which nearly the entire cycle is spent in flight, minimizes the influence of tipping during support. This model permitted Matsuoka to derive a time-optimal state feedback controller that provided stability for hopping in place and for low speed translations.

To test his method for control, Matsuoka built a planar one-legged hopping machine. The machine operated at low gravity by rolling on ball bearings on a table that was inclined 10° from the horizontal in an effective gravity field of 0.17 g. An electric solenoid provided a rapid thrust at the foot, so the support period was short. The machine hopped in place at about 1 hop per second and traveled back and forth on the table.

16.5 Running Machines

Running is a form of legged locomotion that uses ballistic flight phases to obtain high speed. To study running, my coworkers and I have explored a variety of legged systems and implemented some of them in the form of physical machines. To study running in its simplest form, we built a machine that ran on just one leg. The machine hopped like a kangaroo, using a series of leaps. A machine with only one leg allowed us to concentrate on active balance and dynamics while avoiding the difficult task of coordinating many legs. We wanted to know if there were algorithms for walking and running that are independent of gait and that work correctly

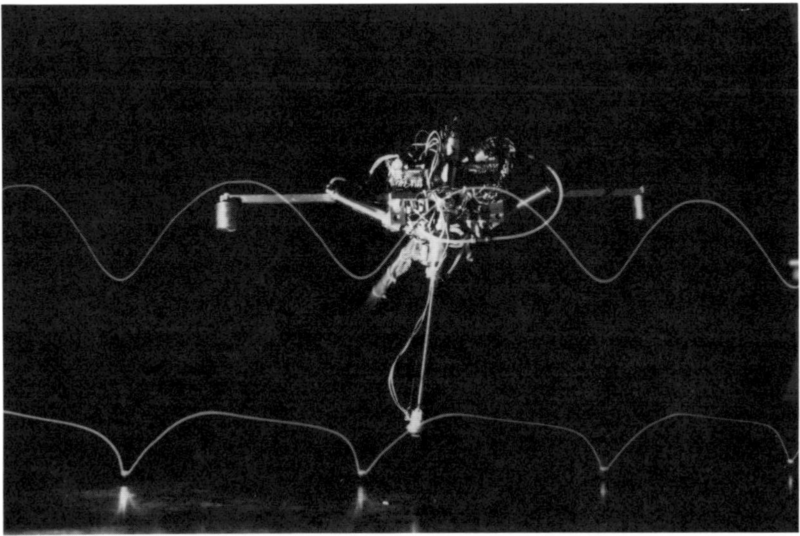

Figure 16.7
Planar hopping machine traveling at about 0.8 m/sec (1.75 mph) from right to left. Lines made by light sources attached to the machine indicate paths of the foot and the hip.

for any number of legs. Perhaps a machine with just one gait could suggest answers to this question.

The first machine we built to study these problems had two main parts: a body and a leg. The body carried the actuators and instrumentation needed for the machine's operation. The leg could telescope to change length and could pivot with respect to the body at a simple hip. The leg was springy along the telescoping axis. Sensors measured the pitch angle of the body, the angle of the hip, the length of the leg, the tension in the leg spring, and contact with the ground. This first machine was constrained to operate in a plane, so it could move only up and down and fore and aft and rotate in the plane. An umbilical cable connected the machine to power and a control computer.

The only way this one-legged machine can run is to hop. The running cycle has two phases. During one phase, called *stance* or *support*, the leg supports the weight of the body and the foot stays in a fixed location on the ground. During stance, the system tips like an inverted pendulum. During the other phase, called *flight*, the center of mass moves ballistically, with the leg unloaded and free to move. (See figures 16.7 and 16.8.)

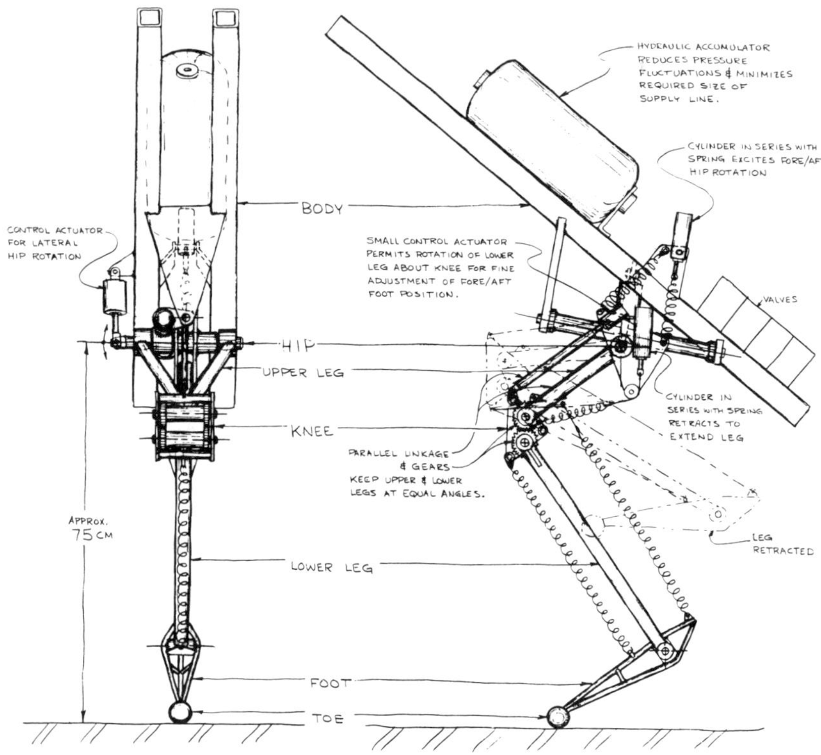

Figure 16.8
Ben Brown and I had this early concept for a one-legged hopping machine that was to operate in three dimensions. This version never left the drawing board.

16.5.1 Control of Running Was Decomposed into Three Parts

We found that a simple set of algorithms can control the planar one-legged hopping machine, allowing it run without tipping over. Our approach was to consider the hopping motion, forward travel, and posture of the body separately. This decomposition lead to a control system with three parts:

• *Hopping.* One part of the control system excites the cyclic hopping motion that underlies running, and regulates the height to which the machine hops. The hopping motion is an oscillation governed by the mass of the body, the springiness of the leg, and gravity. During support, the body bounces on the springy leg, and during flight, the system travels on a ballistic trajectory. The control system delivers a vertical thrust with the leg during each support period to sustain the oscillation and to regulate its amplitude. Some of the energy needed for each hop is recovered by the leg spring from the previous hop.

• *Forward Speed.* A second part of the control system regulates the forward running speed. This is done by moving the leg to a specified forward position with respect to the body during the flight portion of each cycle. The position of the foot with respect to the body when landing has a strong influence on tipping and acceleration behavior during the support period that follows. The body will either continue to travel with the same forward speed, accelerate to go faster, or slow down, depending on where the control system places the foot. To calculate a suitable forward position for the foot, the control system takes account of the actual forward speed, the desired speed, and a simple model of the system's dynamics. The algorithm works correctly when the machine is hopping in place, accelerating to a run, running at a constant speed, and slowing to a stationary hop.

The rule for placing and positioning the foot is based on a kind of symmetry found in running. In order to run at constant forward speed, the instantaneous forward accelerations that occur during a stride must integrate to zero. One way to satisfy this requirement is to shape the running motion so that forward acceleration has an odd symmetry throughout each stride—functions with odd symmetry integrate to zero over symmetric limits. If $x(t)$ is an odd function of time, then $x(t) = -x(-t)$. If $x(t)$ is even, then $x(t) = x(-t)$. A symmetric motion is produced by choosing an appropriate forward position for the foot on each step. In principle, symmetry of this sort can be used to simplify locomotion in systems with any number of legs and for a wide range of gaits.

• *Posture.* The third part of the control system stabilizes the pitch angle of the body to keep the body upright. Torques exerted between the body and leg about the hip accelerate the body about its pitch axis, provided that there is good traction between the foot and the ground. During the support period there is traction because the leg supports the load of the body. Linear feedback control operates on the hip actuator during each support period to restore the body to an upright posture.

Breaking running down into the control of these three functions simplifies locomotion. Each part of the control system acts as though it influences just one component of the behavior, and the interactions that results from imperfect decoupling are treated as disturbances. The algorithms implemented to perform each part of the control task are themselves quite simple, although the details of the individual control algorithms are probably not so important as the framework provided by the decomposition.

Using the three-part control system, the planar one-legged machine hops in place, travels at a specified rate, maintains balance when disturbed, and jumps over small obstacles. Top running speed is about 2.6 mph.

16.5.2 Locomotion in Three Dimensions

The one-legged machine just described was mechanically constrained to operate in the plane, but useful legged systems must balance themselves in three-dimensional space. Can the control algorithms used for hopping in the plane be generalized for hopping in three dimensions? A key to answering this question was the recognition that animal locomotion is primarily a planar activity, even though animals are three-dimensional systems. Films of a kangaroo hopping on a treadmill first suggested this point. The legs sweep fore and aft through large angles, the tail sweeps in counter-oscillation to the legs, and the body bounces up and down. These motions all occur in the sagittal plane, with little or no motion normal to the plane.

Sesh Murthy realized that the plane in which all this activity occurs can generally be defined by the forward vector and the gravity vector. He called this the *plane of motion* (Murthy 1983). For a legged system without a preferred direction of travel, the plane of motion might vary from stride to stride, but it would be defined in the same way. We found that the three-part decomposition could be used to control activity within the plane of motion.

We also found that the mechanisms needed to control the remaining *extraplanar* degrees of freedom could be cast in a form that fit into the original three-part framework. For instance, the algorithm for placing the foot to control forward speed became a vector calculation. One component of foot placement determined forward speed in the plane of motion, whereas the other component caused the plane to rotate about a vertical axis, permitting the control system to steer. A similar extension applied to body posture. The result was a three-dimensional three-part control system that was derived directly from the one used for the planar case.

To explore these ideas, we built a second hopping machine, which is shown in figure 16.9. This machine has an additional joint at the hip to permit the leg to move sideways as well as fore and aft. This machine travels on an open floor without mechanical support. It balances itself as it hops along simple paths in the laboratory, traveling at a top speed of 4.8 mph.

Figure 16.9
Three-dimensional hopping machine used for experiments. The control system operates to regulate hopping height, forward velocity, and body posture. Top recorded running speed was about 2.2m/sec (4.9 mph).

16.5.3 Running on Several Legs

Experiments on machines with one leg were not motivated by an interest in one-legged vehicles. Although such vehicles might very well turn out to have merit, our interest was in getting at the basics of active balance and dynamics in the context of a simplified locomotion problem. In principle, results from machines with one leg could have value for understanding all sorts of legged systems, perhaps with any number of legs. Wallace and Seifert saw merit in vehicles with one leg. Wallace (1942) patented a one-legged hopping tank that was supposed to be hard to hit because of its erratic movements. Seifert (1967) proposed the *Lunar Pogo* as a means of efficient travel on the moon.

Our study of locomotion on several legs has progressed in two stages. For a biped that runs like a human, with strictly alternating periods of support and flight, the one-leg control algorithms apply directly. Because the legs are used in alternation, only one leg is active at a time. One leg is placed on the ground at a time, one leg thrusts on the ground at a time, and one leg exerts a torque on the body at a time. We call this sort of running a *one-foot gait*. Assuming that the behavior of the other leg does not interfere, the one-leg algorithms for hopping, forward travel, and posture can each be used to control the active leg. Of course, to make this workable, some bookkeeping is required to keep track of which leg is active, which leg is idle, and it is required to keep the idle leg "out of the way."

Jessica Hodgins and Jeff Koechling demonstrated the effectiveness of this approach by using the one-leg algorithms to control each leg of a planar biped. The machine can run with an alternating gait, run by hopping on one leg, it can switch back and forth between gaits, and it can run fast. Top recorded speed is 11.5 mph. The biped has also done forward flips and aerials (Hodgins and Raibert, 1987), as shown in figure 16.10.

In principle, this approach could be used to control running on any number of legs, so long as just one touches the ground at a time. Unfortunately, when there are several legs this approach runs into difficulties. There is a conflict between the need to provide balance and the need to move the legs without collisions. For the legs to provide balance the feet

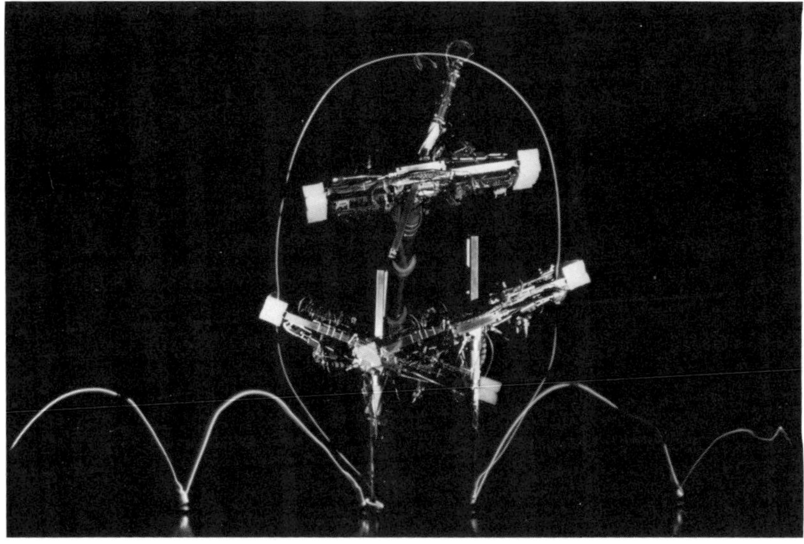

Figure 16.10
Planar biped doing a flip. The three images were made at the touchdown before the flip, the halfway point of the flip, and the lift-off after the flip.

must be positioned so as to sweep under the center of mass during support. This argues for attaching all the legs to the body directly at or below the center of mass. However, the legs must be able to swing without colliding with one another, suggesting separation between the hips (shoulders). In principle, it is not impossible for a quadruped to run using this approach, however they rarely to (Hildebrand, private communication).

An alternative is to use the legs in pairs. Suppose that we introduce a new control mechanism that coordinates the legs of a pair to act like a single equivalent leg—what Ivan Sutherland (1983) called a *virtual leg*. Such coordination requires that the two legs of a pair exert equal forces on the ground, that they exert equal torques on the body, and that the position of each leg's foot with respect to its hip be the same. To control locomotion, the three-part control algorithms described previously specify the behavior of each virtual leg, while the control system ensures that the physical legs move so as to obey the rules required for virtual leg behavior.

Using this approach there are three quadruped gaits to consider: the *trot*, which uses diagonal legs in a pair; the *pace*, which uses lateral legs in a pair; and the *bound*, which uses the front legs as a pair and the rear legs as a pair.

We argue that the quadruped is like a virtual biped, that a biped is like a one-legged machine, and we already know how to control one-legged machines. A control system for quadruped running consists of a controller that coordinates each pair of legs to act like one virtual leg, a three-part control system that acts on the virtual legs, and a bookkeeping mechanism that keeps track. Figure 16.11 shows a four-legged machine that trots, paces, and bounds with this sort of control system. Also see tables 16.2 and 16.3.

16.5.4 Computer Programs for Running

The behavior of the running machines just described was controlled by a set of computer programs that ran on our laboratory computer. These computer programs performed several functions, including

- sampling and filtering data from the sensors,
- transforming kinematic data between coordinate systems,
- executing the three-part locomotion algorithms for hopping, forward speed, and body attitude,
- controlling the actuators,
- reading operator instructions from the console, and
- recording running behavior.

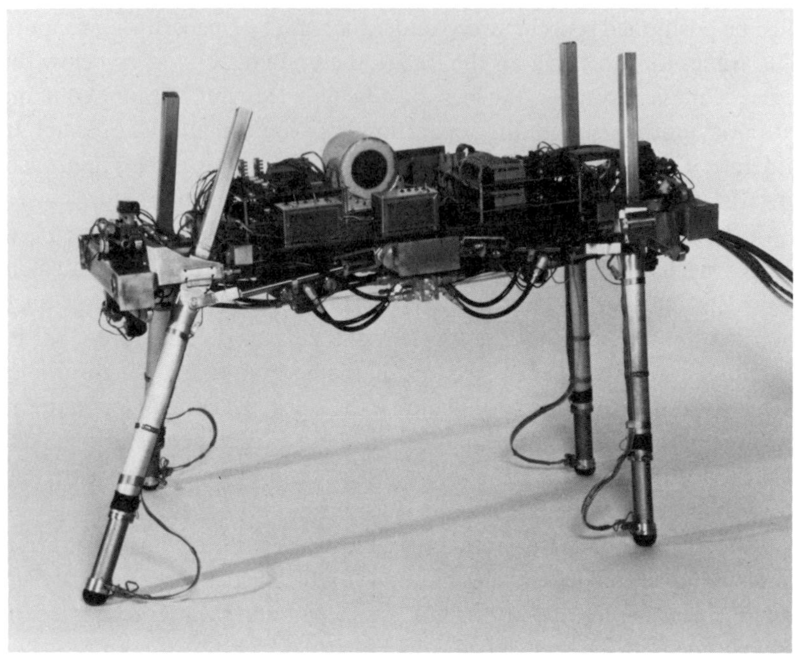

Figure 16.11
Quadruped machine that runs by trotting. *Virtual legs* are used to make trotting like biped running, which in turn is controlled with one-leg algorithms.

The control computer was a VAX-11/780 running the UNIX operating system. In order to provide real-time service with short latency and high-bandwidth feedback, the real-time control programs were implemented as a device that resided within the UNIX kernel. The device driver responded with short latency to a hardware clock that interrupted through the UNIBUS every 8 msec. All sensors and actuators were also accessed through interfaces that connected to the UNIBUS. Each time the clock ticked the running machine driver programs sampled and scaled data from the sensors, estimated joint and body velocities to determine the state of the running machine, executed the three-part locomotion algorithms, and calculated a new output for each actuator.

The programs performing these tasks were written in a mixture of C and assembly language. Assembly language was used where speed was of primary concern, such as in performing the kinematic transformations used to convert between coordinate systems. In some cases tabulated data were

Table 16.2
Details of state machine for biped and quadruped[a]

State[b]	Trigger event	Action
1 LOADING A	A touches ground	Zero hip torque A Shorten B Don't move hip B
2 COMPRESSION A	A air spring shortened	Erect body with hip A Shorten B Position B for landing
3 THRUST A	A air springs lengthening	Extend A Erect body with hip A Keep B short Position B for landing
4 UNLOADING A	A air spring near full length	Shorten A Zero hip torque A Keep B short Position B for landing
5 FLIGHT A	A not touching ground	Shorten A Don't move hip A Lengthen B for landing Position B for landing

a. The state shown in the left column is entered when the event listed in the center column occurs. During normal running states advance sequentially. During states 1–5, leg A is in support and leg B is in recovery. During states 6–10, these roles are reversed. For biped running A refers to leg 1 and B refers to leg 2. For quadruped trotting each letter designates a pair of physical legs.

b. States 6–10 repeat states 1–5, with A and B interchanged.

Table 16.3
Summary of leg laboratory research

1982	Planar one-legged machine hops in place, travels at a specified rate, keeps its balance when disturbed, and jumps over small obstacles.
1983	Three-dimensional one-legged machine runs and balances on an open floor.
1983	Murphy finds passive stability in bounding gait of simulated model.
1984	Data from cat and human runners exhibit symmetries like those used to control running machines.
1984	Quadruped running machine runs with trotting gait. The one-leg algorithms are extended to control this machine.
1985	Planar biped runs with one- and two-legged gaits and changes gait while running. Top speed is currently 11.5 mph.
1986	Monopod runs with new leg design. It uses an articulated structure and a leaf-spring for a foot.
1986	Planar biped does forward flip and aerial.

used to further increase the speed of a transformation. For instance, trigonometric functions, square roots, and higher order kinematic operations were evaluated with linear interpolation among precomputed tabulated data. One such evaluation takes about 25 μsec. The kinematic relationship between the quadruped's fore/aft hip actuator length and the forward position of the foot in body coordinates in an example of a function that was evaluated with a table.

The control programs were synchronized to the behavior of the running machine by a software finite state machine. The state machine made transitions from one state to another when sensory data from the running machine satisfied the specified conditions. For instance, the state machine made a transition from COMPRESSION to THRUST when the derivative of the support leg's length changed from negative to positive. Figure 16.12 and table 16.2 give some detail for the state machine that was used for the biped and quadruped running machines. We found that using a state machine along with properly designed transition conditions aided the interpretation of sensory data by providing noise immunity and hysteresis.

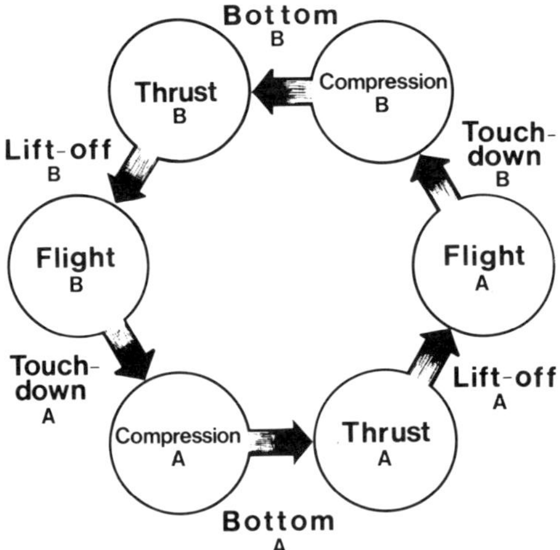

Figure 16.12
Simplified diagram of state machine that synchronizes the control programs to the behavior of the running machine. This state machine is for the biped and quadruped machines, but the one-legged state machines are similar. State transitions are determined by sensory events related to the hopping motion. A different set of control actions are put into effect in each state, as indicated in table 16.2.

Whereas sensory data determine when the state machine makes transitions, the resulting states determine which control algorithms operate to provide control. For instance, when the biped is in the THRUST A state, the control programs extend leg A, exert torque on hip A, shorten leg B, and position foot B.

In addition to the real-time programs that control the running machines, a top-level program was used to control the real-time programs. The top-level program permitted the user to initiate a running experiment, select among control modes, examine or modify the variables and parameters used by the control programs, specify sensor calibration data, mark variables for recording, and save recorded real-time data for later analysis and debugging. Each of these functions was accomplished by one or more system calls to the driver. The top-level program had no particular time constraints, so it was implemented as a time-sharing job that was scheduled by the normal UNIX scheduler.

16.6 Experiments in Animal Locomotion

Earlier in this chapter I said that studying legged machines could help us to understand more about locomotion in animals. Detailed knowledge of concrete locomotion algorithms with well-understood behavior might help us formulate experiments that elucidate the mechanisms used by humans and animals for locomotion. For instance, one might ask if animals control their forward speed as each of the running machines do, by choosing a forward position for the leg during each flight phase, with no adjustments during stance. Is there any decomposition of control? There are several questions like these that I am studying in collaboration with Thomas McMahon of Harvard University. The questions we would like to answer are summarized in the following (each is based on observations or ideas that arose in the course of studying legged machines).

16.6.1 Animal Experiments Motivated by Robot Experiments

Algorithms for Balance. The legged machines and computer simulations we have studied all use a specific algorithm for determining the landing position of the foot with respect to the body's center of mass. The foot is advanced a distance $x_{\hat{f}} = T_s\dot{x}/2 + k_{\dot{x}}(\dot{x} - \dot{x}_d)$. This calculation is based on the expected symmetric tipping behavior of an inverted pendulum.

Do animals use this algorithm to position their feet? To find out, one must measure small changes in the forward running speed, in the angular momentum of the body during the flight phase, and in the placement of the feet. Rather than look at average behavior across many steps, as is done in studies of energetics and neural control, we must look at error terms within each step.

Symmetry in Balance. Symmetry plays a central role in simplifying the control of the dynamic legged robots described earlier. The symmetry of interest specifies that motion of the body in space, and of the feet with respect to the body, are even and odd functions of time during each support period (see figure 16.13). These are interesting motions because they leave the forward and angular motion of the body unaccelerated over a stride, and therefore lead to steady state travel. Do animals run with this sort of symmetry? To find out we are examining film data for several quadrupedal animals. Preliminary measurements show that the galloping and trotting cat sometimes runs with symmetric motions (Raibert 1986b; Raibert 1986c). We would like to know

- How universal is the use of symmetry running by animals, both in terms of different gaits and different animals?
- What is the precision of the observed symmetry?
- Does the symmetry extend to angular motion of the body, as the theory predicts?
- How do asymmetries in the mechanical structure of the body and legs influence motion symmetry?

Virtual Legs. To control the quadruped running machine the control system synchronizes the behavior of pairs of legs that provides support during stance. The synchronization has three parts, requiring that the legs of a pair exert equal vertical forces on the ground, exert equal hip torque on the body, and displace their feet equal distances from the hip or shoulder. The result of such synchronization is called a *virtual leg* (Sutherland and Ullner 1984) because it makes two legs act like one equivalent leg located halfway between the pair. The virtue of virtual legs is that they simplify the control algorithms for balance and dynamic control.

Do animals use virtual legs? To answer this question one must examine the horizontal and vertical forces exerted on the ground by both legs during double support in trotting or pacing. From these measurements one can find the differential force or impulse. One would expect the differential

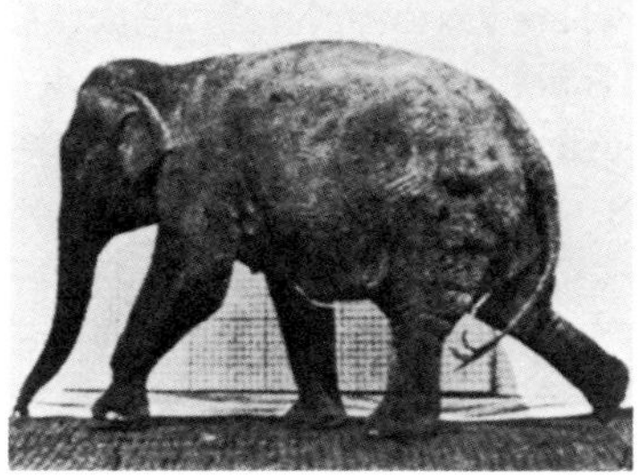

Figure 16.13
Symmetry in animal locomotion. Animals shown in symmetric configuration halfway through the stance phase for several gaits: rotary gallop (top), transverse gallop (second), canter (third), and amble (bottom). In each case the body is at minimum altitude, the center of support is located below the center of mass, the rearmost leg was recently lifted, and the frontmost leg is about to be placed. Photographs from Muybridge (1957).

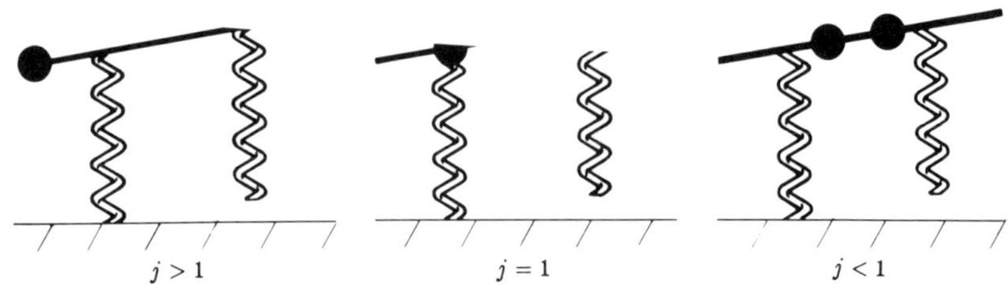

Figure 16.14
The dimensionless moment of inertia, $j = J/(md^2)$, predicts passive stability of the body's pitching motion for a simple simulated model. (Left) For $j > 1$, a vertical force on the left foot causes the right hip to accelerate upward. The model has no passive pitch stability. (Center) For $j = 1$, the system acts as two separate oscillators, with neutral stability. (Right) When $j < 1$, an upward force on the left leg causes the right hip to accelerate downward. The model has passive pitch stability. From Murphy and Raibert (1985).

force to be zero if the system were using a control strategy based on virtual legs. An important manipulation will be to disturb one or both of the feet during stance, and to measure the active force response of the legs. Asymmetries in the distribution of mass in the system may require a somewhat generalized version of the virtual leg, in which a fixed ratio of forces, torques and displacements prevails.

Distribution of Body Mass. The distribution of mass in the body can have a fundamental influence on the behavior of a running system. In earlier work we defined the dimensionless group representing the normalized moment of inertia of the body, $j = J/(\mathrm{md}^2)$ where J is the moment of inertia of the body, m is the mass of the body, and d is half the hip spacing (Murphy and Raibert 1985). As shown in figure 16.14 when $j = 1$ the hips are located at the centers of percussion of the body. Through computer simulations of a simplified model we found that when $j < 1$ the attitude of the body in the sagittal plane can be passively stabilized when running with a bounding gait. However, when $j > 1$ stabilization was not passively obtained.

This finding has implications for the interaction between the mechanical structure of an animal and the control provided by the nervous system. Could it be that bounding animals do not actively control pitching of their bodies, but control only forward running speed and direction? A first step toward answering this question is to measure values of j for a variety of quadrupedal animals, and to relate the measurements to their preferred modes of running, trotting or pacing (no pitching) vs. bounding or galloping (pitching). Does the value of j for a quadruped vary with its trot-to-gallop transition speed? Anatomical measurements like those of Fedak, Heglund, and Taylor (1982) will provide much of the data needed to answer this question.

Yaw Control. How do human runners keep themselves from rotating about the yaw axis? Control of this degree of freedom in the one-legged hopping machine is difficult because it does not have a foot that can exert a torsional torque on the ground. But humans have long feet that might be used to develop substantial torsional traction on the ground about the yaw axis. A first step in exploring this question would be to measure the torsional torque humans exert on the ground during running and to relate the measurements to yaw motions and yaw disturbances of the body.

16.6.2 Philosophy of Interaction between Robotics and Biology

It is interesting to combine the study of animals with the study of machines. Biological systems provide both great motivation by virtue of their striking performance, and guidance with the details of their actions. They are existence proofs that give us a lower bound on what is possible. Unfortunately, biological systems are often too complicated to study—there are many variables, precise measurement is difficult, there are limitations on the experimenter's ability to manipulate the preparation, and perhaps, an inherent difficulty in focusing on the information-level of a problem.

On the other hand, laboratory robots are relatively easy to build. Precisely controlled experiments are possible, as are careful measurements and manipulations, and the "subject" can be redesigned when necessary. However, the behavior of these experimental robot systems is impoverished when compared with the biological counterpart. They are easy to study, but they do not perform nearly so well as biological systems.

Analysis of living systems and synthesis of laboratory systems are complementary activities, each with strengths and weaknesses. Together, these activities can strengthen one another, leading to fundamental principles that elucidate the domain of both problems, independent of the particular implementation. Because machines face the same physical laws and environmental constraints that biological systems face when they perform similar tasks, the solutions they use may embrace similar principles. In solving the problem for the machine, we generate a set of plausible algorithms for the biological system. In observing the biological behavior, we explore plausible behaviors for the machine. In its grandest form, this approach lets the study of robotics contribute to both robotics and biology and lets the study of biology contribute to both biology and robotics.

16.7 The Development of Useful Legged Robots

The running machines described in the previous section are not useful vehicles or even prototypes for such vehicles. They are experimental apparatus used in the laboratory to explore ideas about legged locomotion. Each was designed to isolate and examine a specific locomotion problem, while postponing or ignoring many other problems. Let us now step back and ask what problems remain to be solved before legged robots are transformed into practical machines that do useful work.

16.7.1 Terrain Sensing

Perhaps the deepest problem limiting current walking machines, as well as othe forms of autonomous vehicle, is their inability to perceive the shape and structure of their immediate surroundings. Humans and animals use their eyes to locate good footholds, to avoid obstacles, to measure their own rate and direction of progress, and to navigate with respect to visible landmarks. The problem of giving machines the ability to see has received intensive and consistent attention for the past twenty-five or thirty years. There has been steady progress during that period. Current machines can see well enough to operate in well-structured and partially structured environments, but it is difficult to predict when machines will be able to see well enough to operate autonomously in rough outdoor terrain. I do not expect to see such autonomous machine behavior for at least ten years.

Sensors simpler than vision may be able to provide solutions to certain parts of the problem under certain circumstances. For instance, sonar and laser range data may be used to detect and avoid nearby obstacles. Motion data may be used for measuring speed and direction of travel with techniques that are substantially simpler than those needed to perceive shape in three dimensions.

16.7.2 Travel on Rough Terrain

Complete knowledge of the geometry of the terrain, as might be supplied by vision or these other senses, would not in itself solve the problem of walking or running on rough terrain. A system travelling over rough terrain needs to know or figure out what terrain shapes provide good footholds, which sequence of footholds would permit traversal of the terrain, and how to move so as to place the feet on the available footholds. It will be necessary to coordinate the dynamics of the vehicle with the dynamics of the terrain.

There are several ways that terrain becomes rough and therefore difficult to negotiate:

- not level;
- limited traction (slippery);
- areas of poor or nonexistent support (holes);
- vertical variations:
    - minor vertical variations in available footholds (less than about one-half the leg length);
    - major vertical variations in available footholds (footholds separated vertically by distances comparable to the dimensions of the whole leg);
- large obstacles between footholds (poles);
- intricate footholds (e.g., rungs of a ladder).

The techniques that will allow legged systems to operate in these sorts of terrain will involve the mechanics of locomotion, kinematics, dynamics, geometric representation, spatial reasoning, and planning. Although coarse- and medium-grain knowledge of the terrain will be important, I expect techniques that make legged systems inherently insensitive to fine-grain terrain variations to play an important role too. Ignoring the hard sensing issues mentioned earlier, I believe the perception and control mechanisms required for legged systems to travel on rough terrain will require a substantial research effort, but the important problems can be solved within the next ten years if they are pursued vigorously.

16.7.3 Mechanical Design and System Integration

When these sensing and control problems are solved, it will remain to develop mechanical designs that function with efficiency and reliability. Useful vehicles must carry their own power, control computers, and a payload. A host of interesting problems present themselves, including such matters as energy efficiency, structural design, strength and weight of materials, and efficient control. For instance, the development of materials and structures for efficient storage and recovery of elastic energy will be particularly important for legged vehicles. I expect that early useful legged vehicles can be built with existing mechanical and aerospace technology, but performance will improve rapidly as designs are refined, embellished, and improved.

Acknowledgments

This research was supported by a grant from the System Development Foundation and a contract from the Defense Advanced Research Projects Agency.

References

Ehrlich, A. 1928. Vehicle Propelled by Steppers. Patent Number 1,691,233.

Frank, A. A. 1970. An approach to the dynamic analysis and synthesis of biped locomotion machines. *Medical and Biological Engineering* 8:465–476.

Gurfinkel, V. S., Gurfinkel, E. V., Shneider, A. Yu., Devjanin, E. A., Lensky, A. V., and Shitilman, L. G. 1981. Walking robot with supervisory control. *Mechanism and Machine Theory* 16:31–36.

Hemami, H., and N. Golliday, C. L., Jr. 1977. The inverted pendulum and biped stability. *Mathematical Biosciences* 34:95–110.

Higdon, D. T., and Cannon, R. H., Jr. 1963. On the control of unstable multiple-output mechanical systems. In *ASME Winter Annual Meeting.*

Hirose, S. 1984. A study of design and control of a quadruped walking vehicle. *International J. Robotics Research* 3:113–133.

Hirose, S., and Umetani, Y. 1980. The basic motion regulation system for a quadruped walking vehicle. *ASME Conference on Mechanisms.*

Hodgins, J., and Raibert, M. H., 1987. Planar biped goes head over heels. *ASME Winter Annual Meeting*, Boston.

Hodgins, J., Koechling, J., and Raibert, M. H. 1985. Running experiments with a planar biped. *Third International Symposium on Robotics Research*, Cambridge: MIT Press.

Kato, T., Takanishi, A., Jishikawa, H., and Kato, I. 1983. The realization of the quasi-dynamic walking by the biped walking machine. In *Fourth Symposium on Theory and Practice of Robots and Manipulators*, A. Morecki, G. Bianchi, and K. Kedzior (eds.). Warsaw: Polish Scientific Publishers. 341–351.

Koozekanani, S. H., and McGhee, R. B. 1973. Occupancy problems with pairwise exclusion constraints—an aspect of gait enumeration. *J. Cybernetics* 2:14–26.

Liston, R. A. 1970. Increasing vehicle agility by legs: the quadruped transporter. Presented at *38th National Meeting of the Operations Research Society of America.*

Liston, R. A., and Mosher, R. S. 1968. A versatile walking truck. In *Proceedings of the Transportation Engineering Conference.* Institution of Civil Engineers, London.

Lucas, E. 1894. Huitieme recreation—la machine a marcher. *Recreations Mathematiques* 4:198–204.

Matsuoka, K. 1980. A mechanical model of repetitive hopping movements. *Biomechanisms* 5:251–258.

McGhee, R. B. 1968. Some finite state aspects of legged locomotion. *Mathematical Biosciences* 2:67–84.

McGhee, R. B. 1983. Vehicular legged locomotion. In *Advances in Automation and Robotics*, G. N. Saridis (ed.). JAI Press.

McGhee, R. B., and Frank, A. A. 1968. On the stability properties of quadruped creeping gaits. *Mathematical Biosciences* 3:331–351.

McGhee, R. B., and Jain, A. K. 1972. Some properties of regularly realizable gait matrices. *Mathematical Biosciences* 13:179–193.

McGhee, R. B., and Kuhner, M. B. 1969. On the dynamic stability of legged locomotion systems. In *Advances in External Control of Human Extremities*, M. M. Gavrilovic and A. B. Wilson, Jr. (eds.) Jugoslav Committee for Electronics and Automation, Belgrade, 431–442.

Miura, H. 1986. Biped locomotion robots. Unpublished videotape.

Miura, H., and Shimoyama, I. 1984. Dynamic walk of a biped. *International J. Robotics Research* 3:60–74.

Morrison, R. A. 1968. Iron mule train. In *Proceedings of Off-Road Mobility Research Symposim.* International Society for Terrain Vehicle Systems, Washington, 381–400.

Murphy, K. N., and Raibert, M. H. 1985. Trotting and bounding in a planar two-legged model. In *Fifth Symposium on Theory and Practice of Robots and Manipulators*, A. Morecki, G. Bianchi, and K. Kedzior (eds.). Cambridge: MIT Press, 411–420.

Murthy, S. S., and Raibert, M. H. 1983, 3D balance in legged locomotion: modeling and simulation for the one-legged case. In *Inter-Disciplinary Workshop on Motion: Representation and Perception*, ACM.

Muybridge, E. 1955. *The Human Figure in Motion.* New York: Dover Publications. First edition, 1901 by Chapman and Hall, Ltd., London.

Muybridge, E. 1957. *Animals in Motion.* New York: Dover Publications, First edition, 1899 by Chapman and Hall, Ltd., London.

Raibert, M. H. 1986. *Legged Robots That Balance* Cambridge: MIT Press.

Raibert, M. H. 1986. Symmetry in running. *Science*, 231:1292–1294.

Raibert, M. H., and Brown, H. B., Jr. 1984. Experiments in balance with a 2D one-legged hopping machine. *ASME J. Dynamic Systems, Measurement, and Control* 106:75–81.

Raibert, M. H., Brown, H. B., Jr., and Chepponis, M. 1984. Experiments in balance with a 3D one-legged hopping machine. *International J. Robotics Research* 3:75–92.

Raibert, M. H., Chepponis, M., and Brown H. B. Jr. 1986. Running on four legs as though they were one. *IEEE J. Robotics and Automation*, 2.

Russell, M., Jr. 1983. Odex I: the first functionoid. *Robotics Age* 5:12–18.

Schaefer, J. F., and Cannon, R. H., Jr. 1966. On the control of unstable mechanical systems. *International Federation of Automatic Control.* London, 6c.1–6c.13.

Seifert, H. S. 1967. The lunar pogo stick. *J. Spacecraft and Rockets* 4:941–943.

Shigley, R. 1957. *The Mechanics of Walking Vehicles.* Land Locomotion Laboratory, Report 7, Detroit, Michigan.

Sitek, G. 1976. Big Muskie, *Heavy Duty Equipment Maintenance* 4:16–23.

Snell, E. 1947. Reciprocating Load Carrier. Patent Number 2,430,537.

Sutherland, I. E., and Ullner, M. K. 1984. Footprints in the asphalt. *International J. Robotics Research* 3:29–36.

Urschel, W. E. 1949. Walking Tractor. Patent Number 2,491,064.

Vukobratovic, M., and Stepaneko, Y. 1972. On the stability of anthropomorphic systems. *Mathematical Biosciences* 14:1–38.

Waldron, K. J., Vohnout, V. J., Pery, A., and McGhee, R. B. 1984. Configuration design of the adaptive suspension vehicle. *International J. Robotics Research* 3:37–48.

Wallace, H. W. 1942. Jumping Tank Vehicle. Patent Number 2,371,368.

List of Contributors

Suguru Arimoto
Faculty of Engineering Science
Osaka University
Toyonaka
Osaka, Japan

Haruhiko Asada
Department of Applied Mathematics and Physics
Kyoto University
Kyoto, Japan

Klaus B. Biggers
Center for Engineering Design
Department of Mechanical Engineering
University of Utah
Salt Lake City, UT

Robert Bolles
Artificial Intelligence Center
SRI International
Menlo Park, CA

Michael Brady
Department of Engineering Science
Oxford University
Oxford, England

David Brock
MIT Artificial Intelligence Laboratory
Massachusetts Institute of Technology
Cambridge, MA

Rodney Brooks
MIT Artificial Intelligence Laboratory
Massachusetts Institute of Technology
Cambridge, MA

Paolo Dario
Scuola Superiore S. Anna and Centro E. Piaggio
University of Pisa
Pisa, Italy

Olivier D. Faugeras
INRIA Sophia-Antipolis
Valbonne, France

Martial Hebert
The Robotics Institute
Carnegie-Mellon University
Pittsburgh, PA

John M. Hollerbach
MIT Artificial Intelligence Laboratory
Massachusetts Institute of Technology
Cambridge, MA

Edwin K. Iversen
Center for Engineering Design
Department of Mechanical Engineering
University of Utah
Salt Lake City, UT

Stephen C. Jacobsen
Center for Engineering Design
Department of Mechanical Engineering
University of Utah
Salt Lake City, UT

Takeo Kanade
The Robotics Institute
Carnegie-Mellon University
Pittsburgh, PA

Tomás Lozano-Pérez
MIT Artificial Intelligence Laboratory
Massachusetts Institute of Technology
Cambridge, MA

Matthew T. Mason
Department of Computer Science
Carnegie-Mellon University
Pitttsburgh, PA

Patrick O'Donnell
MIT Artificial Intelligence Laboratory
Massachusetts Institue of Technology
Cambridge, MA

Alex Pentland
Media Lab
Massachusetts Institute of Technology
Cambridge, MA

Marc Raibert
MIT Artificial Intelligence Laboratory
Massachusetts Institute of Technology
Cambridge, MA

Bernard Roth
Department of Mechanical Engineering
Stanford University
Stanford, CA

Kenneth Salisbury
MIT Artificial Intelligence Laboratory
Massachusetts Institute of Technology
Cambridge, MA

Craig C. Smith
Department of Mechanical Engineering
Brigham Young University
Provo, UT

Russell H. Taylor
IBM T. J. Watson Research Center
Yorktown Heights, NY

Daniel E. Whitney
Charles Stark Draper Laboratory, Inc.
Cambridge, MA

Author Index

Acronyms and numbers in square brackets refer to notes at chapter ends. Page numbers in **boldface** refer to chapters written by contributors.

Subject Index

Page numbers in *italics* refer to figures.